Methods in Molecular Biology™

Series Editor
John M. Walker
School of Life Sciences
University of Hertfordshire
Hatfield, Hertfordshire, AL10 9AB, UK

For other titles published in this series, go to
www.springer.com/series/7651

METHODS IN MOLECULAR BIOLOGY™

Protein Blotting and Detection

Methods and Protocols

Edited by

Biji T. Kurien and R. Hal Scofield

Oklahoma Medical Research Foundation, Oklahoma City, OK, USA

Humana Press

Editors
Biji T. Kurien
Oklahoma Medical Research Foundation
Oklahoma City
OK, USA

R. Hal Scofield
Oklahoma Medical Research Foundation
Oklahoma City
OK, USA

ISSN: 1064-3745 e-ISSN: 1940-6029
ISBN: 978-1-61737-943-7 e-ISBN: 978-1-59745-542-8
DOI: 10.1007/978-1-59745-542-8
Springer Dordrecht Heidelberg London New York

Cover illustration: Figure 15 in Chapter 26

Printed on acid-free paper

springer.com

Preface

Amazingly for a procedure so fundamental to present-day biomedical research as well as medical diagnostics, development of the western blot occurred within the memory of many individuals who continue to work in these fields. That is, conversely, there are many investigators and clinicians presently working who recall the days prior to the western blot. The personal stories of how and why the technique was developed are given in this volume by the people intimately involved with this development. These stories date from only roughly 30 years ago. Since then, a very large number of variations on the theme of gel separation of biologic products by different parameters such as size or charge, followed by transfer to solid support and identification, have been developed. The present compilation brings together a large number of these techniques, some of which are adaptations of the original technique in order to solve a problem, whereas others are far flung and vastly different from how the original techniques were envisioned. The goal of our work is to not only compile the vast array of techniques based on western blot, but also give practical methods. We suspect that almost everyone involved in the enterprise has tried to bring a new technique to their laboratory by reproducing methods found in traditional publications. Doing this is commonly fraught with difficulty and may take weeks to accomplish, or may be abandoned as impossible. We hope that investigators will be able to open this volume and rapidly begin to use a technique new to them and their laboratories because the chapters give detailed practical methods, tips, and alternatives. If you are able to open a chapter, and conveniently and quickly perform a new procedure in your laboratory, then we will have accomplished our goals in editing this work.

Oklahoma City, OK

R. Hal Scofield
Biji T. Kurien

Contents

Contributors

EDITA AKSAMITIENE • *Thomas Jefferson University, Philadelphia, PA, USA*
F. JAVIER ALBA • *Universitat Autònoma de Barcelona, Bellaterra, Barcelona, Spain*
MICHAEL BACHMANN • *Carl Gustav Carus University TU Dresden, Dresden, Germany*
GAD BANETH • *Hebrew University of Jerusalem, Rehovot, Israel*
SALVADOR BARTOLOMÉ • *Universitat Autònoma de Barcelona, Bellaterra, Barcelona, Spain*
HOLGER BARTSCH • *Carl Gustav Carus University TU Dresden, Dresden, Germany*
ANTONIO BERMÚDEZ • *Universitat Autònoma de Barcelona, Bellaterra, Barcelona, Spain*
LEE B. BOCKUS • *Oklahoma Medical Research Foundation, Oklahoma City, OK, USA*
W. NEAL BURNETTE • *Molecular Pharmaceutics Corporation, Chapel Hill, NC, USA*
ENGELBERT BUXBAUM • *Ross University School of Medicine, Roseau, Commonwealth of Dominica, West Indies*
Ulrich Canzler • *Carl Gustav Carus University TU Dresden, Dresden, Germany*
GEEN-DONG CHANG • *National Taiwan University, Taipei 106, Taiwan*
NEIL W. CHARTER • *DiscoveRx Corporation, Fremont, CA, USA*
JOON-YONG CHUNG • *National Cancer Institute, Bethesda, MD, USA*
ELLEN W. COLLISSON • *Texas A&M University, College Station, TX, USA*
KARSTEN CONRAD • *Carl Gustav Carus University TU Dresden, Dresden, Germany*
JOAN-RAMON DABAN • *Universitat Autònoma de Barcelona, Bellaterra, Barcelona, Spain*
BELKISYOLÉ ALARCÓN DE NOYA • *Universidad Central de Venezuela, Caracas, Venezuela*
ALBERT DOAN • *DiscoveRx Corporation, Fremont, CA, USA*
PAMELA M. DONOGHUE • *UCD Conway Institute of Biomolecular and Biomedical Research, Dublin, Ireland*
YASER DORRI • *Oklahoma Medical Research Foundation, Oklahoma City, OK, USA*
ANIL D'SOUZA • *Oklahoma Medical Research Foundation, Oklahoma City, OK, USA*
MICHAEL J. DUNN • *UCD Conway Institute of Biomolecular and Biomedical Research, Dublin, Ireland*
E. ANN ELLIS • *Texas A&M University, College Station, TX, USA*
CLAUDIA FRANKE • *Carl Gustav Carus University TU Dresden, Dresden, Germany*
ROBERT FUHLBRIGGE • *Brigham and Women's Hospital, Boston, MA, USA*
NAOKO GOTO-INOUE • *Frontier Research Center, Hamamatsu University, Hamamatsu, Japan*
TANIA VALDES GONZALEZ • *Otsuka Pharmaceutical Co., Ltd., Tokushima, Japan*
DANIEL GRÄFE • *Carl Gustav Carus University TU Dresden, Dresden, Germany*
MARION L. GREASER • *University of Wisconsin-Madison, Madison, WI, USA*
JODY J. GROSS • *Oklahoma Medical Research Foundation, Oklahoma City, OK, USA*

KAI GROßMANN • *Attomol, Lipten, Germany*

TAKAHIRO HAYASAKA • *Frontier Research Center, Hamamatsu University, Hamamatsu, Japan*

KENNETH HENSLEY • *Oklahoma Medical Research Foundation, Oklahoma City, OK, USA*

STEPHEN M. HEWITT • *National Cancer Institute, Bethesda, MD, USA*

ANDREAS HOLZENBURG • *Texas A&M University, College Station, TX, USA*

JOE HORECKA • *DiscoveRx Corporation, Fremont, CA, USA*

XIONSHU HUANG • *Burnham Institute for Medical Research, La Jolla, CA, USA*

SHERRY HUBBELL • *Oklahoma Medical Research Foundation, Oklahoma City, OK, USA*

CHARLES L. JAFFE • *Hebrew University of Jerusalem, Jerusalem, Israel*

JYOTHI JAYARAM • *Texas A&M University, College Station, TX, USA*

CAROL B. JOHNSON • *Texas A&M University, College Station, TX, USA*

ALEXANDER E. KALYUZHNY • *R&D Systems, Inc., Minneapolis, MN, USA*

ALI W. KHALILI • *Oklahoma Medical Research Foundation, Oklahoma City, OK, USA*

TEAYOUN KIM • *Department of Nutrition and Food Sciences, Auburn, AL, USA*

ANATOLY KIYATKIN • *Thomas Jefferson University, Philadelphia, PA, USA*

SETSUKO KOMATSU • *National Institute of Crop Science, Tsukuba, Japan*

JOSEPH KOST • *Ben-Gurion University, Beer-Sheva, Israel*

MARYLA KRAJEWSKA • *Burnham Institute for Medical Research, La Jolla, CA, USA*

STAN (STANISLAW) KRAJEWSKI • *Burnham Institute for Medical Research, La Jolla, CA, USA*

BIJI T. KURIEN • *Oklahoma Medical Research Foundation, Oklahoma City, OK, USA*

FRANÇOISE LASNE • *Agence Française de Lutte contre le Dopage, Châtenay-Malabry, France*

DER-YEN LEE • *National Taiwan University, Taipei 106, Taiwan*

WERNER LEHMANN • *Attomol, Lipten, Germany*

YU-TING LIU • *Graduate School of Medicine, Kyoto University, Kyoto, Japan*

CHIN-YEE LOH • *DiscoveRx Corporation, Fremont, CA, USA*

SANDRA LOSADA • *Universidad Central de Venezuela, Caracas, Venezuela*

GRAEME LOWE • *Monell Chemical Senses Center, Philadelphia, PA, USA*

JOSE LUQUE-GARCIA • *New York University School of Medicine, New York, NY, USA*

MASAHARU MABUCHI • *Millipore Corporation, Danvers, MA, USA*

FRANCESCA MACCARI • *University of Modena and Reggio Emilia, Modena, Italy*

KAZUYA MACHIDA • *University of Connecticut Health Center, Farmington, CT, USA*

SHANNON MAIER • *Oklahoma Medical Research Foundation, Oklahoma City, OK, USA*

IOANNIS N. MANOUSSOPOULOS • *Plant Protection Institute of Patras, Patras, Greece*

ANNAMMA MATHAI • *Sree Chitra Tirunal Institute for Medical Sciences and Technology, Thiruvananthapuram, Kerala, India*

SURESH T. MATHEWS • *Department of Nutrition and Food Sciences, Auburn, AL, USA*

BRUCE J. MAYER • *University of Connecticut Health Center, Farmington, CT, USA*

CIARA A. MCMANUS • *UCD Conway Institute of Biomolecular and Biomedical Research, Dublin, Ireland*

PADMAJA MEHTA • *Oklahoma Medical Research Foundation, Oklahoma City, OK, USA*
BRUCE J. MEYER • *University of Connecticut Health Center, Farmington, CT, USA*
THOMAS A. NEUBERT • *New York University School of Medicine, New York, NY, USA*
OSCAR NOYA • *Universidad Central de Venezuela, Caracas, Venezuela*
INGRID OLSEN • *National Veterinary Institute, Oslo, Norway*
KEITH R. OLSON • *DiscoveRx Corporation, Fremont, CA, USA*
JUN PARK • *Millipore Corporation, Danvers, MA, USA*
ERIC P. PLAISANCE • *Department of Nutrition and Food Sciences, Auburn, AL, USA*
RAFAEL PONT-LEZICA • *Université de Toulouse III, Castanet-Tolosan, France*
VISHNAMPET VENKATARAMAN RADHAKRISHNAN • *Sree Chitra Tirunal Institute for Medical Sciences and Technology, Thiruvananthapuram, Kerala, India*
ROBERT SACKSTEIN • *Brigham and Women's Hospital, Boston, MA, USA*
R. HAL SCOFIELD • *Oklahoma Medical Research Foundation, Oklahoma City, OK, USA*
MITSUOTOSHI SETOU • *Frontier Research Center, Hamamatsu University, Hamamatsu, Japan*
AJAY SHARMA • *Millipore Corporation, Danvers, MA, USA*
AARON G. SMITH • *Texas A&M University, College Station, TX, USA*
SUMI SURENDRAN • *Sree Chitra Tirunal Institute for Medical Sciences and Technology, Thiruvananthapuram, Kerala, India*
TAKAO TAKI • *Otsuka Pharmaceutical Co., Ltd, Tokushima, Japan*
DALIT TALMI-FRANK • *Hebrew University of Jerusalem, Rehovot, Israel*
IRA TARGOFF • *Veterans Affairs Medical Center - Oklahoma City, University of Oklahoma Health Sciences Center, and Oklahoma Medical Research Foundation, Oklahoma City, OK, USA*
MARILYAN TOLEDO • *Universidad Central de Venezuela, Caracas, Venezuela*
HARRY TOWBIN • *Swiss Federal Institute of Technology, Zurich, Switzerland*
SHINYA TOYOKUNI • *Graduate School of Medicine, Kyoto University, Kyoto, Japan*
EDWARD TRIEU • *Oklahoma Medical Research Foundation, Oklahoma City, OK, USA*
MINA TSAGRIS • *Institute of Molecular Biology and Biotechnology and University of Crete, Heraklion, Greece*
STANISLAV VITHA • *Texas A&M University, College Station, TX, USA*
NICOLA VOLPI • *University of Modena and Reggio Emilia, Modena, Italy*
CHAD M. WARREN • *University of Wisconsin-Madison, Madison, WI, USA*
TOM WEHRMAN • *DiscoveRx Corporation, Fremont, CA, USA*
HARALD G. WIKER • *National Veterinary Institute, Oslo, Norway*
OLGA YELIOSOF • *Oklahoma Medical Research Foundation, Oklahoma City, OK, USA*

Chapter 1

Origins of Protein Blotting

Harry Towbin

Summary

The development of protein blotting in its early days is recounted as arising from the need to tackle a specific analytical problem. Combining diverse elements of common methods and simple lab equipment resulted in a procedure of general utility. The expansion of the idea of carrying out immunoassays on membranes as predecessors of microarrays is briefly touched upon.

Key words: Blotting, Western, History, Protein array

Easily persuaded by the editors' invitation to write about the origins of protein blotting, I begin the story in the laboratory of Julian Gordon at the Friedrich Miescher Institute in Basel. In need of a postdoctoral fellow, Julian had hired me in 1978 for the task of creating antibodies against ribosomal proteins. With some experience in ribosomes and none in immunology, I set out to purify our proteins from chicken liver. We believed that it was necessary to get really fresh samples. So, together with a courageous colleague, who later turned to running a motorbike shop, we ventured to a slaughterhouse in the countryside. While we watched their grisly machinery, the friendly workers were captivated by the fog running over the brim of the Dewars as the freshest ever chicken liver dropped into the liquid nitrogen.

Back in the new and spacious laboratory, we struggled with ultracentrifuges, ample amounts of urea, and countless column fractions. We wondered whether the animals we were going to immunize would raise the antibodies we so eagerly sought. How could we ever be sure that we would not get antibodies to some

Biji T. Kurien and R. Hal Scofield (eds.), *Methods in Molecular Biology, Protein Blotting and Detection, vol. 536*

DOI: 10.1007/978-1-59745-542-8_1

contaminants? We were lucky to be in contact with Theo Staehelin at Roche, a pioneer in the field of initiation factors for mammalian protein synthesis who had ample expertise in ribosomes. Theo also told us about the new hybridoma technique. The idea of immunizing with mixtures of proteins and still getting an eternal source of a specific antibody was irresistible. After learning the secrets of the trade from Theo, we happily switched to mice and spared the rabbits and goats we had already injected to familiarize ourselves with time-honored immunological techniques. I was fascinated by precipitation arcs of the Ouchterlony double diffusion test and by the sensitivity of solid phase immunoassays achieved with remarkably simple equipment. Still, the problem of assuring specificity of hybridoma antibodies remained.

A common way of characterizing ribosomal proteins was by electrophoresis on two-dimensional gels. Could one recover the proteins from the gel as the literature described and use them in these sensitive immunoassays? The extraction worked, in principle, but the bulk of homogenized polyacrylamide was deterring. Even today, few researchers take that approach for purifying proteins from gels. We also discarded the idea of letting the antibodies react with proteins in the gel because of the impeded diffusion within the polyacrylamide matrix.

As the three of us later realized, the idea of preparing a replica of a protein gel on a membrane, in close analogy to Southern's DNA blotting, was in many people's mind *(1)*. But how could we copy the proteins to a membrane? It was a lucky coincidence that Julian had an electrophoretic destainer in use. This apparatus, now rarely seen in the laboratory, served to remove excess stain from gels, simply by placing the gel between two grids and applying current at a right angle to the plane of the sheet. The electrically charged dye molecules quickly cleared off the gel. Well, proteins were also charged – would they behave like the dye? The basic setup was quickly put together. Meticulously cleaning the destainer from residual amido black I remember as being the most tedious part of the chore. Pipette tip holders, Scotch Brite scouring pads, and rubber strings were all what was needed to build the sandwich that is still popular for protein blotting. A series of straightforward experiments showed us that the nitrocellulose sheets reliably captured the proteins as they were leaving the gel. From my wife, Marion, I knew about the art of immunohistochemical staining. One could easily test staining procedures by placing little dots of proteins directly on the membrane and running series of dilutions. Developing the first blot from a gel with antibodies thrilled me with bands that darkened within seconds. I felt like a child who reveals secret messages written in invisible ink by holding a sheet of paper over a flame.

After publication of the method *(2)*, it dawned on us that the blotting procedure might have some commercial value, after all.

We learned from the lawyers that we could still claim protection in some countries. We also learned how hard it was to define what was really new in an invention and also that there needed to be an unforeseen element in it. In some way, almost everything in our procedure had some precedence! We stand on the shoulders of giants, as every Google Scholar user knows. Finally, the patent application was written; the fact that Theo Staehelin was at Roche and Julian and myself at Ciba-Geigy, though unusual, was no impediment.

The idea of placing proteins on nitrocellulose sheets by direct spotting, as trivial as it appeared, proved to be stimulating *(3)*. You could easily probe little dots of protein on nitrocellulose or create sandwich tests, for example, for determining antibody subtypes in hybridoma supernatants. The potential of carrying out assays on arrayed protein spots, a bit awkwardly named dot immunobinding, was most clearly recognized by Julian. With the advent of spotting devices, always in the footsteps of DNA technologies, protein arraying is only now gaining popularity. These efforts might all be viewed as aiming at eliminating the cumbersome gel electrophoresis step from western blotting. Also, for those weary of running gels and handling membranes, relief is in sight with a new system that automatically resolves proteins according to isoelectric point in a capillary *(4)*. Still, as the contributions to this volume attest, membranes remain attractive supports, giving room for countless variations, unforeseen applications, and an expanding nomenclature inspired from the compass set by E.M. Southern.

References

1. Rocco, R.M. (2006) Western blots [commentary]. In: Rocco, R.M., ed. Landmark Papers in Clinical Chemistry, Elsevier, Amsterdam, 381–383.
2. Towbin, H., Staehelin, T., and Gordon, J. (1979) Electrophoretic transfer of proteins from polyacrylamide gels to nitrocellulose sheets: procedure and some applications. *Proc Natl Acad Sci USA* 76, 4350–4354.
3. Hawkes, R., Niday, E., and Gordon, J. (1982). A dot-immunobinding assay for monoclonal and other antibodies. *Anal Biochem* 119, 142–147.
4. O'Neill, R.A., Bhamidipati, A., Bi X., Deb-Basu, D., Cahill, L., Ferrante, J., et al. (2006) Isoelectric focusing technology quantifies protein signaling in 25 cells. *Proc Natl Acad Sci USA* 103, 16153–16158.

Chapter 2

Western Blotting: Remembrance of Things Past

W. Neal Burnette

Summary

Western blotting sprung from the need to develop a sensitive visual assay for the antigen specificity of monoclonal antibodies. The technique employed SDS-PAGE of protein antigens, electrophoretic replica transfer of gel-resolved proteins to unmodified nitrocellulose sheets, probing the immobilized antigens with hybridomas, and detection of antibody–antigen complexes with radiolabeled staphylococcal protein A and autoradiography. The simplicity and relevance of the method has led to its expansive application as an immunodiagnostic and a ubiquitous research tool in biology and medicine.

Key words: Western blotting, SDS-PAGE, Electroblotting, Unmodified nitrocellulose, Immobilized replica, Antigen specificity, Antibody–antigen complex, Protein A

Paraphrasing Plato, Jonathan Swift once famously observed that "Necessity is the Mother of Invention" *(1)*. Such necessity was the antecedent of western blotting. The fact that similar techniques arose within the same time frame indicates the temporal pressure of an unfilled demand in biology and medicine – a common exigency to provide a tool by which to visualize specific antigens.

The requirement that impelled the development of western blotting *(2)* in my laboratory came to light in 1977, when I moved to Robert Nowinski's RNA tumor virus group at the Fred Hutchinson Cancer Research Center. This was just at the time when monoclonal antibodies were first described by Köhler and Milstein *(3)*, and Bob's group was developing monoclonal reagents as probes to assess the structural and immunologic nature of retrovirus proteins *(4)*. It quickly became clear that there was no simple, objectively visual way to easily screen the vast numbers of

Biji T. Kurien and R. Hal Scofield (eds.), *Methods in Molecular Biology, Protein Blotting and Detection, vol. 536*

DOI: 10.1007/978-1-59745-542-8_2

generated clones for their specificity toward individual structural polypeptides comprising the retrovirus envelope and core.

Although the main focus of my work at the time was in other areas of retroviral research, I had a methodological background in electrophoretic antigen assessment; therefore, I agreed to undertake the effort in the Nowinski group to develop new and streamlined techniques to facilitate screening of the hybridomas for antigen specificity. Having been trained as a postdoc in Tom August's lab at Albert Einstein College of Medicine in radioimmunoassays, immunoprecipitation, and SDS-polyacrylamide gel electrophoresis (SDS-PAGE), I attempted to conceive of ways in which these methods might be combined. RIAs had great sensitivity, but lacked the ability to give a simple picture of specificity, especially in complex protein mixtures. Conversely, immunoprecipitation required radiolabeling of diverse antigen species and, while it provided reasonable sensitivity and definition of specificity when linked to SDS-PAGE and autoradiography, it was plagued by significant background that led to substantial uncertainty and was not easily adaptable to high-throughput screening.

Launching into this project, essentially on my own and without benefit of knowledge of others who might be engaged in similar work, I attempted a wide array of techniques, hoping that I would stumble upon something useful or, at least, something that might light the pathway to proceed further. Here, I was trying through trial-and-error to fulfill another Swiftian dictum: "Discovery consists of seeing what everybody has seen and thinking what nobody else has thought" *(1)*. In retrospect, some of the things I tried verged on the laughable. Nevertheless, the early work furnished me with the recognition that purified, radiolabeled (in this case, radioiodinated) staphylococcal protein A *(5)* provided a more functionally stable and "universal" imaging agent for detection of antigen–antibody complexes than did "second antibody" reagents.

As incongruous as it might seem in the hindsight of nearly 30 years, I struggled with how to apply the monoclonal antibodies (as well as monospecific antisera) to gel-separated antigens. The "Eureka" moment occurred while I was concomitantly performing other experiments that employed "Northern" blots *(6)*, an effulgent clarity of vision that an immobilized "replica" of the PAGE-resolved proteins was to be an intrinsic element. Initially, I attempted passive transfer by placing gels in direct contact with derivatized, and later unmodified, nitrocellulose sheets. After overcoming problems associated with nonspecific binding of immunoglobulin and protein A reagents to the nitrocellulose by the use of a blocking agent (I employed immunoglobulin-depleted, purified bovine serum albumin), it became apparent that capillary transfer was slow, inefficient, and resulted in unacceptable diffusional band-spreading of the gel-resolved antigens.

A second Archimedean moment occurred at this point, when I came across an old electrophoretic gel destainer that I had not used for years. Perhaps, I reasoned, if I could work fast enough or keep temperatures low enough to minimize band diffusion *within* the parent gel, and find electrophoretic conditions and nitrocellulose pore size to prevent driving the proteins out of the gel and *through* the paper, I might be able to make better "replicas" of the gel-resolved antigens.

It only took about a week from this point to work out the "final" parameters of the basic electroblotting technique, and another few weeks to work on adaptations that could increase resolution and sensitivity in complex mixtures (e.g., cell culture, blood, tissue, and other clinical samples) using isotachophoresis in a first dimension, then applying such cylindrical gels to the SDS-PAGE slab gels. During this period, a manuscript was prepared and a discussion with Bob Nowinski ensued wherein the name "western blotting" was conceived. It was just at this time that the publication of Towbin et al. *(7)* appeared. Although the basic technique described by these investigators was similar, I believed that many of the simplifying and "universalizing" aspects of western blotting (e.g., unmodified nitrocellulose, radiolabeled protein A detection, 2-D separations, etc.) were sufficiently important to warrant submission of my manuscript. I also became aware at this time of the publication by Renart et al. *(8)*; however, the technique described in their paper employed conditions with which I had experimented (e.g., derivatized paper, passive capillary transfer, second antibody, etc.) and found wanting from the perspectives of simplicity, ease of use, resolution, sensitivity, and specificity.

The manuscript was submitted to *Analytical Biochemistry* and was rejected without, it seemed, any recourse to resubmission. It was interesting to note that the rejection appeared to me to be based not on any technical criticisms or its ostensible similarity to the methods of Towbin et al. *(7)*, but rather on the reviewers' sentiment of the pedestrian nature of the contribution and, particularly, to the flippant and frivolous whimsy in the name "western blotting."

As previously documented *(9)*, preprints of the rejected manuscript had been sent to colleagues, who subsequently provided them to others, and they to others until, eventually (even in this preelectronic era of written communications), it seemed as though this unpublished article had received wider distribution than many published ones. I only became aware of this subsequent to my move to the Salk Institute at the end of 1979. It was there that I was tracked down and spent a good part of every work day fielding telephonic questions about the technique and providing readable copies of the preprint – the original I had sent to a few colleagues had undergone many cycles of photocopy replication as it wended its way from lab to lab, the later generations being difficult to read. After about a half year of operating this private "journal club," I called the editor-in-chief of *Analytical Biochemistry*; he agreed that the situation was untenable, that the general immunoblotting technique (as well as the name "western blotting") was becoming widely accepted, and that the initial rejection of my manuscript was probably unfortunate. Therefore, I resubmitted the paper (with only very minor changes); it was accepted immediately, and finally published a few months later *(2)*.

For those who have felt the sting of journal rejection, it is worth noting that this paper has entered a small pantheon of the most highly cited scientific articles, all of which were initially rejected for publication *(10)*. Humility is an oft-reinforced virtue in science; it is humbling to realize that this little paper on western blotting far transcended the sum of journal citations for all of my other published research efforts. Nevertheless, it is a source of immense satisfaction to have made – along with Towbin et al. *(7)* – a lasting contribution to the methodological armamentarium of biological and medical scientists.

To complete the analogy hinted in the title of this review, I wish to thank the editors of this volume for providing me, like the proffered "madeleine" in Proust's *À la recherche du temps perdu (11)*, the occasion for this reminiscence.

References

1. Swift, J. (1726) *Travels into Several Remote Nations of the World, in Four Parts. By Lemuel Gulliver, First a Surgeon, and then a Captain of several Ships.* Benjamin Motte, London.
2. Burnette, W.N. (1981) "Western blotting": electrophoretic transfer of proteins from sodium dodecyl sulfate-polyacrylamide gels to unmodified nitrocellulose and radiographic detection with antibody and radioiodinated protein A. *Anal Biochem* 112, 195–203.
3. Köhler, G. and Milstein, C. (1975) Continuous cultures of fused cells secreting antibody of predefined specificity. *Nature* 256, 495–497.
4. Nowinski, R.C., Stone, M.R., Tam, M., Lostrom, M.E., Burnette, W.N., et al. (1980) Mapping of viral proteins with monoclonal antibodies: analysis of the envelope proteins of murine leukemia viruses, in *Monoclonal Antibodies: Hybridomas: A New Dimension in Biological Analyses* (Kennett, R.H., McKearn, T.J., and Bechtol, K.B., eds.), Plenum, New York, pp. 295–316.
5. Langone, J.J. (1978) [^{125}I]protein A: a tracer for general use in immunoassay. *J Immunol Methods* 24, 269–285.
6. Alwine, J.C., Kemp, D.J., and Stark, G.R. (1977) Method for detection of specific RNAs in agarose gels by transfer to diazobenzyloxymethyl-paper and hybridization with DNA probes. *Proc Natl Acad Sci USA* 74, 5350–5354.
7. Towbin, H., Staehelin, T., and Gordon, J. (1979) Electrophoretic transfer of proteins from polyacrylamide gels to nitrocellulose sheets: procedure and some applications. *Proc Natl Acad Sci USA* 76, 4350–4354.
8. Renart J., Reiser, J., and Stark, G.R. (1979) Transfer of proteins from gels to diazobenzyloxymethyl-paper and detection with antisera: a method for studying antibody specificity and antigen structure. *Proc Natl Acad Sci USA* 76, 3116–3120.
9. Burnette, W.N. (1991) Western blotting (Citation Classic). *Curr Contents* 34, 8.
10. Campanario, J.M. (1996) Have referees rejected some of the most-cited articles of all times? *J Am Soc Inf Sci* 47, 302–310.
11. Proust, M. (1913) *À la recherche du temps perdu (vol. I)*. Bernard Grasset, Paris.

Chapter 3

Introduction to Protein Blotting

Biji T. Kurien and R. Hal Scofield

Summary

Protein blotting is a powerful and important procedure for the immunodetection of proteins following electrophoresis, particularly proteins that are of low abundance. Since the inception of the protocol for protein transfer from an electrophoresed gel to a membrane in 1979, protein blotting has evolved greatly. The scientific community is now confronted with a variety of ways and means to carry out this transfer.

Key words: Western blotting, Sodium dodecyl sulfate polyacrylamide gel electrophoresis, Nitrocellulose membrane, Polyvinylidene difluoride membrane

1. Introduction

The transfer of macromolecules (proteins or nucleic acids) to microporous membranes is referred to as "blotting," and this term encompasses both "spotting" (manual sample deposition) and transfer from planar gels. Proteins that are resolved on sodium dodecyl sulfate polyacrylamide gel electrophoresis (SDS PAGE) gels are typically transferred to adsorbent membrane supports under the influence of an electric current in a procedure that is known as western blotting (WB) or protein blotting *(1, 2)*. Nucleic acids are routinely transferred from agarose gels, to a membrane support, through capillary action (Southern blotting). Protein blotting evolved from DNA (Southern) blotting *(3)* and RNA (northern) blotting *(4)*. The term "western blotting" was coined to describe *(5)* this procedure to retain the "geographic" naming tradition initiated by Southern's paper *(3)*. The blotted

Biji T. Kurien and R. Hal Scofield (eds.), *Methods in Molecular Biology, Protein Blotting and Detection, vol. 536*

DOI: 10.1007/978-1-59745-542-8_3

proteins form an exact replica of the gel and have proved to be the starting step for a variety of experiments. The subsequent employment of antibody probes directed against the membrane-bound proteins (immunoblotting) has revolutionized the field of immunology (**Fig. 1**). Dot blotting refers to the analysis of proteins applied directly to the membrane rather than after transfer from a gel.

The utility of the high resolving power of SDS PAGE *(6)* was limited in purpose, owing to the fact that the separated proteins in the gel matrix were difficult to access with molecular probes, until the advent of protein blotting. Protein transfer with subsequent immunodetection has found wide application in the fields of life sciences and biochemistry. This procedure *(1, 2)* is a powerful tool to detect and characterize a multitude of proteins, especially those proteins that are of low abundance. It offers the following specific advantages: (a) wet membranes are pliable and are easy to handle compared with gels, (b) easy accessibility of the proteins immobilized on the membrane to different ligands, (c) only small amount of

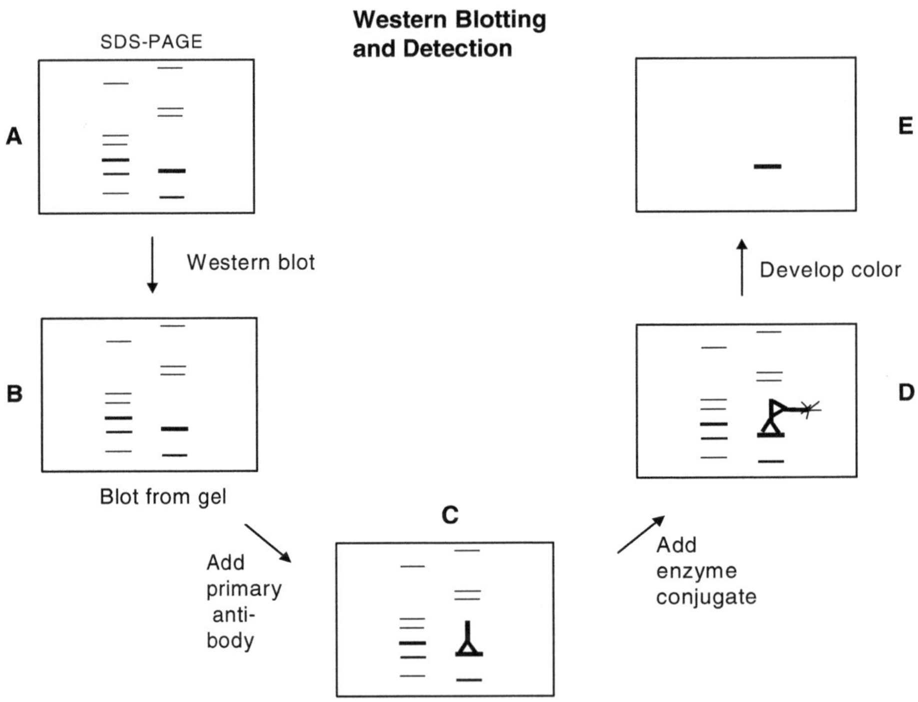

Fig. 1. Schematic of western blotting and detection procedure: (**A**) Unstained SDS PAGE gel prior to western blot. The bands shown are hypothetical. (**B**) Exact replica of SDS PAGE gel obtained as a blot following western transfer. (**C**) Primary antibody binding to a specific band on the blot. (**D**) Secondary antibody conjugated to an enzyme (alkaline phosphatase or horse radish peroxidase) binding to primary antibody. (**E**) color development of specific band (reproduced from *(10)* with permission from Elsevier).

reagents is required for transfer analysis, (d) multiple replicas of a gel are possible, (e) prolonged storage of transferred patterns, prior to use, becomes possible, and (f) the same protein transfer can be used for multiple successive analyses *(7–9)*.

Protein blotting has been evolving constantly, since its inception, and now the scientific community is faced with a multitude of ways and means of transferring proteins *(10)*. Nonetheless, western blot sensitivity is dependent on efficiency of blotting or transfer, retention of antigen during processing, and the final detection/amplification system used. Results are compromised if there are deficiencies in any of these steps *(11)*.

1.1. Blotting Efficiency

The efficient transfer of proteins from a gel to a solid membrane support depends greatly on the nature of the gel, the molecular mass of the proteins being transferred, and the membrane used. Running the softest gel, in terms of acrylamide and cross-linker that yields the required resolution, is the best option. Transfer becomes more complete and faster with the use of thinner gels. However, the use of ultrathin gels may cause handling problems, and a 0.4-mm thickness represents the lower practical limit *(12)*. Proteins with a high molecular mass blot poorly following SDS PAGE, resulting in low levels of detection on immunoblots. However, the efficiency of transfer of such proteins has been facilitated with heat, special buffers, and partial proteolytic digestion of the proteins prior to transfer *(11, 13–17)*.

2. Immobilizing Supports for Protein Transfer

A wide range of solid phases are available for immobilization, ranging from the truly solid phase such as glass or plastic to latex and cellulose that are porous. The most common phases used for blotting comprise microporous surfaces and membranes such as cellulose, nitrocellulose (NC), polyvinylidine diflouride, cellulose acetate, polyethane sulfone, and nylon. The unique properties of microporous surfaces that make them suitable for traditional assays such as western blotting are (a) large volume-to-surface area ratio, (b) high binding capacity, (c) short- and long-term storage of immobilized molecules, (d) ease of processing by allowing a solution phase to interact with the immobilized molecule, (e) lack of interference with the detection strategy, and (f) reproducibility. These properties are useful for the high-thoroughput assays used in the postgenomic era as well *(2, 4, 14, 18, 19)*.

Typically, these microporous surfaces are used in the form of membranes or sheets with a thickness of 100 μm and possessing an average pore size that ranges from 0.05 to 10 μm in diameter.

The interaction of biomolecules with each of these membranes is not completely understood, except for the fact that it is generally known to be noncovalent *(20, 21)*.

Regardless of the type of membrane used, it must be borne in mind that exceeding the protein binding capacity of the membrane used tends to reduce the signal obtained in immunoblotting. Excess protein, weakly associated with the membrane, is readily accessible to react with the primary antibody or any other ligand in solution (e.g., lectin). However, the resulting antibody–protein complexes will easily wash off during further processing of the membrane. Such a scenario would not have prevailed if the protein had initially made good contact with the membrane *(18)*.

2.1. Nitrocellulose Membranes

Nitrocellulose (NC) is perhaps the most versatile of all the surfaces mentioned earlier for the immobilization of proteins, glycoproteins, or nucleic acids *(3, 4, 19)*. In addition to traditional blotting, NC is used in high-throughput array, immunodiagnostic as well as mass spectrometry-coupled proteomic applications, filtration/concentration, ion exchange, and amino acid sequencing in addition to traditional blotting procedures. It was Southern who first demonstrated (in 1975) the usefulness of NC to capture nucleic acids. Towbin in 1979 *(1)* and Burnette in 1981 *(5)* showed that NC could also be used for proteins.

This unique polymer derived from cellulose has been used as the most common immobilization surface in biological research for over 65 years. Since high-throughput methodologies for proteomics and genomics rely heavily on traditional concepts of molecular immobilization followed by hybridization binding or analysis, NC continues to be useful in postgenomic era technology *(19)*.

2.1.1. Synthesis of Nitrocellulose from Cellulose

Treatment of cellulose with nitric acid results in the hydroxyl moieties on each sugar unit of cellulose being substituted by nitrate groups, resulting in NC. Organic solvents readily dissolve dry NC resulting in the formation of a lacquer. When the solvents are evaporated the polymer is deposited as a thin film. By including a nonsolvent such as water in the lacquer pores, nonsolvent can be introduced into the film to create a microporous membrane. Pore formation is a consequence of differential evaporation of the nonsolvent and the solvent. Therefore, pore size and porosity can be readily controlled by the amount of the nonsolvent in the lacquer *(2)*. The pore size of 0.45 μm refers to the average effective diameter of the irregular long and tortuous channels that traverse the membrane. The pores of 0.45 μm in NC membranes account for about 80% of the filter's volume reaching an average density of $450 \times 10^6/\text{cm}^2$ *(18)*. In the blotting process, the membrane needs to be porous to allow it to be saturated with buffer and will permit the required flow of current or liquid for electro and convection blotting.

2.1.2. Immobilization Mechanism

Even though the exact mechanism by which biomolecules interact with NC is unknown, several lines of evidence suggest that the interaction is noncovalent and hydrophobic. One evidence favoring hydrophobic interaction is the fact that since most proteins at pH values above 7 are negatively charged, it is surprising that NC which is also negatively charged can bind proteins efficiently. An additional fact is that nonionic detergents (such as Triton X-100) are effective in removing bound antigens from NC *(8)*.

High concentrations of salt and low concentrations of methanol increase immobilization efficiency *(22)*. NC is unique, when compared with other microporous membranes, in its ability to distinguish between single- and double-stranded nucleic acids, small and large proteins, short and long nucleic acids, and complexed versus uncomplexed molecules *(22)*.

It can be stained with amido black *(4)*, Coomassie Brilliant Blue (CBB) *(1)*, aniline blue black, Ponceau S, fast green, or toluidine blue. Amido black staining can detect a 25-ng dot of bovine serum albumin readily with acceptable background staining. The background staining tends to be higher with CBB while Ponceau S gives a very clean pattern but with slightly less sensitivity than amido black.

2.1.3. Disadvantages of NC

One clear disadvantage of NC is the fact that it cannot be stripped and reprobed multiple times owing to its fragile nature. It also has a tendency to become brittle when dry. In addition, small proteins tend to move through NC membranes and only a small fraction of the total amount actually binds. Using membranes with smaller pores can obviate this *(12)*. Gelatin-coated NC has been used for quantitative retention *(10, 23)*. In supported NC (e.g., Hybond-C Extra), the mechanical strength of the membrane has been improved by incorporating a polyester support web, thereby making handling easier.

2.2. Polyvinylidene Difluoride

Polyvinylidene difluoride (PVDF) is a linear polymer with repeating –(CF_2-CH_2)– units. The use of "di" in polyvinylidene *di*fluoride is redundant (including its use here) and its use needs to be discouraged *(2)*. Polyvinylidene fluoride or polyvinylidene difluoride refers to the same membrane first made available for protein blotting by Millipore in June of 1986. The product was renamed as Immobilon-P™ Transfer Membrane after being initially referred to as Immobilon™ PVDF transfer membrane to differentiate it from other PVDF and non-PVDF-based blotting membranes referred to collectively as Immobilon family and marketed by Millipore. Immobilon-P^{SQ} membrane with a 0.2-μm pore size suitable for proteins with a molecular weight less than 20,000 (to prevent blow through) and Immobilon-FL membrane optimized for all fluorescence applications also form part of the Immobilon family of PVDF membranes, added recently. Sequelon *(24)*, a PVDF-based sequencing membrane, sold by Milligen/BioSearch, a Millipore subsidiary is advantageous because of high protein binding capacity, physical strength, and chemical stability.

2.2.1. Immobilization Mechanism

Proteins transferred to the Immobilon-P membrane during western transfer are retained well on the membrane surface throughout the immunodetection process via a combination of dipole and hydrophobic interactions. The antigen binding capacity of the membrane is 170 μg/cm^2 for bovine serum albumin and this is proportionate with the binding capacity of NC. In addition, the Immobilon-P membrane has very good mechanical strength and like Teflon™ (a related fluorocarbon polymer) it is compatible with a range of chemicals and organic solvents [acetonitrile, trifluoroacetic acid, hexane, ethylacetate, and trimethylamine *(2, 25)]*.

Blotting mechanics are not different from those seen with NC, except that it is necessary to prewet the membrane in either methanol or ethanol before using with aqueous buffers. This is because PVDF is highly hydrophobic and there is no added surfactant in PVDF.

2.2.2. Advantages of PVDF

One of the advantages of electroblotting proteins onto PVDF membranes is that replicate lanes from a single gel can be used for various purposes such as N-terminal sequencing, proteolysis/peptide separation/internal sequencing along with western analysis. Proteins blotted to PVDF membranes can be stained with amido black, India ink, or silver nitrate *(26)*. These membranes are also amenable to staining with CBB, thus allowing excision of proteins for N-terminal protein sequencing, a procedure first demonstrated by Matsudaira in 1987 *(25)*.

2.3. Activated Paper

Activated paper (diazo groups) binds proteins covalently but is disadvantageous in that the coupling method is incompatible with many gel electrophoresis systems. Linkage is through primary amines, and therefore systems that use gel buffers without free amino groups must be used with this paper. In addition, the paper is expensive and the reactive groups have a limited half-life once the paper is activated.

2.4. Nylon Membranes

Nylon-based membranes are thin and smooth surfaced as NC but with much better durability. Two kinds of membranes are available commercially: Gene Screen and Zetabind (ZB). ZB is a nylon matrix (polyhexamethylene adipamine or Nylon 66) modified by the addition of numerous tertiary amino groups during the manufacturing process (extensive cationization). It has excellent mechanical strength and also offers the potential of very significant (yet reversible) electrostatic interactions between the membrane and polyanions. Nylon shows a greater protein binding capacity compared with NC (480 μg vs. 80-μg BSA bound/cm^2). In addition, nylon offers the advantages of more consistent transfer results and a significantly increased sensitivity compared with other membranes *(7, 18)*. This effect is possible owing to the extra potential difference created by the positive charge of ZB.

2.4.1. Disadvantages of Nylon

The high binding capacity of these membranes, however, produces higher nonspecific binding. Another problem with using nylon membranes is that they bind strongly to the commonly used anionic dyes such as CBB, amido black 10B *(18)*, aniline blue black, Ponceau S, fast green, or toluidine blue. SDS, dodecyl trimethylammonium bromide, or Triton X-100 at low concentrations (0.1% in water) remove the dyes from the membrane while simultaneously destaining the transferred proteins, with SDS being the best. Destaining of this membrane is thus not possible, unlike NC, and therefore the background remains as high as the signal *(8)*. On account of these problems, NC membranes have remained the best compromise for most situations. However, an immunological stain and India ink have been used to detect proteins on ZB *(27, 28)* and NC membranes.

Nylon membranes, especially the positively charged ZB membranes, have been found very useful in binding the negatively charged DNA. As a consequence it has been used more for DNA blotting than for protein blotting.

3. Buffers Used in Transfer Protocols

Commonly used buffers for western blotting are (a) Towbin system buffer [25 mM Tris, 192 mM glycine, 20% methanol (v/v), none to 0.01% SDS *(1)]* and (b) CAPS buffer system [CAPS: 10 mM 3-(cyclohexyl-amino)-1-propanesulfonic acid, 10% methanol (v/v), pH 11] for transfer to PVDF popularized by Matsudaira *(24)* for use prior to in situ blot sequencing. Transfer buffers without SDS are better, in general, when using Immobilon-P, since proteins have been reported to pass through the plane of the membrane in the presence of SDS *(29, 30)*. However, for proteins that have a tendency to precipitate, SDS should be in the buffer (<0.01%) during the transfer, and then one must fine-tune transfer time, current, etc. The Towbin system is used widely for applications that require immunodevelopment while the low ionic strength buffer system of Matsudaira *(25)* allows rapid transfer (ca. 10 min) and prevents introduction of additional Tris and glycine that is detrimental to sequence analysis using PVDF membranes.

Methanol, introduced originally by Towbin, is typically present in the transfer buffer and aids in stripping SDS from proteins transferred from denaturing SDS-containing polyacrylamide gels. It stabilizes the geometry of the gel during the transfer process, and tends to increase the binding capacity of NC for protein as well as helps proteins to bind better to NC membrane *(5, 8, 18, 31)*. Methanol can be eliminated completely from transfer buffer when using Immobilon-P membranes as well as NC. Ten to fifteen percent methanol is suggested for general protein transfer (standard Towbin buffer used 20% methanol). Methanol shrinks the gel, and therefore when transferring high molecular weight proteins (>150,000) best results are obtained without added methanol. Nonmethanolic transfer is also advised when enzyme activity needs to be preserved as well as when transferring conformation-sensitive antibodies. PAGE gels tend to swell in low ionic strength buffers in the absence of methanol. The "bands" may become distorted if this swelling is allowed to occur during protein transfer. Preswelling of the gel by incubating it in transfer buffer for 30 min to 1 h prior to transfer has been shown to prevent this problem *(5, 8)*.

4. Settings (Current/Voltage) for Protein Transfer

Some of the issues to be considered before electrotransfer include deciding on whether to use constant voltage or constant current and the use of tank of semidry electroblotting units. The use of constant voltage provides the best driving force (that is, potential difference) during transfer *(2)*. The buffer composition changes as salts are eluted from the gels, resulting in an increase in current and a drop in resistance *(8, 18)*. However, joule heating can cause an accompanying rise in current. Ohm's law states that voltage (V) = current (I) × resistance (R). A transfer using constant voltage leads to an increase in current and a decrease in resistance while a transfer using constant current leads to decrease in voltage as well as resistance ($I = V/R$). When current reaches over 500 mA in a constant voltage setting, heating can be a problem in tank buffer systems and the use of cooling elements has been recommended in such a scenario. However, constant voltage transfer can be efficiently carried out using heated buffer, from which methanol was omitted, to transfer high molecular weight proteins *(17, 32)*. Semidry blotters have been used to rapidly transfer proteins electrophoretically without excessive heat, using small volumes of buffer, short electrode distances, and planar electrodes that also serve as heat sinks *(33)*.

Low molecular weight proteins are preferentially eluted from the gel into the plane of the blotting membrane when a planar gel having electrophoretically resolved protein is exposed to a current perpendicular to its surface. As a result, large molecular weight proteins will be undertransferred under conditions optimized for transfer of low molecular weight polypeptides. On the other hand, a prolonged transfer will help the movement of large molecular weight species with accompanying loss of smaller species consequent to "blow through." A second sheet of membrane as a "backup" is useful to capture proteins that span a large molecular weight range. The use of gradient electric fields to reduce overall current use and allow the quantitative transfer of a wide range of proteins has been suggested *(18)*. Another approach involves a two-step electrotransfer beginning with elution of low molecular weight proteins at low current (1 mA/cm^2) for an hour followed by transfer at high current density (3.5–7.5 mA/cm^2), which aids the elution of high molecular weight proteins *(34)*. Recent work has shown the utility of heated buffer to transfer high molecular weight proteins rapidly *(17, 32)*.

5. Techniques to Transfer Proteins from Gel to Membrane

Transfer of proteins from SDS-PAGE or native gels to nitrocellulose or PVDF membranes has been achieved by (a) simple diffusion, (b) vacuum-assisted solvent flow, and (c) "western" blotting or electrophoretic elution *(4, 12, 35–39)*.

5.1. Simple Diffusion

Diffusion blotting was originally developed for transferring proteins separated by isoelectric focusing on thin gels to membranes and this was later expanded to other gel systems *(32, 40–46)*. In this procedure a membrane is placed on the gel surface with a stack of dry filter papers on top of the membrane. A glass plate and an object with a certain weight are usually placed on this assembly to enable the diffusion process. However, since there is no quantitative transfer of protein this protocol has not gained widespread acceptance. A waning interest in diffusion transfer was resuscitated when it was demonstrated that up to 12 blots can be obtained from a single gel by sandwiching it between two membranes sequentially (*see* Chapter "Non-electrophoretic bi-directional transfer of a single SDS-PAGE gel with multiple antigens to obtain twelve immunoblots") (**Fig. 2**) *(31)*.

Nonelectrophoretic membrane lifts from SDS-PAGE gels for immunoblotting, obtained by using this method, are very useful for identification of proteins by mass spectrometry *(47, 48)*. The gel can be stained with Coomassie following diffusion blotting. The antigens on the blot are detected by immunostaining, and the immunoblotted target band can be compared with the Coomassie-stained gel by superimposing the blot and the stained gel,

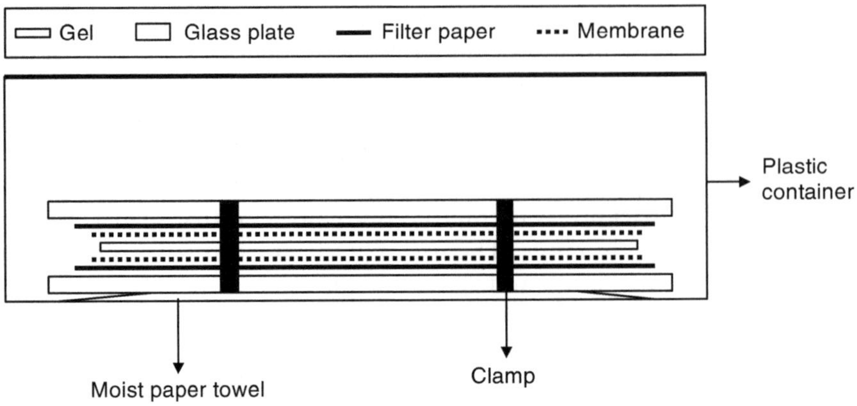

Fig. 2. Bidirectional, nonelectrophoretic transfer of proteins from SDS-PAGE gels to NC membranes to obtain up to 12 blots. The PAGE gel is sandwiched between two membranes, filter paper, and glass plates and incubated at 37°C for varying periods of time to obtain up to 12 blots (reproduced from *(10)* with permission from Elsevier).

allowing the identification of the band to be excised for tryptic digestion and subsequent matrix-assisted laser desorption time of flight mass spectrometric analysis. The main advantage of diffusion blotting compared with electroblotting is that several transfers or imprints can be obtained from the same gel and different antisera can be tested on identical imprints.

Subsequently, quantitative information regarding protein transfer during diffusion blotting was obtained using ^{14}C-labeled proteins. A 3-min diffusion blotting was shown to allow a transfer of 10% compared with electroblotting. Diffusion blotting of the same gels carried out multiple times for prolonged periods at 37°C causes the gel to shrink. This can be overcome by using gels cast on plastic supports *(44, 45)*.

Zymography or activity gel electrophoresis has also been studied with regard to the utility of diffusion. This involves the electrophoresis of enzymes (either nucleases or proteases) through discontinuous polyacrylamide gels containing enzyme substrate (either type III gelatin or β-casein). Following electrophoresis, SDS is removed from the gel by washing in 2.5% Triton X-100. This allows the enzyme to renature, and the substrate to be degraded. Staining of the gel with CBB (in the case of proteins) allows the bands of enzyme activity to be detected as clear bands of lysis against a blue background *(49)*. An additional immunoblotting analysis using another gel is often required in this procedure to examine a particular band that is involved. Diffusion blotting has been used to circumvent the use of a second gel for this purpose *(45)*. The activity gel was blotted onto PVDF for immunostaining and the remaining gel after blotting was used for routine “activity staining.” Since the blot and the activity staining are derived from the same gel, the signal localization in the gel and the replica can be easily aligned for comparison.

Diffusion blotting transfers 25–50% of the proteins to the membrane compared with electroblotting *(45)*. However, the advantage of obtaining multiple blots from the same gel could outweigh the loss in transfer and actually be compensated for by using sensitive detection techniques. The gel remains on its plastic support, which prevents stretching and compression; this ensures identical imprints and facilitates more reliable molecular mass determination. If only a few imprints are made, sufficient protein remains within the gel for general protein staining. These advantages make diffusion blotting the method of choice when quantitative protein transfer is not required.

5.2. Vacuum Blotting

This method was developed *(50)* as an alternative to diffusion blotting and electroblotting. The suction power of a pump connected to a slab gel dryer system was used to drive the separated polypeptides from the gel to the nitrocellulose membrane. Both low and high molecular weight proteins could be transferred using this method. Since small molecular weight proteins (±14,000) are not well adsorbed by the 0.45 μm membrane nitrocellulose, membranes with a small pore size (0.2 or 0.1 μm) should be used when using low molecular weight proteins.

The gel can dry out if the procedure is carried out over 45 min and in such a scenario enough buffer should be used. In some instances low-concentration polyacrylamide gels stick to the membrane following transfer. Rehydrating the gel helps detaching the nitrocellulose membrane from the gel remnants.

5.3. Electroblotting

Electroblotting is the most commonly used procedure to transfer proteins from a gel to a membrane. The main advantages are the speed and the completeness of transfer compared with diffusion or vacuum blotting. Electroelution can be achieved either by (a) complete immersion of a gel-membrane sandwich (**Fig. 3**) in a buffer

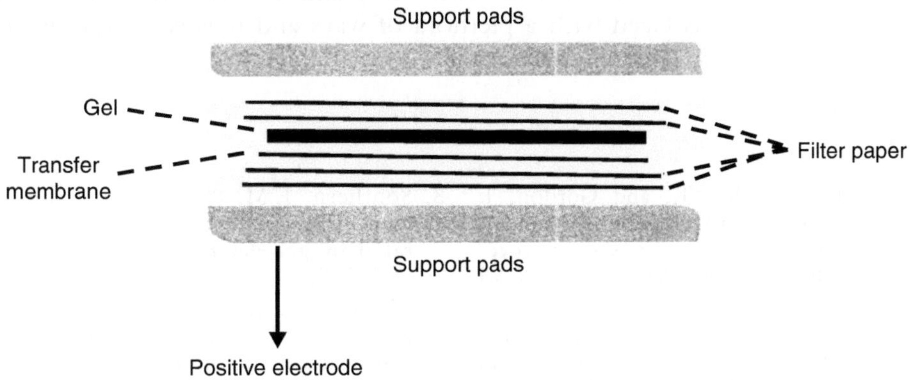

Fig. 3. The western blot transfer assembly (reproduced from *(10)* with permission from Elsevier).

(wet transfer) or by (b) placing the gel-membrane sandwich between absorbent papers soaked in transfer buffer (semidry transfer).

The transfer conditions as such are dependent on gel type, the immobilization membrane, the transfer apparatus used as well as the protein themselves. SDS gels, urea gels *(4)*, lithium dodecyl sulfate-containing gels, nondenaturing gels, two-dimensional gels, and agarose gels have been used for protein blotting (electrophoretic) *(18)*. The electric charge of the protein should be determined and the membrane should be placed on the appropriate side of the gel. When using urea gels the membrane should be placed on the cathode side of the gel *(4)*. Proteins from SDS PAGE gels are eluted as anions and therefore the filter should be placed on the anode side of the gel.

5.3.1. Wet Transfer

In this procedure, the sandwich is placed in a buffer tank with platinum wire electrodes. A large number of different apparatuses are available to efficiently transfer proteins (or other macromolecules) transversely from gel to membrane. Most of these, however, are based on the design of Towbin et al. *(1)*, that is, they have vertical stainless steel/platinum electrodes in a large tank.

5.3.2. "Semidry" Transfer

In semidry transfer, the gel-membrane sandwich is placed between carbon plate electrodes. *Semidry* or *horizontal* blotting uses two plate electrodes (stainless steel or graphite/carbon) for uniform electrical field over a short distance, and sandwiches between these up to six gel/membrane/filter paper assemblies, all well soaked in the transfer buffer. The assembly is clamped or otherwise secured on its side, and electrophoretic transfer is effected in this position, using as transfer buffer only the liquid contained in the gel and filter papers or other pads in the assembly.

The advantages to this procedure over the conventional upright protocol are that (a) gels can be blotted simultaneously, (b) electrodes can be cheap carbon blocks, and (c) less power is required for transfer (and therefore a simpler power pack).

As will be seen in the following chapters, protein blotting has been evolving constantly and now the scientific community is faced with a plethora of ways and means of transferring and detecting proteins.

References

1. Towbin, H., Staehelin, T., and Gordon, J. (1979) Electrophoretic transfer of proteins from polyacrylamide gels to NC sheets: procedure and applications. *Proc Natl Acad Sci USA* 76, 4350–4354.
2. LeGendre, N. (1990). Immobilon-P transfer membrane: applications and utility in protein biochemical analysis. *Biotechniques* 9 (6 Suppl), 788–805. Review.
3. Southern, E.M. (1975). Detection of specific sequences among DNA fragments separated by gel electrophoresis. *J Mol Biol* 98, 503–517.
4. Alwine, J. C., Kemp, D. J., Stark, G.R. (1977) Method for detection of specific RNAs in agar gels by transfer to diazobenzyloxymethyl-paper and hybridization with DNA probes. *Proc Natl Acad Sci USA* 74, 5350–5354.

5. Burnette, W.N. (1981) "Western blotting": electrophoretic transfer of proteins from sodium dodecyl sulfate–polyacrylamide gels to unmodified NC and radiographic detection with antibody and radioiodinated protein A. *Anal Biochem* 112, 195–203.
6. Laemmli, U.K. (1970) Cleavage of structural proteins during assembly of the head of bacteriophage T4. *Nature* 227, 680–685.
7. Kost, J., Liu, L-S., Ferreira, J., and Langer, R. (1994) Enhanced protein blotting from PhastGel media to membranes by irradiation of low-intensity. *Anal Biochem* 216, 27–32.
8. Gershoni, J.M. and Palade, G.E. (1982) Electrophoretic transfer of proteins from sodium dodecyl sulfate-polyacrylamide gels to a positively charged membrane filter. *Anal Biochem* 124, 396–405.
9. Gershoni, J.M. (1988) Protein blotting: a manual. *Methods Biochem Anal*33, 1–58. Review.
10. Kurien, B.T. and Scofield, R.H. (2006) Western blotting. *Methods* 38, 283–293.
11. Karey, K.P. and Sirbasku, D.A. (1989) Glutaraldehyde fixation increases retention of low molecular weight proteins (growth factors) transferred to nylon membranes for western blot analysis. *Anal Biochem* 178, 255–259.
12. Harlow, E. and Lane, D. (1988) Immunoblotting. In: Antibodies. A Laboratory Manual, Harlow, E. and Lane, D. (eds.). Cold Spring Harbor Laboratory, New York, p. 485.
13. Renart, J., Reiser, J., and Stark, G.R. (1979) Transfer of proteins from gels to diazobenzyloxymethyl paper and detection with anti-sera: a method for studying antibody specificity and antigen structure. *Proc Natl Acad Sci USA*, 76, 3116–3120.
14. Elkon, K.B., Jankowski, P.W., and Chu, J.L. (1984) Blotting intact immunoglobulins and other high-molecular-weight proteins after composite agarose-polyacrylamide gel electrophoresis. *Anal Biochem* 140, 208–213.
15. Gibson, W. (1981). Protease-facilitated transfer of high-molecular-weight proteins during electrotransfer to NC. *Anal Biochem* 118, 1–3.
16. Bolt, M.W. and Mahoney, P.A. (1997) High efficiency blotting of proteins of diversesizes following sodium dodecyl sulfate-polyacrylamide gel electrophoresis. *Anal Biochem* 247, 185–192.
17. Kurien, B.T. and Scofield, R.H. (2002) Heat mediated, ultra-rapid electrophoretic transfer of high and low molecular weight proteins to NC membranes. *J Immunol Methods* 266, 127–133.
18. Gershoni, J.M. and Palade, G.E. (1983) Protein blotting: principles and applications. *Anal Biochem* 131, 1–15.
19. Thornton,D.J.,Carlstedt,I.,andSheehan,J.K. (1996) Identification of glycoproteins on nitrocellulose membranes and gels. *Mol Biotechnol* 5, 171–176.
20. Tonkinson, J.L. and Stillman, B. (2002) NC: a tried and true polymer finds utility as a post-genomic substrate. *Front Biosci* 7, c1–c12. Review.
21. Lauritzen, E., Masson, M., Rubin, I., Bjerrum, O.J., and Holm, A. (1993) Peptide dot immunoassay and immunoblotting: electroblotting from aluminum thin-layer chromatography plates and isoelectric focusing gels to activated NC. *Electrophoresis* 14, 852–859.
22. Masson, M., Lauritzen, E., and Holm, A. (1993) Chemical activation of NC membranes for peptide antigen–antibody binding studies: direct substitution of the nitrate group with diaminoalkane. *Electrophoresis* 14, 860–865.
23. Too, C.K., Murphy, P.R., and Croll, R.P. (1994) Western blotting of formaldehyde-fixed neuropeptides as small as 400 daltons on gelatin-coated NC paper. *Anal Biochem* 219, 341–348.
24. Coull, J.M., Dixon, J.D., Laursen, R.A., Koester, H., and Pappin, D.J.C. (1989) Development of membrane supports for the solid-phase sequence analysis of proteins and peptides. In: Methods in Protein Sequence Analysis, Witmann-Liebold, B. (ed.). Springer, Berlin, pp. 69–78.
25. Matsudaira, P. (1987) Sequence from picomole quantities of proteins electroblotted onto polyvinylidene difluoride membranes. *J Biol Chem* 262,10035–10038.
26. Pluskal, M.F., Przekop, M.B., Kavonian, M.R., Vecoli, C., and Hick, D.A. (1986) *Biotechniques* 4, 272–282.
27. Kittler, J.M., Meisler, N.T., Viceps-Madore, D., Cidlowski, J.A., Thanassi, J.W. (1984) A general immunochemical method for detecting proteins on blots. *Anal Biochem* 137, 210–216.
28. Hughes, J.H., Mack, K., and Hamparian, V.V. (1988) India ink staining of proteins on nylon and hydrophobic membranes. *Anal Biochem* 173, 18–25.
29. Tovey, E.R. and Baldo, B.A. (1989) Protein binding to NC, nylon and PVDF membranes in immunoassays and electroblotting. *J Biochem Biophys Methods* 19, 169–183.
30. Xu, Q.Y. and Shively, J.E. (1988) Microsequence analysis of peptides and proteins. VIII. Improved electroblotting of proteins

onto membranes and derivatized glass-fiber sheets. *Anal Biochem* 170, 19–30.

31. Nielsen, P.J. (1982) The phosphorylation of ribosomal protein S6 in rat tissues following cycloheximide injection, in diabetes, and after denervation of diaphragm. A simple immunological determination of the extent of S6 phosphorylation on protein blots. *J Biol Chem* 257, 12316–12321.
32. Kurien, B.T. and Scofield, R.H. (1997) Multiple immunoblots after non-electrophoretic bidirectional transfer of a single SDS-PAGE gel with multiple antigens. *J Immunol Methods* 205, 91–94.
33. Kyhse-Andersen, J. (1984) Electroblotting of multiple gels: a simple apparatus without buffer tank for rapid transfer of proteins from polyacrylamide to nitrocellulose. *J Biochem Biophys Methods* 10, 203–209.
34. Otter, T., King, S.M., and Witman, G.B. (1987) A two-step procedure for efficient electro transfer of both high-molecular weight (greater than 400,000) and low-molecular weight (less than 20,000) proteins. *Anal Biochem* 162, 370–377.
35. Harper, D.R., Kit, M.L., and Kangro, H.O. (1990) Protein blotting: ten years on. *J Virol Methods* 30, 25–39. Review.
36. Egger, D. and Bienz, K. (1994) Protein (western) blotting. *Mol Biotechnol* 1, 289–305.
37. Wisdom, G.B. (1994) Protein blotting. *Methods Mol Biol* 32, 207–213.
38. Kurien, B.T. and Scofield, R.H. (2003) Protein blotting: a review. *J Immunol Methods* 274, 1–15. Review.
39. Kurien, B.T. and Scofield, R.H. (2005) Blotting techniques, In: Encyclopedia of Analytical Science, Second Edition (Worsfold, P.J., Townshend, A., and Poole, C.F., eds.). Elsevier, Oxford, p. 425.
40. Reinhart, M.P. and Malamud, D. (1982) Protein transfer from isoelectric focusing Gels: the native blot. *Anal Biochem* 123, 229–235.
41. Jagersten, C., Edstrom, A., Olsson, B., and Jacobson, G. (1988) Blotting from PhastGel media after horizontal sodium dodecyl sulfate-polyacrylamide gel electrophoresis. *Electrophoresis* 9, 662–665.
42. Kazemi, M. and Finkelstein R.A. (1990) Checkerboard immunoblotting (CBIB): an efficient, rapid, and sensitive method of assaying multiple antigen/antibody cross-reactivities. *J Immunol Methods* 128, 143–146.
43. Heukeshoven, J. and Dernick, R. (1995). Effective blotting of ultrathin polyacrylamide gels anchored to a solid matrix. *Electrophoresis* 16, 748–756.
44. Olsen, I. and Wiker, H.G. (1998) Diffusion blotting for rapid production of multiple identical imprints from sodium dodecyl sulfate polyacrylamide gel electrophoresis on a solid support. *J Immunol Methods* 220, 77–84.
45. Chen, H. and Chang, G.D. (2001) Simultaneous immunoblotting analysis with activity gel electrophoresis in a single polyacrylamide gel. *Electrophoresis* 22, 1894–1899.
46. Bowen, B., Steinberg, J., Laemmli, U.K., and Weintraub, H. (1980) The detection of DNA-binding proteins by protein blotting. *Nucleic Acids Res* 8, 1–20.
47. Kurien, B.T. and Scofield, R.H. (2000) Association of neutropenia in systemic lupus erythematosus with anti-Ro and binding of an immunologically cross-reactive neutrophil membrane antigen. *Clin Exp Immunol* 120, 209–217.
48. Kurien, B.T., Matsumoto, H., and Scofield, R.H. (2001) Purification of tryptic peptides for mass spectrometry using polyvinylidene fluoride membrane. *Indian J Biochem Biophys* 38, 274–276.
49. Bischoff, K.M., Shi, L., and Kennelly, P.J. (1998) The detection of enzyme activity following sodium dodecyl sulfate-polyacrylamide gel electrophoresis. *Anal Biochem* 260, 1–17. Review.
50. Peferoen, M., Huybrechts, R., and De Loof, A. (1982) Vacuum-blotting: a new simple and efficient transfer of proteins from sodium dodecyl sulfate-polyacrylamide gels to NC. *FEBS Lett* 145, 369–372.

Chapter 4

Simultaneous Immunoblotting Analysis with Activity Gel Electrophoresis and 2-D Gel Electrophoresis

Der-Yen Lee and Geen-Dong Chang

Summary

Diffusion blotting method can couple immunoblotting analysis with another biochemical technique in a single polyacrylamide gel, however, with lower transfer efficiency as compared with the conventional electroblotting method. Thus, with diffusion blotting, a protein blot can be obtained from a sodium dodecyl sulfate polyacrylamide gel for zymography assay, from a native polyacrylamide gel for electrophoretic mobility shift assay (EMSA), or from a two-dimensional (2-D) polyacrylamide gel for large-scale screening and identification of a protein marker. Therefore, a particular signal in zymography, EMSA, and 2-D gel can be confirmed or identified by simultaneous immunoblotting analysis with a corresponding antiserum. These advantages make diffusion blotting desirable when partial loss of transfer efficiency can be tolerated or can be compensated by a more sensitive immunodetection reaction using enhanced chemiluminescence substrates.

Key words: Diffusion blotting, Zymography, Electrophoretic mobility shift assay, Autoantigen

1. Introduction

Diffusion blotting is originally developed for isoelectric focusing (IEF) gels or ultrathin gels *(1–4)*, which requires laying a blotting membrane on the gel surface and a stack of dry filter paper on top of the blotting membrane. Usually a glass plate and an object carrying certain weight is further stacked on the filter to facilitate the diffusion process. It is more suitable using diffusion blotting than the electrophoretic blotting for certain applications such as electrophoresis on gels bound on plastic sheets *(3, 4)*, multiple blotting from a single gel *(1, 4, 5)*, and simultaneous

Biji T. Kurien and R. Hal Scofield (eds.), *Methods in Molecular Biology, Protein Blotting and Detection, vol. 536*

DOI: 10.1007/978-1-59745-542-8_4

immunoblotting analysis with activity gel electrophoresis such as proteolytic zymography and electrophoretic mobility shift assay (EMSA) *(6)*.

We have previously demonstrated several applications of diffusion blotting *(6)*. They are (1) multiple blotting, (2) combined immunoblotting analysis with gel staining, (3) combined immunoblotting analysis with proteolytic zymography, and (4) combined immunoblotting analysis with EMSA from a single gel. In activity gel electrophoresis, several positive signals are frequently observed, which requires an additional immunoblotting analysis in another gel to examine a particular protein involved. A method of simultaneous immunoblotting analysis with EMSA has been reported as "shift-western blotting" *(7)*. In this method, a nitrocellulose filter and an anion-exchange membrane are stacked together for electroblotting of proteins and the radioactive DNA, respectively, following native gel electrophoresis. The protein–DNA complex was detected by autoradiography of the DNA blot and by immunoblot analysis of the protein blot *(7–10)*. Simultaneous immunoblotting analysis can also be done with activity gel electrophoresis on the same gel with diffusion blotting. Because the blot and the activity staining are derived from the same gel, the localization of signals in the gel and the replica can be easily aligned for comparison. In this chapter we describe protocols for combined immunoblotting analysis with proteolytic zymography in identifying nephrosin in carp tissue extracts (**Fig. 1**) and combined multiple immunoblotting analysis with silver staining of mouse

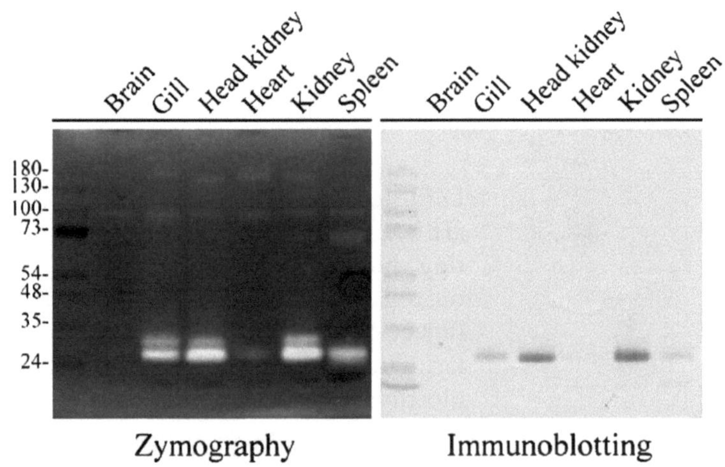

Fig. 1. Simultaneous immunoblotting analysis for nephrosin and proteolytic zymography assay of carp tissue extracts. Four microliters of carp tissue extracts was analyzed by SDS-PAGE in a gel containing 0.2% gelatin. After electrophoresis, the proteins were blotted by diffusion blotting and immunodetected by an antinephrosin antiserum (*right panel*). The remaining gel was treated for the proteolytic zymography assay (*left panel*).

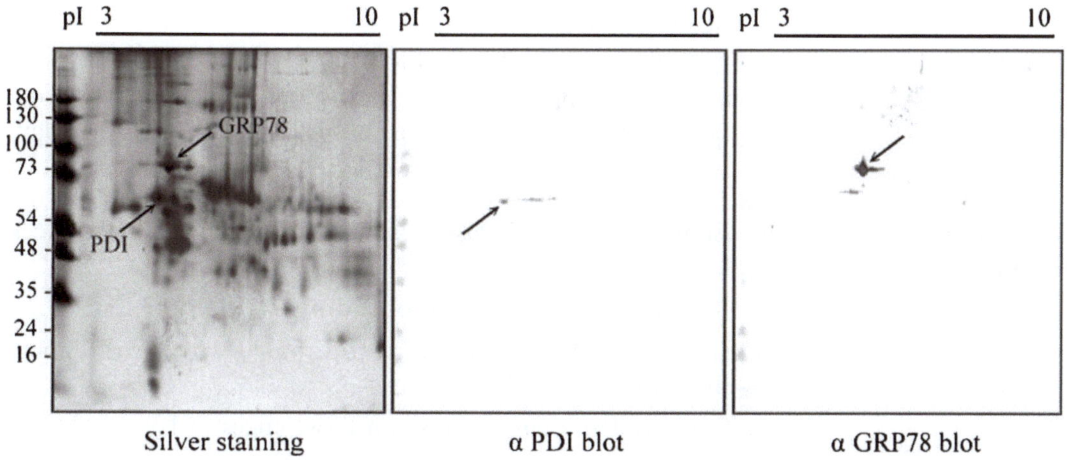

Fig. 2. Simultaneous multiple immunoblotting analysis and silver staining of mouse liver microsomal proteins resolved by 2-DE. Forty micrograms of mouse liver microsomal proteins was resolved by a 7 cm × 8 cm, 2-D polyacrylamide gel. After 2-D gel electrophoresis, the gel was blotted with two pieces of PVDF membrane for 2 h. One membrane was then probed with an anti-PDI and the other with an anti-GRP78 serum. The remaining gel was subjected to silver staining to reveal each protein spot. Theoretical pI/M_w of PDI and GRP78 are 4.77/57058.49 and 5.07/72332.96, respectively

microsomal proteins resolved by 2-D gel electrophoresis in identifying endoplasmic reticulum markers such as protein disulfide isomerase (PDI) and glucose-regulated protein 78 (GRP78) (**Fig. 2**). Potentially, simultaneous immunoblotting analysis with silver staining on a 2-D gel can be applied in large-scale screening and identification of protein markers such as antoantigens and tumor markers *(11–13)* in the future.

2. Materials

2.1. Preparation of Carp Tissue Extracts for Zymography (14)

1. Fresh carp tissues: brains, gills, head kidneys, heart, kidneys, spleens.
2. TE buffer: 20 mM Tris–HCl, pH 8, 5 mM EDTA.
3. 2× Sodium dodecyl sulfate (SDS) sample buffer: 0.1 M Tris–HCl, pH 7.8, 8% SDS, 24% glycerol, and 0.002% bromophenol blue.

2.2. Preparation of Mouse Liver Microsomal Fraction (15) for IEF

1. Fresh ICR mouse livers.
2. Extraction buffer: 20 mM Tris–HCl, pH 8, 0.15 M NaCl, 0.25 M sucrose, and 1 mM dithioerythreitol (DTE).
3. 10% Triton X-100 solution.
4. 20% Trichloroacetic acid (TCA) solution.

5. Acetone is kept at –20°C.
6. Rehydration buffer contains 1.5% 3-10 IPG buffer (Amersham Biosciences Ltd., Piscataway, NJ, USA) and 100 mM DTE in FOCUS™ extraction buffer III (Geno Technology, Inc., Maryland Heights, MO, USA) (*see* **Note 1**).

2.3. Isoelectric Focusing

1. Ettan IPGphor Focusing system (Amersham Biosciences Ltd.).
2. Rehydration buffer (*see* **Subheading 2.2, item 6**).
3. IPG strip: Immobiline™ DryStrip, pH 3–10, 7 cm (Amersham Biosciences Ltd.).
4. Strip holder, 7 cm (Amersham Biosciences Ltd.).
5. IPG Cover Fluid (Amersham Biosciences Ltd.).
6. SDS equilibration buffer: 0.1 M Tris–HCl, pH 7.8, 1% SDS, 0.002% bromophenol blue.

2.4. SDS-PAGE (16)

1. Hoefer SE-250 and SE-260 Mighty small II gel system (Amersham Biosciences Ltd.).
2. Acrylamide/bis-acrylamide solution (50% T; 3% C) contains 48.5 g of acrylamide and 1.5 g of bis-acrylamide in 100 mL of aqueous solution.
3. Gel buffer: 1.5 M Tris–HCl, pH 8.45.
4. Stacking gel buffer: 0.1 M Tris–HCl, pH 7.8, 0.4% SDS.
5. Prepare 10% ammonium persulfate solution by dissolving 0.1 g of ammonium persulfate into 1 mL of deionized water (*see* **Note 2**).
6. TEMED.
7. Glycerol.
8. 2% Gelatin (from porcine skin): Dissolve gelatin in deionized water heated in a water bath.
9. Prestained protein ladder (PageRuler™; Fermentas, Inc., MD, USA).
10. Chromatography paper (3MM Chr; Whatman, Inc., Florham Park, NJ, USA).
11. Marker strip: Blot one end of a 3-mm-wide strip of 3MM chromatography paper with 5 μL of prestained marker and leave it air-dried for 5 min (*see* **Note 3**).

2.5. Diffusion Blotting (6)

1. Millipore Immobilon-P PVDF membrane: 8.6 cm × 6.5 cm and 8.6 cm × 8.8 cm.
2. Chromatography paper (3MM Chr; Whatman).
3. Methanol.

2.6. Staining for Gelatin Zymography (6)

1. Tris/Triton solution: 20 mM Tris–HCl, pH 8, 2% Triton X-100.
2. Incubation buffer: 20 mM Tris–HCl, pH 8, 0.2 mM $CaCl_2$, 0.1 mM $ZnCl_2$.
3. Coomassie blue R-250 solution: 0.1% Coomassie Brilliant Blue R-250, 40% methanol, and 7% acetic acid.
4. Destaining solution: 0.1% Triton X-100 and 7% acetic acid.

2.7. Staining for 2-DE (17)

1. 20% TCA solution.
2. Sensitizing solution: Prepare 12.5 mM sodium thiosulfate and 0.8 M sodium acetate in 30% ethanol as stock solution and add 0.2 mL of glutardialdehyde (25%) to 40 mL of the stock solution prior to use (*see* **Note 4**).
3. Silver nitrate solution: Prepare 0.25% silver nitrate as stock solution and add 20 μL of formaldehyde (37%) to 40 mL of the stock solution prior to use (*see* **Note 4**).
4. Developing solution: Prepare 2.5% sodium carbonate as stock solution and add 10 μL of formaldehyde to 40 mL of the stock solution prior to use (*see* **Note 4**).
5. 4% Acetic acid.

2.8. Immunoblotting (18)

1. Phosphate buffered saline (PBS).
2. PBST: 0.1% Tween-20 in PBS.
3. Blocking solution: 3% skim milk in PBST.
4. Primary antibody solution: Anti-nephrosin, anti-GRP78, and anti-PDI were raised in guinea pigs in our laboratory. They are used at a titer of 1:2,000 in PBST containing 0.3% BSA.
5. Secondary antibody solution: 0.2 μg/mL horse radish peroxidase (HRP)-conjugated anti-guinea pig IgG (Jackson Lab, Inc., West Grove, PA, USA) in 0.3% BSA in PBST.
6. $NiCl_2$-DAB solution: Prepare 0.5 mg of 3′,3′-diaminobenzidine (DAB) and 0.5 mL of 1% $NiCl_2$ in 10 mL of PBS prior to use.
7. Hydrogen peroxide (35% H_2O_2).

3. Methods

3.1. Preparation of Carp Tissue Extracts for Zymography (14)

1. Brains, gills, head kidneys, hearts, kidneys, and spleens are freshly obtained from a male carp sacrificed by decapitation.
2. At a ratio of 10 mL of cold TE buffer to 1 g of carp tissues, carp tissues are homogenized in an ice bath and the homogenates

are centrifuged at 27,000 × *g* at 4°C for 30 min. The supernatant fractions are stored at –20°C.

3. Add 20 μL of 2× SDS sample buffer to 20 μL of supernatant fractions, and boil them at 100°C for 5 min. The samples are ready for SDS-PAGE.

3.2. Preparation of Mouse Liver Microsomal Fraction (15) for IEF

1. Livers are isolated from ICR mice (killed by CO_2-induced asphyxiation) and immediately frozen on dry ice.
2. Twenty grams of mouse livers is homogenized with the addition of 30 mL of cold extraction buffer in an ice bath and the homogenate is centrifuged at 27,000 × *g* at 4°C for 30 min. The supernatant fraction contains cytosolic proteins and microsomes except most mitochondria, nuclei, large organelles, cells, and cell debris.
3. The supernatant fraction is centrifuged at 100,000 × *g* for 1 h and the microsomes are pelleted. The pellet is suspended in 60 mL of extraction buffer and centrifuged at 100,000 × *g* for 1 h to remove excess cytosolic proteins. After a second wash with 60 mL of extraction buffer, the pellet is suspended in 5 mL of extraction buffer and stored at –20°C in 0.6-mL aliquots.
4. Add 100 μL of 10% Triton X-100 to 100 μL of microsomal fraction (about 2 mg of proteins) to lyse microsomes by vigorous vortexing until the solution becomes clear.
5. Add 200 μL of 20% TCA to the lysate to precipitate proteins by vigorous vortexing for few seconds, and centrifuge at 6,000 × *g* for 30 s. Discard the supernatant.
6. Disperse the pellet in 200 μL of deionized water. Add 1 mL of –20°C precooled acetone, and centrifuge at 6,000 × *g* for 30 s. Discard the supernatant.
7. Repeat **step 6** three times. Finally, disperse the pellet in 20 μL of acetone with repeated pipetting. Dry the pellet in a chemical hood for 2–3 h.
8. Solubilize the pellet in 100 μL of rehydration buffer and centrifuge at 14,000 × *g* for 20 min. Transfer the supernatant into a new microfuge tube, and the sample is ready for IEF.

3.3. Isoelectric Focusing

1. IEF is performed with the Ettan IPGphor Focusing system.
2. Ten microliters of sample (about 40 μg of proteins) was diluted with 115 μL of rehydration buffer and centrifuged at 14,000 × *g* for 20 min. Apply supernatant to the strip holder channel and avoid formation of air bubbles.
3. Remove the protective cover foil from an IPG strip. Position the IPG strip with the gel side down and direct the acidic (anodic) end of the strip toward the anodic end (+) of the

strip holder. Hold the basic (cathodic) end of the IPG strip and lower the acidic end of the strip onto the solution. Push the IPG strip toward the anodic end of the strip holder and lower the strip onto the solution at the same time. Slide the IPG strip back and forth along the surface of the solution to coat the entire strip. Be careful not to trap air bubbles under the IPG strip.

4. Add IPG Cover Fluid dropwise into both ends of the strip holder until the entire IPG strip is covered. Place the cover on the strip holder to maintain good contact between the gel and electrodes, and put the holder onto the platform of IPGphor. Position anodic (+) end of the strip holder on anode (+) and cathodic end (–) of the strip holder on cathode (–) and then cover the lid of IPGphor.
5. After rehydration at 30 V for 10 h, start the IEF until the total voltage-hour reaches 20,000–30,000.
6. Transfer the IPG strip to a tray and soak the IPG strip in the SDS equilibration buffer twice, each for 10 min, with gentle agitation (*see* **Note 5**).

3.4. SDS-PAGE

3.4.1. SDS-PAGE of Sample for Gelatin Zymography (8)

1. Modified Tricine SDS-PAGE is used to resolve the sample for gelatin zymography, and electrophoresis is performed with Hoefer SE-250 Mighty small II gel system.
2. Prepare 7.5% separation gel by mixing 4.5 mL of acrylamide/bis-acrylamide solution, 10 mL of gel buffer, 4 g of glycerol, 8.5 mL of deionized water, 3 mL of 2% gelatin, 0.1 mL of 10% ammonium persulfate solution, and 10 μL of TEMED. Cast gel within 8 cm × 10 cm × 0.75 mm gel cassette, leave space for a stacking gel, and overlay with deionized water.
3. Prepare stacking gel by mixing 1 mL of acrylamide/bis solution, 3 mL of stacking gel buffer, 1 g of glycerol, 7 mL of deionized water, 0.1 mL of 10% ammonium persulfate solution, and 10 μL of TEMED. Then insert a 10-well comb into the stacking gel solution immediately.
4. Apply 4 μL of each sample and 5 μL of prestained marker to each well and complete the electrophoresis at 150 V for 60 min with water cooling system.

3.4.2. SDS-PAGE of Sample for 2-DE

1. Modified Tricine SDS-PAGE is used to resolve the sample for 2-DE and the electrophoresis is performed with Hoefer SE-260 Mighty small II gel system.
2. Prepare 7.5% separation gel by mixing 4.5 mL of acrylamide/bis solution, 10 mL of gel buffer, 4 g of glycerol, 11.5 mL of deionized water, 0.1 mL of 10% ammonium persulfate solution, and 10 μL of TEMED. Cast the gel within 10.5 cm × 10 cm × 1.0 mm

gel cassette, leave space for a stacking gel, and overlay with deionized water.

3. Prepare stacking gel by mixing 1 mL of acrylamide/bis solution, 3 mL of stacking gel buffer, 1 g of glycerol, 7 mL of deionized water, 0.1 mL of 10% ammonium persulfate solution, and 10 μL of TEMED. Leave space for positioning an IPG strip and overlay with deionized water.
4. Position an equilibrated IPG strip onto the stacking gel and leave space near the acidic end of the IPG strip for insertion of the marker strip. Insert the marker strip onto the stacking gel and start SDS gel electrophoresis as soon as possible. Electrophoresis is complete at 150 V for 120 min with water cooling.

3.5. Diffusion Blotting

1. PVDF membranes from different sources are equally efficient to retain proteins blotted by diffusion blotting. However, Millipore Immobilon-P PVDF membrane is used here.
2. Wet PVDF membranes with 100% methanol and rinse them in deionized water three times before this procedure.
3. After electrophoresis, remove excess cathode buffer from the gel assembly. Lever the glass plate up with one spacer and remove the glass plate. Most of the time, the gel remains firmly attached on the aluminum plate.
4. Press one piece of moistened PVDF membrane between two pieces of chromatography paper to remove excess water. Cover the gel surface with PVDF membrane. Remove air bubbles, if any, between the gel and the PVDF membrane with a narrow weighing spatula by gently scratching the PVDF membrane from one end to the other.
5. If a second blot is required, the gel can be overlaid with another PVDF membrane on the other gel surface. With the aluminum plate side up, peel the gel from the aluminum plate, set the PVDF membrane down on a piece of chromatography paper, and make the other side of the gel facing up. Cover the gel surface with another piece of PVDF membrane. Remove air bubbles between the gel and the second PVDF membrane.
6. Cover with a piece of chromatography paper on the PVDF membrane and place a piece of glass plate of similar size on top to maintain the gel–membrane contact. Diffusion blotting should be done satisfactorily in 1–2 h (*see* **Note 6**).
7. After diffusion blotting, peel the PVDF membrane and rinse it with deionized water for three times. The PVDF membrane can be processed for immunoblotting immediately (*see* **Note 7**).

Meanwhile, the gel can be stained to reveal the protein profiles.

3.6. Staining for Gelatin Zymography

1. Activation of proteases within the electrophoresed gelatin gel occurs in conditioning solution after renaturation of protease by removing bound SDS. However, different proteases should be treated in different incubation conditions.
2. After diffusion blotting, soak the gelatin gel in Tris/Triton solution with constant agitation for 20 min, twice. Rinse the gel with deionized water and develop protease activities by agitating the gel in incubation buffer at 37°C for 3 h.
3. Rinse the gel with deionized water and agitate the gel in Coomassie blue R-250 solution for 1–3 h until the entire gel becomes deep blue. Then soak the gel in destaining solution with constant agitation for few hours until the appearance of transparent bands.

3.7. Staining for 2-DE

1. Modified silver staining protocol for 2-DE gels is based on the description of Heukeshoven and Dernick *(17)*.
2. Wash a 2-DE gel with deionized water twice, each for 15 min, and then agitate the gel in 20% TCA solution for 1 h to fix proteins within the gel after diffusion blotting.
3. Wash the gel with deionized water three times, each for 5 min, to remove excess TCA.
4. Soak the gel in sensitizing solution with constant agitation for 30 min.
5. Wash the gel with deionized water twice, each for 5 min, to remove excess sensitizing solution.
6. Soak the gel in silver nitrate solution with constant agitation for 20 min.
7. Wash the gel with deionized water twice, each for 20 s. After washing the gel, agitate it in developing solution for several seconds to few minutes under observation until staining is satisfactory.
8. Stop developing reaction by agitating the gel within refilled 4% acetic acid for 5 min. After washing the gel in deionized water three times, each for 5 min, keep the gel in deionized water.

3.8. Immunoblotting (18)

1. Blotted PVDF membranes can be processed directly for immunoblotting after washing with deionized water. For stored membrane, wet the membrane with 100% methanol and then briefly rinse the membrane with deionized water, three times.

2. Agitate a blotted PVDF membrane in blocking solution for 1–2 h. When blocking is complete, rinse the membrane with PBS once.
3. Pour the primary antibody solution into the container and agitate the membrane for 1–2 h. Wash the membrane with PBST twice, each for 5 min, and PBS once for 5 min.
4. Pour the secondary antibody solution into the container and agitate the membrane for 1–2 h. Wash the membrane with PBST twice, each for 5 min, and PBS once for 5 min.
5. Submerge and agitate the membrane within 10 mL of $NiCl_2$-DAB solution for about 10 s, and add 10 μL of H_2O_2 to activate HRP activity. Rock the container until the image becomes apparent, pour the waste into a reservoir containing bleach, and rinse the membrane with tap water. Flush the membrane under tap water for few minutes and preserve the membrane after being air-dried.

4. Notes

1. Some proteins are very resistant to reduction, so we have used up to 100 mM DTE or dithiothreitol routinely.
2. The ammonium persulfate solution would expire after 2 weeks.
3. Marker strips should be prepared fresh, and avoid prolonged drying because some proteins would absorb onto the paper permanently.
4. Glutardialdehyde and formaldehyde are highly volatile and should not be prepared in stock solution.
5. The IPG strip can be immediately subjected to second dimension of SDS-PAGE or be frozen at –20°C for later use.
6. The transfer time for diffusion blotting should be adjusted for gels with different thickness. We routinely blot gel of 0.75 mm for 1 h and gel of 1 mm for 2 h.
7. After three washes with deionized water, the membrane can be air-dried and stored for later use.

Acknowledgements

This work was supported by grants (to G.D.C.) from the National Science Council, Republic of China.

References

1. Reinhart, M.P. and Malamud, D. (1982) Protein transfer from isoelectric focusing gels: the native blot. *Anal Biochem.* 123, 229–235.
2. De Keyser, F., Verbruggen, G., Veys, E.M., Nimmegeers, J., Schatteman, L., Goethals, K., et al. (1990) "Microgel diffusion blotting" for sensitive detection of antibodies to extractable nuclear antigens. *Clin Chem.* 36, 337–339.
3. Heukeshoven, J. and Dernick, R. (1995) Effective blotting of ultrathin polyacrylamide gels anchored to a solid matrix. *Electrophoresis* 16, 748–756.
4. Olsen, I. and Wiker, H.G. (1998) Diffusion blotting for rapid production of multiple identical imprints from sodium dodecyl sulfate polyacrylamide gel electrophoresis on a solid support. *J Immunol Methods.* 220, 77–84.
5. Kurien, B.T. and Scofield, R.H. (1997) Multiple immunoblots after non-electrophoretic bidirectional transfer of a single SDS-PAGE gel with multiple antigens. *J Immunol Methods.* 205, 91–94.
6. Chen, H. and Chang, G.D. (2001) Simultaneous immunoblotting analysis with activity gel electrophoresis in a single polyacrylamide gel. *Electrophoresis* 22, 1894–1899.
7. Demczuk, S., Harbers, M., and Vennstrom, B. (1993) Identification and analysis of all components of a gel retardation assay by combination with immunoblotting. *Proc Natl Acad Sci USA* 90, 2574–2578.
8. Reddy, S.V., Alcantara, O., and Boldt, D.H. (1998) Analysis of DNA binding proteins associated with hemin-induced transcriptional inhibition. The hemin response element binding protein is a heterogeneous complex that includes the Ku protein. *Blood* 91, 1793–1801.
9. Iwata, T., Sato, S., Jimenez, J., McGowan, M., Dey, A., Ibaraki, N., et al. (1999) Osmotic response element is required for the induction of aldose reductase by tumor necrosis factor-alpha. *J Biol Chem.* 274, 7993–8001.
10. Bouet, J.Y., Surtees, J.A., and Funnell, B.E. (2000) Stoichiometry of P1 plasmid partition complexes. *J Biol Chem.* 275, 8213–8219.
11. Le Naour, F. (2007) Identification of tumor antigens by using proteomics. *Methods Mol Biol.* 360, 327–334.
12. Canellea, L., Bousquet, J., Pionneaua, C., Deneuxc, L., Imam-Sghiouara, N., Carona, M., et al. (2005) An efficient proteomics-based approach for the screening of autoantibodies. *J Immunol Methods.* 299, 77–89.
13. Matsuo, K., Xiang, Y., Nakamura, H., Masuko, K., Yudoh, K., Noyori, K., et al. (2006) Identification of novel citrullinated autoantigens of synovium in rheumatoid arthritis using a proteomic approach. *Arthritis Res Ther.* 8, R175–R187.
14. Hung, C.H., Huang, H.R., Huang, C.J., Huang, F.L., and Chang, G.D. (1997) Purification and cloning of carp nephrosin, a secreted zinc endopeptidase of the astacin family. *J Biol Chem.* 272, 13772–13778; Correction 272, 19632.
15. Nigam, S.K., Goldberg, A.L., Ho., S., Rohde, M.F., Bush, K.T., and Sherman, M.Y. (1994) A set of endoplasmic reticulum proteins possessing properties of molecular chaperones includes Ca^{2+}-binding proteins and members of the thioredoxin superfamily. *J Biol Chem.* 269, 1744–1749.
16. Schägger, H. and von Jagow, G.V. (1987) Tricine-sodium dodecyl sulfate-polyacrylamide gel electrophoresis for the separation of proteins in the range from 1 to 100 kDa. *Anal Biochem.* 166, 368–379.
17. Heukeshoven, J. and Dernick, R. (1988) Improved silver staining procedure for fast staining in PhastSystem Development Unit I. Staining of sodium dodecyl sulfate gels. *Electrophoresis* 9, 28–32.
18. Harlow, E. and Lane, D. (1988) Antibodies, Cold Spring Harbor Laboratory, New York, pp. 471–510.

Chapter 5

Diffusion Blotting for Rapid Production of Multiple Identical Imprints from Sodium Dodecyl Sulfate Polyacrylamide Gel Electrophoresis on a Solid Support

Ingrid Olsen and Harald G. Wiker

Summary

A very simple and fast method for diffusion blotting of proteins from precast SDS-PAGE gels on a solid plastic support was developed. Diffusion blotting for 3 min gave a quantitative transfer of 10% compared with 1-h electroblotting. For each subsequent blot from the same gel a doubling of transfer time is necessary to obtain the same amount of protein onto each blot. The relative transfer of low and high molecular weight components was similar in diffusion and electroblotting. However, both methods do give a higher total transfer of the low molecular weight proteins compared with the large proteins. The greatest advantage of diffusion blotting is that several blots can be made from each lane, thus enabling testing of multiple antisera on virtually identical blots. The gel remains on the plastic support, which prevents it from stretching or shrinking. This ensures identical blots and facilitates more reliable molecular weight determination. Furthermore, the proteins remaining in the gel can be stained with Coomassie Brilliant Blue or other methods for exact and easy comparison with the developed blots. These advantages make diffusion blotting the method of choice when quantitative protein transfer is not required.

Key words: SDS-PAGE Diffusion blotting, Electrophoresis, Western blotting

1. Introduction

Electrophoretic transfer of protein is the most widely applied method for western blotting after SDS-PAGE. When using precast gels supported on plastic films, removal of the film became necessary to allow electrophoretic transfer. Ultrathin gels (0.1–0.2 mm) are difficult to separate from the supporting material, and a method for diffusion blotting that gave

Biji T. Kurien and R. Hal Scofield (eds.), *Methods in Molecular Biology, Protein Blotting and Detection, vol. 536*

DOI: 10.1007/978-1-59745-542-8_5

efficient transfer was subsequently developed *(1–3)*. Furthermore, a method for diffusion blotting from 1.5-mm slab gels has been described *(4)*. We wished to investigate diffusion blotting as an alternative method for the commercially available precast 0.5-mm SDS-PAGE Excel gels (GE Healthcare Bio-Sciences, Little Chalfont, UK) on plastic supports and obtain quantitative data concerning the efficiency of the method *(5)*. The efficiency of diffusion blotting was compared with 1-h electroblotting, and it was found that the transfer rate was fast initially and that it declined with time. Compared with electroblotting, 10% of the proteins was transferred during the first 3 min, 20% was transferred after 10 min, and 45–65% was transferred after 3 h. Diffusion blotting also allows for several blots to be made from one gel. A doubling of transfer time from each blot is necessary to obtain similar amount of proteins on each blot *(5)*. After blotting, the proteins in the gel can be stained by standard methods such as Coomassie Brilliant Blue or silver staining. Diffusion blotting is a simple method, which is performed with minimal equipment. The gel can remain on the solid matrix, which maintains the integrity of the gel and permits the generation of multiple blots (**Fig. 1**). Several antisera can thus be tested on identical runs and these can be compared with protein staining of the same gel. The many advantages of diffusion blotting make this an alternative to electroblotting when ultimate sensitivity is not required.

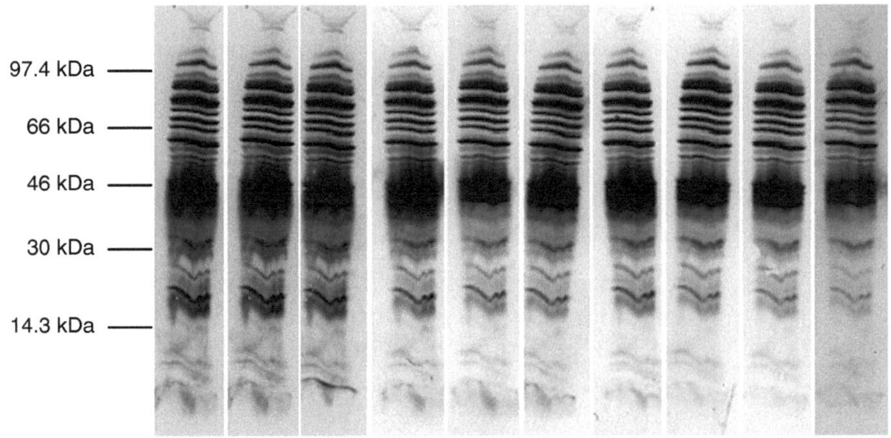

Fig. 1. Multiple blots of the same lane. *Mycobacterium avium* subspecies *avium* proteins (5-μg total protein) were separated on a SDS-PAGE 8–18% precast gel and transferred to a nitrocellulose membrane by diffusion blotting. The transfer time was kept constant at 3 min for each imprint. The blots were stained with polyclonal anti-*M. avium* antibodies and HRP-labeled anti-rabbit immunoglobulin with diaminobenzidine as substrate. The picture shows the individual blots 1–10 and the position of molecular mass markers at the left. Note reproduction of artifacts such as curving of the bands.

2. Materials

1. Prestained molecular mass standard from any supplier.
2. Nitrocellulose membrane from Millipore, Bedford, MA.
3. 3MM Chr paper (Whatman, Maidstone, UK).
4. Tris/glycine transblot buffer (pH 8.0): 25 mM Tris base, 190 mM glycine, and 20% (v/v) methanol. Adjustment of pH is not required (*see* **Note 1**).
5. Glass plate.
6. Weight.

3. Methods

1. After SDS-PAGE put the gel on a flat surface leaving the solid support in place (for blotting after 2D gel electrophoresis, *see* **Note 2**).
2. Use gloves when handling membrane. Cut appropriate number of nitrocellulose membranes of the exact same size as the gel. Cut upper right corner corresponding to the cut corner on the precast SDS-PAGE Excel gels to help correct orientation.
3. Soak the nitrocellulose membrane in the transblot buffer until wet.
4. Remove excess buffer by holding two corners of the membrane with your fingers and carefully allowing the edge to touch a sheet of filter paper.
5. Carefully lay the membrane onto the gel. Avoid air bubbles. Do not remove the membrane after it has been in contact with the gel because some protein is transferred immediately and this may result in double bands.
6. Wet three sheets of 3MM paper in the transblot buffer. Remove excess buffer by squeezing the sheets between two fingers and place the papers on top of the membrane.
7. Cover the gel, membrane, and filter paper assembly with a glass plate and a weight of approximately 1.5 kg (6 g/cm^2).
8. Leave for 3 min (*see* **Note 3**).
9. Repeat **steps 3–6** for every blot to be made and double the transfer time for each blot to obtain the same amount of protein on each blot.

10. Cut a mark in the nitrocellulose for each band in the prestained standard (*see* **Note 4**).

4. Notes

1. Instead of transblot buffer we have often used PBS pH 7.4 and it appears to work equally well.
2. The method is described for precast 0.5-mm SDS-PAGE gels from GE Healthcare (previously Amersham Biosciences), Piscataway, NJ, USA. These gels are also used in a two-dimensional electrophoresis system where Immobiline DryStrip for Isoelectrofocusing is used in the first dimension and the precast gels are used in the second dimension. The method works equally well following two-dimensional gel electrophoresis, where exact pattern reproducibility is difficult to achieve. Creation of multiple identical imprints by diffusion blotting will eliminate problems with gel-to-gel variation and facilitate correct identification of protein spots as well as easy comparison of different antisera.
3. The time can be changed according to the sensitivity of the detection system. Three minutes will give a transfer of approximately 10% of what you see for electroblotting while 30% is transferred after 30 min *(5)*.
4. The prestained standard is visible after blotting but will weaken considerably after blocking and development. Marking the bands with a knife will help in locating the molecular mass markers. If protein staining of the gel is performed, molecular mass determination can be done on the corresponding band identified on the gel.

References

1. Heukeshoven, J. and Dernick, R. (1995) Effective blotting of ultrathin polyacrylamide gels anchored to a solid matrix. *Electrophoresis* 16, 748–756.
2. Braun, W. and Abraham, R. (1989) Modified diffusion blotting for rapid and efficient protein transfer with PhastSystem. *Electrophoresis* 10, 249–253.
3. Jagersten, C., Edstrom, A., Olsson, B., and Jacobson, G. (1988) Blotting from PhastGel media after horizontal sodium dodecyl sulfate-polyacrylamide gel electrophoresis. *Electrophoresis* 9, 662–665.
4. Kurien, B.T. and Scofield, R.H. (1997) Multiple immunoblots after non-electrophoretic bidirectional transfer of a single SDS-PAGE gel with multiple antigens. *J Immunol Methods* 205, 91–94.
5. Olsen, I. and Wiker, H.G. (1998) Diffusion blotting for rapid production of multiple imprints from sodium dodecyl sulfate polyacrylamide gel electrophoresis on a solid support. *J Immunol Methods* 220, 77–84.

Chapter 6

Tryptic Peptide Purification Using Polyvinylidene Difluoride Membrane for Matrix-Assisted Laser Desorption Ionization Time of Flight Mass Spectrometry

Biji T. Kurien and R. Hal Scofield

Summary

Matrix-assisted laser desorption ionization time of flight mass spectrometry (MALDI TOF MS) is extremely sensitive to minor impurities in tryptic peptide digests, resulting in suppression of the signal obtained. Therefore, it becomes necessary to purify the sample, especially those samples that fail to yield good mass spectra. Here, we describe a simple protocol using polyvinylidene difluoride (PVDF) membrane for purifying tryptic peptides prior to mass spectrometric analysis. The tryptic digest is spotted on a PVDF membrane, air-dried, and washed. The membrane is then extracted with trifluoroacetic acid/acetonitrile and the extract is then subjected to MALDI TOF MS. Using this procedure, we were able to identify a cross-reactive D1 autoantigen on the surface of neutrophils that bound antibodies targeting Ro 60 autoantigen in systemic lupus erythematosus.

Key words: MALDI TOF mass spectrometry, PVDF, D1 autoantigen, Systemic lupus erythematosus, Ro 60 autoantigen

1. Introduction

Mass spectrometry provides a highly sensitive method for the analysis and identification of small molecules, peptides, proteins, nucleic acids, and polysaccharides *(1–7)*. Proteins excised following sodium dodecyl sulfate polyacrylamide gel electrophoresis (SDS-PAGE) are subjected to in-gel tryptic digestion and matrix-assisted laser desorption ionization time of flight mass spectrometry (MALDI TOF MS) analysis *(8–15)*. The sensitivity of mass spectrometry is very high with detection thresholds at subpicomolar levels, which makes this procedure vulnerable to the slightest contamination.

Biji T. Kurien and R. Hal Scofield (eds.), *Methods in Molecular Biology, Protein Blotting and Detection, vol. 536*

DOI: 10.1007/978-1-59745-542-8_6

Owing to sample contamination, we have experienced difficulties in obtaining good mass spectra using this procedure. Severe signal suppression due to residual acrylamide monomer *(16)*, trifluoroacetic acid *(17)*, and nonionic detergents *(18)* has been shown in earlier studies. An automated online ionic detergent precolumn has been used to remove SDS from trypsin-digested protein samples *(19)*. In this report we show that tryptic peptides, derived from an autoantigen (located on neutrophil cell surface) bound by sera from patients with systemic lupus erythematosus (SLE), can be purified prior to MALDI TOF MS by using PVDF when little or no signal can be obtained as a result of contaminants in the sample.

Anti-Ro 60 autoantibodies *(20, 21)* from patients with SLE were observed to bind intact normal neutrophils in an in vitro fluorescence-activated cell sorter assay *(22)*. A 60,000 molecular weight protein, part of the Ro ribonucleoprotein complex, is the normal target of these autoantibodies. The Ro 60 autoantigen, noncovalently associates with at least one of four short uridine-rich human cytoplasmic RNAs and is targeted by up to 40% of patients with SLE *(20–22)*. To determine the identity of the antigen on the surface of the neutrophils bound by these antibodies, we purified human neutrophil membranes by nitrogen cavitation *(23)*. The purified neutrophil membranes were solubilized and passed over an anti-Ro 60 affinity chromatography column. The proteins eluted from the column were analyzed on SDS PAGE and transferred nonelectrophoretically to nitrocellulose membrane *(24)*. The membrane was immunoblotted with anti-Ro 60 sera. The residual proteins on the gel were stained with Coomassie blue *(25)*, and the immunoblotted band was compared with the Coomassie blue-stained gel and identified. The target protein band was excised and subjected to in-gel tryptic digestion. The tryptic peptides were extracted, purified using PVDF membrane, and subjected to MALDI TOF MS. This protocol allowed us to identify a cross-reactive D1 antigen on the neutrophil membranes bound by anti-Ro 60 antisera from SLE patients.

2. Materials

All reagents and materials were purchased from either Fisher Scientific, Dallas, TX, USA or Sigma Chemical Company, St Louis MO, USA. Prepare all solutions using ultrapure water (prepared by purifying deionized water to attain a sensitivity of 18 MΩ cm at 25°C) and analytical grade reagents. Prepare and store all reagents at room temperature (unless indicated otherwise). Diligently follow all waste disposal regulations when disposing waste materials.

1. 60% acetonitrile containing 0.1% trifluoroacetic acid.
2. 70% acetonitrile containing 5% trifluoroacetic acid.
3. Ferulic acid.
4. Nitrocellulose membrane (Gelman Sciences/Fisher Scientific, Dallas, TX, USA).
5. Polyvinylidene difluoride (PVDF) membrane (Millipore, Bedford, MA, USA).
6. Savant Speed-Vac Concentrator (Savant Instruments, Inc., Farmingdale, New York, USA).
7. PGC microcentrifuge tubes; 1.7 mL (PGC Scientific, Frederick, MD, USA).
8. Super-Mixer (Curtin Matheson Scientific, USA).

3. Methods

All procedures were performed at room temperature (unless indicated otherwise).

1. Cut a piece each of nitrocellulose and PVDF membranes and place them on top of a Whatman #3 filter paper cut slightly larger than the size of the membranes. Mark a small circle to spot the sample (*see* **Note 1**).
2. Spot a 5-μL aliquot of the 70% acetonitrile/5% TFA sample (obtained after extraction of trypsin-digested gel slice) on each of the membrane surfaces (*see* **Note 2**).
3. Air-dry the membranes for 30 min.
4. Excise the dry spot from each of the membranes and place them individually in a PGC microcentrifuge tube (*see* **Note 3**) containing 1 mL of water and shake gently for 15 min.
5. Decant the water carefully and repeat **step 4** once more.
6. Add 1 mL of water to the tube and vortex for 1 min on a Super-Mixer.
7. Decant the water carefully and repeat **step 6** once more.
8. Add 150 μL of 60% acetonitrile/0.1% TFA to each membrane and shake the mixture for 15 min (*see* **Note 4**).
9. Save the supernatant and repeat the extraction once more.
10. Vacuum-dry the combined supernatants using a Savant Speed-Vac Concentrator.
11. Dissolve the resulting pellet in 20 μL of 70% acetonitrile/5% trifluoroacetic acid.
12. Mix this acetonitrile preparation 1:1 with ferulic acid.

13. Spot 0.5 μL of this mixture on a stainless steel grid and subject it to MALDI-TOF MS.
14. Enter the atomic weights of the peaks obtained into the European Molecular Biology Laboratory (EMBL) protein and peptide group database for protein identification (*see* **Note 5**).

4. Notes

1. A membrane 2 in. x 2 in. size would suffice for one sample. Always use one sample per membrane and wash each membrane separately. Make sure that the tip does not press against the membrane. Use a pencil to mark the spot.
2. Subject the Coomassie-stained band containing the candidate antigen to in-gel digestion with trypsin *(22)*. Following this, add 150 μL 60% acetonitrile/0.1% TFA to the gel slice and shake the mixture for 15 min. Save the supernatant and repeat extraction once more. Vacuum dry the combined supernatants. Dissolve the resulting pellet in 70% acetonitrile/5% trifluoroacetic acid.
3. It is very important to use PGC tubes for the assay, since contaminants are minimal when these tubes are used.
4. Nitrocellulose is not compatible with acetonitrile and hence should not be used. No useful signal was obtained when nitrocellulose was used. There is no need to prewet the PVDF membrane, since the peptide mixture is in acetonitrile and readily spotted on the membrane.
5. We did not obtain usable signal when we analyzed the sample prior to PVDF membrane purification. It appears that impurities in the peptide mixture extracted from the trypsin-digested gel slice quenched the signal when the sample was analyzed prior to PVDF membrane purification. The tryptic peptides that were spotted on the PVDF membrane bound to the membrane, whereas the impurities did not bind. The contaminants were washed away leaving behind the purer peptide pool that was then later extracted from the membrane.

Acknowledgements

This work was supported by NIH grant ARO1844 and Oklahoma Center for the Advancement of Science and Technology to RHS. We also express our gratitude to Drs. Hiroyuki Matsumoto and

Sadamu Kurono (Department of Biochemistry and Molecular Biology, Oklahoma City, OK, USA) for their help with mass spectrometry and analysis.

References

1. Want, E.J., O'Maille, G., Smith, C.A., Brandon, T.R., Uritboonthai, W., Qin, C., et al. (2006) Solvent-dependent metabolite distribution, clustering, and protein extraction for serum profiling with mass spectrometry. Anal Chem. 78, 743–752.
2. Fenselau, C. (1997) MALDI MS and strategies for protein analysis. Anal Chem. 69, 661A–665A. Review.
3. Yates, J.R. III (1998) Database searching using mass spectrometry data. Electrophoresis 19, 893–900. Review.
4. Kinumi, T., Tobin, S.L., Matsumoto, H., Jackson, K.W., and Ohashi, M. (1997) *Eur Mass Spectrom.* 3, 367.
5. Fernandez, J., Gharahdaghi, F., and Mische, S.M. (1998) Routine identification of proteins from sodium dodecyl sulfate-polyacrylamide gel electrophoresis (SDS-PAGE) gels or polyvinyl difluoride membranes using matrix assisted laser desorption/ionization-time of flight-mass spectrometry (MALDI-TOF-MS). *Electrophoresis* 19, 1036–1045.
6. Qin, J., Fenyo, D., Zhao, Y., Hall, W.W., Chao, D.M., Wilson, C.J., et al. (1997) A strategy for rapid, high-confidence protein identification. *Anal Chem.* 69, 3995–4001.
7. Borchers, C., Peter, J.F., Hall, M.C., Kunkel, T.A., and Tomer, K.B. (2000) Identification of in-gel digested proteins by complementary peptide mass fingerprinting and tandem mass spectrometry data obtained on an electrospray ionization quadrupole time-of-flight mass spectrometer. *Anal Chem.* 72, 1163–1168.
8. Moritz, R.L., Eddes, J.S., Reid, G.E., and Simpson, R.J. (1996) S-pyridylethylation of intact polyacrylamide gels and in situ digestion of electrophoretically separated proteins: a rapid mass spectrometric method for identifying cysteine-containing peptides. *Electrophoresis* 17, 907–917.
9. Fountoulakis, M. and Langen, H. (1997) Identification of proteins by matrix-assisted laser desorption ionization-mass spectrometry following in-gel digestion in low-salt, nonvolatile buffer and simplified peptide recovery. *Anal Biochem.* 250, 153–156.
10. Kurien, B.T., Matsumoto, H., and Scofield, R.H. (2001) Purification of tryptic peptides for mass spectrometry using polyvinylidene fluoride membrane. *Indian J Biochem Biophys.* 38, 274–276.
11. Hellman, U. (2000) Sample preparation by SDS/PAGE and in-gel digestion. *EXS* 88, 43–54.
12. Bantscheff, M., Weiss, V., and Glocker, M.O. (1999) Identification of linker regions and domain borders of the transcription activator protein NtrC from *Escherichia coli* by limited proteolysis, in-gel digestion, and mass spectrometry. *Biochemistry* 38, 11012–11020.
13. Shevchenko, A., Jensen, O.N., Podtelejnikov, A.V., Sagliocco, F., Wilm, M., Vorm, O., et al. (1996) Linking genome and proteome by mass spectrometry: large-scale identification of yeast proteins from two dimensional gels. *Proc Natl Acad Sci USA.* 93, 14440–14445.
14. Li, G., Waltham, M., Anderson, N.L., Unsworth, E., Treston, A., and Weinstein, J.N. (1997) Rapid mass spectrometric identification of proteins from two-dimensional polyacrylamide gels after in gel proteolytic digestion. *Electrophoresis* 18, 391–402.
15. Qiu, Y., Benet, L.Z., and Burlingame, A.L. (1998) Identification of the hepatic protein targets of reactive metabolites of acetaminophen in vivo in mice using two-dimensional gel electrophoresis and mass spectrometry. *J Biol Chem.* 273, 17940–17953.
16. Alvarez, E.J., Larsen, B.S., Coldren, C.D., and Rice, J. (2000) Effect of residual acrylamide monomer from two-dimensional gels on matrix-assisted laser desorption/ionization peptide mass mapping experiments. *Rapid Commun Mass Spectrom.* 14, 974–978.
17. Apffel, A., Fischer, S., Goldberg, G., Goodley, P.C., and Kuhlmann, F.E. (1995) Enhanced sensitivity for peptide mapping with electrospray liquid chromatography-mass spectrometry in the presence of signal suppression due to trifluoroacetic acid-containing mobile phases. *J Chromatogr A.* 712, 177–190.
18. Gharahdaghi, F., Kirchner, M., Fernandez, J., and Mische, S.M. (1996) Peptide-mass profiles of polyvinylidene difluoride-bound proteins by matrix-assisted laser desorption/ionization time-of-flight mass spectrometry in the presence of nonionic detergents. *Anal Biochem.* 233, 94–99.

19. Vissers, J.P., Hulst, W.P., Chervet, J.P., Snijders, H.M., and Cramers, C.A. (1996) Automated on-line ionic detergent removal from minute protein/peptide samples prior to liquid chromatography-electrospray mass spectrometry. *J Chromatogr B Biomed Appl.* 686, 119–128.
20. Harley, J.B., Scofield, R.H., and Reichlin, M. (1992) Anti-Ro in Sjogren's syndrome and systemic lupus erythematosus. *Rheum Dis Clin North Am.* 18, 337–358. Review.
21. Harley, J.B., Alexander, E.L., Bias, W.B., Fox, O.F., Provost, T.T., Reichlin, M., et al. (1986) Anti-Ro (SS-A) and anti-La (SS-B) in patients with Sjogren's syndrome. *Arthritis Rheum.* 29, 196–206.
22. Kurien, B.T. and Scofield, R.H. (2000) Association of neutropenia in systemic lupus erythematosus with anti-Ro and binding of an immunologically cross-reactive neutrophil membrane antigen. *Clin Exp Immunol.* 120, 209–217.
23. Moore, K.L., Stults, N.L., Diaz, S., Smith, D.F., Cummings, R.D., Varki, A., et al. (1992) Identification of a specific glycoprotein ligand for P-selectin (CD62) on myeloid cells. *J Cell Biol.* 118, 445–456.
24. Kurien, B.T. and Scofield, R.H. (1997) Multiple immunoblots after non-electrophoretic bidirectional transfer of a single SDS-PAGE gel with multiple antigens. *J Immunol Methods* 205, 91–94.
25. Kurien, B.T. and Scofield, R.H. (1998) Heat mediated quick Coomassie blue protein staining and destaining of SDS-PAGE gels. *Indian J Biochem Biophys.* 35, 385–389.

Chapter 7

Affinity Immunoblotting for Analysis of Antibody Clonotype Distribution in a Lupus Patient Developing Anti-Ro 60 Over Time

Biji T. Kurien and R. Hal Scofield

Summary

We describe a sensitive and specific method to analyze specific antibody clonotype changes in a patient with systemic lupus erythematosus who developed autoantibodies to the Ro 60 autoantigen under observation. Patient sera collected over several years were separated by flatbed isoelectric focusing and analyzed by affinity immunoblotting utilizing Ro 60-coated nitrocellulose membrane. When the Ro 60-coated nitrocellulose was laid over the surface of the IEF gel, the antibodies present on the surface of the acrylamide gel bound the Ro antigen on the nitrocellulose. Tween-20 was used to prevent non-specific binding. The bound IgG clonotypes were detected using alkaline phosphatase conjugated anti-IgG. The patient's sera demonstrated an oligoclonal response to the Ro 60 autoantigen that increased in complexity and affinity over time.

Key words: Affinity immunoblotting, Clonotype distribution, Systemic lupus erythematosus, Ro 60 autoantigen, Flatbed IEF

1. Introduction

Isoelectric focusing (IEF) is a very useful method for investigating the heterogeneity of antibody and immunoglobulin (Ig) clonotypes *(1)*. Antigen-specific antibody clonotype patterns can show whether changes in cell population happen during ongoing immune responses as a response to regulatory influences. Affinity IEF can determine whether changes in hybridoma cell lines can

Biji T. Kurien and R. Hal Scofield (eds.), *Methods in Molecular Biology, Protein Blotting and Detection, vol. 536*

DOI: 10.1007/978-1-59745-542-8_7

occur with time *(2)*. Previously, it was customary to study these changes by immobilizing the separated antibody clonotypes after IEF and incubating them with radioactive antigen. In one method, radiolabeled hapten was allowed to diffuse into a gel before precipitation of Ig with sodium sulfate followed by detection of hapten-specific clonotype distribution by autoradiography *(3)*. In another study Ig was precipitated in the gel with sodium sulfate immediately after completion of the focusing run and was cross-linked with glutaraldehyde followed by the addition of labeled antigen or anti-Ig *(4)*. Subsequently it was shown that fixation with glutaraldehyde could decrease the antigen-binding ability of certain Igs *(5)*. However, the previous study was unable to define optimal cross-linker (glutaraldehyde or suberimidate) concentration, since certain antibodies could not be fixed at cross-linker concentrations that substantially inactivated others. Another drawback of these methods is the excess time needed to diffuse antigen into the gel and to rinse the unbound antigen out of the gel, which can take several days especially when using radioactive probes.

One method for immobilizing focused antibodies involved the use of nitrocellulose membranes. Focused antibodies were transferred electrophoretically or nonelectrophoretically to nitrocellulose, and labeled antigen was used to detect clonotypes that were antigen specific *(6)*. Yet another method involved lying the gel with the focused antibodies with agarose containing antigen-coated sheep erythrocytes *(7)*. In this method, antibodies diffuse into the RBC-containing gel, bind the antigen-coated cells, and lyse the cells following complement addition.

Here, we adapt a method *(2)* to study the 60,000 molecular weight Ro autoantigen. The antigen was first passively immobilized on nitrocellulose membrane and placed in contact with an IEF gel that contained autoantibodies [derived from a systemic lupus erythematosus (SLE) patient who developed antibodies to the Ro 60 autoantigen over time] focused according to its isoelectric point. Following diffusion-mediated transfer to the membrane the antibody clonotypes that are not antigen specific are removed by washing, while the antigen-specific antibody clonotypes are detected using alkaline phosphatase conjugated anti-Ig.

SLE is a complex, chronic disorder characterized by the production of antibodies to self-antigens, including the Ro (or SS-A) ribonucleoprotein complex. Antibodies to the Ro 60 autoantigen occur in up to 40% of patients with SLE *(8)*. The epitopes of the Ro 60 autoantigen bound by SLE patients have been previously characterized *(9, 10)*. Even though anti-Ro 60 sera were commonly observed to bind to short peptides, it was not found to bind the denatured antigen well. Furthermore, the antibodies that were found to bind to octapeptides were also found to bind the native protein *(10)*.

There have been instances of some SLE autoantibodies appearing and disappearing, at times in association with specific disease manifestations, therapy, or generalized clinical disease activity. For instance, antibodies to native DNA are associated with renal disease, and the detection of these autoantibodies may be an indication of disease exacerbation *(11)*. Antibodies to the P autoantigen (ribosomal P antigens) can appear with an increase of neurologic or renal disease. Autoantibodies such as anti-Ro, on the other hand, occur in some normal subjects as well as in SLE patients before onset of disease *(8)*, and develop only rarely during the course of SLE.

This investigation was carried out following the identification of an SLE patient who developed antibodies to the Ro 60 autoantigen after about 10 years of illness. As shown in **Fig. 3** anti-Ro 60 clonality increased in complexity, and affinity to the Ro 60 antigen also increased as the response developed.

2. Materials

Prepare all solutions using ultrapure water (prepared by purifying deionized water to attain a sensitivity of 18 MΩ cm at 25°C) and analytical grade reagents. Prepare and store all reagents at room temperature (unless indicated otherwise). Diligently follow all waste disposal regulations when disposing waste materials. We do not add sodium azide to the reagents.

1. 25% Glycerol (v/v): Add 25 mL glycerol to 75 mL of distilled water. Mix well.
2. 5× Acrylamide (26.5% T, 3% C): Add about 25 mL water to a 100-mL graduated cylinder or a glass beaker. Weigh 12.84 g acrylamide and 0.4098 g bis acrylamide and transfer to the cylinder (*see* **Note 1**). Add a spatula of AG 501-X8 (D) mixed-resin beads (Bio-Rad, Hercules, CA, USA) and stir using a magnetic stir bar on a magnetic plate for about 30 min. Make up to 50 mL (after removing the stir bar) with water and filter through a 0.45 μm Corning filter (*see* **Note 2**). Store at 4°C, with bottle wrapped in a aluminum foil (*see* **Note 3**).
3. 10% Tween-20: Add 90 mL of distilled water into a glass beaker. Add 10 mL Tween-20 and mix.
4. 2% Ammonium persulfate: Weigh 0.02 g ammonium persulfate and dissolve in 1 mL of distilled water (*see* **Note 4**).
5. *N,N,N,N'*-tetramethyl-ethylenediamine (Sigma Chemical Company, St Louis, MO, USA). Store at 4°C (*see* **Note 5**).
6. Alkaline phosphatase buffer: Weigh 6.1 g of Tris, 2.9 g sodium chloride, and 0.51 g magnesium chloride·$6H_2O$, and make it

to 500 mL with water after adjusting pH to 9.3 with HCl (*see* **Note 1**). Store at 4°C.

7. Nitro blue tetrazolium (NBT)/5-bromo-4-chloro-3-indolyl phosphate (BCIP): Dissolve 1 g NBT in 20 mL of 70% dimethylformamide (DMF). Disolve 1 g BCIP in 20 mL of 100% DMF. Add 33 μL of BCIP and 66 μL of NBT to 10 mL of alkaline phosphatase buffer just before adding to membrane.
8. Nitrocellulose membrane (Fisher Scientific, Dallas, TX, USA).
9. 0.5 M sodium bicarbonate solution, pH 9.5.
10. Phosphate buffered saline (PBS), pH 7.4.
11. PBS containing 0.05% Tween-20 (PBST).
12. Ampholytes: pH 3–10 and pH 8–10.5 (Amersham/GE Healthcare, Piscataway, NJ, USA).
13. Ro 60 autoantigen (Immunovision, Springdale, AK, USA).
14. Glass plates: Two 5 in. × 4 in. glass plates.
15. Medium binder clips (1¼ in.).
16. Small binder clips (¾ in.).
17. Gasket with three edges, about 3-mm wide, to serve as spacer between the plates.
18. LKB-2117 Multiphor apparatus for IEF (LKB Instruments, Rockville, MD, USA).
19. Model 3000/300 power supply (Bio-Rad).
20. pH 3 and pH 10 solutions (Serva Electrophoresis GmbH, Heidelberg, Germany).
21. Helium gas.
22. Sample applicator strip (Bio-Rad).
23. Paper wicks.

3. Methods

All procedures are performed at room temperature unless otherwise specified.

1. The night before focusing cut a piece of nitrocellulose membrane according to the size that would fit a small pipet box lid (yellow tip box). Add antigen (Ro 60) at 10 μg/mL in sodium bicarbonate, pH 9.5, and incubate this with the membrane overnight with shaking.

2. Pipet 5.6 mL of distilled water into a conical flask. Add 2 mL of 25% glycerol followed by 2.1 mL of the 5× acrylamide solution. Then add 300 μL of pH 3–10 ampholytes followed by 100 μL of pH 8–10.5 ampholytes.
3. Degas this solution by bubbling helium through it for 15 min. Rinse the metal end of degassing tube first with water and wipe dry with Kimwipes.
4. While the solution is degassing, set up the gel apparatus. Soak the gasket in water for few minutes. Mop dry with Kimwipes.
5. Take one glass plate and lay the gasket on top of the glass plate around the edges so that it will seal the bottom and two sides of the plates. Lay the other glass plate on top of the gasket. Clamp the clips around the edges of the plates (bottom, the left side, and the right side; *see* **Fig. 1**). Stand the gel upright using the base of the clips (*see* **Fig. 1**) to pour the gel. Prepare a 2% ammonium persulfate solution freshly.
6. After degassing is complete, the metal end of the degassing tube is cleaned with water. To the degassed solution, add 100 μL of 10% Tween-20 and mix gently. Then add 100 μL of 2% APS. Have a Pasteur pipette ready for pouring the gel. Add 10 μL of TEMED and mix gently. Pipet the gel mixture into the Pasteur pipette and transfer into the gel apparatus quickly. Attempts should be made to avoid bubbles. Fill up the gel apparatus to the top. Polymerization should begin within minutes. However, let the assembly stand for 2 h without disturbance.
7. Turn the cooling unit ON and set it at 4°C in preparation for focusing.
8. After 2 h carefully remove one of the glass plates and gasket. The gel will remain on one of the glass plates.

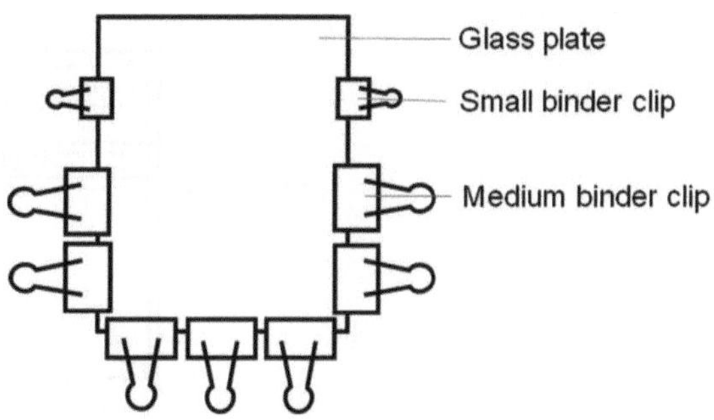

Fig. 1. Gel assembly for isoelectric focusing.

9. Lie the glass plate on top of the IEF unit, with the gel side facing up (wipe off water on top of the unit beforehand). Place the smaller cover in place and press down slightly so as to make imprints for the wicks. Cut two wicks to the size of the gel (be as close as possible). Soak the top wick in Serva pH 3 solution and the bottom wick in Serva pH 10 solution. Dab off excess solution and place where imprints were made by the cover (*see* **Fig. 2**).
10. Put the smaller cover back on, making sure that connection is made with both wicks. Connect red and black wires. Put on a larger cover and make connections to power supply (red = +ve; black = −ve).
11. Prefocus by setting constant voltage of 200 V for 20 min, and then increase voltage to 400 V for another 20 min. Prepare samples for application.
12. Turn OFF power supply, disconnect wires, and remove the covers. Take applicator strip and lay on top of gel 1–2 cm below the top wick. Make sure that the strip is stuck to the gel well. The strip can hang over the gel a little (*see* **Fig. 2**; *see* **Note 6**).
13. Apply the samples, being careful not to spill over into other wells. Replace the covers and make the connections. Turn the power supply ON to 12-W constant power. Focus for approximately 1–2 h. When focusing, the voltage will rise, and the current will drop. The rate at which these two parameters change is much faster in the beginning than in the end. The run is complete when the voltage is between 1,800 and 2,000 V and the current 3–5 A. When the change appears to be very slow or not at all, turn off the unit (*see* **Note 7**).
14. Thirty minutes prior to end of the run, rinse nitrocellulose membrane three times with PBS, pH 7.4, and shake with PBST for 30 min and rinse three times with PBS.

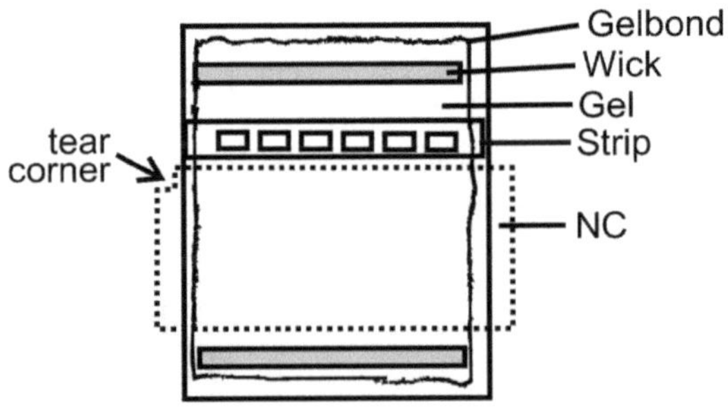

Fig. 2. The membrane–gel assembly following flat bed IEF. The gel bond is shown in this figure. However, we did not use the gel bond to support the gel. The gel was directly in contact with the glass plate.

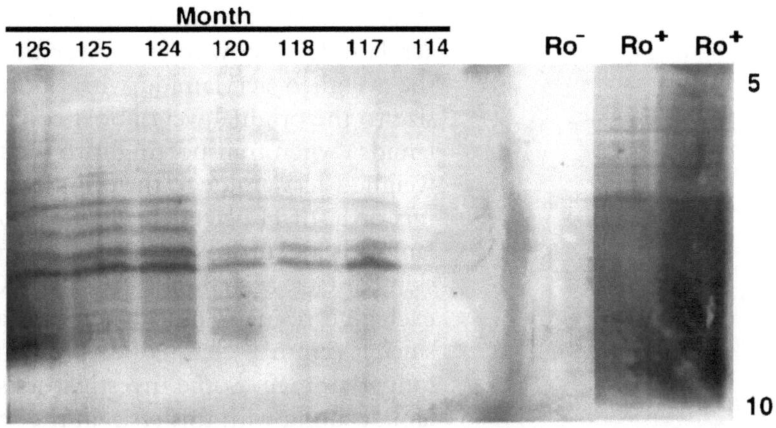

Fig. 3. Affinity immunoblotting of the patient's sera obtained in different months following first observation, showing anti-Ro 60-specific IgG clonotypes. Sera from an anti-Ro 60-negative SLE patient (Ro$^-$) and from two typical anti-Ro 60-positive patients (Ro$^+$) are shown for comparison. The pH range of the IEF gel is shown on the right (reproduced from *(14)* with permission from Wiley and Sons, Inc.).

15. After the run is complete, transfer the focused protein from the gel to the membrane. Take the gel off the flatbed and remove applicator strip. Place the nitrocellulose membrane between two Kimwipes and dab dry. Place the membrane over bottom half of gel above the bottom wick. The membrane may cover the applicator strip area.
16. Place the gel in a Tupperware container with a moist towel; cover and place in an oven at 37°C for 20 min.
17. Take the membrane off the gel. Rinse with 200 mL of deionized water 2–3 times (*see* **Note 8**).
18. Rinse three times with PBS. Wash for 20–30 min in PBST and rinse three times with PBS.
19. Add appropriate alkaline phosphatase conjugate (10 mL) to the membrane and shake for 1 h.
20. Develop bands with NBT/BCIP (*see* **Fig. 3**).
21. Scan the membrane and save results.

4. Notes

1. Having water at the bottom of the cylinder helps to dissolve the Tris relatively easily, allowing the magnetic stir bar to go to work immediately. If using a glass beaker, the Tris can be dissolved faster if the water is warmed to about 37°C. However, the downside is that care should be taken to bring the solution to room temperature before adjusting pH.

2. Wear a mask when weighing acrylamide. To avoid exposing acrylamide to coworkers, cover the weigh boat containing the weighed acrylamide with another weigh boat (similar size to the original weigh boat containing the weighed acrylamide) when transporting it to the fume hood. Transfer the weighed acrylamide to the cylinder inside the fume hood and mix on a stirrer placed inside the hood. Unpolymerized acrylamide is a neurotoxin and care should be exercised to avoid skin contact. Mixed resin AG 501 –X8 (D) (anion and cation exchange resin) is used when acrylamide solution is made, since it removes charged ions (e.g., free radicals) and allows longer storage. Some investigators store the prepared acrylamide along with this resin in the refrigerator. However, we filter them out before storage. The used mixed resin should be disposed as hazardous waste. Manufacturer's warning states that this resin is explosive when mixed with oxidizing substances. The resin contains a dye that changes from blue-green to gold when the exchange capacity is exhausted.
3. The acrylamide solution can be stored at 4°C for 1 month. Acrylamide hydrolyzes to acrylic acid and ammonia. The acrylamide mixture, buffer, and water can be prepared in large batches, frozen in aliquots (for greater day-to-day reproducibility), and used indefinitely (*see* **Ref.** *12*). Remove the required amount, bring to room temperature, and add the other ingredients for polymerization. However, in our laboratory we make the acrylamide solution fresh about every month when we cast our own gels.
4. We find that it is best to prepare this fresh each time.
5. We find that storing at 4°C reduces its pungent smell.
6. Large well = 10 μL, medium wells = 5 μL, small well = 1 μL. Only lay one size of wells on to the gel. Cut if necessary. Strips may be used again.
7. During the end of the run, the gel must be watched carefully in case a fire starts. Many times gel will burn near the applicator strip. If this happens, turn off the unit. The gel can still be used if it had been focused for a long time. The bands are usually below the strip.
8. Rinsing the membrane with deionized water 2–3 times will help remove a bulk of the nonspecific antibodies and help reduce the amount of TBST used subsequently and also reduce the number of washes. This wash helps to reduce nonspecific binding of NBT/BCIP to the strip. The water, owing to its low ionic strength compared with TBST, will be able to remove contaminants much better than TBST. Water is much cheaper than TBST, in terms of money and labor. Other investigators

have found no reduction in detection of specific signals due to washing with water *(13)*.

Acknowledgement

This work was supported by NIH grant ARO1844 and Oklahoma Center for the Advancement of Science and Technology to RHS.

References

1. Awed, Z.L., Williamson, A.R., and Askonas, B.A. (1968) Isoelectric focusing in polyacrylamide gel and its application to immunoglobulins. *Nature* 219, 66–67.
2. Knisley, K.A. and Rodkey, L.S. (1986) Affinity immunoblotting. High resolution isoelectric focusing analysis of antibody clonotype distribution. *J Immunol Methods* 95, 79–87.
3. Askonas, B.A., Williamson, A.R., and Wright, B.E. (1970) Selection of a single antibody-forming cell clone and its propagation in syngeneic mice. *Proc Natl Acad Sci USA* 67, 1398–1403.
4. Keck, K., Grossberg, A.L., and Pressmann, D. (1973) Specific characterization of isoelectrofocused immunoglobulins in polyacrylamide gel by reaction with 125 I-labeled protein antigens or antibodies. *Eur J Immunol* 3, 99–102.
5. Nicolotti, R.A., Briles, D.E., Schroer, J.A., and Davie, J.M. (1980) Isoelectric focusing of immunoglobulins: improved methodology. *J Immunol Methods* 33, 101–115.
6. Friedenson, B. and Soong, C.J. (1984) A simple general method of exactly comparing different binding activities and antigenic properties of antibodies by obtaining duplicate copies from a single isoelectric focusing gel. *J Immunol Methods* 67, 235–242.
7. Phillips, J.M. and Dresser, D.W. (1973) Antibody isoelectric spectra visualized by antigen-coated erythrocytes. *Eur J Immunol* 3, 738–740.
8. Harley, J.B., Scofield, R.H., and Reichlin, M. (1992) Anti-Ro in Sjögren's syndrome and systemic lupus erythematosus.*Rheum Dis Clin North Am*18, 337–358. Review.
9. Scofield, R.H. and Harley, J.B. (1991) Autoantigenicity of Ro/SSA antigen is related to a nucleocapsid protein of vesicular stomatitis virus. *Proc Natl Acad Sci USA* 88, 3343–3347.
10. Huang, S-C., Yu, H., Scofield, R.H., and Harley, J.B. (1995) Human anti-Ro autoantibodies bind peptides accessible to the surface of the native Ro autoantigen. *Scand J Immunol* 41, 220–228.
11. Lloyd, W. and Schur, P.H. (1981) Immune complexes, complement, and anti-DNA in exacerbations of systemic lupus erythematosus (SLE). *Medicine* 60, 208–217.
12. Harlow,E.andLane,D.(1988)Electrophoresis. Appendix I. In: Harlow, E. and Lane, D., eds. Antibodies. A Laboratory Manual. Cold Spring Harbor Laboratory, New York, p. 638.
13. Wu, M., Stockley, P.G., and Martin, W.J., II. (2002) An improved western blotting technique effectively reduces background. *Electrophoresis* 23, 2373–2376.
14. Scofield, R.H., Zhang, F., Kurien, B.T., Anderson, C.J., Reichlin, M., Harley, J.B., and Stafford, H.A. (1996) Development of the anti-Ro autoantibody response in a patient with systemic lupus erythematosus. *Arthritis Rheum* 39, 1664–1668.

Chapter 8

Non-electrophoretic Bi-directional Transfer of a Single SDS-PAGE Gel with Multiple Antigens to Obtain 12 Immunoblots

Biji T. Kurien and R. Hal Scofield

Summary

Protein blotting is an invaluable technique in immunology to detect and characterize proteins of low abundance. Proteins resolved on sodium dodecyl sulfate (SDS) polyacrylamide gels are normally transferred electrophoretically to adsorbent membranes such as nitrocellulose or polyvinylidene diflouride membranes. Here, we describe the nonelectrophroretic transfer of the Ro 60 (or SSA) autoantigen, 220- and 240-kD spectrin antigens, and prestained molecular weight standards from SDS polyacrylamide gels to obtain up to 12 immunoblots from a single gel and multiple sera.

Key words: Nonelectrophoretic transfer, Immunoblots, Bidirectional transfer, Nitrocellulose membrane, Autoantibodies, Autoantigens

1. Introduction

Electrophoretic protein transfer to a microporous membrane support with subsequent immunodetection *(1)* has made a tremendous impact in the field of immunology. Proteins separated on sodium dodecyl sulfate (SDS) polyacrylamide gel electrophoresis (SDS PAGE) are transferred electrophoretically to mainly nitrocellulose or polyvinylidene difluoride *(1–4)*. Protein transfer from gels to membranes has been achieved in three ways: (a) simple diffusion *(3)*, (b) vacuum-assisted solvent flow *(5, 6)*, and (c) electrophoretic elution *(1)*. Only electrophoretic elution has been used widely owing to a variety of reasons, including efficiency,

Biji T. Kurien and R. Hal Scofield (eds.), *Methods in Molecular Biology, Protein Blotting and Detection, vol. 536*

DOI: 10.1007/978-1-59745-542-8_8

simplicity, and length of transfer. Bidirectional transfer procedure was demonstrated in 1980, when Smith and Summers *(7)* transferred DNA and RNA from polyacrylamide gels to nitrocellulose in 36 h to obtain two blots. Here, we demonstrate that proteins can be efficiently transferred nonelectrophoretically from SDS-PAGE gels to nitrocellulose membranes. A similar procedure has been used in 1982 *(8)* to transfer proteins from thin (0.5 mm) native isoelectric focusing gels to nitrocellulose membranes to obtain two blots in 1 h. Diffusion-mediated transfer of immuno globulins from one side of a native gel after isoelectric focusing to an antigen-coated nitrocellulose sheet has been achieved in a similar fashion *(9, 10)*. We were able to transfer and immunologically detect a 60,000 molecular weight autoantigen (Ro 60) as well as the spectrin antigens (molecular weight >200,000) using this method. When prestained molecular weight standards were transferred, only markers up to 118,000 molecular weight could be visualized (these proteins were not visualized immunologically).

Thus, we have obtained up to 12 immunoblots from a single gel using multiple antigens (high, intermediate, and low molecular weight proteins) and multiple sera. Subsequently several investigators have also obtained similar results and also in a quantitative manner (*see* Chapters "Simultaneous Immunoblotting Analysis with Activity Gel Electrophoresis and 2-D Gel Electrophoresis" and "Diffusion Blotting for Rapid Production of Multiple Identical Imprints from Sodium Dodecyl Sulfate Polyacrylamide Gel Electrophoresis on a Solid Support").

2. Materials

Prepare all solutions using ultrapure water (prepared by purifying deionized water to attain a sensitivity of 18 MΩ cm at 25°C) and analytical grade reagents. Prepare and store all reagents at room temperature (unless indicated otherwise). Diligently follow all waste disposal regulations when disposing waste materials. We do not add sodium azide to the reagents.

2.1. SDS Polyacrylamide Gel Components

1. Resolving gel buffer: 1.5 M Tris–HCl, pH 8.8. Add about 100 mL water to a 1-L graduated cylinder or a glass beaker (*see* **Note 1**). Weigh 181.7 g Tris and transfer to the cylinder. Add water to a volume of 900 mL. Mix and adjust pH with HCl (*see* **Note 2**). Make up to 1 L with water. Store at 4°C.
2. Stacking gel buffer: 0.5 M Tris–HCl, pH 6.8. Weigh 60.6 g Tris and prepare a 1 L solution as in previous step. Store at 4°C.

3. Thirty percent acrylamide/Bis solution (29.2:0.8 acrylamide: Bis): Weigh 29.2 g of acrylamide monomer and 0.8 g Bis (cross-linker) and transfer to a 100 mL graduated cylinder containing about 40 mL of water. Add a spatula of AG 501-X8 (D) mixed-resin beads (Bio-Rad, Hercules, CA, USA) and mix for about 30 min. Make up to 100 mL with water and filter through a 0.45 μm Corning filter (*see* **Note 3**). Store at 4°C, in a bottle wrapped with aluminum foil (*see* **Note 4**).
4. Ammonium persulfate: 10% solution in water (*see* **Note 5**).
5. *N,N,N,N′*-tetramethyl-ethylenediamine (TEMED) (Sigma Chemical Company, St. Louis, MO, USA). Store at 4°C (*see* **Note 6**).
6. SDS-PAGE running buffer: 0.025 M Tris-HCl, pH 8.3, 0.192 M glycine, 0.1% SDS (*see* **Note** 7).
7. SDS lysis buffer (5×): 0.3 M Tris–HCl (pH 6.8), 10% SDS, 25% β-mercaptoethanol, 0.1% bromophenol blue (BPB), 45% glycerol. Leave one aliquot at 4°C for current use and store remaining aliquots at –20°C (*see* **Note 8**).
8. BPB solution: Dissolve 0.1 g BPB in 100 mL water.

2.2. Immunoblotting Components

1. Nitrocellulose membranes (Gelman Sciences/Fisher Scientific, Dallas, TX, USA).
2. Western blot transfer buffer: 0.025 M Tris, 0.192 M glycine, and 20% methanol (*see* **Note 9**).
3. Tris buffered saline (TBS; 10×): 1.5 M NaCl, 0.1 M Tris–HCl, pH 7.4.
4. TBS containing 0.05% Tween-20 (TBST).
5. Blocking solution: 5% milk in TBS (*see* **Note 10**). Store at 4°C.
6. Diluent solution: 5% milk in TBST (*see* **Note 10**). Store at 4°C.
7. Mini PROTEAN® 3 System glass plates (catalog number 1653311) (Bio-Rad).
8. Medium binder clips (1¼ in.).
9. Plastic container.
10. Wypall X-60 reinforced paper (Kimberly-Clark, Neenah, WI, USA).
11. Nitro blue tetrazolium (NBT)/5-bromo-4-chloro-3-indolyl phosphate (BCIP): Dissolve 1 g NBT in 20 mL of 70% dimethylformamide (DMF). Disolve 1 g BCIP in 20 mL of 100% DMF. Add 33 μL of BCIP and 66 μL of NBT to 10 mL of alkaline phosphatase buffer just before adding to membrane.

2.3. Antigens and Conjugates

1. Purified red blood cell spectrin, antispectrin polyclonal antibody, and antihemoglobin antibody (Sigma).
2. Benchmark prestained molecular weight standards (Gibco BRL, Bethesda, MD, USA).
3. Purified bovine Ro 60: Purify Ro 60 as reported *(11, 12)*, or it can be purchased from Immunovision, Springdale, AK, USA.
4. Prepare human erythrocyte membranes according to Dodge et al. *(13)*.

3. Methods

Carry out all procedures at room temperature unless otherwise specified.

3.1. 10% Sodium Dodecyl Sulfate Polyacrylamide Gel Electrophoresis

1. Mix 2.5 mL of resolving buffer, 3.33 mL of acrylamide mixture, and 4 mL water in a 50 mL conical flask. Degas with helium for 10 min. Add 100 µL of SDS, 80 µL of ammonium persulfate, and 10 µL of TEMED, and cast gel within a 7.25 cm × 10 cm × 1.5 mm gel cassette. Allow space for stacking the gel and gently overlay with isobutanol or water (*see* **Note 11**).
2. Prepare the stacking gel by mixing 1.25 mL of resolving buffer, 0.67 mL of acrylamide mixture, and 3 mL water in a 50 mL conical flask. Degas with helium for 10 min. Add 100 µL of SDS, 40 µL of ammonium persulfate, and 5 µL of TEMED. Insert a 10-well gel comb immediately without introducing air bubbles.
3. Heat aliquots of bovine Ro 60, RBC membrane, and human spectrin antigens at 95°C for 5 min. Do not add lysis buffer to the prestained protein standard or subject it to heat. Centrifuge the heated samples at 3,000 × *g* for 30 s to bring down the condensate. Load increasing amounts of Ro antigen (1–4 µg) on one gel and same amounts of spectrin (3 µg/lane) or RBC membrane antigens on two other gels along with protein standards (10 µL/well; 2 µg/marker/lane). Add protein standards in every other lane (alternating with spectrin) in the gel with spectrin. Electrophorese at 15 mA until the sample has entered the gel and then continue at 25 mA till the dye front (from the BPB dye in the samples) has reached the bottom of the gel (*see* **Note 12**).
4. Following electrophoresis, pry the gel plates open with the use of a spatula. The gel remains on one of the glass plates. Rinse the gel with water and transfer carefully to a container with western blot transfer buffer.

5. Cut a nitrocellulose membrane to the size of the gel and immerse in methanol. Rinse twice in distilled water and once with transfer buffer.

3.2. Nonelectrophoretic Transfer

1. Immediately following SDS-PAGE, when the dye front reaches the end of the gel, turn off the power supply. Separate the gel plates with the help of a spatula or similar tool. Remove the stacking gel.
2. Rinse the gel (still supported by the bottom glass plate) carefully with deionized water to remove traces of SDS-PAGE running buffer.
3. Excise the gel with spectrin antigen such that there is one lane with the protein markers and one with the spectrin antigen.
4. Leave the gels to air-dry for 5–10 min (*see* **Note 13**).
5. Gently lay one nitrocellulose membrane, cut to the shape of the gel, on top of the gel (*see* **Note 14**).
6. Gently lift the gel–membrane sandwich from the glass plate and place it on a Whatman no. 3 filter (place membrane side directly on the filter paper and the exposed gel side on top) cut to the size of the gel.
7. Place a second nitrocellulose membrane, cut to the shape of the gel, on top of the gel, followed by a Whatman no. 3 filter paper cut similarly (*see* **Note 14**).
8. Place the nitrocellulose–gel–filter paper sandwich between two mini-PROTEAN® 3 System glass plates and secure with clamps.
9. Place this assembly in a prewarmed humidified plastic container (**Fig. 1**) and incubate at 37°C for 30 min (*see* **Note 15**). Remove the membranes for immunoblotting (*see* **Note 16**) (**Fig. 2**).

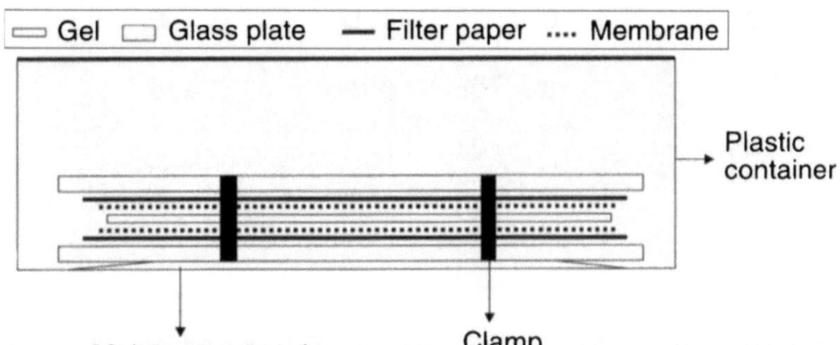

Fig. 1. Gel membrane assembly for the nonelectrophoretic transfer of proteins from SDS-PAGE gels to nitrocellulose membranes to obtain up to 12 blots. The polyacrylamide gel is sandwiched between two membranes, filter paper, and glass plates and incubated at 37°C for varying periods of time to obtain up to 12 blots (reproduced from *(10)* with permission from Elsevier).

10. Repeat this procedure with another set of nitrocellulose membranes and incubate the assembly at 37°C for 2 h to obtain two more blots from the same gel (*see* **Note 17**).
11. Use the same gel further to obtain blots by repeating this procedure to obtain a total of 12 blots. Incubate the gel with the respective membranes for a period of 3 or 4 h to obtain the third and fourth sets of blots. Obtain the fifth and sixth sets of blots (**Fig. 3**) by incubating with the respective membranes for a period of 9 or 36 h, respectively (*see* **Note 18**).

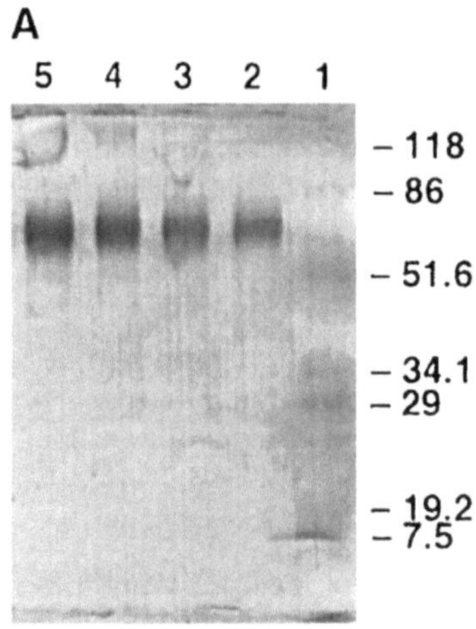

Fig. 2. Ro 60 immunoblots obtained following nonelectrophoretic transfer for 30 min or 2 h. (**a**) One of two Ro 60 blots obtained by the first incubation of nitrocellulose membranes on either side of the gel at 37°C for 30 min. (**b**) A blot from the second set (blots 3 and 4) obtained from the same gel following incubation at 37°C for 2 h (reproduced from *(6)* with permission from Elsevier).

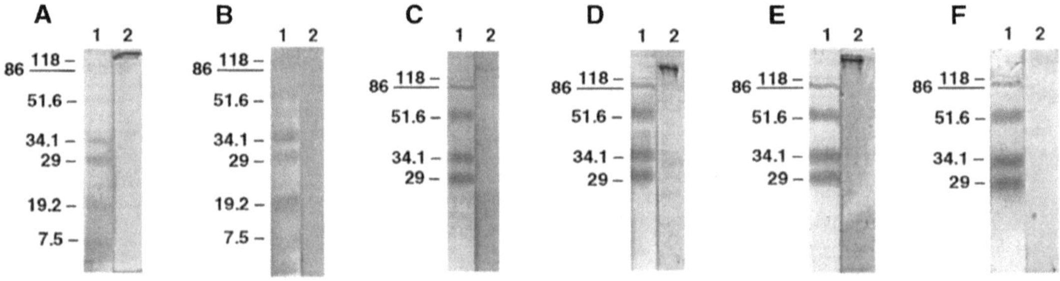

Fig. 3. Immunoblots obtained using spectrin as antigen. Lane 1 in **A–F** shows prestained SDS-PAGE molecular weight standards. Lane 2 shows spectrin probed with either preimmune (**B, C, F**) or antispectrin rabbit sera (**A, D, E**). (**A**) shows one of the first two blots obtained and has been probed with antispectrin. (**B**) shows the second of the two blots of the first set obtained from the reverse side of the gel and was probed with preimmune sera. (**C, D**) show the fifth set (blots 9 and 10) of immunoblots probed with preimmune and antispectrin sera, respectively. (**E, F**) show the sixth set (blots 11 and 12) of blots probed with antispectrin and preimmune sera, respectively (reproduced from *(6)* with permission from Elsevier).

12. Cut excess membrane to smoothen edges and also cut the spectrin-containing membrane into individual lanes (*see* **Note 19**).
13. Block the membranes with blocking solution for 1 h.
14. Add appropriate antisera to the membranes (antispectrin, anti-Ro 60, or control sera) and incubate for 2 h.
15. Rinse membrane strips with deionized water 2–3 times (*see* **Note 20**).
16. Wash 5× with TBST, 5 min each time.
17. Add anti-human IgG or anti-rabbit IgG alkaline phosphatase conjugate (1:5,000 dilution, diluted in diluent) and incubate for 1 h.
18. Wash as in **steps 15** and **16**.
19. Add 500 μL of nitroblue tetrazolium/BCIP substrate and let bands develop (*see* **Note 8**).
20. Rinse 2–3 times with deionized water (*see* **Note 20**).
21. Wash with TBST and arrange strips on paperboard inserts (*see* **Note 21**).

4. Notes

1. Having water at the bottom of the cylinder helps to dissolve Tris relatively easily, allowing the magnetic stir bar to go to work immediately. If using a glass beaker, Tris can be dissolved faster provided the water is warmed to about 37°C. However, the downside is that care should be taken to bring the solution to room temperature before adjusting pH.
2. Concentrated HCl (12 *N*) can be used at first to narrow the gap from the starting pH to the required pH. From then on it would be better to use a series of HCl (e.g., 6 N and 1 N) with lower ionic strengths to avoid a sudden drop in pH below the required pH.
3. Wear a mask when weighing acrylamide. To avoid exposing acrylamide to coworkers, cover the weigh boat containing the weighed acrylamide with another weigh boat (similar size to the original weigh boat containing the weighed acrylamide) when transporting it to the fume hood. Transfer the weighed acrylamide to the cylinder inside the fume hood and mix on a stirrer placed inside the hood. Unpolymerized acrylamide is a neurotoxin and care should be exercised to avoid skin contact. Mixed resin AG 501–X8 (D) (anion and cation exchange resin) is used when acrylamide solution is made,

since it removes charged ions (e.g., free radicals) and allows longer storage. Some investigators store the prepared acrylamide along with this resin in the refrigerator. However, we filter them out before storage. The used mixed resin should be disposed as hazardous waste. Manufacturer's warning states that this resin is explosive when mixed with oxidizing substances. The resin contains a dye that changes from blue-green to gold when the exchange capacity is exhausted.

4. The acrylamide solution can be stored at 4°C for 1 month. Acrylamide hydrolyzes to acrylic acid and ammonia. The acrylamide mixture, buffer, and water can be prepared in large batches, frozen in aliquots (for greater day-to-day reproducibility), and used indefinitely (*see* **Ref.** *14*). Remove the required amount, bring to room temperature, and add the other ingredients for polymerization. However, in our laboratory we make the acrylamide solution fresh about every month when we cast our own gels.
5. We find that it is best to prepare this fresh each time.
6. We find that storing at 4°C reduces its pungent smell.
7. Simple method of preparing running buffer: Prepare 10× native buffer (0.25 M Tris, 1.92 M glycine). Weigh 30.3 g Tris and 144 g glycine, mix, and make it to 1 L with water. Dilute 100 mL of 10× native buffer to 990 mL with water and add 10 mL of 10% SDS. Care should be taken to add SDS solution last, since it makes bubbles.
8. SDS precipitates at 4°C. Therefore, the lysis buffer needs to be warmed prior to use.
9. Dilute 100 mL of 10× native buffer to 800 mL with water and add 200 mL of methanol. Avoid adding methanol directly to the 10× buffer, since it precipitates its ingredients. Even in such a scenario the precipitate can be redissolved by the addition of 800 mL water.
10. Add 100 mL of 10× TBS to a 1 L graduated cylinder and make it to about 800 mL with water. Transfer 50 g skim milk powder into the cylinder and stir until dissolved. Make to 1 L with water. Separate 500 mL as the blocking solution. To the remaining 500 mL add 250 μL of Tween-20 (cut end of blue tip to aspirate Tween-20 easily), dissolve, and use it as the diluent.
11. The gel cassette was sealed at the base using 1% agarose. Overlay the resolving gel with water for gels having acrylamide concentration lower than 8% and use isobutanol (or isobutanol saturated with water) for gels of 10% or greater (*see* **Ref.** *15*). This overlay prevents contact with atmospheric oxygen (which inhibits acrylamide polymerization) in addition to helping to level the resolving gel solution.

12. Centrifuging the samples prior to the run helps remove insoluble debris, which could produce streaks in the protein lanes (revealed when stained with Coomassie Blue). Add a drop of 0.1% BPB to the upper chamber buffer. This helps to form a much stronger dye front during the electrophoretic run.
13. Membrane contact with the gel is much better when the gel is not moist. Therefore, it is important to dry the gel for 5–10 min. The membrane will now stick well to the gel and the gel will peel off the bottom glass plate by just lifting the membrane.
14. Hold the two top corners of the membranes with each hand. Lower the bottom part of the membrane first on the lower part of the gel and gently release the membrane little by little to lay the complete membrane on the gel. This will prevent trapping of bubbles in between the gel and the membrane. A 10 mL pipette was used to roll out the air bubbles from the gel–membrane sandwich prior to placing in transfer cassette. In the case of the gel with spectrin, cut the membrane to fit the two lanes of the gel.
15. The humid chamber consisted of a closed plastic container with a moist Terri Wipes paper towel (Kimberly-Clark, Neenah, WI, USA) at the bottom. The container must be big enough to contain the nitrocellulose–gel–filter paper assembly encased within the glass plates.
16. The second set of two blots was also obtained following incubation with the gel for 1 h (*see* **Fig. 3b, c**).
17. When removing the nitrocellulose membranes from the gel for immunoblotting, it would be common to find that the gel comes up stuck to one of the two membranes. To remove this membrane from the gel, place a fresh, dry nitrocellulose membrane on top of the gel and gently lift the gel. The gel becomes stuck to this fresh membrane, thus releasing the other membrane.
18. The gel dries, inspite of placing in humid chamber, when incubated for longer time periods (36 h). Therefore, it is best to use the blots obtained after 12-h incubation with the membrane.
19. Cut a tiny wedge from the bottom left side of the marker lane and the main membrane sheet for orientation purposes. Also, in the case of the membranes with spectrin (*see* **Subheading 3.1, item 3**), excise the spectrin lane from the protein marker lane after matching each spectrin lane with its specific protein marker lane with pencil marks.
20. Rinsing the membrane strips with deionized water 2–3 times will help remove a bulk of the nonspecific antibodies and help reduce the amount of TBST used subsequently and also

reduce the number of washes. This wash helps to reduce nonspecific binding of NBT/BCIP to the strip. The water, owing to its low ionic strength compared with TBST, will be able to remove contaminants much better than TBST. Water is much cheaper than TBST, in terms of money and labor. Other investigators have found no reduction in detection of specific signals due to washing with water *(16)*.

21. We use paper boards placed in between stacks of ELISA plates in packages of ELISA plates (Costar, Cambridge, MA, USA) for this purpose.

Acknowledgement

This work was supported by NIH grant ARO1844 and Oklahoma Center for the Advancement of Science and Technology to RHS.

References

1. Towbin, H., Staehelin, T., and Gordon, J. (1979) Electrophoretic transfer of proteins from polyacrylamide gels to NC sheets: procedure and applications. *Proc Natl Acad Sci USA* 76, 4350–4354.
2. Kurien, B.T. and Scofield, R.H. (2006) Western blotting. *Methods* 38, 283–293.
3. Renart, J., Reiser, J., and Stark, G.R. (1979) Transfer of proteins from gels to diazobenzyloxymethyl paper and detection with anti-sera: a method for studying antibody specificity and antigen structure. *Proc Natl Acad Sci USA* 76, 3116–3120.
4. Gershoni, J.M. and Palade, G.E. (1983) Protein blotting: principles and applications. *Anal Biochem* 131, 1–15.
5. Peferoen, M., Huybrechts, R., and De Loof, A. (1982) Vacuum-blotting: a new simple and efficient transfer of proteins from sodium dodecyl sulfate-polyacrylamide gels to NC. *FEBS Lett* 145, 369–372.
6. Kurien, B.T. and Scofield, R.H. (1997) Multiple immunoblots after non-electrophoretic bidirectional transfer of a single SDS-PAGE gel with multiple antigens. *J Immunol Methods* 205, 91–94.
7. Smith, G.E. and Summers, M.D. (1980) The bidirectional transfer of DNA and RNA to nitrocellulose or diazobenzyloxymethyl-paper. *Anal Biochem* 109, 123–129.
8. Reinhart, M.P. and Malamud, D. (1982) Protein transfer from isoelectric focusing gels: the native blot. *Anal Biochem* 123, 229–235.
9. Knisley, K.A. and Rodkey, L.S. (1986) Affinity immunoblotting. High resolution isoelectric focusing analysis of antibody clonotype distribution. *J Immunol Methods* 95, 79–87.
10. Scofield, R.H., Zhang, F., Kurien, B.T., Anderson, C.J., Reichlin, M., Harley, J.B., et al. (1996) Development of the anti-Ro autoantibody response in a patient with systemic lupus erythematosus. *Arthritis Rheum* 39, 1664–1668.
11. Deutscher, S.L., Harley, J.B., and Keene, J.D. (1988) Molecular analysis of the 60 kD human Ro ribonucleoprotein. *Proc Natl Acad Sci USA* 85, 9479–9483.
12. Ben-Chetrit, E., Gandy, B.J., Tan, E.M., and Sullivan, K.F. (1989) Isolation and characterization of a cDNA clone encoding the 60 kDa component of the human SS-A/Ro ribonucleoprotein autoantigen. *J Clin Invest* 83, 1284–1292.
13. Dodge, J.T., Mitchell, C., and Hanahan, D.J. (1963) The preparation and chemical characteristics of hemoglobin-free ghosts of human erythrocytes. *Arch Biochem Biophys* 100, 119–130.
14. Harlow, E., and Lane, D. (1988) Electrophoresis. Appendix I. In: Harlow, E. and

Lane, D. Ed. Antibodies. A Laboratory Manual. Cold Spring Harbor Laboratory, New York, p. 638.

15. Harlow, E., and Lane, D. (1988) Electrophoresis. Appendix I. In: Harlow, E. and Lane, D. Ed. Antibodies. A Laboratory Manual. Cold Spring Harbor Laboratory, New York, p. 636.

16. Wu, M., Stockley, P.G., and Martin, W.J. II (2002) An improved Western blotting technique effectively reduces background. *Electrophoresis* 23, 2373–2376.

Chapter 9

Slice Blotting

Graeme Lowe

Summary

Slice blotting is a technique for recording the spatial distribution of extracellular signaling molecules released from thin slices of living tissue. Slices are positioned on the surface of a membrane that can trap secreted substances diffusing from the tissue. The pattern of membrane-bound antigens is subsequently visualized by immunoblotting.

Key words: Slice, Immunoblot, Secretion

1. Introduction

A variety of functional imaging technologies can be applied to visualize communication between cells in diverse structured systems, ranging from neuronal circuits of the brain to the developing embryo. Many techniques are indirect, inferring intercellular communication by detecting intracellular signals or gene expression. Methods for directly imaging extracellular signals are limited in the kinds of molecules that can be detected. Slice blotting fills a gap in the spectrum of available methods, allowing investigators to map patterns of secretion for a broad range of signaling compounds.

The basic steps of slice blotting are relatively simple and easily implemented in most laboratories without a major investment in specialized equipment (*see* **Fig. 1**). Tissue slices are cut using a vibrating blade slicer and incubated in vitro on a blotting membrane for a set time period, under physiological conditions determined by the experimenter. During the incubation period, secreted molecules escaping the tissue matrix by diffusion are

Biji T. Kurien and R. Hal Scofield (eds.), *Methods in Molecular Biology, Protein Blotting and Detection, vol. 536*

DOI: 10.1007/978-1-59745-542-8_9

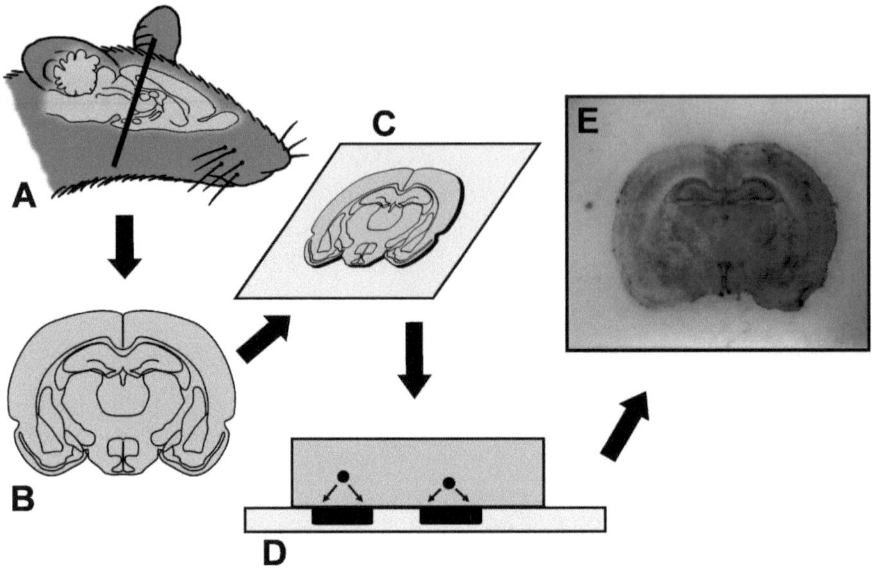

Fig. 1. Steps in slice blotting. A rat brain is extracted (**A**) and in vitro slices are cut (**B**). The slices are placed onto the surface of porous nitrocellulose membranes (**C**) and incubated in saline solution for a fixed time period. Immunoreactive signaling molecules released from slices are bound to the membranes (**D**), which are then developed using western blotting protocols to reveal the distribution of release sites (**E**).

bound and immobilized by the membrane. Afterward, the tissue slice is removed and the membrane is processed using standard western blot methods *(1, 2)* using primary antibodies to probe for particular antigens, and secondary antibodies for staining.

The slice blotting protocols described here were used to detect and visualize secretion patterns of neuropeptide-Y (NPY) from in vitro rat brain slices (midbrain and brainstem) stimulated by high potassium *(3)*. Potassium treatment caused strong depolarization and action potential firing of NPY-containing nerve terminals, releasing peptide that was collected on nitrocellulose membranes. Similar protocols were applied to detect unstimulated release of a chemotactic guidance molecule, semaphorin III and glial cell line-derived neurotrophic factor (GDNF), from the rat olfactory system *(3, 4)*. With appropriate modifications, these basic protocols should be adaptable for use in many other signaling systems.

2. Materials

2.1. Brain Slice Preparation

1. Slicing solution, either sucrose-based: 240 mM sucrose, 2.5 mM KCl, 10 mM Na-HEPES, 1 mM $CaCl_2$, 4 mM $MgCl_2$, 10 mM glucose, 0.2 mM ascorbic acid, pH 7.3; or glycerol-based (same solution with 240 mM glycerol replacing sucrose).

2. High Mg^{2+} artificial cerebrospinal fluid (Mg-ACSF): 122 mM NaCl, 3.75 mM KCl, 26 mM $NaHCO_3$, 1 mM $CaCl_2$, 3.3 mM $MgCl_2$, 0.5 mM NaH_2PO_4, 7.5 mM glucose.
3. Standard ACSF: 122 mM NaCl, 3.75 mM KCl, 26 mM $NaHCO_3$, 2 mM $CaCl_2$, 1 mM $MgCl_2$, 0.5 mM NaH_2PO_4, 10 mM glucose.
4. High K^+ ACSF (same as standard ACSF, but 50 mM Na^+ replaced with 50 mM K^+).
5. Regulated gas supplies: (a) pure O_2, (b) 95% O_2, and (c) 5% CO_2.
6. Fritted-glass gas dispersion tubes for bubbling solutions.
7. Vibratome™ or other vibrating blade tissue slicer.
8. Slice interface chamber: This is a closed 1 L plastic vessel containing a rack of nylon mesh wells where slices can be incubated prior to blotting. There is an entry port for a gas line (95% O_2, 5% CO_2) to enter and bubble the solution, and a small exit port.
9. Heated water bath.

2.2. Blotting

1. Nitrocellulose blotting membrane (e.g., Amersham Hybond™ ECL™, 0.45-μm pore size, 100-μm thickness, GE Healthcare Life Sciences, Piscataway, NJ, USA). Using thin membranes facilitates rapid O_2 diffusion to both sides of the slice.
2. Plastic pipette tips (1 mL capacity), razor blade.
3. Acrylic block 5 cm × 5 cm × 1 cm, with fine slots cut into top surface (3 mm deep, 0.2 mm wide).
4. Fine surgical forceps.
5. Pipettor with 200 μL pipette tips.
6. 500 mL dish or beaker.

2.3. Immunostaining

1. Tris-buffered saline-Tween 20 (TBST): 0.9% NaCl, 100 mM Tris–HCl, 0.1% Tween 20 (Sigma-Aldrich, St. Louis, MO, USA), pH 7.5.
2. Goat serum (Vector Laboratories, Burlingame, CA, USA).
3. Primary antibody (e.g., rabbit anti-NPY antibody, Sigma-Aldrich).
4. Biotinylated secondary antibody (e.g., biotinylated anti-rabbit IgG made in goat, Vector Laboratories).
5. Vectastain™ Elite™ ABC (Avidin Biotinylated Enzyme Complex) Kit (Vector Laboratories).
6. Vector™ VIP Peroxidase Substrate Kit (Vector Laboratories).

3. Methods

3.1. Brain Slice Preparation

Techniques for preparing in vitro brain slices have been well documented, and different laboratories may employ slightly different solutions and procedures. The slice blotting procedure yields best results when tissue is optimally healthy, and cell damage or death associated with slicing is kept to a minimum (*see* **Note 1**). The protocol described here is a variant of a widely used method that replaces sodium chloride with equimolar sucrose in the slicing solution, to reduce cell damage caused by chloride influx *(5)*. A similar strategy using glycerol as the replacement solute to maintain osmolarity has also been applied successfully to improve slice health *(6)*. Lowered Ca^{2+} and elevated Mg^{2+} are included to suppress depolarization and Ca^{2+} entry associated with NMDA receptor activation by glutamate released when slicing brain tissue. Ascorbic acid is added to scavenge free radicals during slicing.

1. Freeze a volume of slicing solution half-filling the bath vessel used to prepare slices, and layer chilled (4°C) unfrozen slicing solution on top of the frozen solution (or, use active cooling system of the slice bath if it is available on the slicer). Bubble with O_2 gas.
2. Fill slice interface chamber with Mg-ACSF, bubbled with 95% O_2, 5% CO_2 gas, and warm it to 32°C in a water bath.
3. Decapitate an anesthetized rat, remove the brain, and block off the portion to be sliced. Using a cyanoacrylate adhesive, attach the blocked brain to a slicing stage and submerge in the chilled, bubbled bath of the slicer.
4. Cut 300–400 μm-thick slices.
5. Transfer freshly cut slices to nylon mesh wells of slice interface chamber. Slices should be floated on the air–water interface of each well to maximize their contact with humidified O_2/CO_2 gas. This can be done by momentarily raising the nylon mesh bottom above the air–water interface.
6. Remove the interface chamber from the water bath, and allow it to cool slowly to room temperature (23°C) over a 30–60 min period.
7. Replace the solution in interface chamber with standard ACSF.

3.2. Blotting

1. Cut sheets of nitrocellulose membrane into small rectangles. The size of membrane to use depends on the size of the slices (e.g., 1–1.5 cm × 1.5–2 cm for rat brain).
2. Place a nitrocellulose rectangle onto the cut end of a vertically oriented 1 mL plastic pipette tip, or similar vertical support

(e.g., a short acrylic rod). The support is smaller in diameter than the rectangle. This allows a drop of solution to be pipetted onto the membrane without the liquid spreading and spilling off the edges.

3. Pick up a slice from the interface chamber using a truncated 1 mL pipettor tip and deposit it onto the nitrocellulose rectangle. Pipette off excess solution until the slice settles flat onto the membrane surface. Using forceps, pick up the membrane and lay it onto absorbent tissue or filter paper, with the slice surface up – this will remove remaining solution by drawing it through the filter, causing the slice to adhere to the membrane. It is important to make sure that there are no kinks in the slice.
4. Immediately submerge the membrane and attached slice in a bath of standard ACSF bubbled with 95% O_2, 5% CO_2 gas. Many membrane rectangles can be suspended in the same bath by holding them edgewise in thin slots of a machined acrylic block. They might also be submerged in multiwell plates with nylon mesh bottoms, similar to the way individual slices were kept in the interface chamber.
5. To stimulate or modulate release or secretion of signaling molecules, the incubating bath solution can be changed. For example, incubating in high K^+ ACSF will depolarize and drive the release of neurotransmitters or neuromodulators from presynaptic terminals (*see* **Note 2**). Pharmacological agents can be added to modulate secretion. Incubation period is variable, but 10–40 min exposure to high K^+ was sufficient to register immunodetectable NPY release from rat brain (*see* **Note 3**).
6. After blotting, remove slices from nitrocellulose membranes with a gentle squirt of ACSF from a 200 μL pipette tip. Rinse membranes in water, air-dry, and store at –20°C until staining. Tissue slices can be retained and further processed using other procedures if necessary.

3.3. Immunostaining

Conventional western blotting procedures can be applied to the nitrocellulose slice blots to visualize patterns of bound antigens.

1. Rinse nitrocellulose membranes for 30 min in TBST.
2. Incubate for 30 min in TBST + 3% goat serum (blocking serum).
3. Incubate for 30 min in TBST + 1.5% goat serum + primary antibody (e.g., 1:2,000 dilution of rabbit anti-NPY antibody, Sigma-Aldrich).
4. Rinse three times, 10 min, in TBST.
5. Incubate for 60 min in TBST + 1.5% goat serum + biotinylated secondary antibody (e.g., 7.5 μg/mL biotinylated anti-rabbit IgG, made from goat, Vector Laboratories).

6. Rinse three times, 10 min, in TBST.
7. Incubate for 60 min in ABC solution. Prepare the ABC solution 30 min ahead of this step (i.e., two drops of solution A + two drops of solution B, per 5 mL TBST).
8. Rinse three times, 10 min, in TBST.
9. Incubate for up to 30 min in VIP solution (i.e., three drops of Reagent 1 + three drops of Reagent 2 + three drops of Reagent 3 + three drops of hydrogen peroxide). The darkening due to peroxidase reaction can be monitored visually.
10. Rinse three times, 10 min, in TBST, after a strong contrast blot pattern is obtained.
11. Record blot patterns immediately by photographic or digital imaging. Nonspecific staining on membranes (*see* **Note 2**) tends to darken over time, reducing contrast. Darkening can be slowed by storing membranes at low temperature (–20°C in 80% glycerol).

4. Notes

1. Slice blotting can be sensitive to nonspecific background staining that may arise when the tissue is not in optimal health. Neurons in some brain regions may be more easily damaged by slicing due to their intrinsic structure, such as having widespread dendritic arbors. In these cases, it is expected that injured cells may more readily lyse during slice handling and blotting, which would contribute to elevated background staining. Some degree of background is expected because direct contact with the nitrocellulose membrane is likely to rupture a thin layer of surviving cells at the surface of the slice, leading to nonspecific release of contents. If slices become sticky (e.g., they cannot be easily lifted off the membrane by a gentle pulse of saline solution), major cell damage is indicated, and slices should be discarded. Cell condition can be checked by examining slices directly under a differential interference contrast microscope.
2. To verify a signal, it is important to set up a series of control blots in which the blotting conditions are varied. For example, in brain slices treated with high K^+, controls would correspond to incubation in a low K^+ ACSF solution. Other controls to determine the level of nonspecific background staining include omitting primary antibody, or preabsorbing primary antibody with antigen before applying to a blot.
3. Slice blotting integrates signal over time, so weak signals can be amplified by incubating slices on a blotting membrane for extended time periods.

References

1. Kurien, B.T. and Scofield, R.H. (2006) Western blotting. *Methods* 38, 283–293.
2. Kurien, B.T. and Scofield, R.H. (2003) Protein blotting: a review. *J Immunol Methods* 274, 1–15.
3. Lowe, G. (1999) Slice blotting: a method for detecting the release of immunoreactive substances from living brain tissue. *J Neurosci Methods* 90, 117–127.
4. Paratcha, G., Ibanez, C.F., and Ledda, F. (2006) GDNF is a chemoattractant factor for neuronal precursor cells in the rostral migratory stream. *Mol Cell Neurosci.* 31, 505–514.
5. Aghajanian, G.K. and Rasmussen, K (1989) Intracellular studies in the facial nucleus illustrating a simple new method for obtaining viable motoneurons in adult rat brain slices. *Synapse* 3, 331–338.
6. Ye, J.H., Zhang, J., Xiao, C., Kong, J.Q. (2006) Patch-clamp studies in the CNS illustrate a simple new method for obtaining viable neurons in rat brain slices: glycerol replacement of NaCl protects CNS neurons. *J Neurosci Methods* 158, 251–259.

Chapter 10

Localizing Proteins by Tissue Printing

Rafael F. Pont-Lezica

Summary

The simple technique of making tissue prints on appropriate substrate material has made possible the easy localization of proteins, nucleic acids, carbohydrates, and small molecules in a tissue-specific mode. Plant tissues can be used to produce prints revealing a remarkable amount of anatomical detail, even without staining, which might be used to record developmental changes over time. In this chapter we will focus on the protocols for the localization of proteins and glycans using antibodies or lectins, probably the most frequently used application, but the localization of other molecules is reported and the sources indicated.

Key words: Immunodetection, Tissue blotting, Protein localization, Western blotting

1. Introduction

Fifty years ago, the first series of film printing for the localization of several enzymes (protease, amylase, RNAse, and DNAse) was realized by placing cryostat sections of various organs on gelatin, starch, or gelatin-nucleic acids *(1)*. Those films were then stained for the substrate, giving a negative image. Since both substrates and enzymes were macromolecules with slow diffusion rates, the images obtained were clear. The availability of several membranes such as nitrocellulose, Nytran, Genescreen, and Immobilon [polyvinylidene difluoride (PVDF)] designed to bind proteins and nucleic acids opens a new era in tissue blotting, and a book edited by Reid et al. *(2)* included numerous protocols for the visualization of enzyme activities, protein and glycan localization,

Biji T. Kurien and R. Hal Scofield (eds.), *Methods in Molecular Biology, Protein Blotting and Detection, vol. 536*

DOI: 10.1007/978-1-59745-542-8_10

and gene expression, in plants and some animal tissues. Nitrocellulose membrane adsorbs relatively large quantities of proteins that are tightly bound, whereas it usually does not retain salts and hydrophilic small molecules. This type of membrane has been used for transferring proteins separated by sodium dodecyl sulphate–polyacrylamide gel electrophoresis to membranes for immunological detection. The procedure involves the electro elution of the negatively charged proteins from the gel to the membrane, followed by the visualization of the targeted protein with specific antibodies. The basic principle of tissue printing is that much of the contents of the cells at the surface of a freshly cut tissue section can be transferred to an adhesive or absorptive surface with little or no diffusion, by simple contact *(3)*. The obtained imprints reveal anatomical details that are difficult to see without fixing, embedding, and then sectioning; however, the resolution of anatomical prints is inferior to that of classical fixed and stained sections. The procedure is particularly useful with big samples such as sections of fruits, tubers, or stems, revealing the distribution of particular proteins in the whole organ *(4, 5)*.

A digest of different applications of tissue printing for the detection of particular proteins, glycans, mRNA, enzyme activities, and microorganism in plants is given in **Table 1**.The steps to make a print and to show the localization of proteins using specific antibodies, as well as the necessary precautions to avoid nonspecific staining in plant tissues, will be described in this chapter.

Table 1
Different applications of tissue prints for the detection of enzyme activity, specific proteins, virus, gene expression, and carbohydrates in plant tissues

Detected	Tissue	Assay	Film	References
Enzyme activity				
Amylase	Barley aleurone, pole bean hypocotyl, and cotyledons	Negative stain I_2.KI	Starch	*(6, 7)*
β-Glucosidase	Oil palm	Naphtyl-glucoside	NC	*(8)*
Catalase	Various	Negative stain I_2	KI starch–H_2O_2	*(9)*
Cinnamyl alcohol dehydrogenase	Tomato, poplar	NBT, $NADP^+$, PMS, conyferyl alcohol	NC	*(10)*
Endo-β-mannanase	Coffea beans	Gel-locust bean gum	Paper	*(11)*

(continued)

Table 1
(continued)

Detected	Tissue	Assay	Film	References
Laccase	Sycamore cells	4-Methylcatechol, syringaldmine	NC	*(12)*
Myrosinase	Turnip and horse radish	Sinigrin–glucose oxidase–peroxidase system	PVDF	*(13)*
Peroxidase	Pea epicotyls	α-Phenilene diamine	NC	*(14)*
Polyphenol oxidases	Tomato			*(15)*
Protease	Pea, mung bean	Negative stain	Gelatin-India ink	*(16, 17)*
Tyrosinase	Mushrooms	Tropolone & L-dopa	NC	*(18)*
Proteins				
Allergens	Pollen	Immunodetection	NC	*(19)*
Antifreeze proteins	Winter rye	Immunodetection	NC	*(20)*
Cardosin A	Artichoke	Immunodetection	NC	*(21)*
Cellulase	Abscission zones various plants	Immunodetection	NC	*(22, 23)*
Cys-rich proteins	Soybean stem and pod	Fluorescent S-corboxymethylation	NC	*(24)*
Endoxylanase (fungal)	Tobacco	Immunodetection	Nylon	*(25)*
Expansins	Rice	Immunodetection	NC	*(26)*
Extensin, HRGPs	Soybean, bean, maize	Immunodetection	NC	*(27–29)*
Ferritin	Transgenic rice	Immunodetection	NC	*(30)*
Glutamine synthase	Potato	Immunodetection	NC	*(31)*
GRPs	Bean, soybean	Immunodetection	NC	*(32–34)*
Laccase	Sycamore	Immunodetection	NC	*(12)*
Lectins	Potato tuber, Canavalia seeds	Immunodetection	NC	*(5, 35)*
Lipid transfer proteins	Sunflower	Immunodetection	NC	*(36)*
Lipoxigenase	Tomato fruit	Immunodetection	NC	*(37)*
Natriuretic peptide	Ivy, potato	Immunodetection	NC	*(38)*
Polygalacturonase	Tomato fruit	Immunodetection	NC	*(4)*

(continued)

Table 1 (continued)

Detected	Tissue	Assay	Film	References
Virus				
Bromovirus	Cowpea	Immunodetection	NC	*(39)*
Pox virus	Plump	Immunodetection	NC	*(40)*
Several viruses	Plump, Malus	Immunodetection	NC	*(41)*
Soybean MV	Soybean leaves	Immunodetection	NC	*(42)*
Tobacco MV	Tobacco leaves	Immunodetection	NC	*(43, 44)*
Virus and mycoplasma	Sybean leaves	Immunodetection	NC	*(45)*
Zucchini yellow mosaic potyvirus	Melon leaves	Immunodetection	NC	*(46)*
Gene expression				
Dirigent proteins	Forsythia	RNA–RNA hybridization	Nylon	*(47)*
Arabinan	Lemon	Immunodetection	NC	*(48)*
β-Conglycinin	Soybean seeds	RNA–RNA hybridization	NC	*(49)*
Extensin, HRGP	Soybean seeds, tomato, petunia, tobacco	RNA–RNA hybridization	NC, nylon	*(33, 49)*
Pectin Me-esterase	Tomato	RNA–RNA hybridization	NC	*(50)*
SAUR	Soybean shypocotyls	RNA–RNA hybridization	NC	*(51)*
Carbohydrates				
Glycoconjugates	Soybean seed, potato	Periodic acid–DNS hydrazine	PVDF	*(52)*
Man- and Gal-containing proteins	Soybean seed and pod	Fluorescent lectins	NC	*(52)*
Pectic-galactan	Tomato	Immunodetection		*(53)*

2. Materials

2.1. Doing a Tissue Print

1. Whatman no. 1 filter paper (Fisher Scientific Company, Pittsburgh, PA, USA).
2. Blotting membrane: nitrocellulose (type BA-85, Schleicher & Schuell, St. Marcel, France), nylon (Zetaprobe, Bio-Rad,

Richmond, CA, USA), Immobilon P, 0.45 μm (Millipore Corp, Bedford, MA, USA).

3. Double-edged razor blades, forceps, rubber gloves, paper to protect the membrane, acrylic sheet, marking pen.
4. Hand lens or microscope for viewing the specimen, and the biological material (**Fig. 1a**).

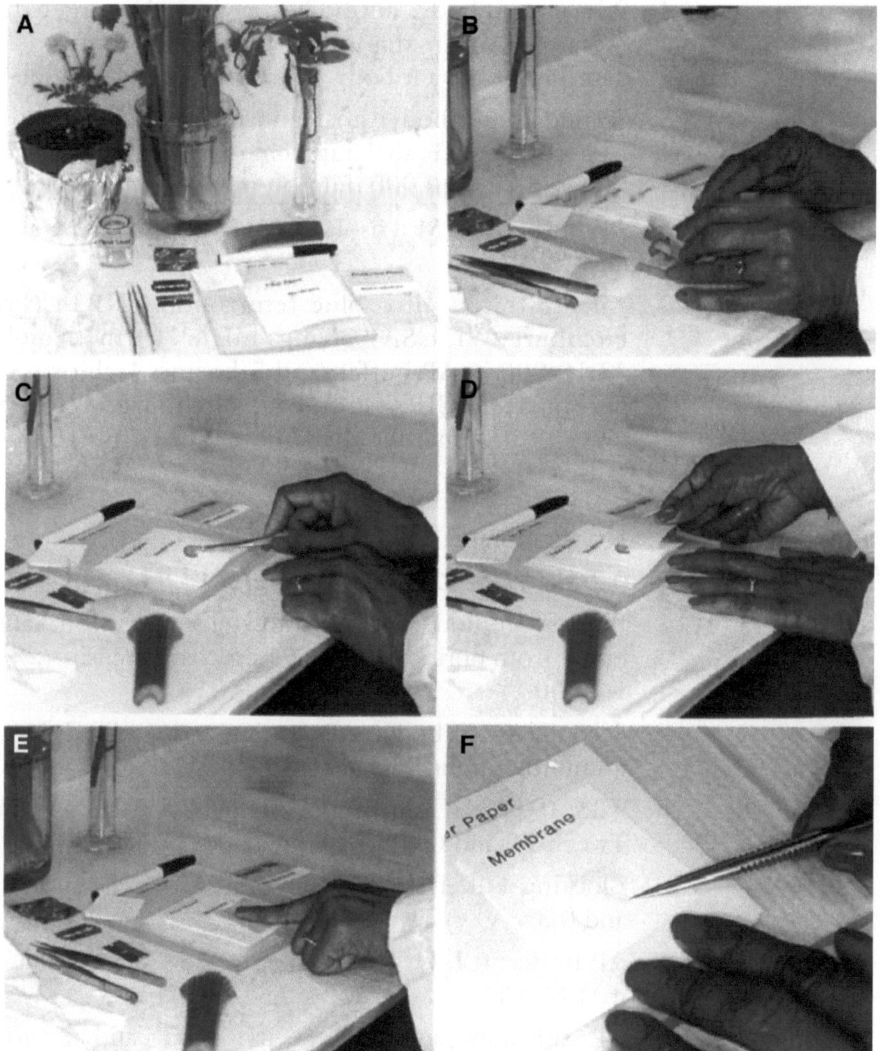

Fig. 1. Detailed description of the steps involved in tissue printing. (**A**) Materials required, (**B**) cutting the tissue, (**C**) placing the section on the membrane, (**D**) protecting the section and the print, (**E**) applying pressure and (**F**) removing the section (reprinted from *(2)* with permission from Elsevier).

2.2. Detection of Total Proteins in a Tissue Blot

1. Nitrocellulose (NC) membrane, 0.45-µm pore size (Schleicher & Schuell, St. Marcel, France) (another membrane may be used).
2. Tris buffered saline (TBS)–Tween-20 (TBST): 20 mM Tris–HCl (pH 7.5), 0.15 M NaCl, and 0.05% Tween-20.
3. India ink staining: 1-µL Pelikan India ink/mL of PBST.

2.3. Revealing Proteins with Antibodies

1. Nitrocellulose membrane, 0.45-µm pore size (Schleicher & Schuell, St. Marcel, France) (another membrane may be used).
2. TBST.
3. Blocking buffer: 0.25% gelatin and 0.025% bovine serum albumin (BSA) in TBST.
4. Primary antibody: antibodies against the targeted protein (potato lectin in this case), in blocking buffer (1:15,000 dilution for the antibody against the deglycosylated lectin).
5. Secondary antibody: goat anti-rabbit alkaline phosphatase (AP) conjugate $F(ab)^2$ fragment (Sigma Chemical Company, MO, USA), 1:20,000 dilution in blocking buffer.
6. AP buffer: 0.1 M Tris–HCl (pH 9.5), 0.1 M NaCl, and 5 mM $MgCl_2$.
7. AP substrates: nitro blue tetrazolium (NBT) (Promega, Fitchburg, WI, USA), 50 mg/mL in 70% methanol diluted 33:10,000 in AP buffer, and 5-bromo-4-chloro-3-indolyl phosphate (BCIP) (Promega, Fitchburg, WI, USA), 50 mg/mL in dimethyl formamide diluted 66:10,000 in AP buffer.

2.4. Localizing Cell Wall Proteins in Plants

1. NC membrane.
2. 0.2 M $CaCl_2 \sim 2H_20$.
3. Primary antibody: specific antiextensin polyclonal antibodies raised from purified soybean seed coat extensin *(27)* diluted 1:15,000.
4. Secondary antibody: AP conjugated antibody anti-rabbit immunoglobulin IgG (Fc).
5. TBS: 0.9% NaCl in 20 mM Tris–HCl (pH 7.4) plus 0.3% Tween-20 and 0.05% sodium azide (NaN_3).
6. Blocking buffer: 0.25% (w/v) BSA, 0.25% (w/v) gelatin, and 0.3% (v/v) Tween-20 in TBS.
7. AP buffer: 0.1 M Tris–HCl (pH 9.5), 0.1 M NaCl, and 5 mM $MgCl_2$.
8. AP substrate solution: 66-µL NBT and 33-µL BCIP (Promega, Fitchburg, WI, USA) in 10-mL AP buffer.
9. 3 MM Whatman filter paper (Fisher Scientific Company, Pittsburgh, PA, USA).

3. Methods

3.1. Doing a Tissue Print

1. Place several layers of filter paper on a smooth, hard surface, and place a blotting membrane on top. Use a double-edged razor blade to cut a tissue sample (**Fig. 1b**) (*see* **Note 1**).
2. Using forceps transfer the tissue section to the membrane (*see* **Note 2**). Several successive sections can be printed on the same piece of membrane (**Fig. 1c**).
3. Place a small piece of nonabsorbent paper over the section to protect the membrane from fingerprints (**Fig. 1d**) (*see* **Note 3**).
4. Apply the appropriate amount of pressure to the section for the type of print desired. A chemical print requires only light pressure, but a physical print requires several times as much. The proper pressure also varies with the tissue used (**Fig. 1e**).
5. Gently remove the protective paper and the section with forceps, and air-dry the print with warm air, and observe (**Fig. 1f**).
6. The prints may be illuminated from the top or from the one side by white or UV light and may also be viewed with transmitted light.

3.2. Detection of Total Proteins in a Tissue Blot

It is useful to have an image stained for total proteins in the blot, which will indicate the pattern of protein distribution in the tissue, as well as some indications on the different tissues present in the section. We have used India ink staining *(54)* for total protein on nitrocellulose membranes since Coomasie Blue stain gives a strong background. The same tissue slice can be printed several times, showing the tissues with higher amount of proteins (**Fig. 2**).

1. Wash the dry nitrocellulose filter twice with TBST for 5 min with constant shaking.
2. Incubate the NC membrane with India ink staining mix until the image appears clearly (*see* **Note 4**).
3. Wash the print with water 5 min, and dry it.

3.3. Revealing Proteins with Antibodies

Chemical tissue prints result from the molecules that transfer from the freshly cut cells of a tissue section to the surface of a synthetic membrane, where they are retained and immobilized. The principal steps for tissue printing are (a) the release of the protein of interest from the tissue, (b) the contact–diffusion transfer of the protein to the recipient membrane, and (c) the retention and binding of the protein into the synthetic matrix. The protein print is a mirror image of the tissue and can be used to detect and localize specific proteins. Retention and binding of a protein into the recipient matrix depend on the chemistry of the membrane and its ability to interact electrostatically and hydrophobically with the protein of interest. NC membranes have a high binding

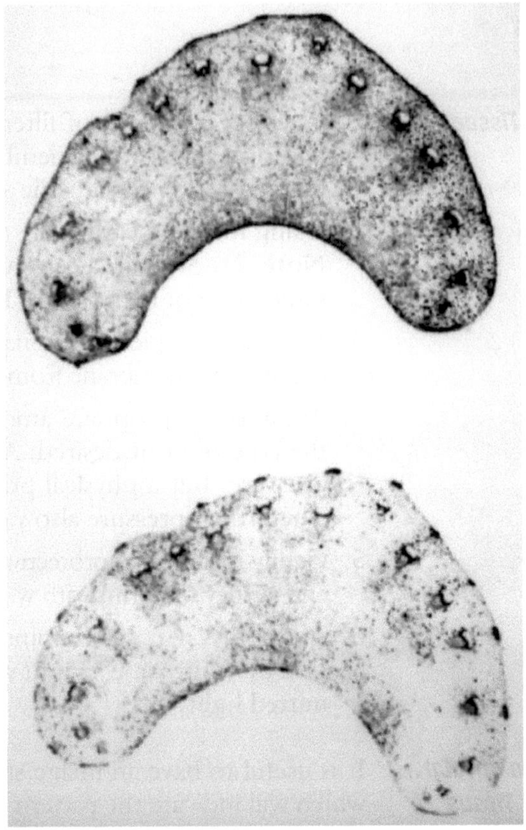

Fig. 2. Tissue print of a cross section from a celery stalk stained for total proteins with India ink. The same section was printed a first time (*upper image*) and a second time (*lower image*). Most of the proteins from the cut cells were transferred on the first print, and the second one shows mainly the apoplastic proteins abundant in some tissues.

capacity for proteins and should be tried first for printing a new type of protein molecule. Adding methanol to the transfer buffer increases the capacity and affinity on NC for proteins, presumably by promoting hydrophobic interactions (*see* **Note 5**).

Methods for detecting the protein of interest bound to the imprinted membranes must be specific and sensitive. Best results are achieved by using antibodies raised against the targeted protein. An enzyme-conjugated secondary antibody raised against the primary antibody is commonly used to visualize the binding between the primary antibody and the antigen (*see* **Note 6**). To reduce background from nonspecific cross reactions, the primary antibody is combined either with 1% (v/v) normal serum from the species in which the secondary antibody was raised or with a low concentration of sodium dodecyl sulfate or Tween-20.

The following procedure was used to localize potato lectin by means of polyclonal antibodies against the native and deglycosylated potato tuber lectin *(5)*.

1. Cut a section of the tissue about 1-mm thick with a new razor blade, and gently wipe the surface with a filter paper to absorb excess liquid. Put the freshly cut surface on the membrane, and press for 10–15 s (*see* **Note 7**).
2. Transfer the printed membrane to TBST, and wash away the unbound material two times for 5 min each in a shaker.
3. Block the unoccupied site of the membrane by shaking the membrane in a blocking buffer for 30 min at room temperature (RT).
4. Transfer the membrane to the first antibody solution (antilectin serum), and incubate overnight in a refrigerator or for 2 h at RT. For controls either use preimmune serum at the same dilution or skip the first antibody step (*see* **Note 8**).
5. Wash the membrane in TBST three times for 10 min each.
6. Incubate the prints with the secondary antibody (goat anti-rabbit alkaline phosphatase in this case) for 2 h at RT.
7. Wash the membrane several times with TBST (five washes of 5 min each) (*see* **Note 9**).
8. Incubate the prints in AP substrates at RT until the reaction product is observed; treated and control prints should be incubated under the same conditions (*see* **Note 10**). The result of this tissue printing method is shown in **Fig. 3**.

3.4. Localizing Cell Wall Proteins in Plants

The presence of a cell wall is one of the outstanding features distinguishing plant cells from those of animals. The cell wall is not an organelle with relatively constant functions; rather, it is subject to continuous developmental processes that govern cell size, division, shape, and function. In addition to the well-known polysaccharides present in the cell wall, proteins are important modulators of cell wall structure and function *(55)*. One of the best characterized cell wall proteins is extensin, a basic hydroxyproline-rich glycoprotein important for cell wall structure. It is a difficult protein to isolate, because high proportion of it becomes insolubilized in the wall, but it can be extracted with a solution with high salt concentration. A modification of the tissue print method was developed *(27)* to transfer cell wall proteins to NC, soaking the membrane previously in 0.2 M $CaCl_2$ and then printing the tissue.

1. Soak the NC membrane in 0.2 M $CaCl_2$ for 30 min, and dry on 3-MM Whatman paper.
2. Cut fresh tissue into sections of 0.3–3-mm thick with a new razor blade, previously washed in distilled water for 3 s and

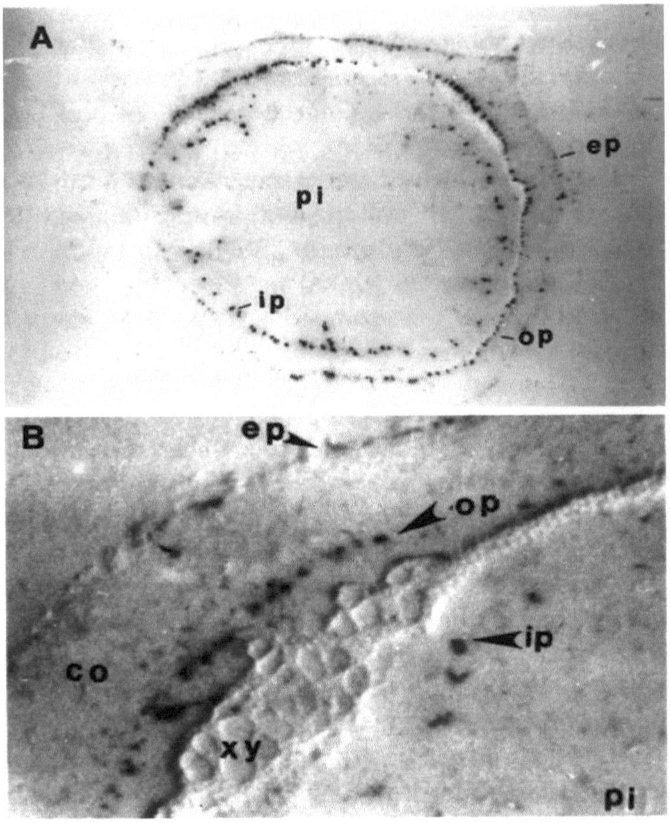

Fig. 3. Tissue prints of potato stem incubated with antibodies raised in rabbit against the deglycosylated lectin, followed by goat anti-rabbit AP. (**A**) Transverse section. (**B**) Magnification of a stem sector with lateral illumination showing greater detail of lectin localization. The lectin is present only in the outer (op) and inner phloem (ip). The stain at the epidermis (ep) is not a positive reaction but a transfer of natural pigments (*xy* xylem, *co* cortex, *pi* pith parenchyma) (reprinted from *(52)* with permission from Elsevier).

dried on Kimwipes. Then transfer each section to the NC membrane as indicated.

3. Block the NC with blocking buffer for 1–3 h at RT with constant shaking.
4. Add the primary antibody to the desired dilution in the blocking buffer, and incubate the membrane for 1–3 h at RT with shaking.
5. Wash the NC three times for 30 min each in TBS with agitation.
6. Soak the NC membrane in secondary antibody (AP-conjugated anti-IgG) diluted 1:20,000 in blocking buffer for 1–3 h with agitation.
7. Wash the NC membrane with AP buffer, and add the AP substrate solution. Develop the tissue print until a color signal appears, and stop the reaction by washing the membrane in distilled water.

4. Notes

1. Use gloves for all the manipulations to avoid finger prints on the membrane. It may be necessary to gently preblot the section on a separate piece of filter paper before printing to remove excess tissue exudates from cut cells and to ensure an accurate print.
2. To avoid double images be careful when blotting and removing the tissue section from the membrane.
3. When printing a thin section (200–300 μm), place a piece of membrane on top of the section to prevent the nonabsorbent paper from marking the membrane under the section.
4. The incubation time for India ink staining can be very variable, according to the amount of protein present in the tissue. For a seed blot from a legume, 15-min incubation will give a very good image. For other tissues such as stem or petiole, it may be necessary to incubate for several hours or even overnight.
5. In some instances NC membranes do not bind the protein of interest at all or else bind it only weakly. Try next a membrane that can react chemically with the protein, and covalently bind it, such as Immunodyne Immunoaffinity membrane (Pall Corp, Cortland, NY, USA). The type of buffer, salt concentration, and pH of the incubation mixture are important for protein binding and must be determined empirically.
6. Detection of the alkaline phosphatase-conjugated second antibody on tissue prints from plant organs was selected over the peroxidase-conjugated second antibody procedure because the substrates used for detecting the peroxidase, such as *O*-phenylenediamine and hydrogen peroxide, will react with endogenous plant peroxidase activity in the tissue sections *(14)*, making the immunoblotting reaction.
7. The same tissue surface can be reprinted several times; the successive images will be weaker, but a good imprint can be found among these when the proteins are very abundant in a particular tissue.
8. To be sure that the antibody used for tissue printing is specific for the targeted protein, it is important to test it against a total protein extract of the tissue previously by SDS-PAGE, followed by a western blot. Only the targeted protein should be stained, and use the highest dilution to obtain the best results.
9. The last wash is made with TBS without detergent to avoid interference with the alkaline phosphatase reaction.
10. Occasionally some tissues show cross reaction with the goat serum. To avoid such a reaction, two procedures are available:

(a) block the prints with blocking buffer containing 1:3,000 dilution of normal goat serum, or (b) use a secondary antibody raised in a different animal.

Acknowledgements

The author is grateful to the Université Paul Sabatier (Toulouse III, France) and the CNRS for support.

References

1. Daoust, R. (1957) Localization of deoxyribonuclease in tissue sections. A new approach to the histochemistry of enzymes. *Exp Cell Res.* 12, 203–211.
2. Reid, P.D., Pont-Lezica, R.F., del Campillo, E., and Taylor, R. (eds.) (1992) *Tissue Printing.* Academic Press, San Diego, CA.
3. Varner, J. and Ye, Z. (1994) Tissue printing. *FASEB J.* 8, 378–384.
4. Tieman, D.M. and Handa, A. (1989) Immunocytolocalization of polugalacturonase in ripening tomato fruit. *Plant Physiol.* 90, 17–20.
5. Pont-Lezica, R.F., Taylor, R., and Varner, J. (1991) *Solanum tuberosum* aggultinin accumulation during tuber development. *J Plant Physiol.* 137, 453–458.
6. Jacobsen, J. and Knox, R. (1973) Cytochemical localization and antigenicity of α-amylase in barley aleurone tissue. *Planta* 112, 213–224.
7. Nolan, M. and David, B. (1984) A starch-agar gel method for the localization of starch hydrolyzing enzymes in the cotyledons and hypocotyls of beans. *Am J Bot.* 71, 137–141.
8. Spruce, J., Mayer, A.M., and Osborne, D.J. (1987) A simple histochemical method for locating enzymes in plant tissue using nitrocellulose blotting. *Phytochemistry* 26, 2901–2903.
9. Pearse, A.G.E. (1980) Substrate film techniques, In: *Histochemistry, Theoretical and Applied*, Churchill Livingstone, Edinburgh, pp. 253–269.
10. Roth, R., Boudet, A., and Pont-Lezica, R.F. (1997) Lignification and cinnamyl alcohol dehydrogenase activity in developing stems of tomato and polar: a spatial and kinetic study through tissue printing. *J Exp Bot.* 48, 247–254.
11. da Silva, E., Toorop, P., van Aelst, A., and Hilhorst, H. (2004) Abscisic acid controls embryo growth potential and endosperm cap weakening during coffee *(Cofea arabica* cv. Rubi) seed germination. *Planta.* 220, 251–261.
12. Driouich, A., Lainé, A., Vian, B., and Faye, L. (1992) Characterization and localization of laccase forms in stem and cell cultures of sycamore. *Plant J.* 2, 13–24.
13. Hara, M., Eto, H., and Kuboi, T. (2001) Tissue printing for myrosinase activity in roots of turnip and japanese radish and horseradish: a technique for localizing myrosinases. *Plant Sci.* 160, 425–431.
14. Cassab, G.I., Lin, J.J., Lin, L.S., and Varner, J.E. (1988) Ethylene effect on extensin and peroxidase distribution in the subapical region of pea epicotyls. *Plant Physiol.* 88, 522–524.
15. Maki, H. and Morohashi, Y. (2006) Development of polyphenol oxidase activity in the micropylar endosperm of tomato seeds. *J Plant Physiol.* 163, 1–10.
16. Yomo, H. and Taylor, M.P. (1973) Histochemical studies on protease formation in the cotyledons of germinating bean seeds. *Planta* 112, 35–43.
17. Harris, N. and Chrispeels, M.J. (1975) Histochemical and biochemical observations on storage protein metabolism and protein body autolysis in cotyledons of germinating beans. *Plant Physiol.* 56, 292–299.
18. Moore, B., Kang, B., and Flurkey, W. (1989) Histochemical localization of mushroom tyrosinase in whole sections on nitrocellulose. *Histochemistry* 90, 379–381.

19. Singh, M.B. and Knox, R. (1985) Grass pollen allergens: antigenic relationships detected using monoclonal antibodies and dot immunobinding assay. *Int Arch Allergy Appl Immunol* 78, 300–304.
20. Antikainen, M., Griffith, M., Zhang, J., Hon, W., Yang, D., and Pihakaski-Maunsbach, K. (1996) Immunolocalization of antifreeze proteins in winter rye leaves, crowns, and roots by tissue printing. *Plant Physiol.* 110, 845–857.
21. Ramalho-Santos, M., Pissarra, J., Verissimo, P., Pereira, S., Salema, R., Pires, E., et al. (1997) Cardosin a, an abundant aspartic proteinase, accumulates in protein storage vacuoles in the stigmatic papillae of *Cynara cardunculus* L. *Planta.* 203, 204–212.
22. del Campillo, E., Reid, P., Sexton, R., and Lewis, L. (1990) Occurrence and localization of 9.5 cellulase in abscising and nonabscising tissues. *Plant Cell* 2, 245–254.
23. Reid, P.D., del Campillo, E., and Lewis, L.N. (1990) Anatomical changes and immunolocalization of cellulase during abscission as observed on nitrocellulose tissue prints. *Plant Physiol.* 93, 160–165.
24. Pont-Lezica, R.F. and Varner, J.E. (1989) Histochemical localization of cysteine-rich proteins by tissue printing on nitrocellulose. *Anal Biochem.* 182, 334–337.
25. Bailey, B.A., Taylor, R., Dean, J.F., and Anderson, J.D. (1991) Ethylene biosynthesis-inducing endoxylanase is translocated through the xylem of *Nicotiana tabacum* cv xanthi plants. *Plant Physiol.* 97, 1181–1186.
26. Cho, H. and Kende, H. (1997) Expansins and internodal growth of deepwater rice. *Plant Physiol.* 113, 1145–1151.
27. Cassab, G.I. and Varner, J.E. (1987) Immunocytolocalization of extensin in developing soybean seed coats by immunogold-silver staining and by tissue printing on nitrocellulose paper. *J Cell Biol.* 105, 2581–2588.
28. Hood, K., Baasiri, R., Fritz, S., and Hood, E. (1991) Biochemical and tissue print analyses of hydroxyproline-rich glycoproteins in cell walls of sporophytic maize tissues. *Plant Physiol.* 96, 1214–1219.
29. Ye, Z.H., Song, Y.R., Marcus, A., and Varner, J.E. (1991) Comparative localization of three classes of cell wall proteins. *Plant J.* 1, 175–183.
30. Goto, F., Yoshihara, T., Shigemoto, N., Toki, S., and Takaiwa, F. (1999) Iron fortification of rice seed by the soybean ferritin gene. *Nature Biotechnol.* 17, 282–286.
31. Pereira, S., Carvalho, H., Sunkel, C., and Salema, R. (1992) Immunocytolocalization of glutamine synthase in mesophyll and phloem of leaves of *Solanum tuberosum* L. *Protoplasma.* 167, 66–73.
32. Keller, B., Sauer, N., and Lamb, C.J. (1988) Glycine-rich cell wall proteins in bean: gene structure and association of the proteins with the vascular system. *EMBO J.* 7, 3625–3633.
33. Ye, Z.H., and Varner, J. (1991) Tissue-specific expression of cell wall proteins in developing soybean tissues. *Plant Cell.* 3, 23–37.
34. Ryser, U. and Keller, B. (1992) Ultrastructural localization of a bean glycine-rich protein in unlignified primary walls of protoxylem cells. *Plant Cell* 4, 773–783.
35. Gabius, S., Hellmann, K.P., Hellmann, T., Brinck, U., and Gabius, H.J. (1989) Neoglycoenzymes: a versatile tool for lectin detection in solid phase assays and glycohistochemistry. *Anal Biochem.* 182, 447–451.
36. Gonorazky, A., Regente, M., and de la Canal, L. (2005) Stress induction and antimicrobial properties of a lipid transfer protein in germinating sunflower seeds. *J Plant Physiol.* 162, 618–624.
37. Kausch, K. and Handa, A. (1997) Molecular cloning of a ripening-specific lipoxygenase and its expression during wild-type and mutant tomato fruit development. *Plant Physiol.* 113, 1041–1050.
38. Maryani, M., Morse, M., Bradley, G., Irving, H., Cahill, D., and Gehring, C. (2003) *In situ* localization associates biologically active plant natriuretic peptide immuno-analogues with conductive tissue and stomata. *J Exp Bot.* 54, 1553–1564.
39. Fujita, Y., Fujita, M., Mise, K., Kobori, T., Osaki, T., and Furusawa, I. (2000) Bromovirus movement protein conditions for the host specificity of virus movement through the vascular system and affects pathogenicity in cowpea. *Mol Plant Microbe Interact.* 13, 1195–1203.
40. Dicenta, F., Martinez-Gomez, P., Bellanger, I., and Audergon, J. (2000) Localization of plum pox virus in stem and petiole tissues of apricot cultivars by immuno-tissue printing. *Acta Virol.* 44, 323–328.
41. Knapp, E., da Camara Machado, A., Puhringer, H., Wang, Q., Hanzer, V., Weiss, H., et al. (1995) Localization of fruit tree viruses by immuno-tissue printing in infected shoots of *Malus* sp. and *Prunus* sp. *J Virol Methods.* 55, 157–173.
42. Mansky, L.M., Andrews, R.E., Durand, D.P., and Hill, J.H. (1990) Plant virus localization in leaf tissue by press blotting. *Plant Mol Biol Rep.* 8, 13–17.

43. Holt, C.A. and Beachy, R.N. (1991) *In vivo* complementation of infectious transcripts from mutant tobacco mosaic virus cDNAs in transgenic plants. *Virology* 181, 109–117.
44. Jung, G.L. and Hahne, G. (1992) A simple method to increase resolution in whole leaf blotting. *Plant Sci.* 82, 125–132.
45. Lin, N.S., Hsu, Y.H., and Hsu, H.T. (1990) Immunological detection of plant viruses and mycoplasma-like organism by direct tissue blotting on nitrocellulose membranes. *Phytopathology.* 80, 824–828.
46. Polston, J.E., Bubrick, P., and Perring, T.M. (1991) Detection of plant virus coat proteins on whole leaf blots. *Anal Biochem.* 196, 267–270.
47. Burlat, V., Kwon, M., Davin, L., and Lewis, N. (2001) Dirigent proteins and dirigent sites in lignifying tissues. *Phytochemistry* 57, 883–897.
48. Willats, W., Marcus, S., and Knox, J. (1998) Generation of monoclonal antibody specific to (1–5)-á-l-arabinan. *Carbohydr Res.* 308, 149–152.
49. Varner, J.E., Song, Y.R., Lin, L.S., and Yuen, H. (1989) The role f cell wall proteins in the structure and function of cell walls, In: *The molecular Basis of Plant Development* (Goldberg, R., ed.), Alan R Lis, New York, pp. 161–168.
50. Harriman, R.W.F., Tieman, D.M., and Handa, A.K. (1992) Localization of pectin methylesterase mRNA in tomato fruit, In: *Tissue Printing* (Reid, P.D., et al., eds.), Ademic Press, San Diego, CA, pp. 121–123.
51. McClure, B. and Guilfoyle, T. (1989) Tissue print hybridization. A simple technique for detecting organ- and tissue-specific gene expression. *Plant Mol Biol.* 12, 517–524.
52. Pont-Lezica, R.F. (1992) Lectin and glycan recognition, In: *Tissue Printing* (Reid, P.D., et al., eds.), Academic Press, San Diego, CA, pp. 71–79.
53. Jones, L., Seymour, G., and Knox, J. (1997) Localization of pectic galactan in tomato cell walls using a monoclonal antibody specific to (1,4)-â-d-galactan. *Plant Physiol.* 113, 1405–1412.
54. Hancock, K. and Tsang, V.C.W. (1983) India ink staining for protein on nitrocellulose paper. *Anal Biochem.* 133, 157–162.
55. Jamet, E., Canut, H., Boudart, G., and Pont-Lezica, R.F. (2006) Cell wall proteins: a new insight through proteomics. *Trends Plant Sci.* 11, 33–39.

Chapter 11

Dot-Immunobinding Assay

S. Sumi, A. Mathai, and V.V. Radhakrishnan

Summary

Dot-immunobinding assay (Dot-Iba) is a simple and highly reproducible immunodiagnostic method. Antibody or antigen is dotted directly onto nitrocellulose membrane (NCM) discs. The diagnostic material to be checked can be incubated on this disc. Presence of antigen–antibody complex in NCM discs can be directly demonstrated with enzyme-conjugated antiglobulins and substrate. Development of a purple-pink colored, insoluble substrate product in the NCM will be considered a positive result in the assay. This assay allows the processing of multiple specimens at a time and the entire operational procedures require only 4–6 h. Dot-Iba is rapid, and the technical steps involved in the assay are much simpler than in the other immunoassays such as enzyme-linked immunosorbant assay in detecting circulating antigen and antibody in clinical samples. The Dot-Iba showed an overall sensitivity of 60% for tuberculous meningitis diagnosis and no false positive results were encountered. Hence, this assay is highly specific for the diagnosis of paucibacillary diseases such as extrapulmonary tuberculosis. Dot-Iba is best suited to laboratories in developing world where there are constraints in laboratory resources.

Key words: Dot-immunobinding assay, Dot-Iba, Nitrocellulose discs, Circulating antibody

1. Introduction

In most of the developing world, infectious diseases continue to remain as one of the major public health problems. Most of these infectious diseases are potentially curable with appropriate antimicrobials. Any delay in diagnosis and treatment in patients will invariably lead to irreversible sequelae and complications. Development of newer techniques other than bacteriological methods has not only become essential but also is relevant for the early and rapid diagnosis of paucibacillary conditions as in tuberculous meningitis. The newly developed assay should be more sensitive

Biji T. Kurien and R. Hal Scofield (eds.), *Methods in Molecular Biology, Protein Blotting and Detection, vol. 536*

DOI: 10.1007/978-1-59745-542-8_11

and also possess operational advantages over the existing diagnostic methods. The newly designed assay should be specific, cost-effective, and feasible for application in the developing world where there are constraints in laboratory resources *(1–5)*.

1.1. Advantages of Dot-Iba

1. NCM discs have superior binding capacity than polystyrene microtiter plates.
2. The results can be read visually and there is no need of any expensive equipment such as microtiter ELISA readers.
3. The technical procedures involved in the assay are user-friendly, and laboratory personnel can be trained to undertake the assay.
4. Large number of specimens can be handled together.
5. Only limited laboratory space is required and there is no need for extraordinary instrumentation.
6. The NCM discs following the assay can be safely stored at 4°C and can be retained for review of previous results in the same patient.
7. The results can be made available within 6 h after the receipt of the specimen in the laboratory. Hence, this assay is rapid and will be helpful in making decision at the bedside management of patients.
8. The Dot-Iba is best suited to laboratories in developing world where there is a definite limitation in the laboratory resources and technical expertise.

1.2. Limitations of Dot-Iba

The Dot-Iba will yield positive results if the circulating antigen concentration (as in specimens like CSF) is less than 50 ng/mL. So it is possible that the assay may yield false negative results in patients with tuberculous meningitis who received a course of antituberculosis chemotherapy.

2. Materials

1. Punching machine.
2. 1-cm-Diameter circular nitrocellulose membranes (NCM).
3. Flat bottom microtiter plates (NUNC, Denmark).
4. Phosphate buffered saline (PBS).
5. Bovine serum albumin (BSA).
6. 0.15 M PBS in Tween-20 (PBST).
7. Anti-rabbit IgG–biotin conjugate (Sigma-Aldrich, MO, USA).

8. Extr–Avidin alkaline phosphatase (Sigma).
9. Ortho dianisidine tetrazotized (0.25 mg/mL).
10. β-naphthyl acid phosphate (0.25 mg/mL).
11. 0.06 M sodium borate buffer (pH 9.7).
12. Methanol.
13. Acetic acid.
14. Double-distilled water.

3. Methods

1. Cut 1-cm-diameter circular NCM with a clean "punching" machine.
2. NCM discs should be placed properly in the "flat bottom" microtiter plates with fine forceps (*see* **Note 1**).
3. 5 μL of antigen (*see* **Note 2**) should be carefully spotted in the central portion of the NCM disc and incubated for 12 h at 4°C (*see* **Note 3**).
4. Another incubation at 37°C for 1 h (optional).
5. The plates should be washed with PBST.
6. The unbound sites in the NCMs are quenched with 3% BSA in PBST.
7. Incubate the discs with 1:1,000 diluted specific monoclonal or polyclonal antibody for 1 h at 37°C followed by repeated washing in PBST.
8. Subsequently the NCM discs are incubated with 1:1,000 diluted anti-rabbit IgG–biotin conjugate and then Extr–Avidin alkaline phosphatase for 1 h, respectively, and finally washed thoroughly with PBST (*see* **Note 4**).
9. The NCM discs should be completely immersed in substrate containing *o*-dianisidine tetrazotized (0.25 mg/mL), β-naphthyl acid phosphate (0.25 mg/mL) in 0.06 M sodium borate buffer (pH 9.7) for 10 min.
10. Wash in PBST and fix in a solution containing methanol, acetic acid, and distilled water in the proportion 5:1:5.
11. A positive reaction is indicated by the development of purple to purple-pink in the central portion of the NCM discs (**Fig. 1**; *see* **Note 5**).
12. A positive control and negative control should be included in the assay.

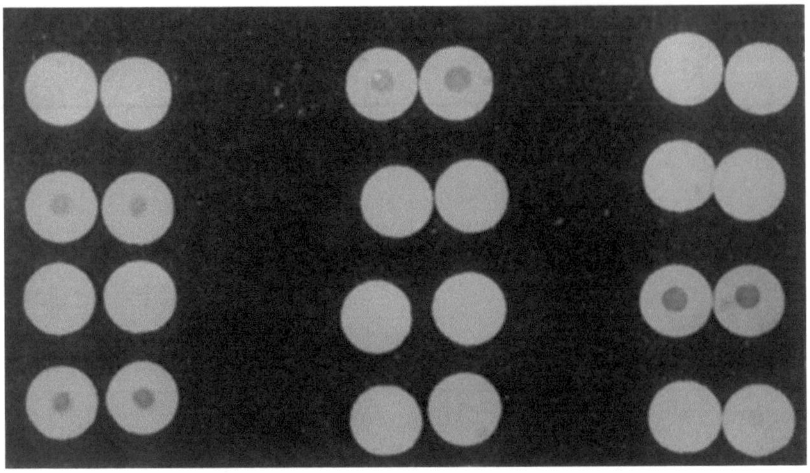

Fig. 1. NCM discs showing positive and negative results in Dot-Iba.

4. Notes

1. Circular discs should be cleanly cut with the help of a punching device. Discs with irregular edges and with wrinkles should not be used in the assay. NCM strips should be handled with a non-tooth and nonserrated forceps ("do not touch" with fingers).
2. Different concentrations (5–500 ng/mL) should be tried for the standardization procedure. In our experience, the dot-Iba gave positive results in all those NCM discs containing 50 ng/mL and above. In other words, the lowest detection limit in the assay is 50 ng/mL.
3. Do not release more than 5 μL of the antigen or antibody over the NCM disc. Also release the reagent steadily and uniformly over the NCM disc. Use a 5-μL micropipette with disposable tips for this purpose. During the release the reagent should not dribble outside the NCM disc.
4. The immunoreagents used in the assay such as antigen and antibodies must be stored in aliquots in optimal amounts, and this should be used during the assay. This will also eliminate repeated effects of freezing and thawing. This will also enhance the shelf life of reagents.
5. As the result of the assay is interpreted by the visual examination of the NCM, the assay should be performed during daylight. However, if the assay needs to be undertaken during the night, then a provision should be made to connect a light source to work bench.

References

1. Radhakrishnan, V.V., Mathai, A., Rao, S.B., and Sehgal, S. (1990) Immunoelectrophoresis for mycobacterial antigens. *Indian J Exp Biol.* 28, 812–815.
2. Sumi, M.G., Mathai, A., Sarada, C., and Radhakrishnan, V.V. (1999) Rapid diagnosis of tuberculous meningitis by a Dot immunobinding assay to detect mycobacterial antigen in cerebrospinal fluid specimens. *J Clin Microbiol.* 37, 3925–3927.
3. Sumi, M.G., Annamma, M., Sarada, C., and Radhakrishnan, V.V. (2000) Rapid diagnosis of tuberculous meningitis by a dot-immunobinding assay. *Acta Neurol Scand.* 101, 61–64.
4. Sumi, M.G., Mathai, A., Sarada, C., Radhakrishnan, V.V., Indulakshmi, R., Sathish, M., Ajaykumar, R., and Manju, Y.K. (2002) A comparative evaluation of dot-immunobinding assay (Dot-Iba) and polymerase chain reaction (PCR) for the laboratory diagnosis of tuberculous meningitis. *Diag Microbiol Infect Dis* 42, 35–38.
5. Mathai, A., Radhakrishnan, V.V., and Molly, M. (1991) Rapid diagnosis of tuberculous meningitis with a dot enzyme immunobinding assay to detect antibody in the cerebrospinal fluid. *Eur J Clin Microbiol Infect Dis.* 10, 440–443.

Chapter 12

Capillary Blotting of Glycosaminoglycans on Nitrocellulose Membranes After Agarose Gel Electrophoresis Separation

Nicola Volpi and Francesca Maccari

Summary

A method for the blotting and immobilizing of several nonsulfated and sulfated complex polysaccharides on membranes made hydrophilic and positively charged by cationic detergent after their separation by conventional agarose gel electrophoresis is illustrated. This new approach to the study of glycosaminoglycans (GAGs) utilizes the capacity of agarose gel electrophoresis to separate single species of polysaccharides from mixtures and the membrane technology for further preparative and analytical uses.

Nitrocellulose membranes are derivatized with the cationic detergent cetylpyridinium chloride and mixtures of GAGs are capillary blotted after their separation in agarose gel electrophoresis. Single purified species of variously sulfated polysaccharides are transferred on derivatized membranes with an efficiency of 100% and stained with alcian blue (irreversible staining) and toluidine blue (reversible staining). This enables a lower amount limit of detection of 0.1 μg. Nonsulfated polyanions, for example hyaluronic acid, may also be transferred to membranes with a limit of detection of approximately 0.1–0.5 μg after irreversible or reversible staining. The membranes may be stained with reversible staining and the same lanes are used for immunological detection or other applications.

Key words: Glycosaminoglycans, Electrophoresis, Blotting, Heparin, Chondroitin sulfate

1. Introduction

Separation of glycosaminoglycans (GAGs), i.e., hyaluronic acid (HA), keratan sulfate, chondroitin sulfate (CS)/dermatan sulfate (DS), and heparan sulfate (HS)/heparin (Hep) by electrophoresis is routine in many laboratories and their characterization to the microgram level form an integral part of biochemical research, especially with respect to obtaining information from unknown purified polysaccharides. This is more important for natural (macro)

Biji T. Kurien and R. Hal Scofield (eds.), *Methods in Molecular Biology, Protein Blotting and Detection, vol. 536*

DOI: 10.1007/978-1-59745-542-8_12

molecules produced by extraction and purification from different animal tissues having several fundamental biological activities, as well as pharmacological properties, making them important drugs for use in clinical and pharmaceutical fields *(1–5)*.

Blotting of molecules on membranes after their separation by electrophoresis takes advantage of the possibility to conduct further analysis on single separated molecules by several approaches, e.g., specific binding and identification with antibodies or recovery of single band, and can be used for preparative, quantitative, and qualitative studies. Clearly, the first step is related to the electrophoretic separation and in the case of GAGs, cellulose acetate *(6)*, nitrocellulose membrane (NC) *(7)*, and agarose gel *(8)* and polyacrylamide gel *(9, 10)* electrophoretic techniques are generally utilized for qualitative and quantitative analyses of mixtures of these macromolecules or single species. However, agarose gel electrophoresis permits the separation and the identification of several GAG species, such as slow (SM heparin) and fast moving heparin (FM heparin) or heparan sulfate, dermatan sulfate, and chondroitin sulfate *(8, 11)*.

GAGs are strongly hydrophilic and negatively charged macromolecules that do not bind well to either polystyrene surfaces or hydrophobic blotting membranes. As a consequence, membranes have been derivatized with cationic detergents to make them hydrophilic and positively charged, like cetylpyridinium chloride (CPC)-treated NC membranes used in this protocol. After their electrophoretic separation, several intact GAGs with high molecular mass, such as hyaluronic acid, chondroitin sulfate (CS), highly-sulfated CS, dermatan sulfate, heparan sulfate, heparin (Hep), and its two components, FM Hep and SM Hep species, were transferred on NC membranes treated with a cationic detergent, CPC *(12)*. Quantitative analysis was performed after visualization of bands by cationic dyes, the recovery of single molecules released from membrane was also examined *(12)*, and the direct and specific recognition of these polysaccharides by antibodies on CPC-treated NC supports has also been described *(13)*.

2. Materials

1. High purity agarose is from Sigma Chemical Co, MO, USA or Biorad, CA, USA (*see* **Note 1**).
2. Barium acetate and NC membranes 0.45 μm, binding capacity of 80–100 μg/cm^2, are from BioRad.
3. 1,2-diaminopropane (PDA) and cresol red are from Merck Eurolab, Leicestershire, UK.

4. Cetylpyridinium chloride (CPC) is from Sigma-Aldrich, MO, USA.
5. Toluidine blue, alcian blue, and Whatman 3MM paper are purchased by Sigma-Aldrich.
6. All the other reagents should be of analytical grade.

2.1. Glycosaminoglycans

1. Different GAGs to be used as standard may be purchased from Sigma-Aldrich (http://www.sigmaaldrich.com) or other specialized companies.
2. Extraction and purification protocols for various GAGs are available in specific scientific articles and monographs *(3, 5, 14–17)*.

2.2. Electrophoresis

1. 40 mM barium acetate buffer, pH 5.8 with 1 M acetic acid. Store at 4°C.
2. 50 mM 1,2-diaminopropane (PDA) buffer: Buffered at pH 9 with glacial acetic acid. Store at 4°C.
3. Cresol red solution: Dissolve 10 mg of the dye in 100 mL of distilled water (final concentration of 0.1 mg/mL). Store at 4°C.

2.3. Blotting on Membranes

1. 1% CPC in 30% 2-propanol. This solution should be always freshly prepared.
2. 150 mM NaCl.
3. Transfer buffer: 100 mM Tris-acetate buffer at pH 7.3. Store at 4°C.

2.4. Membrane Staining

1. Alcian blue solution: Dissolve 50 mg of the dye in 1 ml 8 M guanidine and 19 ml of 18 mM sulfuric acid-0.25% Triton X-100. This staining reagent is always freshly prepared.
2. Reversible staining is always freshly prepared with toluidine blue. 20 mg toluidine blue is dissolved in 100 ml of 3% acetic acid.

3. Methods

3.1. Agarose-Gel Electrophoresis of Glycosaminoglycans

1. Prepare 0.5% agarose solution in 40 mM barium acetate buffer (*see* **Note 2**). Heat the solution in a microwave oven or on a stirrer mixer, mixing continuously until the agarose completely dissolves and a clear solution is obtained. Do not allow the solution to boil.
2. Thoroughly clean a single glass plate of 7 × 8 cm (of approximately 2 mm thickness) with alcohol. After drying with paper,

place it in a holder of approximately 8 × 10 cm and pour the warm agarose solution (*see* **Notes 3** and **4**).

3. Prior to electrophoresis, leave the gel at room temperature (RT) for approximately 30 min. Cut the gel by the side of the glass plate and put it with the same glass plate on a grid made of 1 × 1 cm squares. Make four small wells by using a flat chisel of approximately 5 mm, taking care to leave approximately 5 mm between each well (*see* **Note 5**). Make the wells approximately 2 cm from the edge of the gel.
4. 10 μL of GAGs standard or samples (*see* **Note 6**) may be layered by micropipets into the wells.
5. The electrophoretic run is performed in 50 mM PDA for 150 min at 50 mA by using a Pharmacia Multiphor II (Pharmacia LKB Biotechnology, Uppsala, Sweden) electrophoretic cell instrument (*see* **Note 7**).

3.2. Glycosaminoglycans Blotting

1. Cut NC membranes in portions of 7 × 8 cm.
2. Wet the membrane in freshly prepared 1% CPC in 30% 2-propanol for 5 min. Six milliliters of CPC solution should be used. Shake the membrane manually.
3. After wetting, add 50 mL of 150 mM NaCl and incubate the membrane on a shaker for 15 min.
4. CPC-derivatized membrane is rinsed several times in 150 mM NaCl (*see* **Note 8**) and then equilibrated with continuous shaking in the same NaCl solution until blotting is performed.
5. After agarose-gel electrophoresis, carefully remove the gel from the glass plate.
6. Prepare the blotting sandwich by assembling a Whatman 3MM paper immersed in the buffer reservoir (*see* **Note 9**). Carefully place the agarose gel on the 3MM paper with the wells located parallel to the two buffer reservoirs. The detergent-treated NC membrane is then laid on top of the gel (*see* **Note 10**). Three further wetted filter papers, two wetted sponges, and 5 cm of absorbent paper tissue are carefully laid on top of the NC membrane, ensuring that no bubbles are trapped in the resulting sandwich. The blotting sandwich is stabilized by putting a 500 g weight on the top.
7. The capillary blotting is performed overnight at RT.

3.3. Staining Procedures

1. Irreversible staining is performed by means of alcian blue. After 2 h staining, membrane is destained by rinsing in 150 mM NaCl until background staining disappears (*see* **Note 11**). An example of the results produced is shown in **Fig. 1a**.
2. Irreversible staining with alcian blue can be used for quantitative studies with a high detection sensitivity (see below).

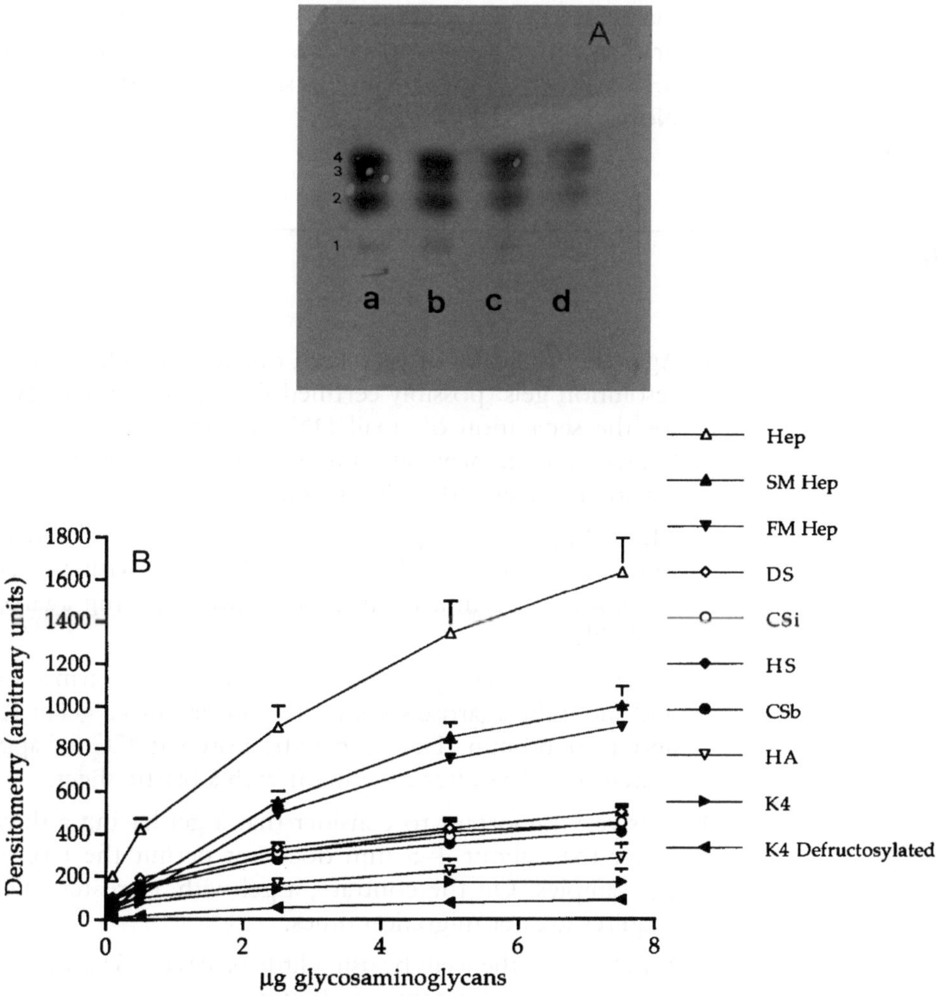

Fig. 1. (**A**) Decreasing amounts (a: 7.5 μg, b: 5.0 μg, c: 2.5 μg, d: 0.1 μg) of mixtures of GAGs composed of (1) slow moving heparin, (2) fast moving heparin, (3) dermatan sulphate, and (4) chondroitin sulfate electrophoretically separated by agarose-gel and blotted on CPC-treated NC membrane. The membrane was stained with alcian blue. (**B**) Quantitation of immobilized sulfated and nonsulfated GAGs on CPC-treated NC membranes after agarose-gel electrophoresis, capillary blotting, staining with alcian blue, destaining, and densitometric analysis. *Hep* Heparin, *SM Hep* Slow moving heparin, *FM Hep* Fast moving heparin, *DS* Dermatan sulfate, *CSi* Highly sulfated chondroitin sulfate, *HS* Heparan sulfate, *CSb* Chondroitin sulfate, *HA* Hyaluronic acid, *K4* Bacterial polysaccharide K4, and *K4 Defructosylated* Defructosylated bacterial polysaccharide K4.

Quantitative analysis of GAGs may be performed with a densitometer connected to a computer by using an image processing and analysis software. The wet membranes should be scanned in the RGB mode and saved in gray scale. An example of the results produced is shown in **Fig. 1b**.

3. GAGs separated by agarose-gel electrophoresis and transferred to a membrane can also be detected by toluidine blue staining (*see* **Note 12**). Reversible staining may be obtained by treating

membranes with toluidine blue solution for 5 min. The membrane is rinsed for 30–60 s in 3% acetic acid in the presence of 0.1% CPC to remove the excess stain, and further utilized (*see* **Notes 13** and **14**).

4. Notes

1. Agarose should be of very high quality, suitable to run high resolution gels (possibly certified for molecular biology, ideal for the separation of small DNA fragments). Agarose from Sigma-Aldrich, product number A4718, or from Bio-Rad, catalog number 161-3101, is suitable.
2. The volume of the agarose solution is strictly related to the dimension of the gel. For a gel of 7 × 8 cm with a thickness of about 4–5 mm a volume of 50 mL (250 mg agarose) is advisable.
3. Carefully eliminate possible air bubbles in the warm solution, and allow the agarose solution to convert into a gel at RT for about 30–60 min. The gel may be stored at 4°C for approximately 4–5 days after covering it with a plastic sheet.
4. It is very important to consider that a gel having a thickness lower than about 4–5 mm does not permit the layering of the samples. On the contrary, a gel with a greater thickness requires greater migration times.
5. Carefully dry the wells by using little pieces of Whatman 3 MM paper of approximately 5 × 20 mm.
6. GAGs standard should be prepared at a concentration of 0.5 mg/mL in distilled water with a final absolute amount of 5 μg loaded on the gel. Extracted GAGs from different matrices should be quantitatively evaluated by means of known assays *(1, 13–15, 17)* before performing the electrophoretic separation. The optimum concentration range of unknown purified GAGs loaded on the gel should be from 2 to 8 μg.
7. Due to the possible variability of the electrophoretic conditions, 2 μL of cresol red solution (0.1 mg/mL) should be added to each standard or sample solution (10 μL) in order to make a more accurate evaluation of the electrophoretic migration. A good migration time and electrophoretic separation is obtained at a cresol red migration of approximately 20–25 mm.
8. Derivatized membranes should be rinsed several times in 150 mM NaCl with vigorous shaking until no foaming is observed. This step permits the complete removal of excess CPC.

9. The Whatman 3MM is immersed in the two buffer reservoirs permitting the migration of the buffer from the tank to the top side of the blotting sandwich. As a consequence, make sure that sufficient buffer volume for a complete GAGs migration (approximately 1 L) is available.
10. Carefully remove possible air bubbles entrapped between the gel and the membrane by using a little glass pestle.
11. It is very important to optimize the irreversible staining period to obtain a good band staining against a clear background. Under the experimental conditions described, an optimum staining time would be 2 h. Furthermore, use several changes of the destaining solution to produce the best results.
12. The sensitivity of staining with toluidine blue is about 10–15 times lower than that with alcian blue (an example is illustrated in **Fig. 2**), but if the membrane is then destained, the same lanes can be used for immunological detection or other applications.
13. After toluidine blue detection, membrane is destained using 3% acetic acid in the presence of 0.1% CPC. Under these conditions, the destaining of the bands is completed within 5 min and without any loss of immobilized molecules.
14. The electrophoretically separated GAGs transferred on NC may also be released and recovered from the cationized

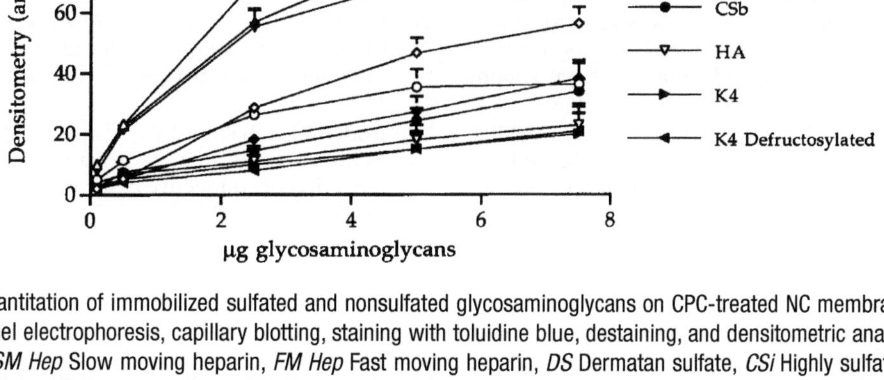

Fig. 2. Quantitation of immobilized sulfated and nonsulfated glycosaminoglycans on CPC-treated NC membranes after agarose-gel electrophoresis, capillary blotting, staining with toluidine blue, destaining, and densitometric analysis. *Hep* Heparin, *SM Hep* Slow moving heparin, *FM Hep* Fast moving heparin, *DS* Dermatan sulfate, *CSi* Highly sulfated chondroitin sulfate, *HS* Heparan sulfate, *CSb* Chondroitin sulfate, *HA* Hyaluronic acid, *K4* Bacterial polysaccharide K4, *K4 Defructosylated* Defructosylated bacterial polysaccharide K4.

membranes at the microgram level for further analysis, such as disaccharide pattern evaluation, molecular mass determination, and characterization of specifically sulfated sequences inside the polysaccharide chains. The immobilized GAGs are efficiently released from the membrane using a nonionic detergent at high ionic strength (for details *see* **ref. *13***).

References

1. Mammen, E.F., Walenga, J.M., and Fareed. J. (eds) (1991) Development of non-heparin glycosaminoglycans as therapeutic agents. *Semin Thromb Hemost* 17, 1–135.
2. Ofosu, F.A., Danishefsky. I., and Hirsh, J. (eds) (1989) Heparin and related polysaccharides. Structure and activities. *Ann N Y Acad Sci* 556, 1–501.
3. Heinegard, D., and Sommarin, Y. (1987) Isolation and characterization of proteoglycans. *Meth Enzymol* 144, 319–73.
4. Crescenzi, V., Dea, I.C.M., Paoletti, S., Stivala, S.S., and Sutherland, I.W. (eds) (1989) *Biomedical and biotechnological advances in industrial polysaccharides*. Gordon and Breach, New York, NY.
5. Volpi, N. (ed) (2006) *Chondroitin sulfate: structure, role and pharmacological activity*. Academic, New York, NY.
6. Cappelletti, R., Del Rosso, M., and Chiarugi, V.P. (1979) A new electrophoretic method for the complete separation of all known animal glycosaminoglycans in a monodimensional run. *Anal Biochem* 99, 311–15.
7. Volpi, N. (1996) Electrophoresis separation of glycosaminoglycans on nitrocellulose membranes. *Anal Biochem* 240, 114–8.
8. Nader, H.B., Takahashi, H.K., Guimaraes, J.A., Dietrich, C.P., Bianchini, P., and Osima, B. (1981) Heterogeneity of heparin: characterization of one hundred components with different anticoagulant activities by a combination of electrophoretic and affinity chromatography methods. *Int J Biol Macromol* 3, 356–60.
9. Rice, K.G., Rottink, M.K., and Linhardt, R.J. (1987) Fractionation of heparin-derived oligosaccharides by gradient polyacrylamide-gel electrophoresis. *Biochem J* 244, 515–22.
10. Lyon, M., and Gallagher, J.T. (1990) A general method for the detection and mapping of submicrogram quantities of glycosaminoglycan oligosaccharides on polyacrylamide gels by sequential staining with azure A and ammoniacal silver. *Anal Biochem* 185, 63–70.
11. Volpi, N. (1993) "Fast moving" and "slow moving" heparins, dermatan sulfate, and chondroitin sulfate: qualitative and quantitative analysis by agarose-gel electrophoresis. *Carbohydr Res* 247, 263–78.
12. Maccari, F., and Volpi, N. (2002) Glycosaminoglycan blotting on nitrocellulose membranes treated with cetylpyridinium chloride after agarose-gel electrophoretic separation. *Electrophoresis* 23, 3270–7.
13. Maccari, F., and Volpi, N. (2003) Direct and specific recognition of glycosaminoglycans by antibodies after their separation by agarose-gel electrophoresis and blotting on cetylpyridinium chloride-treated nitrocellulose membranes. *Electrophoresis* 24, 1347–52.
14. Volpi, N. (ed) (2002) *Analytical techniques to evaluate the structure and function of natural polysaccharides, glycosaminoglycans*. Research Signpost, India.
15. Savolainen, H. (1999) Isolation and separation of proteoglycans. *J Chromatogr B Biomed Sci Appl* 722, 255–62.
16. Beaty, N.B., and Mello, R.J. (1987) Extracellular mammalian polysaccharides: glycosaminoglycans and proteoglycans. *J Chromatogr* 418, 187–222.
17. Roden, L., Baker, J.R., Cifonelli, J.A., and Mathews, M.B. (1972) Isolation and characterization of connective tissue polysaccharides. *Meth Enzymol* XXVIII, 73–140.

Chapter 13

Quantitative Computerized Western Blotting

Dalit Talmi-Frank, Charles L. Jaffe, and Gad Baneth

Summary

Western blotting allows analysis of antibody reactivity against multiple antigens separated according to their molecular weights. The distinction between immune dominant and recessive antigens is often difficult and carried out by qualitative or empirical means. Quantitative computerized western blotting (QCWB) addresses this difficulty by analyzing reactivity to specific antigens and providing a statistically measurable value for each band. This allows differentiation between immunodominant and immunorecessive determinants. QCWB is appropriate for either single time point analysis or longitudinal studies where multiple time points are evaluated and the reactivities against individual bands compared. This technique can be used to study humoral responses to complex antigenic mixtures such as allergens and infectious agents, or to identify serologic markers for early diagnosis of cancer, autoimmune, or infectious diseases, or to monitor patient's clinical status.

Key words: Western blot, Quantitative computerized western blot, Immunodominant antigen, SDS-PAGE, Net intensity, Total lane intensity

1. Introduction

Western blotting is a highly sensitive technique for the detection of proteins or monitoring the presence of antibodies that react with discrete antigens in complex mixtures *(1)*. It can be employed simultaneously against a large number of molecules to test reactivity with a panel of potential antigens *(2, 3)*. The high sensitivity and specificity of this technique makes western blotting suitable for the detection of low concentrations of particular proteins and allows the identification of specific circulating antibodies to allergens, pathogens, tumor, and autoantigens in the sera of humans and animals. Early and sensitive disease diagnosis is a crucial factor in the effective and successful treatment and

Biji T. Kurien and R. Hal Scofield (eds.), *Methods in Molecular Biology, Protein Blotting and Detection, vol. 536*

DOI: 10.1007/978-1-59745-542-8_13

management of various maladies or illnesses. Early diagnosis prior to the development of clinical symptoms, detection of infection in immunosuppressed patients, and the identification of asymptomatic or subclinical carriers are major diagnostic challenges which often require very sensitive and specific assays *(4, 5)*.

Many serologic techniques used for diagnosis are based on reactivity against the whole pathogen, allergen, or tumor. These assays are commonly adjusted to detect high antibody levels, whereas early, subclinical or latent infections and relapses, or diseases are often characterized by low antibody titers against only a few specific antigens in the whole mixture *(5)*. Western blotting allows complex mixtures to be separated into individual components based on molecular weight and posttranslational modification(s), permitting the detection of even low levels of antibody reactivity with discrete antigens present in the mixture. In general, antibody reactions against multiple components are observed when patient serum reactivity is evaluated against complex antigens by western blotting. Often bands that react strongly showing high intensities in a higher percentage of patients are termed immunodominant antigens, whereas other bands that react weakly or moderately in the same or a lower percentage of subjects are termed immunorecessive antigens *(6)*. The distinction between these band definitions is often difficult and frequently done empirically based on single measurements or an overall qualitative impression. Quantitative Computerized Western Blot (QCWB) analysis addresses this difficulty thus allowing for numerical documentation and statistical analysis of reactivity to specific antigenic bands.

Specific immunodominant antigens identified by this method could be targets for developing rapid serodiagnostic tests, as well as effective vaccines. Two parameters can be analyzed using QCWB: first, net band intensity (NBI) that allows the evaluation of antibody reactivity to individual bands and second, total lane intensity (TLI) that measures the total serum antibody reactivity to a whole lane of separated antigens *(6)*. The computerized quantification of band intensity allows a statistically analyzable distinction between immunodominant and immunorecessive bands.

QCWB can be very useful during a longitudinal follow up of the antibody reaction to infection, exposure to foreign proteins, or changes in clinical status when monitoring cancer or autoimmune diseases. The identification and measurement of reactivity to specific antigenic bands at multiple points over time can facilitate the identification of such bands as sensitive sentinels for early disease detection, as prominent markers of patent infection, or as markers of successful treatment. Sera collected from dogs experimentally infected with *Leishmania infantum*, a fatal protozoan parasite of humans and dogs, will be used to illustrate how QCWB can be used to evaluate and compare antibody reactivity against individual antigens in a whole parasite mixture both at a single time point as well as over time during a longitudinal study. Other serological methods such as enzyme-linked immunosorbent assay (ELISA) are compared to QCWB using the same sera to demonstrate the advantages of the latter technique *(6)*.

2. Materials

Sodium dodecylsulfate-polyacrylamide gel electrophoresis (SDS-PAGE) and western blotting are described here essentially as carried out by the authors *(6)*. Other protocols with slightly different reagents can be analyzed by the quantitative computerized technique detailed in the next paragraphs.

2.1. SDS-PAGE and Western Blotting

1. Protein sample – 150 μg/lane/gel.
2. SDS-PAGE equipment.
3. Nitrocellulose paper.
4. Transfer buffer.
5. Blocking solution: Phosphate buffered saline (PBS) containing 0.05% Tween-20 and 5% dried milk.
6. Serum from test subjects and controls.
7. Washing buffer: PBS containing 0.01% Tween20 (PBST).
8. Secondary antibody at 1:1,000 dilution in washing buffer [Horseradish peroxidase (HRP)-conjugated goat anti-canine IgG (Cappel Research Reagents, ICN Biomedicals, Aurora, OH, USA)].
9. 0.0004% 3-3-diaminobenzidine (DAB), Sigma Chemical Company, Missouri, USA.
10. Molecular mass protein standard (Precision Plus Protein Standards, Bio-Rad, Hemel Hempstead, UK).
11. Digital scanner – the 6200C scanner, (Hewlett-Packard Co., Mississauga, Canada).
12. Image analyzing software such as the KODAK 1D program (Eastman Kodak Company, Scientific Imaging Systems, Rochester, NY, USA).
13. Mini-electrophoresis apparatus (Bio-Rad Laboratories, Inc., Hercules, USA).
14. Semidry or wet transfer apparatus (Bio-Rad transfer cell, Bio-Rad).
15. HRP-conjugated goat anti-canine IgG (Cappel research reagents, ICN Biomedicals, Aurora, OH, USA).
16. 3-3-diaminobenzidine (DAB, Sigma, St. Louis, MO, USA).

3. Methods

The methods below describe (1) the fundamentals of western blotting procedure in short; (2) application of an image ana-ly-zing software on a blotted membrane; (3) lane marking and individual band analysis; (4) determination of NBI; (5) defining immunodominant and immunorecessive bands; and (6) total lane intensity.

3.1. Western Blotting

While the full western blotting procedure used here was detailed at length *(6)*, any standard procedure for blotting can be employed (*see* Chapter "Detection of La/SS-B by Western Blot Using Nanogold-Tagged Antibodies and Silver Enhancement").

1. Crude *L. infantum* antigen (CLA, 150 μg/lane/gel) is separated on a 12% SDS-PAGE gel under nonreducing conditions using a minielectrophoresis apparatus.
2. Antigens are transferred to nitrocellulose membrane using either a semidry or wet transfer apparatus.
3. A molecular mass protein standard, preferably prestained, is also run on the same gel and transferred to the membrane.
4. The membranes are cut into individual strips and undergo three steps including blocking, incubation with dog sera, and incubation with an HRP-conjugated goat anti-canine IgG diluted 1:5,000.
5. Development is carried out with 0.0004% DAB.
6. Several rinses (3–4 times) with washing buffer are performed between each of the earlier steps listed in **steps 4** and **5**.
7. After a last wash with distilled water the blots are air dried and at the end of the western blotting procedure dry nitrocellulose membrane strips with dark bands corresponding to serum antibody reactions with protein bands of various molecular weights are obtained.

3.2. Image Formation

In order to acquire an image of highest quality and resolution, blotted membranes should be scanned using a scanner or other imaging device that permits further digital analysis. Any modification of the scanner or imaging device settings that improves the blot appearance or enhances a particular feature of the blot should be avoided. Thus the same settings for brightness, contrast, and gamma values should be strictly maintained during all analysis. Membranes should be scanned immediately following the blotting procedure in order to produce clearly resolved bands and avoid fading of the bands over time. An optimal scanning procedure is the one that keeps the lanes as straight as possible and avoids color artifacts (**Fig. 1A**).

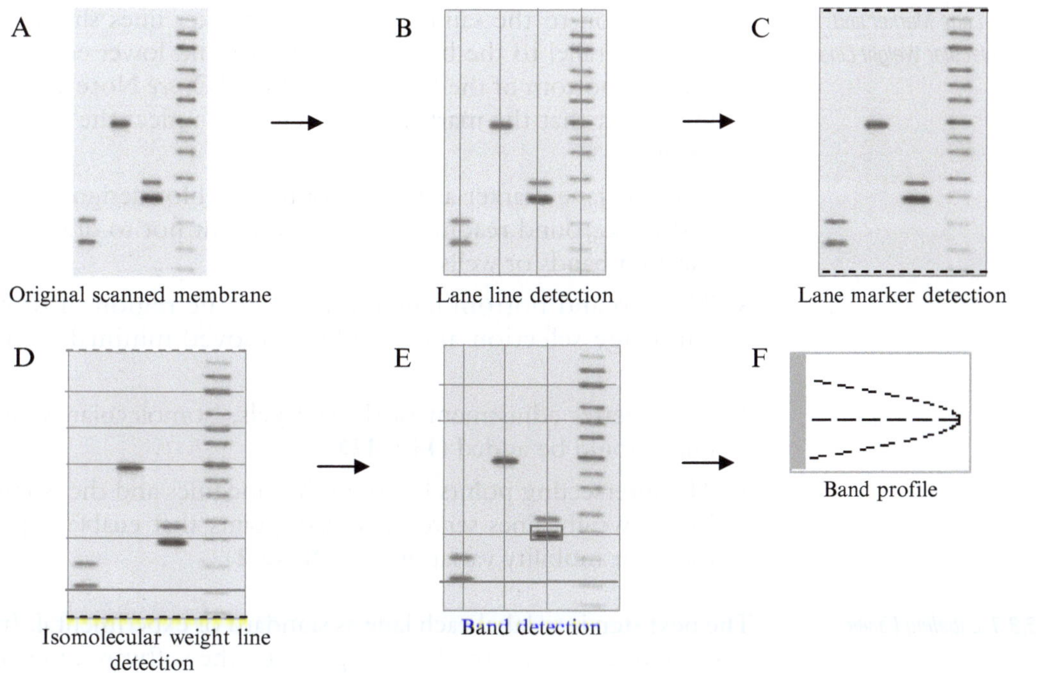

Fig. 1. Flow chart describing the different steps in image analysis of quantitative computerized western blot. Panels (**A**): Scanning of the blot and automatic or manual lane line detection based on pixel peaks of bands. (**B**) Establish the lane markers at the top and bottom of the scanned membrane and designation of the first background reading. (**C**) Marking isomolecular weight lines enabling adjustment of individual and multiple blot membranes and determination of the molecular weight for individual bands. (**D**) Automatic software analysis and definition of rectangles around the detected bands. (**E**) Manual adjustment of the rectangle borders if needed. The software uses an optimized band sensitivity chosen according to the bands of interest. The sensitivity chosen should distinguish the bands of interest from the background while ignoring the extraneous ones. (**F**) Remodeling of profile data if needed.

3.3. Digital Analysis (see Fig. 1)

3.3.1. Defining Sample Lanes

1. Sample lanes on an image are automatically and/or manually found by the Kodak image analyzing program using the lane finding algorithm.
2. The algorithm determines the band location and calculates the spacing and the width of the different lanes from each other on a selected region of the image.
3. The program identifies the location of lanes according to the pixel peaks for bands that are adjacently localized. Therefore, if bands are not well resolved, homogeneous or the transfer is poor the lanes may need to be inserted manually.
4. Lane finding sensitivity can be set to several levels (*see also* **Subheading 3.3.5**) and corrected to fit straight, slanted, or curved lanes. When sensitivity is increased, fainter bands can be identified. However, it is advisable after adjusting the sensitivity to a desired level, to keep it constant throughout the project. It is critical to ensure that the lines defining each lane pass directly through the center of all the bands in each lane (**Fig. 1B**).

3.3.2. Lane Marker and Isomolecular Weight Lines

1. In addition to the sample lane lines, marker lines should be drawn parallel to the bands at the upper and lower edges, or top and bottom of the blotted gel (**Fig. 1C**; *see* **Note 1**). It is important that the marker lines accurately reflect the skew of the gel.
2. The first lane marker at the top of the gel blot designates the first background reading, so it is important not to draw lines through bands or wells.
3. The top and bottom lane markers set the region of original image selection and should be moved minimal or not at all.
4. For a better adjustment of skewed gels, isomolecular weight lines should be added (**Fig. 1D**).
5. The intersecting points between the lane lines and the isomolecular weight lines serve as control points that enable adjustment for mobility variations (*see* **Note 2**).

3.3.3. Labeling Lanes

The next step is to label each lane as standard or experimental. It is possible to either use the formats given by the software or create a new standard lane according to the molecular weights given by the ladder. A standard lane will define the band size in Daltons for proteins, base pairs for DNA, or bases for RNA depending on the type of gel viewed (**Fig. 1B**).

3.3.4. Receiving and Analyzing Band Data

Once the lanes have been marked, bands can be automatically located. The band finding algorithm depends on two parameters that define the band sensitivity and the profile width. These two parameters should be set at the beginning of the project and kept constant. A completely marked image can be seen in **Fig. 2**.

3.3.5. Band Sensitivity

In order to choose the correct sensitivity for a given project, resolution of the individual bands must be considered. If the bands of interest do not stand out from the background, a higher sensitivity should be chosen. Optimally the correct sensitivity should allow the program to find all the bands of interest while minimizing those bands that are irrelevant.

3.3.6. Profile Width and Band Numbering

1. Profile width defines the width of the band used for the analysis out of the total strip width, in case the band is narrower than the strip. It sets the area of the rectangles that surround each band. Optimal profile width encompasses the whole band without getting false readings from the background that surrounds it.
2. Each band is then labeled and numbered, and a horizontal marking line is stretched across it.

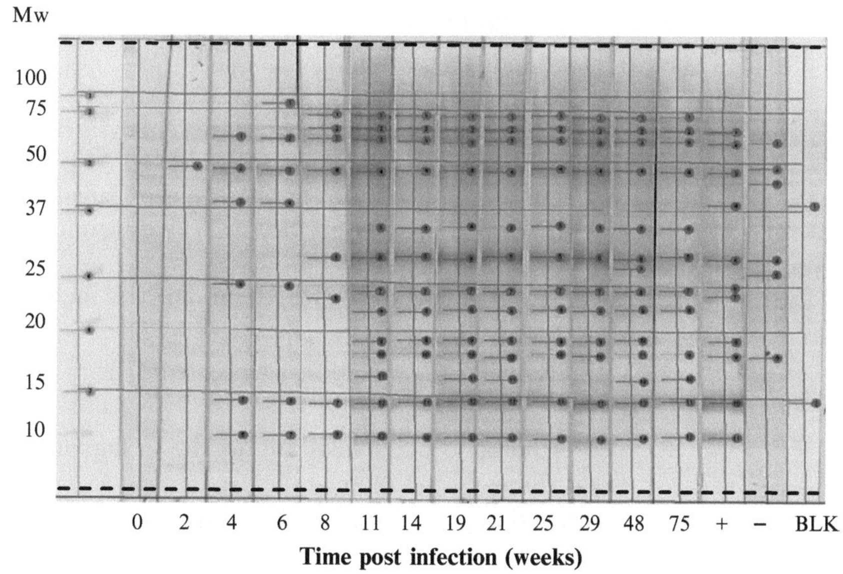

Fig. 2. An example of an image created using the Kodak 1D analyzing software on a typical western blot membrane. The image shows the reaction of sera taken from one dog experimentally infected with *Leishmania infantum* parasites on homologous parasite crude antigen at successive time points during a longitudinal follow up study. The dog was treated from weeks 32 to 45 with allopurinol. The *upper* and *lower long horizontal dashed lines* indicate the lane markers and the *short horizontal lines* ending with a *dot* mark the isomolecular weight lines. Positive and negative control sera are depicted in lanes *plus* and *minus*, respectively, and a *blank lane* depicting reactivity with diluting reagents is marked as BLK.

3. It is important to manually verify that the horizontal line is positioned exactly at the center of the band because the molecular weight is calculated based on the position of this line.
4. A rectangle is marked around the band (**Fig. 1E**).
5. As soon as the central horizontal line is drawn, it can be used to compare the band positions between the standard and experimental lanes, and it is possible to also adjust the position and boundaries of any band while examining the peak intensity viewed in the profile exhibited (**Fig. 1F**).

3.3.7. Data Modeling

After the bands are found and the band profile is created, it is important to determine whether the data needs to be modeled. This is especially important for determining the values of unresolved bands, oversaturated bands, or bands with uneven or high background.

When to Remodel Data?

1. When bands are unresolved (closely spaced) (*see* **Note 3**)
2. If bands are saturated (flat-peaked) (*see* **Note 4**)
3. When faint bands are located adjacent to strong bands (*see* **Note 5**)

4. If the background is uneven or variable
5. If false bands are present

In order to calculate the net intensity, the program uses the pixel values in the band rectangle. Quantitative analysis of overlapping or saturated bands is best approximated by fitting the data to a Gaussian curve *(7)*. For saturated bands, the Gaussian technique better estimates bands that are beyond the dynamic range of the capture device.

Information received from the band analysis includes:

1. Molecular weight
2. Mobility-measured location of the band from the top lane marker
3. NBI – the band pixel value minus the background rectangle pixel values
4. Relative intensity – the percent intensity contribution of a band within a lane

3.4. Applications of Quantitative Computerized Western Blot Analysis

3.4.1. Net Band Intensity

NBI is defined as the sum of background subtracted pixels within the rectangle area formed around the band, reflecting the reaction to it. NBI enables the quantitative comparison of antibody reactivity between specific antigen bands on the same or different blots. It facilitates comparison of antibody reactivity with antigens between individuals or by the same individual at different times, thus allowing one to follow and analyze experimental or natural infections longitudinally. The ability to quantify antibody reactions in a standardized way allows the statistical assessment reactions to treatment confirming whether remission or elimination of infection and decrease of antibody responses have occurred *(8)*. This is especially applicable in the case of diseases where infection does not correlate with clinical signs.

3.5. Immunodominant Bands

NBI can be determined individually or in comparison to other bands allowing one to determine whether specific antigens generate an immunodominant or immunorecessive serological response based on comparative intensity and prevalence (number of individuals reacting with each band).

1. In the experimental model for canine visceral leishmaniasis longitudinal quantification of serologic reactivity (NBI) to *L. infantum* antigens was recorded in six dogs over 75 weeks *(6)*.
2. Computerized quantification was done and after taking the prevalence of the bands into consideration, statistical analysis identified the immunodominant and immunorecessive antigen bands.
3. QCWB could be applied at one time point (**Table 1**) or used to follow changes in seroreactivity against specific antigenic bands over time.

Table 1
Quantitative comparative Western blot analysis of antibodies against *Leishmania* antigens following experimental infection with *L. infantum*

Mean intensity	Band (kDa)	Band category
12[a] ± 4 (5)	12	ID
35.7 ± 22.3(6)	14	ID
15.2 ± 5.2 (6)	24	ID
23.9 ± 9.7 (6)	29	ID
5.8 ± 3.0 (3)	48	ID
14.0 ± 0.1 (4)	68	ID
12.0 ± 8.4 (3)	18	ID
1.2 (1)	19	IR
2.5 ± 1.1 (3)	34–35	IR
13.1 ± 2.4 (2)	71	IR
7.3 (1)	102	IR
------	136	IR

Comparison of the mean net band intensities (n = 6 dogs) at 53 weeks postinfection for each *L. infantum* antigen and categorization into immunodominant or immunorecessive bands. Bands were categorized according to both the mean net band intensity and the number of animals reacting with each band (prevalence). *ID* Immunodominant and *IR* Immunorecessive

[a]Mean intensity ± SEM was determined at 10^6 pixels. Values in parentheses indicate numbers of dogs expressing bands

3.6. Total Lane Intensity

This parameter is defined as the sum of all valid band intensities for each individual lane. TLI is equated to total antibody reactivity against all the components found in a complex crude antigen and roughly parallels serologic reactivity against whole antigen by other serological methods such as ELISA.

1. The serological response against crude *L. infantum* antigen (CLA), following infection of beagles in an experimental canine visceral leishmaniasis model, was measured by TLI and ELISA at different times postinfection and compared.
2. Antibody response measured by TLI showed significantly higher folds increase.
3. At 21 weeks postinfection, TLI was 100 times greater than preinfection, whereas CLA found by ELISA was only 60 times the preinfection value ($p < 0.01$) (**Fig. 3**).

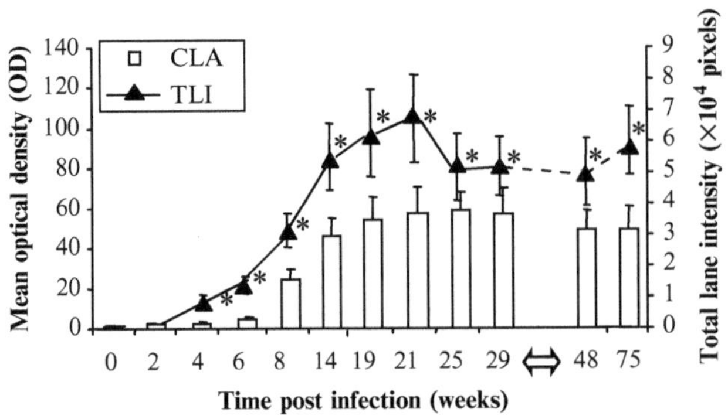

Fig. 3. Comparison of fold increase in antileishmanial antibody reaction with crude leishmanial antigen (CLA) by Quantitative Computerized Western Blotting (QCWB) and Enzyme linked immunosorbent assay (ELISA). Dogs ($n = 6$) were infected intravenously with the parasite *Leishmania infantum* and the increase in total lane intensity (TLI), determined by QCWB, and CLA by ELISA reactivity were measured over a 75-week period. The *arrow* on the *X*-axis indicates the period when the dogs were treated with (32–45 weeks postinfection). *Triangular blocks* and *columns* represent mean values standardized to time zero, with *bars* indicating the standard error. A significant difference ($p < 0.01$) is indicated by *asterisk*.

4. Notes

1. Adding lane markers is only possible at this stage and will not be permitted in later steps when using the KODAK 1D software.
2. It is important to locate these lines correctly for the accurate identification of bands with identical molecular weights that migrate differently on the same or different gels.
3. When the program identifies unresolved bands it might be difficult for it to determine the boundaries of the rectangle, thus not allowing manual insertion of a band close to the one that was found. The way to address this problem is to manually reduce the size of the rectangle confining the band, and then insert the desired neighboring band.
4. Oversaturated bands – when bands are flat peaked they indicate oversaturation, and require adjustment using the Gaussian model.
5. Very faint bands can be difficult to differentiate from the surrounding background and manual insertion may be needed.

References

1. Laemmli, U.K. (1970) Cleavage and structural proteins during the assembly of the head of bacteriophage T4. *Nature* 227, 680–684.
2. Zollner, T.M., Spengler, K., Podda, M., Ergezinger, K., Kaufmann, R., and Boehncke, H. (2001) The Western blot is a highly sensitive and efficient technique in diagnosing allergy to wasp venom. *Clin Exp Allergy* 11, 1754.
3. Belov, L., Meher, H.V., Putaswamy, V., and Miller, R. (1999) Western blot analysis of bile or intestinal fluid from patients with septic shock or systemic inflammatory response syndrome, using antibodies to TNF-α, IL-1α and IL-1β. *Immunol Cell Biol* 77, 34–40.
4. Riera, C., Fisa, R., Udina, M., Gallego, M., and Portus, M. (2004) Detection of *Leishmania*

infantum cryptic infection in asymptomatic blood donors living in an endemic area (Eivissa, Balearic Islands, Spain) by different diagnostic methods. *Trans R Soc Trop Med Hyg* 98, 102–110.

5. Iniesta, L., Fernandez-Barredo, S., Bulle, B., Gomez, M.T., Piarroux, R., Gallego, M., et al. (2002) Diagnostic techniques to detect cryptic leishmaniasis in dogs. *Clin Diagn Lab Immunol* 9, 1137–1141.
6. Talmi-Frank, D., Strauss-AyaliD., , Jaffe, C.L., and Baneth, G. (2006) Kinetics and diagnostic and prognostic potential of quantitative Western blot analysis and antigen-specific enzyme-linked immunosorbent assay in experimental canine leishmaniasis. *Clin Vaccine Immunol* 13, 271–276.
7. Vohradsky, J., and Panek, J. (1993) Quantitative analysis of gel electrophoretograms by image analysis and least squares modeling. *Electrophoresis* 14, 601–612.
8. Hassler, D., Schnaufer, M., Ehrfeld, H., and Muller, E. (2004) Disappearance of specific immune response after successful therapy of chronic Lyme borreliosis. *Int J Med Microbiol* 37, 161–164.

Chapter 14

Cationic Electrophoresis and Eastern Blotting

Engelbert Buxbaum

Summary

Denaturing, discontinuous electrophoresis in the presence of SDS has become a standard method for the protein scientist. However, there are situations where this method produces suboptimal results. In these cases, electrophoresis in the presence of positively charged detergents like cetyltrimethylammonium bromide may work considerably better. Methods for electrophoresis, staining, and blotting of such gels are presented.

Key words: Disk electrophoresis, Cationic electrophoresis, CTAB electrophoresis, Blotting, Eastern blotting, Detergent, CTAB

1. Introduction

Sodium dodecyl sulfate gel electrophoresis (SDS-PAGE) *(1)* and western blotting *(2, 3)* have become indispensable in the protein lab to separate and detect proteins with high resolution. Separation is usually performed on an analytical scale, but preparative equipment is available commercially.

With most proteins, the R_f-value in SDS-PAGE is proportional to size because proteins bind about 1.4 g SDS per gram of protein, equivalent to one molecule of SDS per three amino acids *(4)*. The number of negative charges introduced by the detergent far outweighs the charges on the protein itself, thus the charge/mass ratio, and hence the acceleration in an electrical field, is identical for all proteins. The restriction by the gel matrix, however, increases with the size of the protein.

Biji T. Kurien and R. Hal Scofield (eds.), *Methods in Molecular Biology, Protein Blotting and Detection, vol. 536*

DOI: 10.1007/978-1-59745-542-8_14

There are, however, some situations where SDS-PAGE performs less well:

1. Very hydrophobic proteins (i.e., transmembrane proteins) bind more than the usual 1.4 g SDS per gram of protein, increasing the charge/mass ratio of the protein. Thus these proteins run faster in SDS-PAGE than expected for their molecular mass.
2. Proteins with a large number of positively charged amino acids (e.g., histones) *(5)* also run faster than expected for their molecular weight.
3. Glycoproteins contain a highly variable number of negative charges in their sugar side chains. Thus they do not run as crisp bands in SDS-PAGE, but as broad smears, reducing the achievable resolution of the method.

In the latter two cases, it should be possible to circumvent the problems by using a positively charged detergent in lieu of the negatively charged SDS. Of course, the proteins then run from the positive to the negative electrode. Such attempts have been reported several times in the literature *(6–12)*, but resolution of the gels was usually low, caused by relatively broad protein bands.

The high resolution of SDS-PAGE is the result of band stacking in a discontinuous (multiphasic) buffer system, an effect first described by Ornstein *(13)* and later theoretically elaborated by Jovin and others *(14–18)*. While the protein moves through the stacking gel, it is electrophoretically concentrated from below 1 mg/mL (as present in usual samples) to several 100 mg/mL. Since the amount of protein cannot change due to mass conservation, the only way in which this can happen is by reducing the volume of the protein band, i.e., its height. Changing from an anionic to a cationic detergent requires a change in buffer composition so that stacking is still possible. Suitable buffers fulfilling the theoretical requirements for stacking are listed on http://www.buffers.nichd.nih.gov. With such selected buffer systems high resolution electrophoresis of proteins in cationic detergents is possible *(19)*. As with SDS-PAGE, separation in CTAB (cetyltrimethylammonium bromide)-PAGE is based on protein size, as noted also by others *(6, 7, 10, 11)*.

An additional advantage of CTAB compared to SDS is that it efficiently solubilizes membrane proteins, often without damaging their structure *(8, 19–22)*. Thus one can use the same detergent for electrophoresis that is also used for solubilization and purification, an advantage since extraneous detergent can interfere in PAGE by competing with SDS for binding to the protein.

2. Materials

All chemicals were of the highest purity available and were obtained mostly from Fluka (Buchs, Switzerland). Antibodies were from Accurate (Westbury, NY). Water came from a MilliQ-system (Millipore, Billerica, MA), who also supplied the Immobilon P blotting membranes. Immobiline DryStrip IEF-strips, IPG-buffer, and electrophoresis units were from Pharmacia (now part of GE Healthcare Bioscience, Piscataway, NJ, USA), detergents from Anatrace (Maumee, OH, USA). The blotting tank was from BioRad (Hercules, CA, USA).

2.1. Casting of CTAB Gels

1. 40% acrylamide/Bis (37:1): 1.08 g bisacrylamide and 38.9 g acrylamide made to 100 mL with water (*see* **Note 1**). Stable for months at 4°C, especially when stored over an anion exchanger.
2. 40% acrylamide/Bis (19:1): 2.11 g bisacrylamide and 37.9 g acrylamide made to 100 mL with water. Stable for months at 4°C, especially when stored over an anion exchanger.
3. Potassium hydroxide (KOH; 1 M): 5.611 g KOH made to 100 mL with water. Stable at room temperature (RT) if protected from air.
4. 16.6 M acetic acid (commercial 99.5% glacial acetic acid): Stable at RT. Exact molarity is determined once by titration and noted onto the bottle.
5. 10% CTAB: 10 g CTAB made to 100 mL with water. Store at 37°C to increase solubility.
6. Malachite green (1%): 10 mg/mL malachite green in water, stable at 4°C.
7. Water saturated butanol: *n*-butanol shaken with some water, after phase separation the upper, organic phase is used. Stable at RT.

2.2. For Photopolymerization

1. 100 mM methylene blue: 780 mg methylene blue made to 20 mL with water. Stable for months at 4°C.
2. 100 mM sodium toluene 4-sulfinate (T4S): 356 mg T4S (anhydrous) made to 20 mL with water. Stable for months at 4°C.
3. 1 mM diphenyl iodonium chloride (DPIC): 6.3 mg DPIC made to 20 mL with water. Stable for 1 week at 4°C.

2.3. For Fenton-System

1. 10 mM ferrous sulfate ($FeSO_4$): 27.8 mg $FeSO_4 \cdot 7H_2O$ made to 10 mL with water, make fresh daily.

2. 40 mM ascorbic acid: 70.5 mg ascorbic acid made to 10 mL with water, make fresh daily.
3. 30% hydrogen peroxide: commercially available, store at 4°C (*see* **Note 1**).

The recipes for both photopolymerization *(5)* and a Fenton-system *(8)* are given, obviously only one needs to be prepared.

2.4. Running of CTAB Gels

1. Upper tank buffer: 40 mM (3.56 g/L) β-alanine, 70 mM (2.29 mL/L) acetic acid, 0.1% CTAB. Make fresh each time.
2. Lower tank buffer: 50 mM KOH, 187 mM (3.18 mL/L) acetic acid, 0.1% CTAB. Make fresh each time.
3. Sample buffer (2×): 1.27 mL 1 M KOH (127 mM final), 107 µL acetic acid (187 mM final), 2 mL 10% CTAB (2% final), 100 µL β-mercaptoethanol (βME) (1% final), 1 mL glycerol (10% final), 7.21 g urea (12 M final) and 50 µL 1% basic fuchsin (0.005% final), make to 10 mL with water. Stable at RT for a week.

2.5. Staining of CTAB Gels with Ponceau S

1. Fixative: 100 mL glacial acetic acid and 400 mL methanol made to 1 L with water. Stable at RT.
2. Ponceau S solution: 0.1 g Ponceau S and 10 mL glacial acetic acid made to 1 L with water. The solution is stable at RT and may be reused several times.

2.6. Eastern Blotting

1. Blotting buffer: 10 mL 1 M KOH (10 mM final) and 636 µL acetic acid (11 mM final), water to 1 L. Make fresh.
2. Washing solution: 50 mg SDS (0.05% final) in 100 mL methanol. Stable at RT.
3. Phosphate buffered saline (PBS; 10× stock): 2.0 g KH_2PO_4 (14.7 mM final), 2.0 g KCl (26.9 mM final), 80.0 g NaCl (1,369 mM final), and 21.6 g $Na_2HPO_4 \times 7H_2O$ (or 11.44 g of the anhydride, 80.6 mM final) made to 1 L with water. Store at RT.
4. Blotto: 1 g low fat milk powder (0.1% final), 0.5 mL Tween-20 (0.05% final), and 0.1 g thimerosal (0.01% final) made to 1 L with PBS. Make fresh each day.
5. Blocking solution: 0.5 g low fat milk powder in 10-mL Blotto. Make fresh each time.
6. Horse radish peroxidase (HRP) substrate (20×): 6.01 g Tris (1 M final), 14.6 g NaCl (3.2 M final), 950 mg $NiCl_2 \times 6H_2O$ (80 mM final) dissolved in 40 mL water. Adjust to pH 7.6 with HCl, and add 250 mg DAB (13 mM final) (*see* **Note 1**). Dilute to 50 mL with water, filter and freeze in 1-mL aliquots. Stable for years at −20°C.
7. Chemiluminescence reagent: 400 mg luminol (2.2 mM final) and 9.5 mg *p*-iodophenol (432 µM final) in 19-mL dimethylsulfoxide diluted to 100 mL with water. Immediately before

use take 10 mL and add 4.8 μL 30% hydrogen peroxide (420 nM final, *see* **Note 1**).

8. Primary antibody: 1 μg monoclonal antibody in 10-mL Blotto. Stable for many months at 4°C if the milk powder in Blotto is replaced by BSA (fraction V). For anti-sera the dilution has to be worked out on a case-by-case basis.

2.7. 2D-Electrophoresis

1. IEF sample buffer (2×): 100 μL β-ME (1% final), 100 μL IPG-buffer (same pH-range as IEF-strip, 1% final), 7.21 g urea (12 M final), 0.4 g CHAPS (4% final), and a trace of bromophenol blue made to 10 mL with water. Store in aliquots at –20°C.
2. Equilibration buffer: 3.6 g urea (6 M final), 3-mL glycerol (3% final), 0.2 g CTAB (2% final), 640 μL 1 M KOH (64 mM final), 54 μL glacial acetic acid (94 mM final) and 25 μL 1% basic fuchsin (0.0025% final), make to 10 mL with water.
3. Mounting medium: 50 mg agarose in 10-mL stacking gel buffer (0.5% final), heated to boiling in a microwave oven. Best to prepare fresh for each run (**Table 1**).

Table 1
Mixing table for CTAB-PAGE gels (all volumes in mL)

Solution	5%	7.5%	10%	12.5%	15%	17.5%	20%	Stack
Acrylamide 19:1	–	–	–	–	–	–	–	3.0
Acrylamide 37:1	7.5	11.3	15.0	18.8	22.5	26.3	30.0	–
1 M KOH	2.6	2.6	2.6	2.6	2.6	2.6	2.6	1.91
Glacial acetic acid	0.962	0.962	0.962	0.962	0.962	0.962	0.962	0.161
Urea	10.8 g	10.8 g	10.8 g	10.8 g	10.8 g	10.8 g	10.8 g	5.4 g
10% CTAB	0.6	0.6	0.6	0.6	0.6	0.6	0.6	0.3
100 mM T4S	0.3	0.3	0.3	0.3	0.3	0.3	0.3	0.15
1 mM DPIC	1.5	1.5	1.5	1.5	1.5	1.5	1.5	0.75
10 mM $FeSO_4$	0.048	0.048	0.048	0.048	0.048	0.048	0.048	0.024
40 mM Ascorbate	6.0	6.0	6.0	6.0	6.0	6.0	6.0	3.0
Water to a final volume	60	60	60	60	60	60	60	30

Note that either the reagents for photopolymerization *or* those for the Fenton-system should be used.

3. Methods

3.1. Casting of CTAB Gels

1. Pour the mixture, containing either the reagents for photo-crosslinking or the Fenton-system, into an Erlenmeyer flask with magnetic stirrer. The stirrer is adjusted for vigorous movement without foam production. Then apply a vacuum for at least 10 min to remove dissolved oxygen. This is essential as oxygen inhibits the polymerization reaction. Do not apply the vacuum before starting the stirrer!
2. Add the catalyst: 18 µL 10 mM methylene blue or 2-µL hydrogen peroxide depending on whether the photopolymerization or the Fenton-system is used. Mix by gentle inversion; do not shake oxygen into the solution again. At this stage do not add the catalyst to the stacking gel.
3. Pour the separating gel into the casting sandwich and overlay with water-saturated *n*-butanol. The Fenton-system will polymerize by itself, the photopolymerization system requires exposure to a strong source of white light. A sun-exposed window or a halogen lamp may be used. Polymerization is finished when the interface between gel and butanol becomes prominent. Methylene blue will become colorless during polymerization. Normally polymerization should be complete within 30 min.
4. Pour the butanol from the top of the gel, add catalyst to the stacking gel mixture, and cast the stacking gel. Insert combs immediately and allow the gel to polymerize.

3.2. Running of CTAB Gels

1. Mount the gel in the running chamber according to manufacturers' instruction, add upper and lower tank buffer. Sometimes air bubbles get trapped in the wells; these can be rinsed out using a tuberculin syringe with 27G needle. This process and the loading of samples is aided by malachite green in the stacking gel, which makes the wells easier to see (for SDS-PAGE use phenol red for the same purpose) *(23)*.
2. Load the sample with a 25-µL Hamilton syringe, the needle should have a flat point. Between samples rinse the syringe with upper tank buffer, finally with water.
3. For a standard minigel, electrophoresis is performed at 20 mA per gel (10 mA during stacking), with a maximum voltage of 200 V. Do not forget to *reverse the electrode polarity* compared to SDS-PAGE (*see* **Note 2a–h**).

3.3. Staining of CTAB Gels with Ponceau S

1. Fix the gel for 10-15 min twice on an orbital shaker, then replace the fixative with dye solution for 5 min. The gel can be differentiated by incubation with several changes of fixative (in the same way as is commonly done with CBB-R250), but this procedure is time consuming (*see* **Note 3f**).

3.4. Eastern Blotting

1. Soak the PVDF blotting membrane first in methanol, then in water and finally in blotting buffer (nitrocellulose membranes are soaked directly in buffer).
2. Place one of the sponges and three sheets of filter paper (e.g., Whatman No. 3), all soaked in blotting buffer, onto the red part of the blotting cassette. Place the gel on top and flood it with a small amount of blotting buffer (plastic Pasteur-pipettes are ideal for this). Add the blotting membrane carefully on top without trapping air bubbles between gel and membrane, then place three sheets of filter paper and the second sponge, all soaked in blotting buffer, on top. Close the cassette and place (in the correct orientation, red to red and black to black) into the blotting tank already filled with blotting buffer. In this way, the gel is on the positive and the membrane on the negative side. This orientation is opposite to the one used in western blotting, hence the name "eastern blotting." The entire procedure is best performed on a plastic tea-tray or similar, so that the unavoidable spillage of buffer is contained.
3. Electrotransfer is achieved at 40 V constant voltage (resulting in a current of ~200 mA), no reversal of current is required if the sandwich is assembled as described above. The BioRad blotting tank comes with a plastic vessel for ice to keep the buffer cold during blotting. In my experience 2 h blotting time is sufficient to transfer even high molecular weight proteins at the interface between stacking and separating gel. It is good practice to stain the gel afterward to check for residual protein, at least occasionally.
4. Remove bound CTAB by incubating the membrane twice for 1 h in washing solution.
5. To check whether proteins have been transferred to the membrane incubate the membrane with staining solution for a few minutes and then rinse with running tap water. It is convenient to mark the position of the molecular weight standards at this time with a soft pencil.
6. Block nonspecific protein binding sites on the membrane with 5% low fat milk powder in Blotto for about 15 min. Longer times are often suggested in the literature, but are not required and may reduce sensitivity by covering the bands to be detected with milk proteins. This step also removes any Ponceau that may still be on the membrane.
7. Wash the membrane briefly with Blotto and incubate with the primary antibody. Incubation can be performed for 1 h at 37°C if one is in a hurry. However, if the incubation is performed overnight at 4°C in the cold room on an orbital shaker, sensitivity is higher and the antibody solution can be reused many times. In fact it is quite possible to do ones

entire PhD work with a single 10 mL batch of primary antibody solution.

8. Wash the membrane three times for 15 min with Blotto, incubate with the secondary antibody (HRP-conjugated, diluted as by manufacturers recommendation) for 1 h and wash again three times with Blotto.
9. For chemical staining of the blot *(24)* defrost a 1-mL aliquot of substrate solution, dilute to 20 mL and add 1 μL of 30% H_2O_2 (*see* **Note 1b**). Incubate the membrane with that solution on an orbital shaker until you see well-developed bands. Rinse briefly with water and air-dry. The precipitate of elementary Ni photographs well and is stable for years in a notebook (*see* **Note 4a–h**).

3.5. 2D-Electrophoresis

1. To use an IPGphor horizontal electrophoresis unit mix the sample 1:1 with IEF-sample buffer and place this solution into the ceramic strip holder.
2. Remove the protective plastic cover from the gel and place it gel side down into the sample. The sample volume depends on the length of the strips, for the 7-cm strips compatible with minigels use 125 μL. Add covering solution (obtained from Pharmacia) and place the strip holder onto the IPGphor unit.
3. Allow the sample to soak into the strip at a temperature of 18°C and an applied voltage of 50 V for 8 h. Then increase the voltage to 500, 1,000, and 4,000 V for 30 min each, finally 8,000 V for a total of 25,000 V h. The maximum voltage available on the IPGphor unit is 8,000 V; depending on the conductivity of the sample this value may not be reached. This is no problem, however, since runs can be normalized by the applied V h. It is convenient to run the IEF over night, as it does not need supervision.
4. Remove the strip from the sample holder and place it into a test tube with equilibration buffer. Seal the tube with Parafilm and place it on an end-over-end mixer for 10 min at 37 °C. This allows sufficient time for CTAB binding to the protein bands without diffusional band spreading.
5. Mount the equilibrated strip carefully onto a CTAB-gel without introducing air bubbles. A thin layer of stacking gel buffer on top of the gel can aid this process and is wicked away once the strip is in place. Use gels with one small lane for the molecular weight standards and a wide lane for the IEF-strip. Strip and gel should be hand-warmed to allow mounting the strip with agarose solution (300 μL). Once the agarose has set the gel is ready to be run.

4. Notes

1. *Safety issues.* (a) Acrylamide is neurotoxic, handle with great care. Ready-made acrylamide/bisacrylamide solutions are commercially available which avoid the development of dust during weighing. Store the solution over an anion exchanger to remove any acrylic acid which may form to prevent electroendosmosis during runs. (b) Store hydrogen peroxide solutions in the bottles supplied by the manufacturer which have release valves to prevent the build-up of pressure from decomposition. (c) DAB is a suspected carcinogen. It is safest to add the buffer directly to the bottle and assume that the manufacturer has sold the correct weight. When opening a bottle containing DAB, do *not* wear gloves as the electrostatic charge on the gloves would blow the fine powder directly into your face. Wear eye protection. DAB is also available as tablets for safer handling. (d) Agarose: Some experimentation is required with the settings of the microwave oven so that the agarose dissolves in the mounting medium without boiling over.
2. *Electrophoresis.* (a) Fenton-polymerization results in gels which are somewhat more brittle than those produced by photopolymerization. However, if the Pharmacia multicasting stand is used the nontransparent aluminia back-plates prevent the use of a photopolymerization system, in that case the Fenton-system must be used. The separation achievable does not depend on the polymerization reaction used, however. (b) Both systems are even more sensitive to the presence of oxygen than the TEMED/APS-system used for Laemmli-gels. Proper degassing of the gel mixture is essential. In addition, it is important that the combs prevent access of air to the polymerizing stacking gel. Neither those produced by Pharmacia nor those from BioRad do that however, resulting in ill-formed wells, sometimes so badly that the gel must be discarded. I have solved that problem by having better combs manufactured in a workshop *(19)*. (c) Gel concentration depends on the molecular mass range of the proteins of interest *(25, 26)*. I have found 5–15% gradient gels most convenient for proteins of 10–200 kDa. If the 10-gel multicaster is used, the gradient can be easily formed with a gradient maker, but for single gels the volume required is too small. In that case cast a step gradient by mixing (**Table 2**). These solutions are carefully layered on top of each other; this is made easier when the heavy solution contains 10% glycerol (which does not otherwise interfere with electrophoresis). The total volume of that gradient is 2.35 mL, enough for one mini-gel. With the photopolymerization

Table 2
Volumes of 5 and 15% gel solution for a step gradient gel (1 minigel)

% Desired	5% (μl)	15% (μl)
15.0	0	392
14.0	36	356
13.1	71	320
12.3	107	285
11.3	142	249
10.5	178	214
9.5	214	178
8.6	249	142
7.7	285	107
6.8	320	71
5.9	356	36
5.0	392	0

chemistry you can start polymerization once the gel is cast. (e) Proteins with very high molecular mass (>200 kDa) require low acrylamide concentrations that result in very soft, difficult to handle gels. Add 0.5% agarose to stabilize them without any effect on separation. (f) The buffer system used here was originally described in *(17)* as suitable for cationic electrophoresis. However, a considerable number of other cationic systems with a pH range of 2.5–9.0 are listed on http://www.buffers.nichd.nih.gov for separation of samples with specific requirements. For example the Na/K-ATPase acylphosphate is most stable at acidic pH and destroyed in the presence of potassium. One may try system number 15, 18, 25, or 30. (g) CTAB has a relatively high Krafft-point and precipitates if the temperature drops below 18°C. If lower temperatures are desired during electrophoresis, consider replacing CTAB with a detergent with lower cmc and/or Krafft point like 16-BAC. Note that the stacking properties of buffer systems are temperature dependent, on http://www.buffers.nichd.nih.gov systems for use at room temperature and at 0°C are available. (h) Prestained molecular weight markers, available for SDS-PAGE from several manufacturers, are unsuitable for CTAB-PAGE even after detergent exchange. The bound dye influences the R_f-value and separation is no longer according to molecular mass. It can only be hoped that such standards will also become available for CTAB-PAGE.

3. *Gel staining.* (a) The staining of gels with Ponceau S is sensitive and fast, but staining with CBB-R250 is also possible. Phenol red can be used in the same way as Ponceau S, the same is probably true for other acidic dyes. Phenol red is fluorescent under acidic conditions, making very sensitive detection of proteins possible. (b) Silver staining of CTAB-gels is possible, the method of Heukeshoven and Dernick *(27)* achieves a higher sensitivity than that of Merril *(28)*. You can execute the former method at a constant temperature of 37°C for all steps; replace the glutardialdehyde with formaldehyde for maximum sensitivity. (c) PAS-staining of CTAB-gels *(29)* works, but considerable savings in time and chemicals are possible if the staining is performed on eastern-blots rather than on gels *(30)*. (d) CTAB is a very mild detergent, which can retain the enzymatic activity of enzymes solubilized with it. Try zymograms *(11)*, proteins in gels specifically stained by their enzymatic activity, at least with monomeric enzymes. Another paper in this volume discusses zymograms in detail (*see* Chapter "Simultaneous Immunoblotting Analysis with Activity Gel Electrophoresis and 2-D Gel Electrophoresis"). (e) Gels can be dried after incubation with 1% glycerol between two sheets of cellophane. Span the assembly between two plastic frames and leave on the bench to dry overnight. This is a more convenient approach than drying gels onto filter paper, rehydration is easier should that become necessary. For autoradiography with ^{14}C or ^{35}S replace one of the cellophane sheets by Saran Wrap and remove after drying. (f) Alternatively destain by putting the gel between filter paper (3 sheets of Whatman No. 3 on both sides of the gel) and place it in a tank blotter with blotting buffer. Thirty minutes at 40 V (~200 mA) removes the background stain, while the protein/stain complex is immobile. If need be, proteins become mobile again after incubating the gel in 1 mM KOH for 30 min.
4. *Blotting.* (a) Large membrane proteins like Na/K-ATPase and Mdr1 blot well in a tank blotter but there is no obvious reason why you cannot use other blotting methods (as described in this volume, *see* Chapter "Detection of Calcium Binding by Ro 60 Multiple Antigenic Peptides on PVDF Membrane Using Quin-2") if desired. (b) Before eastern blots can be immunostained, the CTAB needs to be washed off, as the protein/CTAB complexes would unspecifically bind antibodies. Wash PVDF membranes with SDS in methanol. For nitrocellulose 50% methanol, 10% acetic acid is suitable,

but results in weaker bands. (c) After removal of CTAB most of the staining procedures used for western blots can also be used for eastern, for example India ink (permanent) *(31)* or Ponceau S *(32, 33)*. The latter dye is washed off during the blocking step and does not interfere with immune detection. (d) When you use horseradish peroxidase conjugated antibodies for detection, do not use sodium azide as preservative, since it inactivates the enzyme. Thimerosal® (sodium ethylmercurithiosalicylate) is a suitable alternative. (e) In some cases, phosphate or milk powder present in the Blotto may interfere (e.g., anti-phospho-tyrosine antibodies). In those cases use Tris-buffered saline (20 mM Tris–HCl, 137 mM NaCl, 4 mM KCl, pH 7.4). Replace milk powder by fish skin gelatin (mammalian gelatin or bovine serum albumin work less well). (f) Detection of HRP-conjugated secondary antibodies can be done either by forming a colored precipitate directly on the blot *(24)* or by chemiluminescence *(34)*. The direct method is easier to perform and the stained blot can be glued into the notebook for a permanent record. Chemiluminescence is more sensitive by at least one order of magnitude. Note that for chemoluminescence detection the secondary antibody must be more dilute to reduce background. Refer to manufacturers' instruction. Very pure luminol and *p*-iodophenol (should be colourless, not brown) are required; commercially available reagent mixes may also be used. (g) Phosphatase-labeled secondary antibodies with the appropriate detection reagents should also be usable in eastern blotting, however, I have not tested this. (h) Alternatively incubate the membrane with the chemiluminescence reagent *(34)* for several minutes, then wrap in cellophane and place against X-ray film. The light signal is stable for some time, during which you can try different exposure times.

Acknowledgements

This work was supported in part by Kuwait University grant MPB029.

References

1. Laemmli, U. (1970) Cleavage of structural proteins during the assembly of the head of bacteriophage T4. *Nature* 227, 680–685.
2. Towbin, H., Staehelin, T., and Gordon, J. (1979) Electrophoretictransfer of proteins from polyacrylamide gels to nitrocellulose sheets: Procedure and some applications. *Proc Natl Acad Sci USA* 76, 4350–4354.
3. Dunn, S. (1986) Effects of the modification of transfer buffer composition and the

renaturation of proteins in gels on the recognition of proteins on western blots by monoclonal antibodies. *Anal Biochem* 157, 144–153.

4. Shirahama, K., Tsujii, K., and Takagi, T. (1974) Free boundary electrophoresis of sodium dodecyl sulphate-protein polypeptide complexes with special reference to SDS-polyacrylamide gel electrophoresis. *J Biochem* 75, 309–319.
5. Rabilloud, T., Girardot, V., and Lawrence, J.-J. (1996) One- and two dimensional histone separations in acidic gels: Usefulness of methylene blue-driven photopolymerisation. *Electrophoresis* 17, 67–73.
6. Panyim, S., Thitipongpanich, R., and Supatimusro, D. (1977) A simplified gel electrophoretic system and its validity for molecular weight determinations of protein–cetyltrimethylammonium complexes. *Anal Biochem* 81, 320–327.
7. Eley, M., Burns, P., Kannapell, C., and Campbell, P. (1979) Cetyltrimethylammonium bromide polyacrylamide gel electrophoresis: Estimation of protein subunit molecular weights using cationic detergents. *Anal Biochem* 92, 411–419.
8. MacFarlane, D. (1983) Use of benzyldime thyl-*n*-hexadecylammonium chloride ("16-BAC"), a cationic detergent, in an acidic polyacrylamide gel electrophoresis system to detect base labile protein methylation in intact cells. *Anal Biochem* 132, 231–235.
9. Mócz, G., and Bálint, M. (1984) Use of cationic detergents for polyacrylamide gel electrophoresis in multiphasic buffer systems. *Anal Biochem* 143, 283–292.
10. Akin, D., Shapira, R., and Kinkade, J., Jr. (1985) The determination of molecular weights of biologically active proteins by cetyltrimethylammonium bromide-polyacrylamide gel electrophoresis. *Anal Biochem* 145, 170–176.
11. Akins, R., Levin, P., and Tuan, R. (1992) Cetyltrimethylammonium bromide discontinuous electrophoresis: *Mr*-based separation of proteins with retention of enzymatic activity. *Anal Biochem* 202, 172–178.
12. Hartinger, J., Stenius, K., Högemann, D., and Jahn, R. (1996) 16-BAC/SDS-PAGE: A two-dimensional gel electrophoresis system suitable for the separation of integral membrane proteins. *Anal Biochem* 240, 126–133.
13. Ornstein, L. (1962) Disk electrophoresis: I. Background and theory. *Ann N Y Acad Sci* 121, 321–351.
14. Jovin, T. (1973) Multiphasic zone electrophoresis. I. Steady-state moving-boundary systems formed by different electrolyte combinations. *Biochemistry* 12, 871–879.
15. Jovin, T. (1973) Multiphasic zone electrophoresis. II. Design of integrated discontinuous buffer systems for analytical and preparative fractionation. *Biochemistry* 12, 879–890.
16. Jovin, T. (1973) Multiphasic zone electrophoresis. III. Further analysis and new forms of discontinuous buffer systems. *Biochemistry* 12, 890– 898.
17. Jovin, T. (1973) Multiphasic zone electrophoresis. IV Design and analysis of discontinuous buffer systems with a digital computer. *Ann N Y Acad Sci* 209, 477–496.
18. Chrambach, A. (1985) The practice of quantitative gel electrophoresis. VCH, Weinheim.
19. Buxbaum, E. (2003) Cationic electrophoresis and electrotransfer of membrane glycoproteins. *Anal Biochem* 314, 70–76.
20. Pritchard, D., Crawford, C., Duce, I., and Behnke, J. (1985) Antigen stripping from the nematode epicuticle using the cationic detergent cetyltrimethylammonium bromide (CTAB). *Parasite Immunol* 7, 575–585.
21. Freedman, D., Nutman, T., and Ottesen, E. (1988) Enhanced solubilization of immunoreactive proteins from *Brugia malayi* adult parasites using cetyltrimethylammonium bromide. *Exp Parasitol* 65, 244–250.
22. Maki, K., Sagara, J., and Kawai, A. (1991) A cationic detergent, cetyltrimethylammonium bromide (CTAB), selectively dissociates the intermediate filament of the fibroblast. *Biochem Biophys Res Commun* 175, 768–774.
23. Smith, I., Cromie, R., and Stainsby, K. (1988) Seeing gel wells well. *Anal Biochem* 169, 370–371.
24. Lazar, J., and Taub, F. (1992) A highly sensitive method for detecting peroxidase *in situ* hybridization or immunohistochemical assays. In Nonradioactive labeling and detection of biomolecules, Kessler, C., ed., pp. 135–142. Springer, Berlin.
25. Rodbard, D., and Chrambach, A. (1971) Estimation of molecular radius, free mobility, and valence using polyacrylamide gel electrophoresis. *Anal Biochem* 40, 95–134.
26. Stellwagen, N. (1998) Apparent pore size of polyacrylamide gels. *Electrophoresis* 19, 1542–1547.
27. Heukeshoven, J., and Dernick, R. (1988) Improved silver staining procedure for fast staining in phastsystem development unit I. Staining of sodium dodecyl sulfate gels. *Electrophoresis* 9, 28–32.

28. Merril, C., Goldman, D., Sedman, S., and Ebert, M. (1981) Ultrasensitive stain for proteins in polyacrylamide gels shows regional variations in cerebrospinal fluid proteins. *Science* 211, 1437–1438.
29. Segrest, J., and Jackson, R. (1972) Molecular weight determination of glycoproteins by polyacrylamide gel electrophoresis in sodium dodecyl sulfate. *Methods Enzymol.* 28, 54–63.
30. Thornton, D., Holmes, D., Sheehan, J., and Carlstedt, I. (1989) Quantitation of mucus glycoproteins blotted onto nitrocellulose membranes. *Anal Biochem* 182, 160–164.
31. Hancock, K., and Tsang, V. (1983) India ink staining of proteins on nitrocellulose paper. *Anal Biochem* 133, 157–162.
32. Hassan, J., Feighery, C., and Whalan, A. (1987) Staining molecular weight markers on nitrocellulose using Ponceau S. *J Clin Lab Immunol* 24, 104.
33. Moore, M., and Viselli, S. (2000) Staining and quantification of proteins transferred to polyvinylidene difluoride membranes. *Anal Biochem* 279, 241–242.
34. Leong, M., and Fox, G. (1990) Luminescent detection of immunodot and Western blot. *Methods Enzymol* 184, 442–451.

Chapter 15

A Miniaturized Blotting System for Simultaneous Detection of Different Autoantibodies

Ulrich Canzler, Holger Bartsch, Kai Großmann, Werner Lehmann, Karsten Conrad, Biji T. Kurien, Yaser Dorri, R. Hal Scofield, and Michael Bachmann

Summary

Sera of tumor patients frequently contain autoantibodies to tumor-associated antigens. Here we describe a miniaturized immunoblot platform allowing us to screen sera of patients for the presence of autoantibodies to ten autoantigens in parallel.

Key words: Multiparametric assay, Autoantibodies, Tumor-associated antigens

1. Introduction

Autoantibodies to cellular components are not only found in sera of patients with autoimmune diseases: using a variety of different techniques including for example cDNA library screening (SEREX technique), enzyme linked immunosorbant assay, immunoblotting, immunocoprecipitation, and epifluorescence microscopy, but also detected in tumor patients *(1–6)*. The presence of such antibodies are of interest for several reasons: (i) these antibodies indicate that tumor cells are indeed recognized by both, the cellular and humoral arm of the immune system, and (ii) the antibodies against tumor-associated antigens (TAAs) can perhaps be used as early indicators of a developing or already existing tumor or a relapse of a cancer. Many investigators have, therefore, been interested in the use of autoantibodies to TAAs as serological markers for cancer diagnosis [reviewed in *(7)]*. Enthusiasm for

Biji T. Kurien and R. Hal Scofield (eds.), *Methods in Molecular Biology, Protein Blotting and Detection, vol. 536*

DOI: 10.1007/978-1-59745-542-8_15

this approach has been tempered by the low sensitivity when individual antigen–antibody reactions were studied: According to these studies, antibodies to any individual antigen such as for example p53, c-myc, or p62 do not reach levels of sensitivity which could become routinely useful in diagnosis. However, it was proposed that this drawback can be overcome by using a panel of selected TAAs. Indeed, some published data support the idea that a multiparametric assay improves the sensitivity and the specificity for a specific tumor entity when several selected TAAs are analyzed in parallel *(8)*. Here we describe a novel platform, a miniaturized immunoblot system (**Fig. 1**), allowing us to screen sera for the presence of different autoantibodies in parallel.

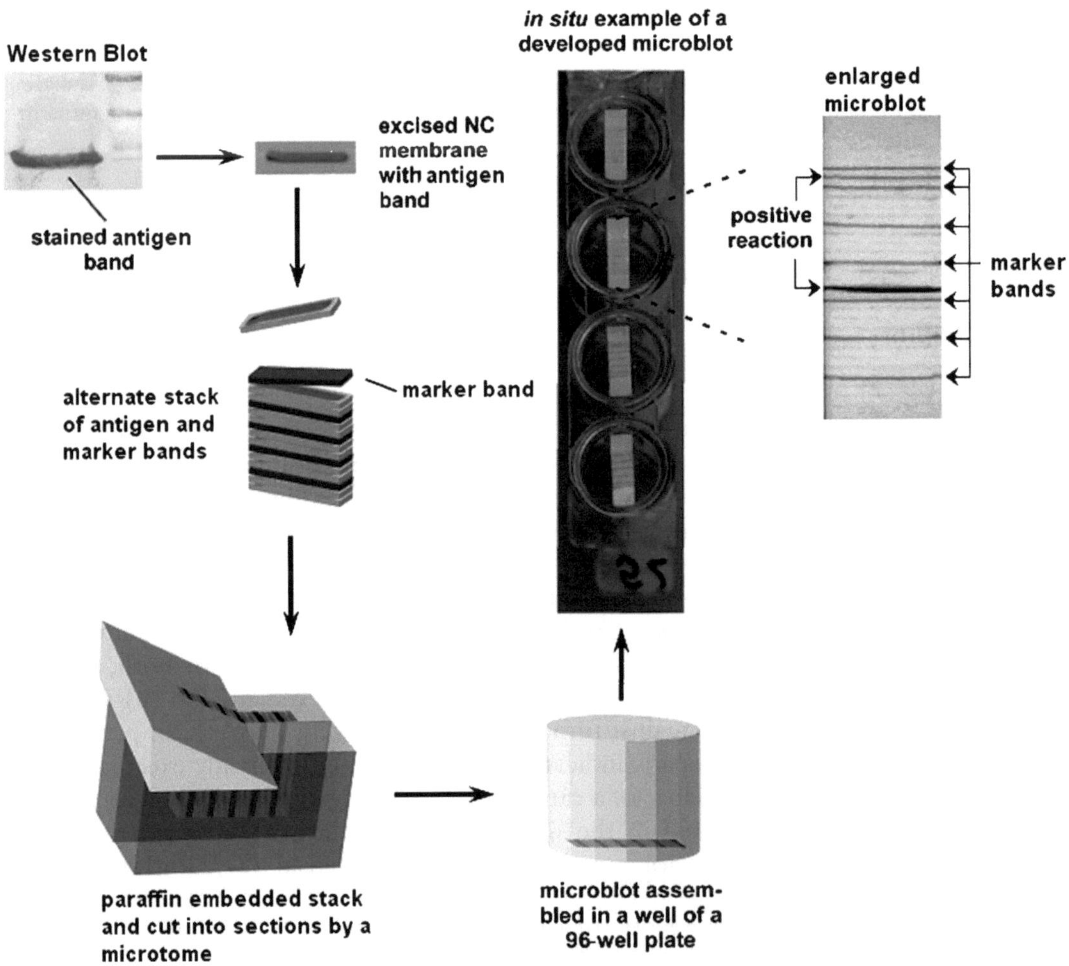

Fig. 1. Schematic view of the microblot manufacturing steps.

2. Materials

2.1. Bacterial Expression and Purification of Antigens

1. Luria-Burtani (LB) medium (10 g/L NaCl, 5 g/L yeast extract, 10 g/L tryptone, pH 7.0), store at 4°C.
2. Isopropyl β-D-1-thiogalactopyranoside (IPTG), 1 M stock solution in water, store in aliquots at –20°C (*see* **Note 1**).
3. Bacterial expression clones (pET28 or equivalent) in BL21(DE3) pLysS bacteria (*see* **Note 2**).
4. Ni-NTA agarose (Qiagen, Hilden, Germany).
5. Lysis buffer (8 M urea, 100 mM NaH_2PO_4, 10 mM Tris–HCl, pH 8).
6. Wash buffer I (10 mM imidazole, 8 M urea, 100 mM NaH_2PO_4, 10 mM Tris–HCl, pH 8.0) and wash buffer II (20 mM imidazole, 8 M urea, 100 mM NaH_2PO_4, 10 mM Tris–HCl, pH 8).
7. Elution buffer (350 mM imidazole, 8 M urea, 100 mM NaH_2PO_4, 10 mM Tris–HCl, pH 8)
8. Suitable antibiotic stock, dependent on the used expression system.

2.2. Western Blotting for Preparing of the Microblot

1. Transfer buffer: Roti-Blot A and Roti-Blot K (Carl-Roth, Karlsruhe, Germany).
2. Nitrocellulose membrane: Porablot NCP, not enforced (Machery Nagel, Düren, Germany) (*see* **Note 3**).
3. 3MM chromatography paper from Whatman, Maidstone, UK.
4. Tris-buffered saline with Tween-20 (TBST) (10×): 1.5 M NaCl, 0.5 M Tris–HCl, pH 8, 1% Tween-20. Dilute 100 mL with 900 mL water for use.
5. Blocking buffer: 5% (w/v) Blocking reagent (Roche, Mannheim, Germany) in TBST.
6. Ponceau S staining solution: 0.2% [w/v] Ponceau S in 0.3% [v/v] trichloroacetic acid.

2.3. Development of Microblot

1. The manufacturing principle of the microblot *(9)* is shown in **Fig. 1**. Briefly, protein antigen bearing lines excised from stained western blots as well as marker bands for software-based data analysis are stacked and embedded in paraffin. Sections of 10 μm are cut by a microtome, paraffin is removed and nitrocellulose slices are mounted to a solvent resistant support membrane by organic solvents. The resulting microblots (6 × 2 mm) are fixed to a plastic holder and integrated into modules of a standard 96-well microtitre plate. Each microblot contains

ten autoantigen bands, seven marker bands, and a conjugate reaction control line.

2. The solutions requested for development of the microblots are provided by Attomol GmbH (Lipten, Germany) upon delivery of your microblot test system. These ready-to-use solutions include a sample diluent and a precipitating 3,3′,5,5′-tetramethylbenzidine (TMB) substrate solution for horseradish peroxidase (HRP), concentrated wash buffer (10×), and sheep anti-human IgG HRP conjugate (27×). The wells containing the microblots are stored at room temperature (RT). The 10× washing buffer has to be stored at RT. The anti IgG HRP conjugate as well as the TMB substrate has to be stored at 4°C. The sample diluent solution should be stored at –20°C.
3. For data collection, a scanner with a suitable depth of focus (e.g., Plustek Optic Pro ST48) and a PC for documentation and analysis is required. The results can be automatically evaluated with the Attomol® Microblot-Analyzer software.
4. For washing of the 96-well microtitre plates, an ELISA washer, adjusted to 400 μL wash volume set to overflow can be used. Adjust the washer needle maximal eccentrically in order to provide the microblot from being touched (*see* **Note 4**). An automatic plate washer from Tecan GmbH (Germany) can be employed.

3. Methods

As patient sera are often limited, it is desirable to evaluate as much parameters with as little serum sample as possible. Therefore, multiparameter assays have been developed. Other problems often confronted with when screening patient sera for specific antigen recognition are limitations concerning the used antigens. Often, these are difficult to purify or, if purchased, very expensive.

One technique that is able to address both problems at one time (limited availability of serum sample and antigen) is the use of miniaturized multiparametric assays as for example the microblot technique used herein. Here, one can use very little amount of antigen, which does not need to be absolutely pure, because there is an additional SDS-PAGE step for separation of the protein of interest from undesired contaminants. In addition, in one well of a 96-well plate the simultaneous screening of up to ten different antigen reactivities can be accomplished. Also, the procedure is related to a normal ELISA assay, which makes it easy to perform. Hence, the microblot is very suitable for a multiparametric evaluation of precious patient sera.

3.1. Preparation of Samples; Bacterial Expression; and Affinity Purification of His_6 Tagged Proteins

1. If not already available, start with cloning of the desired antigens into a bacterial expression system. We have good experience using the pET system provided by Merck KGaA (Darmstadt, Germany), but other systems might work as well.
2. The day before the bacterial expression pick an isolated colony of the desired clone and incubate in 10 mL LB medium supplemented with the appropriate antibiotic in a rotating incubator at 37°C overnight.
3. Transfer 8 mL of the overnight culture to 800 mL of freshly prepared LB medium and continue to incubate at 37°C in a rotating incubator (*see* **Note 5**).
4. Monitor the growth of the bacterial culture by measuring the optical $density_{600}$ (OD_{600}) and continue the culture until the OD_{600} reaches 0.5.
5. Induce the culture by adding 1 mM IPTG.
6. Grow the culture for additional 3 h and keep monitoring OD_{600} at least every hour. If the OD_{600} is not increasing any further, you can stop the culture.
7. Pellet the bacterial suspension at 5,000 × *g* for 20 min at 4°C. Discard the supernatant.
8. Store the cell pellet at –20°C until use or proceed immediately with the purification procedure.
9. Resuspend the pellet in 10 mL lysis buffer, add 1 mM PMSF (*see* **Note 6**).
10. Disintegrate the bacterial suspension by incubating for 15 min at room temperature and periodical mixing. The suspension will become very viscous due to the released nucleic acids.
11. Degrade the nucleic acids by applying ultrasound in short pulses. Keep the protein suspension on ice during the whole procedure (*see* **Note 7**).
12. Centrifuge the lysate for 15 min at 10,000 × *g* (4°C) and save the supernatant for purification.
13. Prepare a Ni-NTA column as follows: fill the bottom of a 5-mL syringe with some sterile glass wool and wet the glass wool with water to remove any air that is trapped. Add 500 μL Ni-NTA agarose beads (approximately 1 mL Ni-NTA slurry) on top of the glass wool and equilibrate the prepared column with 10 mL lysis buffer.
14. Apply the supernatant saved after the centrifugation step onto the column. Keep the flow through and reapply once more. If desired keep an aliquot for control purpose.
15. Wash the column once with 10 mL lysis buffer.

16. Wash four times with 1 mL wash buffer I each, followed by four additional wash steps with 1 mL wash buffer II each. If desired keep an aliquot for control purpose.
17. Elute the protein with elution buffer. Apply six times 500 µL each and save in separate sample tubes.
18. Continue with an SDS-PAGE (*see* **Subheading 3.2**) in order to analyze the efficacy of your protein purification. Load samples of each elution fraction and, if desired samples of your washing steps, flow through and crude extract.

3.2. SDS-PAGE

1. Carry out SDS-PAGE (1 mm thick mini gels) essentially according to Laemmli *(10)*.
2. Load 30 µL (approximately 3 µg protein) of each sample in a well. Include one well for prestained molecular weight markers Page Ruler TM (Fermentas, St. Leon-Roth, Germany). Load either the samples for the control or for the western blot.
3. Stain the control gel in Coomassie Blue staining solution for 1 h and destain until protein bands are clearly visible with a nearly clear background, or continue with western blotting.

3.3. Western Blotting

1. After separation by SDS-PAGE, the samples are transferred to nitrocellulose membranes electrophoretically. These directions assume the use of a semidry system provided by Bio-Rad (TransBlot™ Semi-Dry Transfer Cell, Bio-Rad, Munich, Germany). Two trays with transfer buffer Roti-Blot A and transfer buffer Roti-Blot K, respectively, are prepared with a size slightly bigger than the dimension of the gel.
2. Cut the nitrocellulose membrane and eight pieces of Whatman 3MM paper to the size of the separating gel. Cut one edge of the membrane for later orientation.
3. Four sheets of 3MM are moistened in transfer buffer Roti-Blot A and transferred to the anode plate of the blotting device. On top of this stack, the nitrocellulose membrane is added.
4. The gel unit is disconnected from the power supply and disassembled. The stacking gel is removed and discarded and one corner (corresponding to the membrane) is cut from the separating gel to allow its orientation to be tracked. The separating gel is then laid on top of the nitrocellulose membrane.
5. Another four sheets of 3MM paper are wetted in transfer buffer K and carefully laid on top of the gel, ensuring that no bubbles are trapped in the resulting sandwich. You can remove air bubbles by carefully rolling a glass pipette on top of the stack.
6. The lid is put on top of the stack and the power supply activated. Transfers can be accomplished at 0.8 mA/cm^2 blot size for 1 h.

7. Once the transfer is complete the stack is carefully disassembled. The 3MM paper and the gel can then be discarded. If you want, you can stain the gel in Coomassie Blue staining reagent for 1 h and destain thereafter in order to check for complete transfer of the separated proteins. On the nitrocellulose membrane the colored molecular weight markers should be clearly visible on the membrane.
8. Stain the membrane with Ponceau S staining solution for 5 min at room temperature (RT). Destain the blot with water until the protein band is clearly visible.
9. This protein band together with similarly prepared nine additional protein bands can then be used for the assembly of a stack of protein blots as schematically summarized in **Fig. 1**. This sophisticated assembly of the microblot is made commercially available from the company Attomol GmbH (Lipten, Germany). At present, up to ten different proteins have been used for the assembly of one microblot.

3.4. Detection of Autoantibodies Using Microblots

1. Equilibrate assay reagents and microtitre modules to room temperature (RT), vortex reagents and insert the desired modules into a microplate frame.
2. Prepare 1× washing buffer by adding 900 mL aqua dest. to 100 mL concentrated washing buffer. The ready-to-use buffer is stable for at least 2 weeks if stored at 4°C.
3. Warm sample buffer to 37°C for 15 min in order to solubilize all components. Mix thoroughly. After usage, store the remaining buffer at –20°C.
4. Thaw the needed sera and warm to RT, mix well. Dilute the sera 1:100 with sample buffer and mix well.
5. Fill each well with 100 μL diluted sera sample and cover the microtitre modules. Incubate for 60 min at RT on a rotary shaker.
6. Carefully remove the diluted sera and wash each well six times with 300 μL diluted wash buffer for 200 s at RT (*see* **Note 8**). Remove any remaining liquid by tapping the plate on a pile of filter paper.
7. During the washing steps prepare the HRP conjugate fresh by diluting one to 27 in ready-to-use wash buffer.
8. Fill each well with 100 μL diluted HRP conjugate. Incubate the covered microtitre modules for 60 min at RT on the rotary shaker.
9. Carefully remove the HRP conjugate and wash each well six times with 300 μL ready-to-use wash buffer for 200 s at RT. Remove any remaining liquid by tapping the plate on a pile of filter paper.

10. Fill each well with 100 μL substrate solution. Incubate the covered microtitre modules for 20 min at RT on the rotary shaker. *Protect from sunlight!*
11. Carefully remove the substrate solution and wash each well by applying 300 μL ready-to-use wash buffer. Incubate for 5 min at RT on the rotary shaker. Wash finally with 300 μL aqua dest. per well for 5 min on the rotary shaker. Remove any remaining liquid by tapping the plate on a pile of filter paper.
12. Dry microtitre modules for 2 h at RT. Clean the lower surface of all wells free of lint. Scan the microblots from the bottom of modules (within the microplate frame) by using the scanner (resolution 1,200 dpi). Analyze the immunoreactions on the enlarged scan image. An immunostained microblot is shown in **Fig. 1**. The control band displays a blue staining if the detection reagent does function and if the test was correctly carried out. Differentially stained antigen bands reflect the autoantibody content of the serum sample. The use of the automatic image processing algorithms of the Microblot Analyzer software facilitates the evaluation as it detects marker bands and measures signals in all antigen areas densitometrically. The data are displayed as negative, borderline, or positive results in relation to negative control sera defined cut-off values.

4. Notes

1. Unless stated otherwise, all solutions should be prepared in water that has a conductivity of 0.056 μS/cm and total organic content of less than five parts per billion. This standard is referred to as "water" in this text.
2. There are different expression systems commercially available, some of them for special purpose (e.g., for expression of sequences including rarely used codons in bacteria). You should evaluate the system which works best for your purpose.
3. The usage of the nitrocellulose membrane *porablot NCP*, not enforced from Machery Nagel (Düren, Germany) is mandatory, as the preparation of the microblots is adjusted to that membrane. Other membranes might not work with this system. *Especially the use of reinforced membrane must be avoided!*
4. If there is no automated washer available, use an eight channel pipette for washing instead.
5. Use a flask that has ten times the size of the used expression volume or e.g., Fernbach-Flasks as alternative.

6. There is no need for native purification, as the proteins are run on denaturing SDS-PAGE later on. Normally, the denaturing cleanup procedure is more effective. If purification under native condition is superior, one can use this procedure as well.
7. The settings one will need for this ultrasound step are extremely dependent on your specific device and have to be determined individually.
8. If possible use a commercial automated washer for all washing steps.

Acknowledgements

This work was supported in part by grants from the Medical Faculty of the TU-Dresden and the BMBF (Germany). We like to thank Martina Franke for her excellent technical support.

References

1. Tan, E. M. (1989) Antinuclear antibodies: diagnostic markers for autoimmune diseases and probes for cell biology. *Adv Immunol* 44, 93–151.
2. Cheever, M.A., Disis, M.L., Bernhard, H., Gralow, J.R., Hand, S.L., Huseby, E.S. et al. (1995) Immunity to oncogenic proteins. *Immunol Rev* 145, 33–59.
3. Tan, E.M. (2001) Autoantibodies as reporters identifying aberrant cellular mechanisms in tumorigenesis. *J Clin Invest* 108, 1411–1415.
4. Stockert, E., Jager, E., Chen, Y.T., Scanlan, M.J., Gout, I., Karbach, J. et al. (1998) A survey of the humoral immune response of cancer patients to a panel of human tumor antigens. *J Exp Med* 187, 1349–1354.
5. Soussi, T. (2000) p53 Antibodies in the sera of patients with various types of cancer: a review. *Cancer Res* 60, 1777–1788.
6. Forti, S., Scanlan, M.J., Invernizzi, A., Castiglioni, F., Pupa, S., Agresti, R., et al. (2002) Identification of breast cancer-restricted antigens by antibody screening of SKBR3 cDNA library using a preselected patient's serum. *Breast Cancer Res Treat* 73, 245–256.
7. Conrad K., and Bachmann, M. Tumor-associated autoantibodies. In: Autoantibodies, 2nd Ed (ed. by Y. Shoenfeld, M.E. Gershwin, P.L. Meroni), Elsevier, Amsterdam, 2006, pp. 423–436
8. Zhang, J.Y., Casiano, C.A., Peng, X.X., Koziol, J.A., Chan, E.K., and Tan, E.M. (2003) Enhancement of antibody detection in cancer using panel of recombinant tumor-associated antigens. *Cancer Epidemiol Biomarkers Prev* 12, 136–143.
9. Lehmann, W. (2001) Procedure for production of an array for detection of components of a biological sample (Patent DE 101 10 511).
10. Laemmli, U.K. (1970) Cleavage of structural proteins during the assembly of the head of bacteriophage T4. *Nature* 227, 680–685.

Chapter 16

Transfer and Multiplex Immunoblotting of a Paraffin Embedded Tissue

Joon-Yong Chung and Stephen M. Hewitt

Summary

In the functional proteome era, the proteomic profiling of clinicopathologic annotated tissues is an essential step for mining and evaluations of candidate biomarkers for disease. Previously, application of routine proteomic methodologies to clinical tissue specimens has provided unsatisfactory results. Multiplex tissue immunoblotting is a method of transferring proteins from a formalin-fixed, paraffin-embedded tissue section to a stack of membranes which can be applied to a conventional immunoblotting method. A single tissue section can be transferred to up to ten membranes, each of which is probed with antibodies and detected with fluorescent tags. By this approach, total protein and target signals can be simultaneously determined on each membrane; hence each antibody is internally normalized. Phosphorylation-specific antibodies as well as antibodies that do not readily work well with paraffin-embedded tissue are applicable to the membranes, expanding the menu of antibodies that can be utilized with formalin-fixed tissue. This novel platform can provide quantitative detection retaining histomorphologic detail in clinical samples and has great potential to facilitate discovery and development of new diagnostic assays and therapeutic agents.

Key words: Immunoassay, Histology, Proteomics, Expressional profiling

1. Introduction

Proteomic profiling of tissue specimens, having pathologic and histologic relevance, promises to the development of biomarkers to guide diagnosis and therapy in biomedicine *(1, 2)*. Many of the traditional approaches such as western blots are based on a "grind and bind" means of isolating proteins from tissue. This "grind and bind" approach fails to provide a histomorphologic perspective of protein expression. The only means of gaining a histomorphologic

Biji T. Kurien and R. Hal Scofield (eds.), *Methods in Molecular Biology, Protein Blotting and Detection, vol. 536*

DOI: 10.1007/978-1-59745-542-8_16

understanding of protein expression have been immunohistochemistry and laser capture microdissection-based collection of samples for traditional analysis. Laser capture microdissection does provide the capacity to perform a directed western blot on tissue *(3–5)*, however, it is time consuming and does not provide a global expression view of a target protein. Immunohistochemistry, although providing excellent localization, lacks quantification without sophisticated instrumentation *(6)* and lacks normalization in chromagenic applications. However, formalin-fixed and paraffin-embedded tissue, the gold standard of diagnostic histopathology, is not routinely applicable to the "grind and bind" approach due to the high level of covalently cross-linked proteins arising from formalin fixation. In translational research, there is the great desire to utilize the vast archive of formalin-fixed and paraffin-embedded tissue that has been collected *(7)*. Most research antibodies do not perform well in paraffin-embedded tissues. This failure is thought to be related to protein cross-linking, inadequate deparaffinization, and issues of epitope presentation. This problem has become a bottleneck in translational research *(2)*.

To address these challenges, a number of protein-based arrays have been developed and evaluated in clinical research fields. Although these techniques are generally superior in expression profiling and quantitation of protein changes associated with disease states, each still has significant limitations *(8–10)*. Investigators continue to seek a convenient and reliable proteomic tool, which can detect protein in tissue, while having retention of both quantitative and histomorphologic features. Multiplex tissue immunoblotting meets the criteria. This method provides a level of histomorphologic correlation, but at the same time, presents the proteins on a membrane platform that widens the number of antibodies that can be utilized. Additionally, the total amount of protein present on each membrane can be determined and used for normalization *(11–14)*. As a research tool, this method expands the capacity of a tissue microarray to a protein array with quantitative data that can be normalized and directly composed of different antigens detected on a single stack of membranes *(13–14)*. When applied to a whole section of tissue *(11, 12)*, it allows the ability to profile a tumor for multiple antigens with the use of single paraffin-embedded slide.

As an example of the utility of multiplex tissue immunoblotting, we have quantified the changes in seven proteins in the transition from normal epithelium to invasive tumor *(12)*. This approach allows the quantification of changes in the expression of potential biomarkers in normal, in situ and invasive disease. This approach provided insight into the timing and magnitude of protein changes seen in this transition of from benign to invasive cancer as is useful in the development of novel biomarkers for prevention and screening of cancer. Additionally, this approach can be applied to tissue microarrays, creating a form of protein array.

2. Materials

2.1. Deparaffinization and Enzyme Treatment

1. Xylene or dewaxing reagents.
2. 100%, 95%, and 70% ethanol (EtOH) (molecular biology grade).
3. 50 mM ammonium bicarbonate (NH_4HCO_3 buffer, pH 8.2). Store room temperature (RT).
4. Stock solution of trypsin (200×): Prepare 0.2% trypsin (Sigma, St. Louis, MO, USA) solution in 50 mM ammonium bicarbonate solution (pH 8.2) and immediately freeze in single use (200 μL) aliquots at −20°C.
5. Stock solution of Proteinase-K (400×): Ready-to-use Proteinase-K (DAKO, Glostrup, Denmark), freeze in single use (50 μL) aliquots at −20°C (*see* **Note 1**).
6. Enzyme cocktail solution (prepare freshly before use): 0.001% trypsin plus 1:400-fold diluted the ready-to-use Proteinase K in 50 mM ammonium bicarbonate solution, pH 8.2 (*see* **Note 2**).
7. ProBuffer: One tablet of complete protease inhibitor (Roche Diagnostics, Mannheim, Germany), 0.5 mL phosphatase inhibitor I (Sigma), 0.5 mL phosphatase inhibitor II (Sigma), in 50 mL PBS (pH 7.2). Store at 4°C.
8. Phosphate-buffered saline (PBS, pH 7.2).
9. Plastic or glass coplin jars for slide processing.
10. Incubation chamber for enzyme reaction (5-slides mailer box).

2.2. Transferring from the Tissue Slide to Membrane

1. Transfer buffer (10×): 250 mM Tris (do not adjust pH), 1,900 mM glycine. Store at RT. Dilute 200 mL with 800-mL distilled water for use.
2. A stack of membranes (5 or 10 sheets) (P-Film, 20/20 GeneSystems, Rockville, MD, USA) (*see* **Note 3**).
3. Spacer membrane (GE polycarbonate PVPF membrane, GE Osmonics Labstore, Minnetonka, MN, USA) (*see* **Note 4**).
4. Nitrocellulose membrane (PROTRAN, Schleicher & Schuell, Keene, NH, USA).
5. Absorbent pads (Blot Absorbent Filter paper, Bio-Rad, Hercules, CA, USA).
6. Slide glass (Opticlear® Microscope slide, Kimble, Toledo, OH, USA).
7. Kapak SealPAK pouches (KAPAK, Minneapolis, MN, USA).
8. Impulse sealer (American international electric INC, Whittier, CA, USA).
9. Heat block (Dry bath, Thermolyne, Dubuque, IA, USA).

2.3. Immunoblotting

1. Biotinylation solution (1 μg/mL, EZ-link Sulfo-NHS-Biotin, Pierce, Rockford, IL, USA): Prepare freshly solution before use in PBS.
2. Blot FastStain kit (Chemicon International, Temecula, CA, USA).
3. Tris-buffered saline (TBS, 10×): 1.5 M NaCl, 0.5 M Tris–HCl, pH 8. Store at RT.
4. Tris-buffered saline with Tween-20 (TBST): TBS plus 0.05% (w/v) Tween-20. Store at RT.
5. Primary and secondary antibody dilution buffer: TBST supplemented with 0.5% (w/v) fraction V bovine serum albumin (BSA).
6. Primary antibodies (*see* **Note 5**).
7. Streptavidin linked Cy5 (Amersham Biosciences, Uppsala, Sweden).
8. FITC conjugated anti-rabbit IgG or anti-mouse IgG (Molecular Probes, Eugene, OR, USA).

2.4. Image Scanning and Quantitation

1. Scan Array Express microarray Scanner (Perkin Elmer, Boston, MA, USA).
2. ImageQuant 5.2 software (GE Healthcare, Piscataway, NJ, USA).

3. Methods

This method begins with routinely processed formalin-fixed, paraffin-embedded tissue section on a regular glass slide and utilizing routine laboratory procedures for microtomy. Typically we can obtain five replicate membranes from a 5-μm-thick formalin-fixed, paraffin-embedded tissue section. After transfer is completed, the membranes can be probed using conventional western blotting or immunoblotting method. After transfer, the original slide can be stained with Hematoxylin and Eosin (H&E) for direct correlation of histopathology with the immunoblot results. The residual tissue on the section after transfer should be limited, and may be difficult to interpret. If complex diagnostic features are to be examined, an adjacent H&E section should be utilized. Spatial resolution is determined by the scanning methodology, as well as transfer conditions and abundance of the protein. "Acinar" resolution on the order of 100–200 μm is obtainable with optimized conditions.

3.1. Deparaffinization and Enzyme Treatment

1. Deparaffinize the formalin-fixed, paraffin-embedded tissue section in xylene (3 × 5 min) in glass or plastic coplin jars. Transfer to 100% EtOH (2 × 5 min), 95% EtOH (2 × 5 min), 70% EtOH (1 × 5 min), and then to PBS (2 × 5 min) (*see* **Note 6**).

2. Equilibrate the deparaffinized tissue slide for 5 min with 50 mM ammonium bicarbonate buffer (pH 8.2).
3. Prepare the enzyme cocktail solution by mixing 100 μL of trypsin stock solution, 50 μL of Proteinase-K stock solution, and 20 mL of 50 mM ammonium bicarbonate buffer (pH 8.2).
4. Place the enzyme cocktail solution in an incubation chamber. After the equilibration is completed, incubate the slide for 30 min at 37°C with enzyme cocktail solution (*see* **Note 7**).
5. Pour off the enzyme cocktail solution and rinse the surface of the tissue twice with PBS.
6. Place the slide on a flat surface and then apply immediately 2 mL of proBuffer to the slide for 15 min at RT.
7. Briefly wash the slide with transfer buffer. The slide is ready for transfer.

3.2. Transferring from the Tissue Slide to Membrane

1. After deparaffinization and enzyme treatment are completed, equilibrate the tissue slide in 1 mL of transfer buffer for 15 min.
2. During the equilibration of the tissue slide, prepare a spacer membrane, a stack of membranes (*see* **Note 8**), four absorbent pads, 3MM paper, and a nitrocellulose membrane trap (*see* **Note 9**). Equilibrate all membranes for 5 min in transfer buffer.
3. The pre-equilibrated spacer membrane is laid on the top of the tissue and subsequently the five-membrane stack is laid on the top of the spacer membrane. Gently roll over the sandwich using a disposable pipette to ensure that no air bubbles are present.
4. The pre-wetted nitrocellulose membrane trap is carefully laid on the top of the stack of the membrane, ensuring that no air bubbles are trapped in the resulting sandwich.
5. Place the pre-wetted 3MM paper and two absorbent pads on the top of the nitrocellulose membrane trap.
6. Two additional two absorbent pads are wetted in the transfer buffer and laid on the bottom of the tissue slide. Support the transfer assembly using two slide glasses as shown in **Fig. 1**.
7. Place the transfer assembly unit in a Kapak Seal PAK pouch and then heat seal by an impulse sealer.
8. After the assembly of the transfer unit is completed, the transfer unit is transferred under serial conditions for 1 h at 55°C, for 0.5 h at 65°C, and for 2 h at 80°C using heat block (*see* **Note 10**).

3.3. Immunoblotting

1. Once the transfer is completed, the seal-pak pouch is opened and the membrane stack carefully disassembled, with the top absorbent pads and 3MM paper removed.
2. Remove excess transfer buffer by washing the membranes (3 × 5 min) in PBS.
3. Stain the nitrocellulose and the spacer membranes using the Blot FastStain kit (*see* **Note 11**).

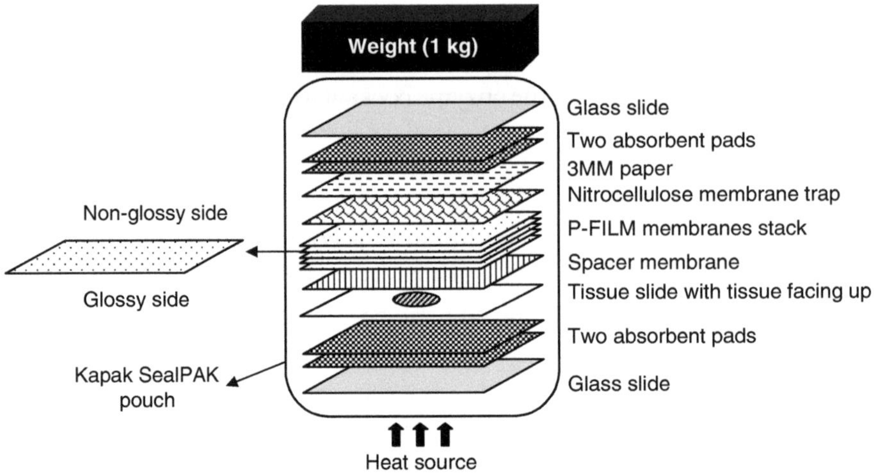

Fig. 1. A schematic diagram of heat-facilitated capillary transfer. Assemble the transfer unit carefully and then seal the assembled transfer unit using a Kapak SealPAK pouch. After the sealing, the transfer unit is moved to the heat block and then applied to a multiple-serial heating system (1 h for 55°C, 0.5 h for 65°C, and 1 h for 80°C). The protein leaves the tissue slide and is deposited onto the membrane stack. A weight applied to the top of the transfer assembly unit helps to ensure a tight connection between the layers of material used in the transfer system.

(a) Prepare 1:7-fold diluted working solutions of reagents A and B with distilled water.

(b) Incubate the nitrocellulose and spacer membranes in 7 mL of reagent A for 10 min, followed by incubation of the both membranes in reagent B for 10 min or until spot visualized.

(c) Move the staining container to 4°C and then let stand for 10–30 min.

4. The five-membrane stack is then incubated in 20-mL biotinylation solution for 10 min at RT on a rocking platform (*see* **Note 12**).
5. The biotinylation solution is discarded and then the membrane is washed three times for 5 min each with TBST.
6. Incubate each membrane by adding 0.5 mL of an appropriately diluted primary antibody in the antibody dilution buffer for overnight (16–18 h) at 4°C on a Kapak SealPAK pouch. Rocking platform is recommended (*see* **Note 13**).
7. The primary antibody is then removed and the membrane is washed three times for 5 min each with TBST.
8. Prepare the mixture of secondary antibodies by mixing 5 μL of streptavidin linked Cy5 and 5 μL of FITC conjugated anti-rabbit IgG or anti-mouse IgG, and 5 mL of the antibody dilution buffer.
9. Add the mixture of secondary antibodies to the membrane and then incubate for 30 min at RT on a rocking platform. The membrane should be protected form light until scan is acquired.

10. The secondary antibodies are discarded and the membrane is washed five times for 10 min each with PBST.

3.4. Image Scanning and Quantitation

1. Dry the membranes after the final wash between two sheets of 3MM paper.
2. Scan each membrane at appropriate wavelength with Scan Array Express Microarray Scanner. Examples of the signals for total protein and pan-cytokeratin are shown in **Fig. 2A** (*see* **Note 14**).
3. Analyze the scanned images using ImageQuant 5.2 software. Example of quantitation of scanned image is shown in **Fig. 2B**.

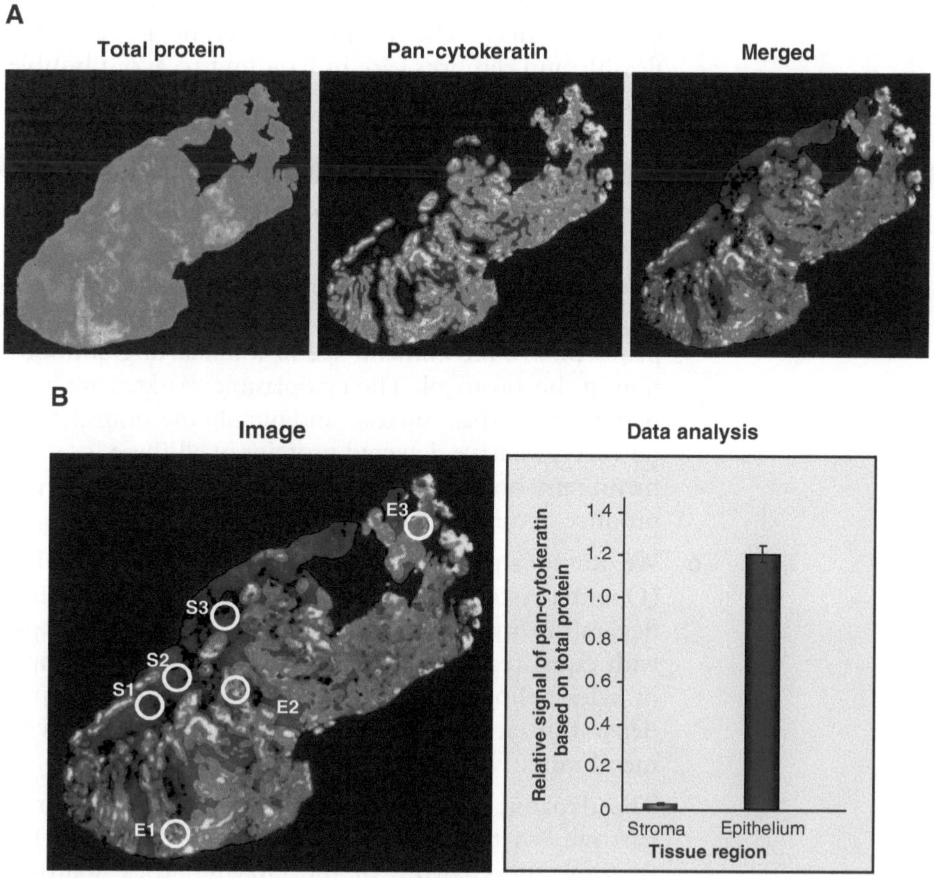

Fig. 2. Representative image and quantitation of whole tissue section by multiplex tissue immunoblotting. (**A**) Prior to immunoblotting, a formalin-fixed and paraffin-embedded tissue section of a gastrointestinal tumor (5 μm thickness) was transferred to the five-membrane stack by heat-facilitated capillary transfer system. The third membrane was incubated in biotinylation solution and then was reacted with anti-pan-cytokeratin antibodies (1:200-fold diluted). After primary antibody incubation is completed, total protein and specific target signals were detected using streptavidin linked Cy5 and FITC conjugated anti-mouse IgG (1:1,000-fold diluted). Membranes were imaged with a microarray scanner. Fluorescent scans are represented in *pseudocolor*, where signal intensity is *white–red–yellow–green–blue–black* from maximum to minimum signal. (**B**) We selected three different representative stroma and epithelium tissue regions based on the H&E slide. We subsequently quantified those areas using ImageQuant 5.2 software. After normalization with total protein level, relative expressional signals were represented as a ratio. The *bar graph* shows the average ± SD of three circle areas (*S* Stroma, *E* Epithelium) (*see Color Plates*).

4. Notes

1. We have found that Dako Proteinase-K is best for this method. Substitution with other Proteinase-K can diminish reproducibility of results.
2. This protocol can be adapted for ethanol-fixed, paraffin-embedded tissue sections. In that case, the enzyme solution should be changed to 0.001% trypsin only for 15 min at 37°C. Overall the final condition of enzyme digestion should be optimized depending on tissue type with a minor change of enzyme concentration and time.
3. When handling membranes, always wear gloves to prevent contamination. The P-Film membrane is very thin and flexible and requires care in handling to avoid bubbles and folds.
4. The spacer is an uncoated polycarbonate PVPF membrane (pore size: 0.4 μm) and is used for filtrations of inappropriately digested proteins and debris during heat-facilitated capillary transfer procedure.
5. The primary antibodies which are suggested for western blotting as well as immunohistochemistry are compatible in this platform. We recommend 1:200-fold starting antibody dilution in this protocol. The cytoplasmic markers are in general better targets than nuclear- and membrane-bound molecules; however, we have detected proteins in all these locations. For membrane-bound targets, thicker tissue sections (10 μm) may produce better results.
6. We found aqueous based *dewaxers* such as AutoDewaxer (Openbiosystems, Huntsville, AL, USA) could be used as a deparaffinizing agent in place of xylene at high temperature with equal results and greater safety *(15)*. The temperature of xylene should be controlled under 65°C. Inappropriate deparaffinization can result in poor protein transfer to the membrane.
7. The dynamic range of the enzyme condition is relatively narrow. For this reason we recommend avoiding repeated freezing and thawing of the enzyme stock solution. The stock solution can be stored up to 6 months in the freezer (−20°C).
8. There are two different features (glossy vs. nonglossy sides) on the membrane. Before pre-soaking the membrane, we marked membrane and case numbers in the margin of nonglossy side for further information such as number of membrane and case using a regular ballpoint pen. The glossy side of the membrane stack should face the tissue on the slide (**Fig. 1**).

9. This protocol is used for a regular tissue slide, and can be adapted for irregular slide size with appropriate membrane size. Cut all membrane and pads with size of approximately 2.2 × 4.5 cm. The cover of cover slip box can be used as a container for incubation chamber, to prevent excess buffer use.
10. The use of a multiserial temperature condition produces an even distribution of proteins across membranes compared to a single temperature, which resulted in uneven transfer or bubble spots. This procedure generated a linear decrease in protein concentration through the membrane stack, with a great correlation coefficient (R^2= 0.985) *(11)*.
11. This step is just a confirmation stage for protein transferred to the membrane stack. The staining kit is very sensitive but is a transient staining. If you want to keep original image you should scan the semi-wet membrane between two transparent films and save the image.
12. Do *not* use a plastic petri dish coated for cell culture. A 20 mL of biotinylation solution can be covered up to 20 membranes (2.2 × 4.5 cm).
13. It is not necessary to block the membranes. The carrier protein (BSA) in the antibody dilution is sufficient to inhibit nonspecific binding. Do not use dry milk as a carrier protein.
14. Excitation at 633 nm induces the Cy5 fluorescence (red emission) for total protein, while excitation at 488 nm induces FITC fluorescence (blue emission). This fluorescence labeling system can be adapted for user purpose.

Acknowledgements

The authors wish to thank Michael R. Emmert-Buck for his advice and insights into applications for this platform.

References

1. Chuaqui, R. F., Bonner, R. F., Best, C. J., Gillespie, J. W., Flaig, M. J., Hewitt, S. M. et al. (2002) Post-analysis follow-up and validation of microarray experiments. *Nat Genet* 32, 509–514.
2. Hewitt, S. M., Dear, J., and Star, R. A. (2004) Discovery of protein biomarkers for renal diseases. *J Am Soc Nephrol* 15, 1677–1689.
3. Simone, N. L., Paweletz, C. P., Charboneau, L., Petricoin, E. F., and Liotta, L. A. (2000) Laser capture microdissection: beyond functional genomics to proteomics. *Mol Diagn* 5, 301–307.
4. Ornstein, D. K., Englert, C., Gillespie, J. W., Paweletz, C. P., Linehan, W. M., Emmert-Buck, M. R., et al. (2000) Characterization of intracellular prostate-specific antigen from laser capture microdissected genign and malignant prostatic epithelium. *Clin Cancer Res* 6, 353–356.
5. Craven, R. A., Totty, N., Harnden, P., Selby, P. J., and Banks, R. E. (2002) Laser capture microdissection and two-dimensional polyacrylamide gel electrophoresis: evaluation of tissue preparation and sample limitations. *Am J Pathol* 160, 815–822.

6. Camp, R. L., Chung, G. G., and Rimm, D. L. (2002) Automated subcellular localization and quantification of protein expression in tissue microarrays. *Nat Med* 11, 1323–1327.
7. Liotta, L. A., and Petricoin, E. (2000) Molecular profiling of human cancer. *Nat Rev Genet* 1, 48–56.
8. Miyaji, T., Hewitt, S. M., Liotta, L. A., and Star, R. A. (2002) Frozen protein arrays: a new method for arraying and detecting recombinant native tissue proteins. *Proteomics* 2, 1489–1493.
9. Lopez, M. F. (2000) Better approaches to finding the needle in a haystack: optimizing proteome analysis through automation. *Electrophoresis* 21, 1082–1093.
10. Molloy, M. P., Phadke, N. D., Maddock, J. R., and Andrews, P. C. (2001) Two-dimensional electrophoresis and peptide mass fingerprinting of bacterial outer membrane proteins. *Electrophoresis* 22, 1686–1696.
11. Chung, J. Y., Braunschweig, T., Baibakov, G., Galperin, M., Ramesh, A., Skacel, M., et al. (2006) Transfer and multiplex immunoblotting of a paraffin embedded tissue. *Proteomics* 6, 767–774.
12. Chung, J. Y., Braunschweig, T., Hu, N., Roth, M., Traicoff, J. L., Wang, Q. H., et al. (2006) Profiling of biomarkers in the normal to tumor transition zone of esophageal squamous cell carcinoma by multiplex tissue immunoblotting. *Cancer Epidemiol Biomarkers Prev* 15, 1403–1408.
13. Chung, J. Y., Braunschweig, T., Tuttle, K., and Hewitt, S. M. (2007) Tissue microarrays as a platform for proteomic investigation. *J Mol Histol* 38, 123–128.
14. Traicoff, J. L., Chung, J. Y., Braunschweig, T., Mazo, I., Shu, Y., Ramesh, A., et al. (2007) Expression of EIF3-p48/INT6, TID1 and Patched in cancer, a profiling of multiple tumor types and correlation of expression. *J BiomedSci* 14, 395–405.
15. Chung, J. Y., Braunschweig, T., and Hewitt, S. M. (2006) Optimization of recovery of RNA from formalin-fixed, paraffin-embedded tissue. *Diagn Mol Pathol* 15, 229–236.

Chapter 17

Multistrip Western Blotting to Increase Quantitative Data Output

Anatoly Kiyatkin and Edita Aksamitiene

Summary

The qualitative and quantitative measurements of protein abundance and modification states are essential in understanding their functions in diverse cellular processes. Typical western blotting, though sensitive, is prone to produce substantial errors and is not readily adapted to high-throughput technologies. Multistrip western blotting is a modified immunoblotting procedure based on simultaneous electrophoretic transfer of proteins from multiple strips of polyacrylamide gels to a single membrane sheet. In comparison with the conventional technique, Multistrip western blotting increases the data output per single blotting cycle up to tenfold, allows concurrent monitoring of up to nine different proteins from the same loading of the sample, and substantially improves the data accuracy by reducing immunoblotting-derived signal errors. This approach enables statistically reliable comparison of different or repeated sets of data, and therefore is beneficial to apply in biomedical diagnostics, systems biology, and cell signaling research.

Key words: Western blotting, Electrophoretic transfer, Gel cutting, Quantitative protein analysis, High-throughput, Blotting errors, Systems biology, Cell signaling

1. Introduction

Qualitative measurement of protein abundance is one of the common tasks in biomedical diagnostics in the search for therapeutic targets and biomarkers of various diseases and disorders *(1–8)*. Quantitative analysis of protein phosphorylation states, recruitment to the specific subcellular compartments, and interaction with other proteins is a paramount goal in systems biology, which explores, predicts, and explains how signaling networks govern cellular behavior by exploiting experimental data-driven mathematical models. To this aim, the cellular response to external stimuli

Biji T. Kurien and R. Hal Scofield (eds.), *Methods in Molecular Biology, Protein Blotting and Detection, vol. 536*

DOI: 10.1007/978-1-59745-542-8_17

in vivo is usually compared to the response obtained under one or more perturbations (e.g., pharmacological inhibitors, exposure to physiochemical stresses or the down- or up-regulation of protein expression, etc). In addition, the variations in the concentration of a ligand and stimulation time provide deeper insight into the spatiotemporal functioning of a specific cell signaling pathway *(9)*. Apparently, these tasks require the generation of large amounts of high-quality data points.

Western blotting used for the immunodetection of the expression levels and/or the modification status of electrophoretically resolved proteins, is a sensitive and widespread technique, which however has several drawbacks *(1, 10)*. It is low-throughput, time-consuming, and sample layout-unfriendly multistep procedure. Each step (sampling, loading, electrophoretical separation and transfer of proteins, immunoblotting, and signal detection) is performed under slightly differing conditions in sequential blotting cycles. This eventually increases data variability, which makes difficult to quantitatively compare the signals obtained from different series of samples *(11)*. Therefore, the improvements of typical western blotting procedure are in great request *(12–18)*.

Multistrip western blotting is a modified immunoblotting procedure based on simultaneous electrophoretic transfer of proteins from multiple strips of polyacrylamide gels to a single membrane sheet. The proposed modification has several advantages over a classic western blotting procedure *(19)*. First, the transfer and the sequential procedures with the blot such as membrane washing, incubation with antibodies, and protein detection are performed under similar conditions. This significantly improves the data accuracy by reducing immunoblotting-derived signal errors. Second, instead of detection of a single protein, many additional proteins of interest that differ in their molecular weight (e.g., ErbB family, GAB1/2, Raf, Shc, ERK1/2, GAPDH, and Grb2) can be synchronously detected in the same sample *per* blotting cycle. Such approach also eliminates the need of stripping and reprobing blots for the detection of house-keeping proteins such as actin, tubulin, GAPDH, etc. The analytical power of western blotting is increased, because one-step analysis of numerous signaling proteins is more productive, saves time as well as costly materials. Third, when the number of samples to be analyzed exceeds the number of wells in a gel, the comparative quantitative protein analysis can be readily achieved by applying Multistrip western blotting technique, which increases the data output *per* single blotting cycle up to tenfold. As a consequence, a large number of data points can be integrated and analyzed on the same graph.

Although the protocol presented here is developed for NuPAGE 4–12% gradient Bis-Tris 10-well mini-gels using the XCell devices (Invitrogen), Multistrip western blotting has wide potential for further uses since it does not require any additional tools and is compatible with any conventional gel electrophoresis as well as protein transfer systems.

2. Materials

2.1. Sample Preparation

1. Sample buffer: 4× NuPAGE LDS Sample Preparation Buffer (pH 8.4) and 10× NuPAGE Sample Reducing Agent (both from Invitrogen, Carlsbad, CA, USA) (*see* **Note 1**).

2.2. SDS-Polyacrylamide Gel Electrophoresis (SDS-PAGE)

1. XCell SureLock Mini-Cell units (Invitrogen) (*see* **Note 2**).
2. NuPAGE Novex 4–12% gradient Bis-Tris Mini-gels (Invitrogen) (*see* **Note 3**).
3. Running buffer: 50 mM 3-(*N*-morpholino)-propanesulfonic acid (MOPS), 50 mM Tris, 0.1% (w/v) sodium-dodecyl sulphate (SDS), 1 mM ethylene-diamine-tetraacetic acid (EDTA, pH 7.7) (Invitrogen and Boston Bioproducts, Worcester, MA, USA). Store at room temperature (RT). Supplement the running buffer in the upper chamber of XCell SureLock Mini-Cell with 0.5 mL NuPAGE Antioxidant (Invitrogen) before electrophoresis (*see* **Note 1**).
4. Prestained molecular weight markers: Precision Plus Protein standards (Bio-Rad, Hercules, CA, USA). Store at –20°C.

2.3. Western Blotting

1. XCell II Blot module (Invitrogen) (*see* **Note 2**).
2. Setup buffer: 20 mM Tris, 154 mM glycine, 0.02% SDS, 20% (v/v) methanol.
3. Transfer buffer: 25 mM bicine, 25 mM Bis-Tris, 1.0 mM EDTA, 0.05 mM chlorobutanol, pH 7.2 (available from Invitrogen), 10% (v/v) methanol. Store at 4°C. Supplement with 0.1% (v/v) NuPAGE antioxidant (Invitrogen) in the transfer apparatus before electrophoretic transfer (*see* **Note 1**).
4. Nitrocellulose membrane from Bio-Rad (*see* **Note 4**).
5. Blotting filter paper 2.45 mm thickness, 320 grade is available from E&K Scientific (Santa Clara, CA, USA) or Colonial Scientific (Richmond, VA, USA).
6. Tris-buffered saline with Triton X-100 (TBST) solution (1×): 10 mM Tris–HCl (pH 8), 150 mM NaCl, 0.5% (v/v) Triton X-100. Store at RT.
7. Blocking buffer: Dissolve 4% (w/v) heat inactivated bovine serum albumin (BSA) (Roche Diagnostics, Indianapolis, IN, USA) in TBST.
8. Primary and secondary antibodies diluted in TBST.

9. Chemiluminescent reagent: SuperSignal West Dura Extended Duration Substrate (Pierce Biotechnology, Rockford, IL, USA).
10. Square dishes with grid (Fisher Scientific).
11. Imaging system: Image Station 440CF (Eastman Kodak Scientific Imaging Systems, New Haven, CT, USA).

3. Methods

3.1. Preparation of Samples for Multistrip Western Blotting

1. Immediately after cell lysis, the supernatant of each cell lysate is mixed with 4× NuPAGE LDS sample buffer and supplemented with 10× NuPAGE Sample Reducing Agent in a ratio of 65:25:10 in labeled Eppendorf tubes. The tubes with samples are then heated for 5 min at 75°C. After cooling to RT, they are ready for separation by SDS-PAGE or can be stored for further use at –80°C.

3.2. Sample Loading and SDS-PAGE

These instructions assume the use of XCell SureLock Mini-Cell apparatus for SDS-PAGE of 10-well mini-gels (*see* **Note 3**). The number of gels to be loaded depends on the number of data series and the number of samples within each series to be analyzed. For example, when the time-course (e.g., 0, 1, 3, 5, 7, 10, 20, 30, and 60 min) of protein X activation in control cells (A) is compared to that in the presence of first perturbation (e.g., inhibitor of protein X) (B) and the second perturbation (e.g., the suppression of protein Y by siRNA) (C), one will have to load three data series (A, B, and C) consisting of nine time-points each (A1, A2,..., A9, etc.) into three gels. There are two alternative ways of loading such array of samples (**Fig. 1**).

1. The upper chamber is filled with 200 mL of running buffer to completely cover the sample wells of a gel. 600 mL of running buffer is poured into the lower chamber.
2. A pipette equipped with prolonged gel loading tip (Fisher) is used to underlay 7 μL of prestained protein molecular weight marker (M) into the first and/or the last gel well.
3. Equal volume of each sample (e.g., 20 μL) (*see* **Note 5**) is loaded into the rest of gel wells. If there are empty wells without loaded sample left, fill them with similar amount of sample buffer. Mark the sequence of loaded samples in a laboratory notebook.
4. Lock the gel tension lever. The electrophoresis unit is then completely assembled by adding the lid on the buffer core, and connected to a power supply. The proteins are electrophoretically separated at 140 V, until the blue dye front

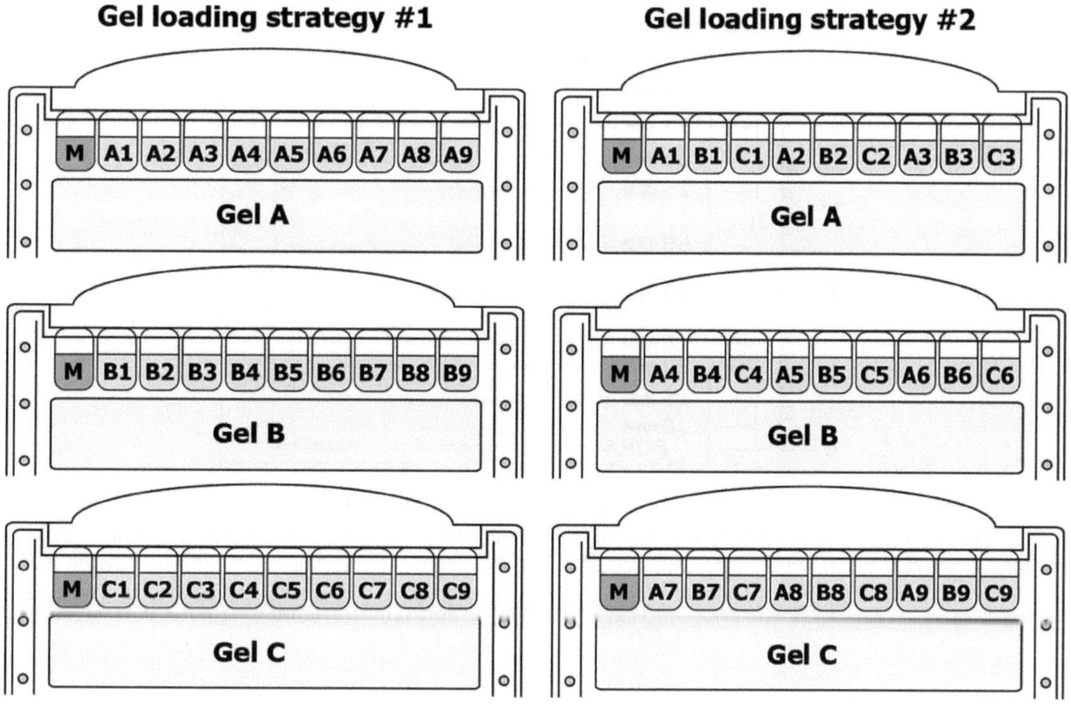

Fig. 1. Gel loading strategies when the number of samples exceeds the number of gel wells.

reaches the bottom of a gel. If running more than two gels, make an interval of 5 min before loading the next tandem of gels and powering on the electrophoresis unit. This will reserve enough time for follow-up steps.

5. At the end of gel run, a gel cassette is removed out of apparatus and gently opened with a gel knife. Note that upon opening the cassette, the gel can be adhered on either side. If the gel remains on a notched side, the sequence of sampling should be rewritten in the laboratory notebook in a reversed order. Discard the plate of gel cassette without the gel.
6. Perform **steps 1–5** with gel cassettes from other electrophoresis units.

3.3. Gel Cutting

1. **Figure 2** illustrates the plate with attached gel after protein separation according to their molecular weight by SDS-PAGE. The prestained marker is visibly separated into the bands corresponding to protein molecular weights of 250, 150, 100, 75, 50, 37, 25, 20, 15, and 10 kDa (**Fig. 2**, *M*).
2. A millimeter-scaled transparent ruler is firmly positioned near the edge lane with separated marker so that zero (0 cm) aligns with the middle of the blue dye front (**Fig. 2**, *BDF*). The distance from the blue dye front to the center of each marker band (**Fig. 2**, *H*) is measured in millimeters and registered in a table (*see* **Note 6**).

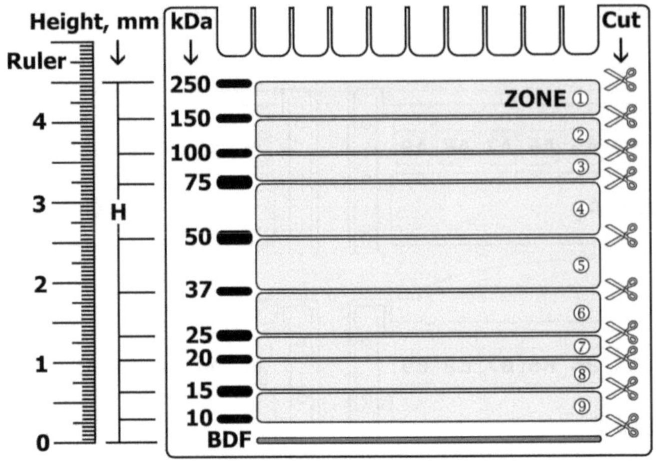

Fig. 2. The identification and cutting of protein migration zones in Multistrip western blotting procedure. *M*, prestained protein molecular weight marker; *BDF*, blue dye front; *H*, the distance from BDF to the center of particular marker band. *Scissor symbols* indicate the cutting lines, which would separate the entire gel into nine strips. Protein migration zones are enumerated by *numbers in circles* (modified from **ref.** *19* with permission from WILEY-VCH Verlag GmbH & Co. KG aA).

3. The distance between two electrophoretically separated marker bands corresponds to the migration range of certain molecular weight proteins. For instance, the distance between H_{250} and H_{150} defines a migration range of electrophoretically separated proteins with molecular sizes between 150 and 250 kDa. This range is termed zone *(1)* in **Fig. 2**. Each sample provides up to nine protein-containing zones that may be simultaneously cut out from a single gel.
4. A regular gel knife is used to cut out the strip, which covers an area with a protein of interest located in the middle, from the gel across its entire width (**Fig. 2**, *scissors symbol*).
5. The number of strips to be cut out from the gel depends on the number of distinct proteins to be detected. Most frequently studied signal transduction proteins migrating in various zones are listed in **Table 1**, which also indicates the appropriate areas that can be cut out of the gel in order to detect these proteins later. For example, it is convenient to separate the activated EGF receptor (165 kDa, zone (1)), from the phosphorylated PLD1 (116 kDa) and the phosphorylated 90 kDa ribosomal S6 kinase (RSK) that migrate in the zones (2) and (3), respectively. Then it is easy to separate the phosphorylated Akt (60 kDa), which is found in the zone (4), from both the activated ERK1 (44 kDa) and ERK2 (42 kDa) kinases that co-migrate in the zone (5), and from the phosphorylated S6 Ribosomal Protein (32 kDa, zone (6)). In this case, according to the **Table 1**, the gel is cut into six strips at 12 mm and 22 mm (for phospho-S6 Ribosomal Protein), at 29 mm (for ERK1/2), at 38 mm (for

Table 1
Migration of various signal transduction proteins in Novex 4–12% gradient NuPAGE mini-gel and their cutting areas, based on the migration of prestained Precision Plus Protein marker (modified from ref. *19* with permission from WILEY-VCH Verlag GmbH & Co. KG aA)

Zone	Protein MW (kDa)	Cut area, H (mm from BDF)	Migration range of signal transduction proteins in a gel			
1	150 – 250	49 ± 1 – 54 ± 1	EGFR, ErbBs, PDGFR, CSFR	c-Met, PLCs, FGFR, c-Kit, SHIP, JAK/Tyk2	SOS, IRS1/4, Filamin	VEGFR, c-Ret, ASK1, RhoGAP
2	100 – 150	44 ± 1 – 49 ± 1	RasGAP, FAK, c-Cbl, Cas, PLD1, PTPα, Vinculin		PI3K-p110, Gab1/2, Pyk2, Vav, STATs, MLK3, Catenin	Cadherin, cPLA2, PKD, eNOS, ERK5
3	75 – 100	38 ± 1 – 44 ± 1	PI3K-p85, FRS-2, GRK, APS	IR, PKCs, RSK, IKKα, IGF-1R		Raf, Grb10, PAK, p70S6K, FKHR, Calpain, Gab3
4	50 – 75	29 ± 1 – 38 ± 1	SHP1/2, Src family, PTEN, Akt, Csk, Grb14, Grb7	SHC, α-Tubulin, SAPK/JNK, SGK, ILK1, PP2A, CaMKII, GSK-3, PTP1B, β-Arrestin	Dok-R, Paxillin, p53, PDK1, Sam68, c-Fos, AMPKα/γ	
5	37 – 50	22 ± 1 – 29 ± 1	Crk, ERK, MEK, Nck, CREB, AMPKβ		β-Actin, PKA, MAPKAPK2, c-Jun, MKKs	Sprouty, p38 MAPK, GAPDH, LAT
6	25 - 37	12 ± 1 – 22 ± 1	14-3-3, Bic	Bcl-2, PP1, S6RP	Bak, GRB2	
7	20 - 25	9 ± 1 – 12 ± 1	Bad, Rac1/cdc42	Caveolin-1, Ras, DAP1, RKIP		
8	15 - 20	5 – 9 ± 1	Survivin, Bmf		TCL1	BID
9	10 - 15	2 – 5				

phospho-Akt), at 44 mm (for phospho-RSK), at 49 mm (for phospho-PLD1), and at 54 mm (for phospho-EGFR) from the BDF (*see* **Note 7**).

6. The gel pieces outside the strips are discarded.
7. The first plate with prepared multiple gel strips is covered with a sheet of moistened filter paper [covering filter paper (CFP)] and placed on the bench top. Similarly, the second gel is cut, covered with another sheet of moistened CFP and placed next to the previously laid plate on the bench top. Repeat above procedure with the rest of the gels.

3.4. Assembly of Gel Strips

During this step, the gel strips that are derived from different gels, are assembled onto a single sheet of filter paper [assembling filter paper (AFP)] for the subsequent electrophoretic protein transfer onto the same piece of nitrocellulose membrane (*see* **Note 8**). The strategy of assembly depends on the quantity of gels used as well on the number of analyzable proteins *per* lane (i.e., the number of precut gel strips comprising of appropriate zones). Here we provide two exemplar cases of gel strip assembly:

(a) When six gels (A, B, C, D, E, and F) are run and five proteins of interest from each sample (e.g., phospho-EGFR from zone (1), phospho-GAB1 migrating between zones (2) and (3), phospho-SHP2 from zone (4), phospho-ERK1/2 from zone (5) and Grb2 as house-keeping protein, which migrates between zones (6) and (7)) are subsequently detected under equal conditions. Guidance for cutting of one out of six gels is provided in left panel of **Fig. 3**

(b) When three gels (A, B, and C) are run and four proteins from each sample (phospho-IRS1 from zone (1), phospho-IR from

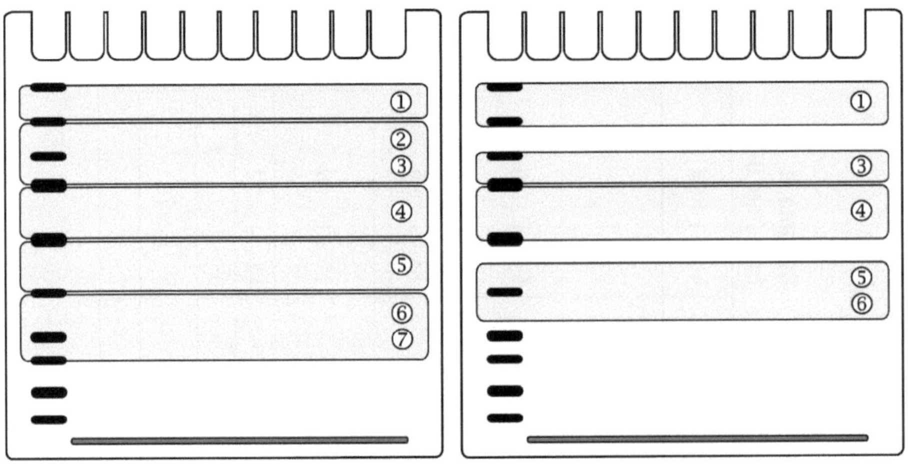

Fig. 3. Example of single gel cutting into five (*left panel*) or four (*right panel*) strips containing distinct protein zones. The strips will be subsequently transferred onto assembling filter papers together with similar strips derived from other gels.

zone (3), phospho-Akt from zone (3) and GAPDH as housekeeping protein, which migrates between zones (5) and (6)) are analyzed. Guidance for cutting of one out of three gels is provided in right panel of **Fig. 3**

1. In case (a), the plate containing gel A strips is flipped and gently lifted so that all strips would stick to the moistened CFP. Use gel knife if the strips do not independently detach from the plate.
2. The first gel strip from the top (possessing zone (1) proteins) is lifted with a gloved hand and carefully transferred onto the AFP #1. The CFP with remaining gel strips is returned onto the plate by flipping it back.
3. Similarly, top gel strips derived from gels B to F are sequentially transferred onto the AFP #1 so that the strips would lay side by side and parallel to each other. AFP #1 is now ready for immediate protein transfer (*see* **Note 9**).
4. **Steps 1–3** are repeated with the strips derived from gels A to F that possess the proteins migrating in zones (2), (3) then (4), (5) and finally (6), (7). This procedure will yield five AFPs (AFP #1 through #5) with collected six gel strips on each (**Fig. 4A**). Now they are ready for electrophoretic protein transfer onto the same membrane.
5. In case (b), **steps 1–2** are performed with gel A, followed by sequential transfer of gel strips derived from gel B and gel C onto the AFP #1. Then, gel strips that contain the proteins migrating in zone (3) are sequentially placed onto the AFP #1 below previously laid triplet of strips as shown in left panel of Fig. 4B. Leave a small gap between triplets.
6. AFP #2 is processed in the same manner so that it would contain triplet of strips with proteins migrating in zone (4) and triplet of strips with zones (5) and (6) (**Fig. 4B**, right panel). After protein transfer, the resulting nitrocellulose membrane is cut into two pieces across the gap between triplets. The pieces are then treated in separate dishes (*see* **Note 10**).

3.5. Western Blotting

Instructions provided below assume the use of XCell II Blot module that is used for protein transfer from one AFP:

1. One side of assembly tray is filled with 500 mL of setup buffer, while another side – with 400 mL of transfer buffer.
2. Four sponge pads are presoaked in setup buffer. Remove air bubbles by squeezing the pads while they are submerged in buffer. A nitrocellulose membrane is cut to the dimensions of AFP and presoaked in transfer buffer for 5 min before using. Three additional sheets of filter paper are briefly moistened in setup buffer immediately before using.

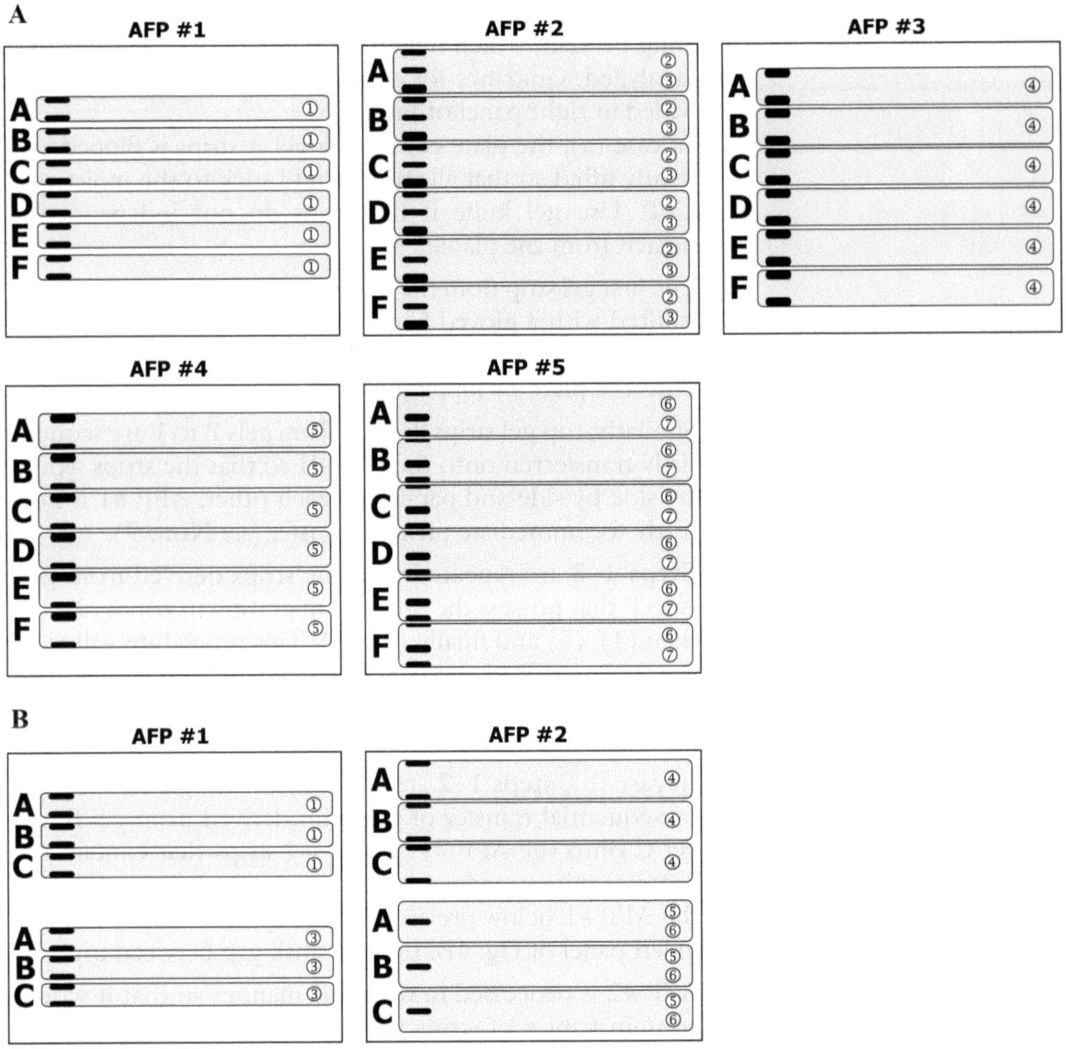

Fig. 4. Assembly of multiple gel strips onto assembling filter paper. (**A**) SDS-PAGE of six gels (A through F) was performed. Five strips were cut out of each gel and combined onto appropriate assembling filter papers (AFP, #1 through #5). (**B**) SDS-PAGE of three gels (A through C) was performed. Four strips were cut out of each gel and combined onto two assembling filter papers (AFP #1 and #2).

3. Two soaked sponge pads are placed into the cathode (−) core of the blot module and covered with one sheet of moistened filter paper. The AFP with collected gel strips is placed on the top. Subsequently, the surface of gel strips is covered with a sheet of nitrocellulose membrane. Remove any trapped air bubbles by rolling a blotting roller over the membrane surface. Two moistened filters are then placed onto the surface of the membrane followed by tandem of soaked sponge pads (*see* **Note 11**).
4. The anode (+) core is placed on the top of the pads. Slide the blot module into the rails on the lower chamber. Lock the gel tension lever.

5. The blot module is filled with transfer buffer until the blotting sandwich is completely submerged. The outer chamber is filled with cold deionized water.
6. The unit is completely assembled by adding the lid on the buffer core, and connected to a power supply. The proteins are electrophoretically transferred at 30 V constant for 90 min.
7. After transfer is stopped, the nitrocellulose membrane is removed out of the blot module and placed into a square Petri dish. Used filter papers and gel strips are discarded.
8. After 3 min rinsing with deionized water, the membrane is incubated with 20 mL of blocking buffer for 1 h at RT on a rotating platform.
9. After the blocking buffer is discarded, the membrane is briefly rinsed with deionized water and blotted with appropriate primary antibody at dilution ratio as recommended by a manufacturer overnight at 4°C on a rotating platform (*see* **Note 12**).
10. The membrane is extensively rinsed with deionized water and washed five times for 7 min each with TBST buffer at RT on a rotating platform.
11. The membrane is incubated with appropriate secondary antibody at dilution ratio as recommended by a manufacturer for 1 h at RT on a rotating platform followed by **step 10** once again.
12. The membrane is incubated with a working solution of chemiluminescent reagent for 5 min and the signal is captured by Imaging system and quantified using KODAK Digital Science software (*see* **Note 13**). To enable side-by-side comparison, the capture time and number of frames should be equal for each separately exposed membrane.

4. Notes

1. Laemmli instead of LDS sample buffer can be used with appropriate running and transfer buffers. Electrophoresis can be performed under reducing as well as nonreducing conditions.
2. Choose the type of apparatus for electrophoresis and protein transfer according to the size of your gels (e.g., mini-, midi-, or maxi-gels).
3. This protocol can be adapted for gels of any percentage, composition, size, and with any number of wells.
4. If desired, PVDF or nylon membranes can be also used.
5. The samples to be loaded can be different or repetitive.

6. The table is designed to track the statistics of marker migration in the gel of selected percentage. In addition, different tables can be created according to the marker type used. The statistics is required for successive Multistrip western blotting procedures if one needs to cut out the gel strip with the protein of interest (with known molecular weight), but no prestained marker has been loaded.
7. If a protein of interest migrates in an intersection of marker-defined zones (e.g., 100 kDa GAB1, which migrates between zones (2) and (3); 25 kDa Grb2, which migrates between zones (6) and (7); 74 kDa c-Raf, which migrates very close to the zone (3); etc.), the cutting area must include both zones or at least should be wider. For example, Shc protein has three isoforms of 46 kDa, 52 kDa, and 66 kDa migrating in the zones (4) and (5), so it can be separated from the proteins that lay in the zone (3) (e.g., p85, a regulatory subunit of PI3K) and between zones (6) and (7) (e.g., GRB2) by cutting the gel into three strips at 9, 22, 38, and 49 mm (*see* **Table 1**).
8. The maximal number of gel strips that can be combined onto a single AFP depends on the overall dimension of the transfer unit, hence on the size (height and width) of AFP. Routinely we use 7 × 10 cm filter, which provides space for maximum of 12 gel strips of 0.6 cm height each. However, regularly we place fewer amounts of gel strips (e.g., six), especially when they are wider and/or the membrane should be cut into two or more pieces after electrophoretic protein transfer.
9. If some pauses occur, regularly wet the surface of gel strips by dropping deionized water.
10. Alternatively, the whole piece of nitrocellulose membrane can be treated with blocking reagent and then incubated with the mixture of primary antibodies (be sure that they do not cross-react) in a single dish.
11. The pads should rise at least 0.5 cm over the rim of the cathode core. If not, place an additional filter paper or a sponge pad in the tank.
12. The primary antibodies can be collected into the tube and reused several times if supplemented with 0.1% sodium azide. If precipitation occurs, filter the solution through 0.22-μm filter unit.
13. The chemiluminescent signal can be visualized by another imaging instrument and quantified using an appropriate software. Alternatively, the signal can be captured on the film followed by densitometric quantification.

Acknowledgements

The authors gratefully acknowledge Drs. Boris N Kholodenko and Jan B Hoek for support. This work was supported by National Institutes of Health Grants GM59570 and AA0125311.

References

1. Soundy, P., Harvey, B. Western blotting as a diagnostic method. In: Walker JM, Rapley R, eds. Medical Biomethods Handbook. Totowa, NJ: Humana Press; 2005:43–62.
2. Omenn, G.S. (2006) Strategies for plasma proteomic profiling of cancers. *Proteomics* 6, 5662–5673.
3. Hueber, W., and Robinson, W.H. (2006) Proteomic biomarkers for autoimmune disease. *Proteomics* 6, 4100–4105.
4. Fardilha, M., Wu, W., Sa, R., et al. (2004) Alternatively spliced protein variants as potential therapeutic targets for male infertility and contraception. *Ann N Y Acad Sci* 1030, 468–478.
5. Ducruet, A.P., Vogt, A., Wipf, P., and Lazo, J.S. (2005) Dual specificity protein phosphatases: therapeutic targets for cancer and Alzheimer's disease. *Annu Rev Pharmacol Toxicol* 45, 725–750.
6. Gaiger, A., Heintel, D., and Jager, U. (2004) Novel molecular diagnostic and therapeutic targets in chronic lymphocytic leukaemia. *Eur J Clin Invest* 34(Suppl 2), 25–30.
7. Sawyer, T.K. (2004) Cancer metastasis therapeutic targets and drug discovery: emerging small-molecule protein kinase inhibitors. *Expert Opin Invest Drugs* 13, 1–19.
8. Krause, D.S., and Van Etten, R.A. (2005) Tyrosine kinases as targets for cancer therapy. *N Engl J Med* 353, 172–187.
9. Kholodenko, B.N. (2006) Cell-signalling dynamics in time and space. *Nat Rev Mol Cell Biol* 7, 165–176.
10. Kurien, B.T., and Scofield, R.H. (2006) Western blotting. *Methods* 38, 283–293.
11. Schilling, M., Maiwald, T., Bohl, S., et al. (2005) Quantitative data generation for systems biology: the impact of randomisation, calibrators and normalisers. *Syst Biol (Stevenage)* 152, 193–200.
12. Bergendahl, V., Glaser, B.T., and Burgess, R.R. (2003) A fast Western blot procedure improved for quantitative analysis by direct fluorescence labeling of primary antibodies. *J Immunol Methods* 277, 117–125.
13. Bolt, M.W., and Mahoney, P.A. (1997) High-efficiency blotting of proteins of diverse sizes following sodium dodecyl sulfate-polyacrylamide gel electrophoresis. *Anal Biochem* 247, 185–192.
14. Kashino, Y., Koike, H., and Satoh, K. (2001) An improved sodium dodecyl sulfate-polyacrylamide gel electrophoresis system for the analysis of membrane protein complexes. *Electrophoresis* 22, 1004–1007.
15. Kurien, B.T., and Scofield, R.H. (2002) Heat-mediated, ultra-rapid electrophoretic transfer of high and low molecular weight proteins to nitrocellulose membranes. *J Immunol Methods* 266, 127–133.
16. Swank, M.W., Kumar, V., Zhao, J., and Wu, G.Y. (2006) A novel method of loading samples onto mini-gels for SDS-PAGE: Increased sensitivity and Western blots using sub-microgram quantities of protein. *J Neurosci Methods* 158, 224–233.
17. Wu, M., Stockley, P.G., and Martin, W.J., 2nd. (2002) An improved western blotting technique effectively reduces background. *Electrophoresis* 23, 2373–2376.
18. Schilling, M., Maiwald, T., Bohl, S., Kollmann, M., Kreutz, C., Timmer, J., et al. (2005) Computational processing and error reduction strategies for standardized quantitative data in biological networks. *FEBS J* 272, 6400–6411.
19. Aksamitiene, E., Hoek, J.B., Kholodenko, B., and Kiyatkin, A. (2007) Multistrip Western blotting to increase quantitative data output. *Electrophoresis* 28, 3163–3173.

Chapter 18

Western Blotting/Edman Sequencing using PVDF Membrane

Setsuko Komatsu

Summary

The western blotting/Edman sequencing technique using polyvinylidene difluoride (PVDF) membrane is one of the most popular technique for determination of primary structure. A protein sample is transferred from a SDS-polyacrylamide gel electrophoresis (PAGE) gel onto a PVDF membrane by electroblotting. The membrane carrying the protein is directly subjected to protein sequencing. If sequencing fails after a few cycles, the PVDF membrane is removed from the sequencer and treated with deblocking solution. If this attempt at sequencing fails, alternative methods such as the Cleveland method are required. Because the resolution of two-dimensional polyacrylamide gel electrophoresis (2D-PAGE) is high, the combined use of 2D-PAGE, western blotting, and Edman sequencing often allows effective sequence determination of crude proteins mixture that could not be easily purified by conventional column chromatography.

Key words: Western blotting, Edman sequencing, PVDF membrane, Blocked protein, Deblocking, Cleveland method

1. Introduction

Edman sequencing of proteins separated on two-dimensional polyacrylamide gel electrophoresis (2D-PAGE) gels became possible with the introduction of protein electroblotting methods that allow efficient transfer of sample from the gel matrix onto supports suitable for gas-phase sequencing or related techniques *(1)*. Picomole amounts of protein are first separated by 2D-PAGE *(2)*, and then electroblotted from 2D-PAGE gels onto polyvinylidene difluoride (PVDF) membrane. The amino acid sequences of

Biji T. Kurien and R. Hal Scofield (eds.), *Methods in Molecular Biology, Protein Blotting and Detection, vol. 536*

DOI: 10.1007/978-1-59745-542-8_18

the electroblotted protein are determined by Edman sequencing. Direct N-terminal sequencing is the most sensitive method (1–5 μg of protein), but, when gaps or ambiguous assignments are seen, verification of the sequence by other means often demands much more material. Proteins are often posttranslationally modified, and N-terminal blockage is one of the more common posttranslational modifications. Proteins can become N-terminally blocked not only in vivo but also in vitro. However, it is possible to prevent in vitro blocking, which is generated during protein extraction, 2D-PAGE, and blotting. The use of very pure reagents during these procedures, the addition of thioglycolic acid as a free radical scavenger to the extraction buffer, electrophoresis and electroblotting buffers, and preelectrophoresis to remove the free radicals from the gel might all be effective in preventing in vitro blocking *(3)*. However, if proteins are blocked in vivo, a chemical or enzymatic deblocking procedure or peptide mapping procedure is required to determine the N-terminal or internal sequence.

2. Materials

1. Sodium dodecyl sulfate (SDS) sample buffer: 60 mM Tris–HCl (pH 6.8), 2% SDS, 10% glycerol, and 5% β-mercaptoethanol *(4)*.
2. Acrylamide for separating gel (acrylamide/BIS = 30:0.135): 30.00 g acrylamide, 0.135 g BIS. Make volume to 100 mL with milli-Q (MQ) water, and keep in the dark (brown bottle).
3. Separating gel buffer (pH 8.8): 12.11 g Tris–HCl to a 1 M final concentration, 0.27 g SDS to a 0.27% final concentration. Dissolve in 80 mL MQ water, adjust pH to 8.8, and make the volume to 100 mL.
4. Acrylamide for stacking gel (acrylamide/BIS = 29.2:0.8): 29.2 g acrylamide, 0.8 g BIS. Make volume to 100 mL with MQ water, and keep in the dark (brown bottle).
5. Stacking gel buffer (pH 6.8): 3.03 g Tris–HCl to a 0.25 M final concentration, 0.20 g SDS to a 0.2% final concentration. Dissolve in 80 mL MQ water, adjust pH to 6.8, and make the volume to 100 mL.
6. SDS-PAGE running buffer: 9 g Tris–HCl, 43.2 g glycine, and 3 g SDS. Dissolve in 3 L MQ water.
7. Bromophenol blue (BPB) solution: Dissolve 0.1 g BPB in 100 mL 10% glycerol.

8. Blotting buffer A: 36.33 g Tris–HCl to a 0.3 M final concentration, 200 mL methanol to a 20% final concentration, and 0.20 g SDS to a 0.02% final concentration. Make volume to 1 L with MQ water, and keep at 4°C.
9. Blotting buffer B: 3.03 g Tris to a 25 mM final concentration, 200 mL methanol to a 20% final concentration, and 0.20 g SDS to a 0.02% final concentration. Make volume to 1 L with MQ water, and keep at 4°C.
10. Blotting buffer C: 3.03 g Tris–HCl to a 25 mM final concentration, 5.20 g ε-aminon-caproic acid to a 40 mM final concentration, 200 mL methanol to a 20% final concentration, and 0.20 g SDS to a 0.02% final concentration. Make volume to 1 L with MQ water, and keep at 4°C.
11. Separating gel solution (amounts for one gel [18%]): 10 mL acrylamide for separating gel, 6.3 mL separating gel buffer (pH 8.8), 120 μL 10% ammonium per sulfate (APS), and 20 μL TEMED.
12. Stacking gel solution (amounts for one gel [5%]): 1 mL acrylamide for stacking gel, 3 mL stacking gel buffer (pH 6.8), 2 mL MQ water, 30 μL 10% APS, and 20 μL TEMED.
13. Hydration buffer: Make this buffer with urea (8 M), CHAPS (2% w/v), DTE (10 mM), Resolyte pH 3.5–10 (2% v/v) and a trace of BPB.

3. Methods

3.1. Two-Dimensional Polyacrylamide Gel Electrophoresis

For 2D-PAGE, there are two methods as immobilized pH gradient (IPG) strip and isoelectric focusing (IEF) tubes as first dimension.

3.1.1. Immobilized pH Gradient as First Dimension

1. A nonlinear IPG strip (pI 3.5–10.0, 18 cm) (GE Healthcare, Piscataway, NJ, USA) is used as the first dimension. It offers high resolution, great reproducibility, and allows high protein loads.
2. Hydration is performed overnight in the preswelling cassette with 25 mL of the hydration buffer.
3. When the rehydration cassette is thoroughly emptied and opened, the strips are transferred to the strip tray. After placing IPG strips, humid electrode wicks, electrodes, and sample cups in position, the strips and cups are covered with low viscous paraffin oil. Samples are applied at the cathodic end of the IPG strips in a slow and continuous manner, without touching the gels.

4. The voltage is linearly increased from 300 to 3,500 V during 3 h, followed by 3 additional hours at 3,500 V, whereupon the voltage is increased to 5,000 V.
5. After IPG, SDS-PAGE in the second dimension is performed using polyacrylamide gels.

3.1.2. Isoelectric Focusing as First Dimension

1. IEF tube gels (pI 3.5–10.0, 11 cm in length and 3 mm in diameter) are used *(2)*. They consist of 8 M urea, 3.5% acrylamide, 2% NP-40, 2% ampholines, APS, and TEMED.
2. The sample overlay buffer consisted of 20 μl of 1/2 lysis buffer.
3. Electrophoresis is carried out at 200 V for 30 min, followed by 400 V for 16 h and 600 V for 1 h.
4. After IEF, SDS-PAGE in the second dimension is performed using polyacrylamide gels.

3.1.3. SDS-PAGE as Second Dimension with IPG and IEF

1. Fix glasses (100 × 140 × 1 mm) with clip, keeping a 1-mm space between the plates.
2. Prepare separating gel solution in a 100-mL beaker. Mix the solutions and fill the plates about 2 cm from top. (Caution: Pour the solutions into the plates immediately after adding 10% APS and TEMED.)
3. Overlay the separating gel solution with 1-mL MQ water.
4. Leave the gel for 40–60 min at room temperature (RT) for polymerization.
5. Remove the overlaid water and pour the following stacking gel solution.
6. Prepare the stacking gel solution in a 100-mL beaker. Mix well, pour on the separating gel.
7. Leave the gel for 20 min at RT for polymerization.
8. The first dimension is applied directly on the top of the stacking gel. The first dimension is overlaid by 1% agarose.
9. The slab gel is assembled for running. A few drops of BPB are added.
10. Run the sample at 35 mA (constant current) until the tracking dye reaches the bottom of the separating gel.
11. Separate the stacking gel, and take out the separation gel.

3.2. Cleveland Peptide Mapping (5)

Prepared samples are separated by 2D-PAGE. Then the gels are stained with Coomassie brilliant blue (CBB), and gel pieces (5–20 pieces) containing protein spots are removed and soak in MQ water in a 2-mL microcentrifuge. Remove MQ water and add 750 μL of electroelution buffer. The protein is electroeluted from the gel pieces using an electrophoretic concentrator run at 2 W constant powers for 2 h. After electroelution, the protein solution is dialyzed against MQ water for 2 days and lyophilized.

3.2.1. Peptide Mapping Protocol

1. Cut out stained protein spots from 2D gels and soak in MQ water.
2. Fill 750 μL of electroelution buffer in the 2-mL Eppendorf tube containing protein spots (5–20 gel pieces). Shake for 30 min.
3. Cut 12- to 15-cm-long seamless cellulose tubing (small size, no. 24, Wako, Osaka, Japan) as space is needed for clipping. Fill a 300-mL beaker with 250 mL MQ water, boil it for 5 min, and keep the tubing membrane in it. Wet the small pieces of cellophane film in a small beaker with MQ water.
4. Close the bottom of the small part of cup with a cellophane film, open the bottom of the large part of cup, connect and twist the tubing membrane, and close the distal end by clipping (**Fig. 1A**).
5. Fix the cup in the electrophoretic concentrator (**Fig. 1B**; Nippon-Eido, Tokyo, Japan). Deposit gel pieces containing proteins on the cellophane film (the small part of the cup) and add 750 μL of the electroelution buffer from the Eppendorf tube. Fill the small part of the cup with electroelution buffer, and then fill the large part of the cup with electroelution buffer in such a way that a layer of buffer joins both the cup parts, allowing movement of protein from the small part of the cup to the tubing membrane. Fill the apparatus with electroelution buffer. The small part of cup containing protein spots should be toward the positive side.
6. Run at 2 W constant power for 2 h.

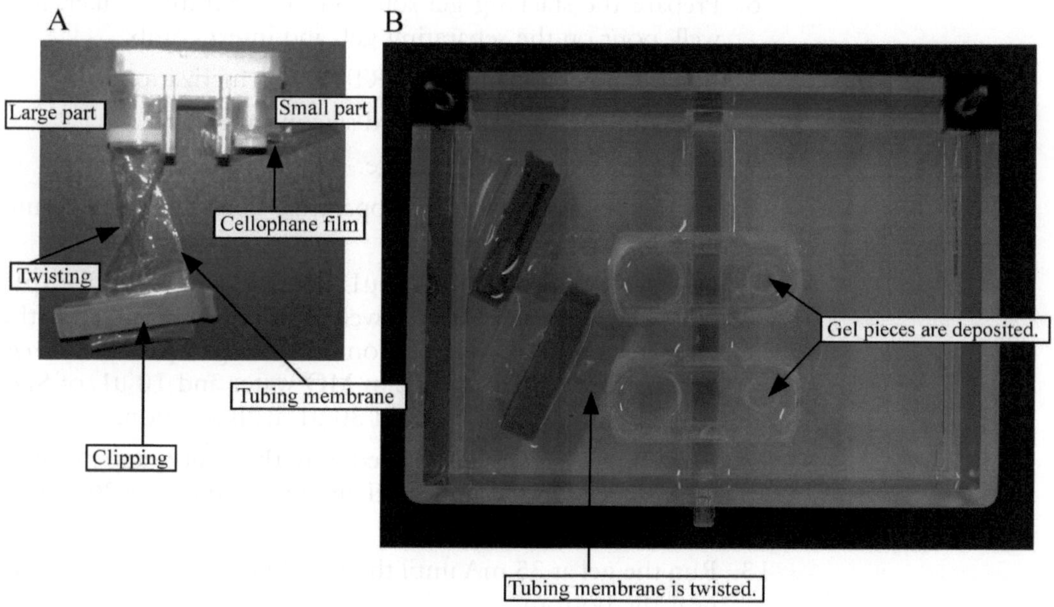

Fig 1. Electrophoretic concentrator. (A) Close-up of the cup with long seamless cellulose tubing. (**B**) The cups fixed to the electrophoretic concentrator.

7. Remove the tubing membrane and clip to close the end. Dialyze in a cold room (4°C). Change the MQ water three times the first day. The next day, change the MQ water, two times.
8. Transfer the protein solution to two to six 2-mL microcentrifuge tubes. Freeze-dry overnight.
9. Dissolve the protein in 20 μL of SDS sample buffer.

3.2.2. V8 Protease Digestion

The protein is dissolved in 20 μL of SDS sample buffer (pH 6.8) and applied to a sample well of an SDS-PAGE gel. The sample solution is overlaid with 20 μL of a solution containing 10 μL of *Staphylococcus aureus* V8 protease (Pierce, Rockford, IL, USA) (0.1 μg/μL) in MQ water and 10 μL of SDS sample buffer (pH 6.8). Electrophoresis is performed until the sample and protease are stacked in the stacking gel. The power is switched off for 30 min to allow digestion of the protein and then electrophoresis is continued. Detailed procedure is as follows:

1. Fix glasses (100 × 140 × 1 mm) with clip, keeping a 1-mm space between the plates.
2. Prepare separating gel solution in a 100-mL beaker. Mix the solutions and fill the plates about 3 cm from top. (Caution: Pour the solutions into the plates immediately after adding 10% APS and TEMED.)
3. Overlay the separating gel solution with 1 mL MQ water.
4. Leave the gel for 40–60 min at RT for polymerization.
5. Remove the overlaid water and pour the following stacking gel solution.
6. Prepare the stacking gel solution in a 100-mL beaker. Mix well, pour on the separating gel, and insert comb.
7. Leave the gel for 20 min at RT for polymerization.
8. Take out the comb, clips, and silicon tubes.
9. Clean the wells with a syringe.
10. Fix the gel plates with the apparatus. Pour SDS-PAGE running buffer.
11. Dissolve the protein in 20 μL of SDS sample buffer (pH 6.8), and apply to a sample well in SDS-PAGE. Overlay the sample with 20 μL of a solution containing10 μL of *S. aureus* V8 protease with 1 μg/μL in MQ water and 10 μL of SDS sample buffer (pH 6.8). Add 30 μL BPB solution.
12. Electrophoresis was performed until the sample and protease were stacked in the upper gel and interrupted for 30 min to digest the protein.
13. Run the gel at 35 mA until the BPB line reaches about 5 mm near the bottom.
14. Disconnect the electricity and take out the plates.

15. Separate two plates with a spatula.
16. Separate the stacking gel and take out the separating gel.

3.3. Western Blotting

Following separation by 2D-PAGE or by the Cleveland method, the proteins are electroblotted onto a PVDF membrane (Pall, Port Washington, NY, USA) (**Fig. 2**) using a semidry transfer blotter (Nippon-Eido, Tokyo, Japan), and detected by CBB staining.

1. Cut the PVDF membrane equal to the size of the gel.
2. Cut the Whatman 3MM filter paper equal to the size of the gel.
3. Wash the PVDF membrane in methanol for a few seconds, and transfer the PVDF membrane to 100-mL blotting buffer C, and shake for 5 min.
4. Wet two Whatman 3MM filter papers in blotting buffers A, B, and C (blotting papers A, B, and C).
5. Place the separating gel in 100-mL blotting buffer C and shake for 5 min.
6. Wet the semidry transfer blotter with MQ water. Place blotting paper A on the blotting plate and then blotting paper B. Remove air bubbles, if any. Place the PVDF membrane on the plate and then the gel and blotting paper C.
7. Connect the power supply. Run the blotting at 1 mA/cm^2 for 90 min.
8. Wash the PVDF membrane in 100-mL MQ water.
9. Stain the PVDF membrane for 2–3 min in CBB stain.
10. Destain the PVDF membrane in 60% methanol for 2-3 min twice.
11. Wash with MQ water and air-dry at RT.

3.4. Deblocking of Blotted Proteins *(6)* (See Note 1)

3.4.1. Acetylserine and Acetylthreonine

Proteins with acetylserine and acetylthreonine at their N-termini separated by 2D-PAGE are electroblotted onto a PVDF membrane. The region of the PVDF membrane carrying the protein spot is excised and treated with trifluoroacetic acid (TFA) at 60°C for 30 min. They are then directly subjected to protein sequencing (*see* **Note 2**).

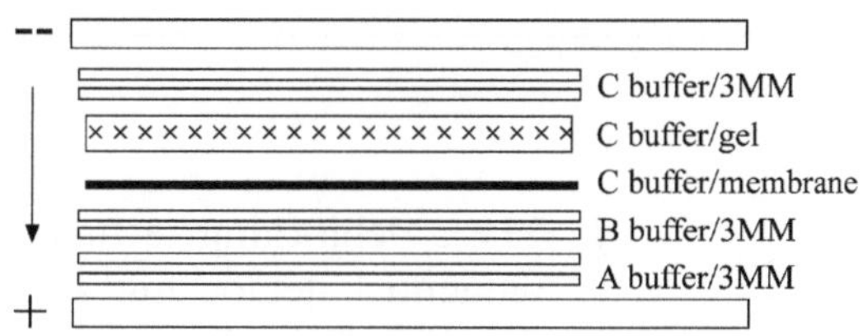

Fig. 2. Western blotting. Following separation by 2D-PAGE or by the Cleveland method, the proteins in the gel are electroblotted onto a PVDF membrane using a semidry transfer blotter. 3MM filter papers are wetted in blotting buffer A, B, and C, and gel and PVDF membrane are wetted in blotting buffer C.

3.4.2. Formyl Group

N-formylated protein separated by 2D-PAGE is electroblotted onto a PVDF membrane. The region of the PVDF membrane carrying the protein spot is excised and treated with 300 μL of 0.6 M HCl at 25°C for 24 h. The membrane is washed with MQ water, dried, and applied to the protein sequencer.

3.4.3. Pyroglutamic Acid

1. Proteins with pyroglutamic acid at their N-termini separated by 2D-PAGE are electroblotted onto a PVDF membrane.
2. The region of the PVDF membrane carrying the protein spot is excised and treated with 200 μL of 0.5% (w/v) polyvinylpyrrolidone (PVP)-40 in 100 mM acetic acid 37°C for 30 min (*see* **Note 3**).
3. The PVDF membrane is washed at least ten times with 1 mL of MQ water.
4. The PVDF membrane is soaked in 100 μL of 0.1 M phosphate buffer (pH 8) containing 5 mM dithiothreitol and 10 mM EDTA acid.
5. Pyroglutamyl peptidase (5 μg) is added and the reaction solution is incubated at 30°C for 24 h.
6. The PVDF membrane is washed with MQ water, dried, and applied to the protein sequencer *(6)*.

3.5. N-Terminal and Internal Amino Acid Sequence Analysis and Homology Search of Amino Acid Sequence

1. The stained protein spots or bands are excised from the PVDF membrane and applied to the upper glass block of the reaction chamber in a gas-phase protein sequencer, Procise 494 or cLC (Applied Biosystems, Foster City, CA, USA), and/or PPSQ (Shimazu, Kyoto, Japan). Edman degradation is performed according to the standard program supplied by Applied Biosystems and/or Shimazu. The released phenylthiohydantoin (PTH) amino acids are separated by an online high-performance liquid chromatography (HPLC) system and identified by retention time.
2. The amino acid sequences obtained are compared with those of known proteins in the Swiss-Prot, PIR, Genpept, and PDB databases with the Web-accessible search program FastA.

4. Notes

1. These deblocking techniques may be combined to allow the sequential deblocking and sequencing of unknown proteins that have been immobilized onto PVDF membranes. A protein on the PVDF membrane is directly subjected to gas-phase sequencing. If sequencing fails at this step, the PVDF membrane

is removed from the sequencer, and performed to remove acetyl group, formyl group, and then pyroglutamic group.

2. The advantage of this method is that the deblocking can be easily and rapidly done, although overall sequencing yields obtained by this procedure were low compared with acylamino acid-releasing enzyme (AARE) digestion. N-acetylated protein: N-acetylated proteins are enzymatically deblocked with AARE after on-membrane digestion with trypsin to generate the N-terminal peptide fragment. This tryptic digestion is required since AARE can only remove the acetylamino acid from a short peptide *(6)*.
3. PVP-40 is used to unbind pyroglutamic acid from the PVDF membrane, while the rest of the protein stays bound to it.

References

1. Eckerskorn, C., Mewes, W., Goretzki, H., and Lottspeich, F. (1988) A new siliconized-gas fiber as support for protein-chemical analysis of electroblotted proteins. *Eur J Biochem* 176, 509–512.
2. O'Farrell, P. F. (1975) High resolution two-dimensional electrophoresis of proteins. *J Biol Chem* 250, 4007–4021.
3. Hirano, H., Komatsu, S., Kajiwara, H., Takagi, Y., and Tsunasawa, S. (1993) Micro-sequence analysis of the N-terminally blocked proteins immobilized on polyvinylidene difluoride membrane by Western blotting. *Electrophoresis* 14, 839–846.
4. Laemmli, U. K. (1970) Cleavage of structural proteins during the assembly of the head of bacteriophage T4. *Nature* 227, 680–685.
5. Cleveland, D. W., Fischer, S. G., Krischer, M. W., and Laemmli, U. K. (1977) Peptide mapping by limited proteolysis in sodium dodecyl sulphate and analysis by gel electrophoresis, *J Biol Chem* 252, 1102–1106.
6. Hirano, H., Komatsu, S., Nakamura, A., et al. (1991) Structural homology between semi-dwarfism-related proteins and glutelin seed protein in rice (Oryza sativa L.). *Theor Appl Genet* 83, 153–158.

Chapter 19

Blotting from PhastGel to Membranes by Ultrasound

Joseph Kost

Summary

Ultrasound-based approach for enhanced protein blotting is proposed. Three minutes of ultrasound exposure (1 MHz, 2.5 W/cm^2) was sufficient for a clear transfer of proteins from a polyacrylamide gel (PhastGel) to nitrocellulose or Nylon 66 Biotrans membrane. The proteins evaluated were prestained sodium dodecyl sulfate-polyacrylamide standards (18,500–106,000 Da) and ^{14}C-labeled Rainbow protein molecular weight markers (14,300–200,000 Da).

Key words: Ultrasound, Protein blotting, Mass transfer, PhastGel, Polyacrylamide gel, Nitrocellulose membranes, Nylon 66 Biotrans membranes

1. Introduction

1.1. Ultrasound

Ultrasound, or sound of frequency higher than 20 kHz, is inaudible to the human ear. Irradiation with ultrasound is widely used in medical applications (imaging and physiotherapy) *(1–3)*, sonochemical processing *(4)*, ultrasonic cleaning of surfaces *(3)*, as the basis for underwater sonar ranging *(3)*, and in many food-processing applications *(5)*. There are three distinctly different biomedical applications of ultrasound that can be identified in terms of their frequency ranges:

1. High-frequency or diagnostic ultrasound (2–10 MHz)
2. Medium-frequency or therapeutic ultrasound (0.7–3.0 MHz)
3. Low-frequency or power ultrasound (5–100 kHz)

Enhanced membrane permeation (a phonophoretic effect) of ultrasound on cells has been widely reported *(6)*. Ultrasound has

Biji T. Kurien and R. Hal Scofield (eds.), *Methods in Molecular Biology, Protein Blotting and Detection, vol. 536*

DOI: 10.1007/978-1-59745-542-8_19

been used successfully to induce transfer of genetic material into live animal *(7–9)* and plant cells *(10)*. Ultrasound has been shown to facilitate the delivery of drugs from liposomes *(11)*, polymers *(12)*, across the skin *(2)* as well as enhance the transdermal non-invasive extraction for continuous detection of glucose *(13)*. At sufficiently high acoustic power inputs, ultrasound is known to rupture cells and ultrasonication is a well-established laboratory technique of cell disruption *(14)*.

1.2. Mass Transfer Enhancement

Ultrasound has the potential for enhancing mass transfer within a cell. At certain ultrasound intensities, intracellular microstreaming has been reported inside animal and plant cells *(15)*. Similarly, rotation of organelles and induced circulation within vacuoles of plant cells have been associated with ultrasound *(16)*.

Ultrasound-enhanced diffusion of nutrients through gels has been used to explain improved dehydrogenation of hydrocortisone by gel-entrapped cells of Arthrobacter simplex *(17)*. Since mass transfer can be a significant limitation in many bioprocessing situations *(18)*, mass transfer enhancing effect of ultrasound has many potential applications.

1.3. Protein Blotting

We previously proposed the application of ultrasound in order to accelerate protein blotting *(19)*. The very pronounced effect of ultrasound on blotting performance is shown in **Fig. 1**. Three minutes of ultrasound exposure (1 MHz, 2.5 W/cm^2) was sufficient for a very clear transfer of prestained SDS-PAGE standards from a polyacrylamide gel (PhastGel) to nitrocellulose (NC) or nylon 66 (BR) membranes. In control experiments, which were performed following similar procedures without turning the ultrasonic generator on, no protein blotting could be seen. For comparable blotting results, 30 min for electroelution or 240 min for the elution by convection blotting procedure was required.

The kinetics of protein transfer from gel to BR or NC membranes, enhanced by (a) ultrasound, (b) electroelution, or (c) convection are shown in **Fig. 2**. As seen, the highest rates of

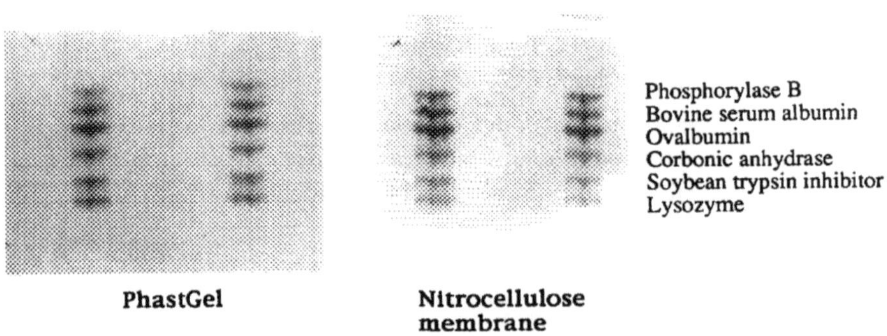

Fig. 1. Low-range SDS-PAGE standards were electrophoretically loaded on polyacrylamide gels (PhastGel) and ultrasonically (3 min in TB-1) transferred to nitrocellulose (Trans-Blot) membrane *(19)*.

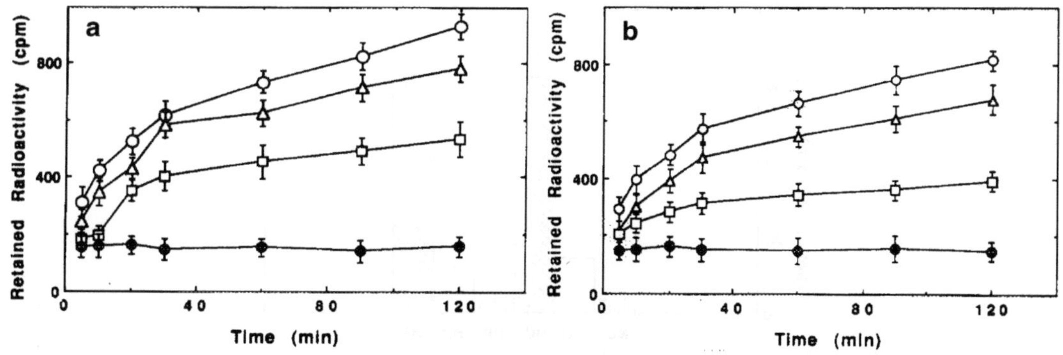

Fig. 2. Retained radioactivity of ^{14}C-methylated protein markers vs. blotting time for 2 h of ultrasonic blotting (*open circles*), 2 h of electroelution (*triangles*), 12 h of convection blotting (*rectangular*) and dry unused membrane controls (*filled circles*) in (**A**) nylon 66 (BR) membranes and (**B**) nitrocellulose (NC) membranes. The transfer solution in all experiments was TB-1 *(19)*.

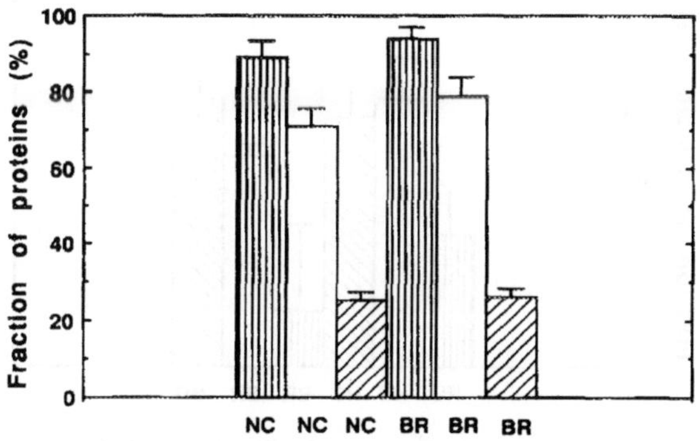

Fig. 3. Fraction of proteine, out of the initial amount loaded to the PhastGel, transferred and retained in BR and NC membranes, following blotting in transfer solution TB-1 by ultrasound for 2 h (*vertical lines*), electroelution (*open bars*), and 12 h of convection blotting (*tilted lines*) *(19)*.

protein transfer from gels to membranes were for systems exposed to ultrasound. The proteins blotted covered a span of molecular weights from 14,300 to 200,000.

The fraction of protein retained in the gel and the fraction transferred from the gel to NC or BR membranes for the different blotting procedures are displayed in **Figs. 3** and **4**. At least 85% of the initial protein amount loaded on the gel was transferred to the membranes when blotted by ultrasound. We could not detect differential efficiencies between transfers of high and low molecular weight proteins. For ultrasonic transport, the highest fraction of proteins was eluted from PhastGels and retained on the membranes, while the fractions of protein lost during blotting were much smaller than those lost during the electrophoretic or convectional methods (**Fig. 5**). The lower fraction of protein lost during ultrasonic transfer might be due to the smaller volumes of transfer buffer required for this procedure relative to those required for electroelution or convectional blotting.

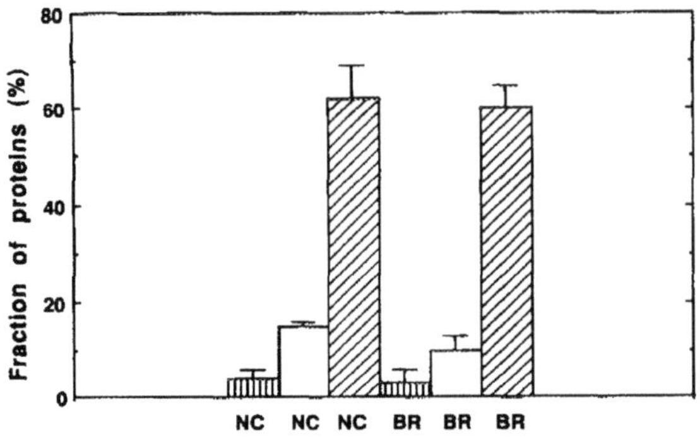

Fig. 4. Fraction of proteins, out of the initial amounts loaded to the PhastGel, retained in PhastGel following blotting in transfer solution TB-1 by ultrasound for 2 h (*vertical lines*), electroelution for 2 h (*open bars*), and 12 h of convection blotting (*tilted lines*) *(19)*.

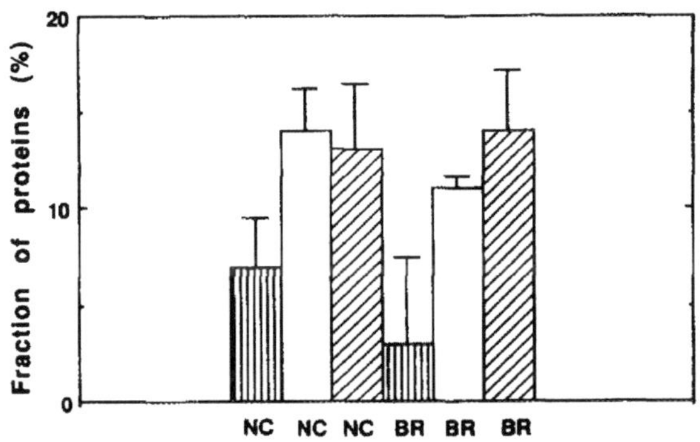

Fig. 5. Fraction of proteins, out of the initial amount loaded to the PhastGel, transferred from the gel and lost following blotting in transfer solution TB-1 by ultrasound for 2 h (*vertical lines*), electroelution for 2 h (*open bars*), and 12 h of convection blotting (*tilted lines*) *(19)*.

2. Materials

1. Transfer buffer-1 (TB-1): 3.05 g/L of Tris(hydroxymethyl) aminomethane (Tris), 19.2 g/L of glycine, and 200 mL/L of methanol.
2. Transfer buffer-2 (TB-2): 3.05 g/L of Tris and 19.2 g/L of glycine.
3. Transfer buffer-3 (TB-3): 3 M sodium chloride and 0.3 M sodium citrate.
4. Waterproof tape (Scotch Super 33+, 3M, Huchinson, MN, USA).
5. Filter paper (Shleicher &Schuell, Keene, NH, USA).
6. Aquasonic gel 100 (Parker Laboratories, Fairfield, NJ, USA).
7. Ultrasound (Sonopuls 434, Enraf Nonius, The Netherlands).

8. Nitrocellulose membranes (Trans-Blot 0.45 μm, Bio-Rad Laboratories, Richmond, CA)
9. Nylon 66 membranes (Biotrans 0.45 μm, ICN Biomedicals, Costa Messa, CA, USA)

3. Methods

3.1. Effect of Transfer Buffer Composition on BSA Blotting by Ultrasound Irradiation

1. BSA blotting with ultrasound and electroelution were performed in different transfer buffer solutions (**Fig. 6**) in order to evaluate the effect of transfer buffer solution composition on blotting.
2. No effect of transfer buffer composition on BSA blotting by ultrasound irradiation could be detected, while for BSA electroelution, TB-1 was more efficient that TB-2 (*see* **Note 1**).
3. Although the therapeutic ultrasound applied in these experiments is routinely used on humans in physiotherapy, its possible effects on the protein biological activity were examined. **Figure 7** shows that 10-min exposure of phospholipase A2 (PLA2), from porcine pancreas, to ultrasound at intensities comparable to those used for ultrasonic blotting, did not affect its biological activity (*see* **Note 2**).

3.2. Protocol for Protein Transfer Using Ultrasonic Irradiation

Protein transfer from PhastGels to membranes by ultrasonic irradiation was performed as follows:

1. A sheet of dry filter paper (equal in size to the PhastGel used in electrophoresis) was placed on a glass plate.
2. Protein-loaded PhastGel was placed on top of the filter paper.

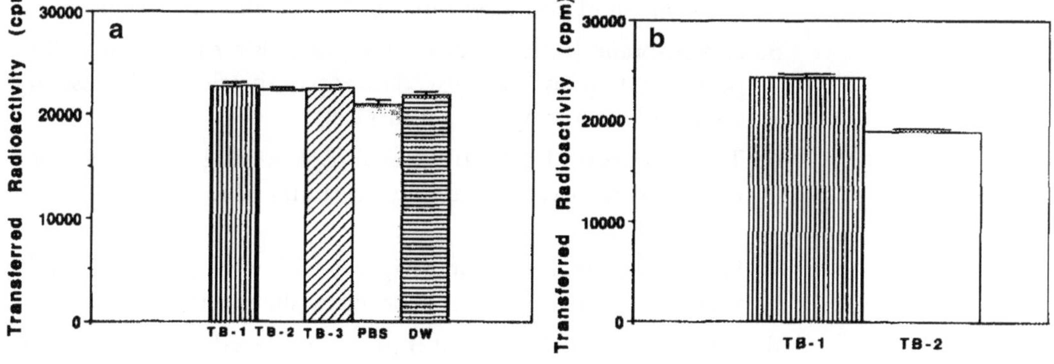

Fig. 6. Amount of ^{125}I-BSA transferred to NC membranes (**A**) ultrasonic blotting for 2 h in TB-1, TB-2, TB-3, PBS, and distilled water (**B**) Electroelution for 2 h in TB-1 and TB-2 *(19)*.

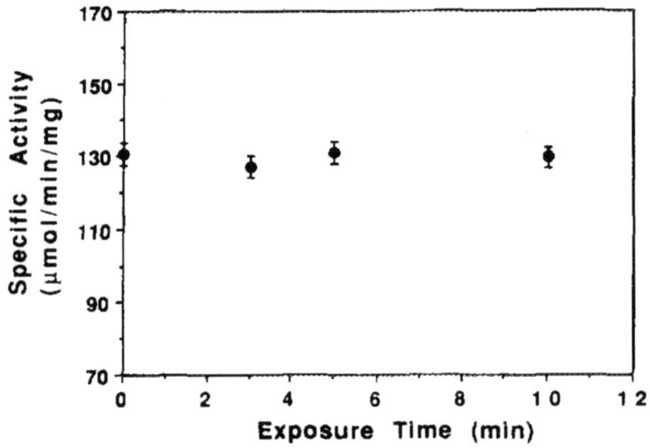

Fig. 7. Specific activity of phospholipase A2 solution subjected to ultrasound at 1 MHz, 2.5 W/cm² *(19)*.

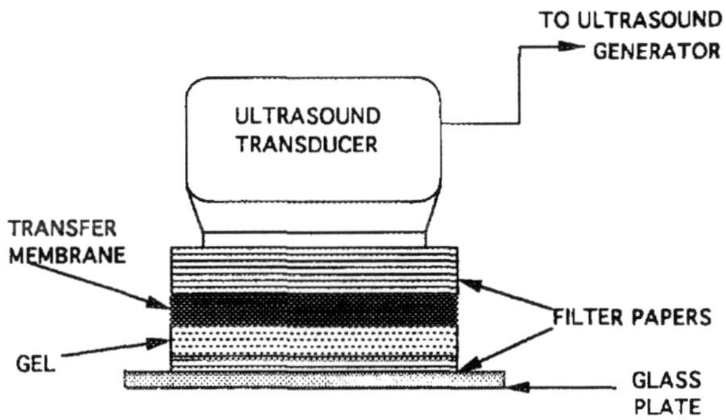

Fig. 8. Schematic presentation of the ultrasonic transfer procedure *(19)*.

3. The assembly was secured in place by waterproof tape which was placed around the edges of the electrophoresis gel, taped to the glass plate.
4. The gel and filter paper were firmly held together so the transfer solution could permeate through the gel and not around it.
5. A membrane (nitrocellulose (NC) or nylon 66 Biotrans (BR), pore size 0.45 μm) was placed on top of the electrophoresis gel, completely covering the gel surface.
6. The air trapped between the electrophoresis gel and the membrane was removed, by rolling a clean metallic cylinder over the membrane.
7. Seven to ten pieces of filter paper, which were prewetted in the transfer buffer, were placed, sheet by sheet on the membrane.
8. Aquasonic gel 100 (about 5 mL) was placed on top of the filter papers as a coupling medium for ultrasound transmission.
9. Ultrasound at 1 MHz, 2.5 W/cm²) was applied, placing the ultrasonic probe on top of the Aquasonic gel (**Fig. 8**).

4. Notes

1. The higher efficiency of TB-1 was expected as methanol was reported to increase the binding capacity of proteins, with moderate molecular weight molecules like BSA, strengthening the hydrophobic interactions between protein and membrane *(20)*. The insensitivity of the ultrasound blotting to this hydrophobic interaction might be due to the smaller amounts of buffer used for the ultrasonic blotting (prewetted filter papers with the transfer solution vs. gels inserted into buffer tank containing transfer solution) and/or the large enhancement of the ultrasound, which might overcome the hydrophobic interaction.
2. The procedure is fast and not as charge specific as electroblotting. We believe that after optimization, the required times could even be shorter and potentially applied to all proteins.

References

1. Moore, C., and Promes, S. B. (2004) Ultrasound in pregnancy, *Emerg Med Clin North Am* 22, 697–722.
2. Mitragotri, S. (2005) Healing sound: the use of ultrasound in drug delivery and other therapeutic applications, *Nat Rev Drug Discov* 4, 255–260.
3. Suslick, K. S. (1998) Ultrasound: its chemical, physical and biological effects, VCH, New York, NY.
4. Mason, J. T. (1998) Chemistry with ultrasound, Elsevier, London.
5. Povey, M. J. W., and Mason, T. J. (1998) Ultrasound in food processing, Blackie, London.
6. Williams, A. R. (1983) Ultrasound: biological effects and potential hazards, Academic, New York, NY, pp. 177–253.
7. Unger, E. C., McCreery, T., and Sweitzer, R. H. (1997) Ultrasound enhances gene expression of liposomal transfection, *Invest Radiol* 32, 723–727.
8. Miller, D. L., and Song, J. (2003) Tumor growth reduction and DNA transfer by cavitation-enhanced high intensity focused ultrasound, *Ultrasound Med Biol* 29, 887–893.
9. Deshpande, M. C., and Prausnitz, M. R. (2007) Synergistic effect of ultrasound and PEI on DNA transfection in vitro, *J Control Release* 118, 126–135.
10. Joersbo, M., and Brunstedt, J. (1992) Sonication: a new method for gene transfer to plants. *Physiol Plant* 85, 230–234.
11. Shroeder, A., Avnir, Y., Weisman, S., Tzemach, D., Najajreh, Y., Gabizon, A., et al. (2007) Controlling liposomal drug release by low-frequency ultrasound: mechanism and feasibility. *Langmuir* 23, 4019–4025.
12. Kost, J., and Langer, R. (2001) Responsive polymeric delivery systems. *Adv Drug Deliv Rev* 46, 125–148.
13. Kost, J., Mitragotri, S., Gabbay, R., Pishko, M., and Langer, R. (2000) Non-invasive blood glucose measurement using ultrasound, *Nat Med* 6, 347–350.
14. Chisti, Y., and Moo-Young, M. (1998) Disruption of microbial cells for intracellular products. *Enzyme Microb Technol* 8, 194–204.
15. Bar, R. (1998) Ultrasound-enhanced bioprocesses: cholesterol oxidation by Rhodococcus erythropolis. *Biotechnol Bioeng* 32, 655–663.
16. Nyborg, W. L. (1998) Ultrasonic microstreaming and related phenomena. *Br J Cancer* 45, 156–160.
17. Zabaneh, M. and Bar, R. (1991) Ultrasound-enhanced bioprocess. II: dehydrogenation of hydrocortisone by Arthrobacter simplex. *Biotechnol Bioeng* 37, 998–1003.
18. Chisti, Y. (2003) Sonobioreactors: using ultrasound for enhanced microbial productivity. *Trends Biotechnol* 21, 89–93.
19. Kost, J., Liu, L -S., Ferreira, J., and Langer, R. (1994) Enhanced protein blotting from PhastGel media to membranes by irradiation of low-intensity. *Anal Biochem* 216, 27–32.
20. Issaq, H. J., Atamna, I. Z., Muschil, G. M., and Janini, G. M. (1991) *Chromatographia* 32, 155.

Chapter 20

Ultrarapid Electrophoretic Transfer of High and Low Molecular Weight Proteins Using Heat

Biji T. Kurien and R. Hal Scofield

Summary

An ultrarapid method for the electrophoretic transfer of high and low molecular weight proteins to nitrocellulose membranes following sodium dodecyl sulfate (SDS) polyacrylamide gel is described here. The transfer was performed with heated (70–75°C) normal transfer buffer from which methanol had been omitted. Complete transfer of high and low molecular weight antigens (molecular weight protein standards, a purified protein, and proteins from a human tissue extract) could be carried out in 10 min for a 7% (0.75 mm) SDS polyacrylamide gel. For 10 and 12.5% gels (0.75 mm) the corresponding time was 15 min. A complete transfer could be carried out in 20 min for 7, 10, and 12.5% gels (1.5 mm gels). The permeability of the gel is increased by heat, such that the proteins trapped in the polyacrylamide gel matrix can be easily transferred to the membrane. The heat mediated transfer method was compared with a conventional transfer protocol, under similar conditions. The conventional method transferred minimal low molecular weight proteins while retaining most of the high molecular weight proteins in the gel. In summary, this procedure is particularly useful for the transfer of high molecular weight proteins, very rapid, and avoids the use of methanol.

Key words: SDS-PAGE, Western blotting, Nitrocellulose, High molecular weight proteins

1. Introduction

Proteins transferred from sodium dodecyl sulfate polyacrylamide gel electrophoresis (SDS-PAGE) gels to nitrocellulose or polyvinylidene difluoride membranes have been achieved by (a) simple diffusion *(1)*; (b) vacuum-assisted solvent flow *(2)*; and (c) electrophoretic elution *(3)*. There is considerable interest in diffusion

Biji T. Kurien and R. Hal Scofield (eds.), *Methods in Molecular Biology, Protein Blotting and Detection, vol. 536*

DOI: 10.1007/978-1-59745-542-8_20

mediated transfer of proteins, since it was showed that *(1)* multiple immunoblots can be generated following nonelectrophoretic bidirectional transfer of a single SDS-PAGE gel with multiple antigens. The lifts from SDS-PAGE gels, for immunoblotting, using this method are particularly useful in identification of proteins by mass spectrometry *(4, 5)*. However, electrophoretic elution is still the method of choice for protein transfer to membranes used widely in most laboratories.

Electrophoretic transfer of proteins, resolved by SDS-PAGE, to nitrocellulose is a fundamental step prior to detection of specific proteins with specific antibodies *(6)*. The protein transfer procedure normally takes about 2–4 h at about 70 V or an overnight transfer at 30 V. High molecular weight proteins are often stubbornly resistant to transfer *(7)* in spite of these prolonged runs and this problem is accentuated when higher percentage gels are used. Prolonged electrotransfer (16–20 h) at high current density coupled with inclusion of 0.01% sodium dodecyl sulfate, to enhance protein elution, has been used to efficiently transfer high-molecular weight proteins *(8)*.

In order to determine the efficiency of heat-mediated electroblotting, we have used purified Ro 60 autoantigen, prestained molecular weight standards, and a human cell extract using gels of two different thicknesses (0.75 mm and 1.5 mm) and gels with three different amounts of acrylamide (7, 10, and 12.5%). In addition, transfer of proteins from a 4–20% gradient SDS-PAGE gel and a 12.5% (1.5 mm) regular gel was also investigated and compared to that obtained using a conventional transfer method under similar conditions.

Prestained high and low molecular weight protein standards (**Figs. 1** and **2**; Lane 4) and bovine Ro 60 (**Figs. 1** and **2**; Lane 2) could be transferred completely to nitrocellulose. All the protein markers (5 μl of the marker) could be transferred in about 10 min from a 0.75 mm, 7% gel (**Fig. 1A**, Lane 4). All the protein markers, ranging from 184 to 9 kDa, could be transferred to membranes from 7, 10, and 12.5% gels (**Fig. 1B**, **C**, Lane 4) in 15 min. The posttransfer polyacrylamide gels were clean, without any sign of residual nontransferred protein markers. It took 20 min to transfer all the protein markers in the case of the 1.5 mm gels (7, 10, and 12.5% gels) (**Fig. 2**, Lane 4).

Figure 3 shows the Fast Green stained nitrocellulose membranes following transfer of protein using our method (**Fig. 3**, right panel) and a conventional method (**Fig. 3**, left panel). The protein transfer was found to be efficient with heat mediated transfer, while even the low molecular weight proteins can be barely seen following 20 min of transfer using a conventional method. Similarly only low levels of proteins could be

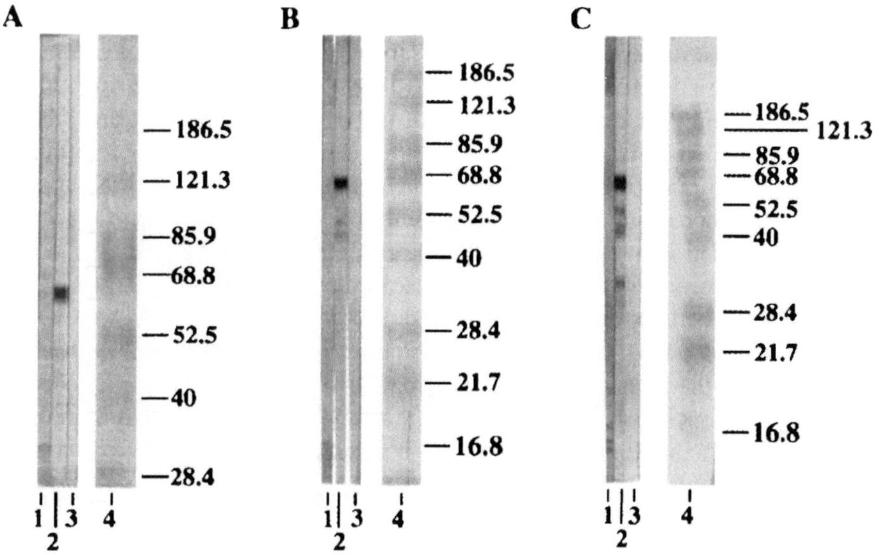

Fig. 1. Western blot transfer and immunoblotting of bovine Ro 60 and prestained molecular weight standards using 7, 10, and 12.5% gels (0.75 mm gels). (**A–C**) Proteins on a 7, 10, and 12.5% SDS-PAGE gels, respectively, transferred using heat. Lane 1 – Anti-Ro 60 negative control; lane 2 – anti-Ro 60 positive control; lane 3 – conjugate control; Lane 4 – prestained molecular weight standards (reproduced from *(8)* with permission from Elsevier).

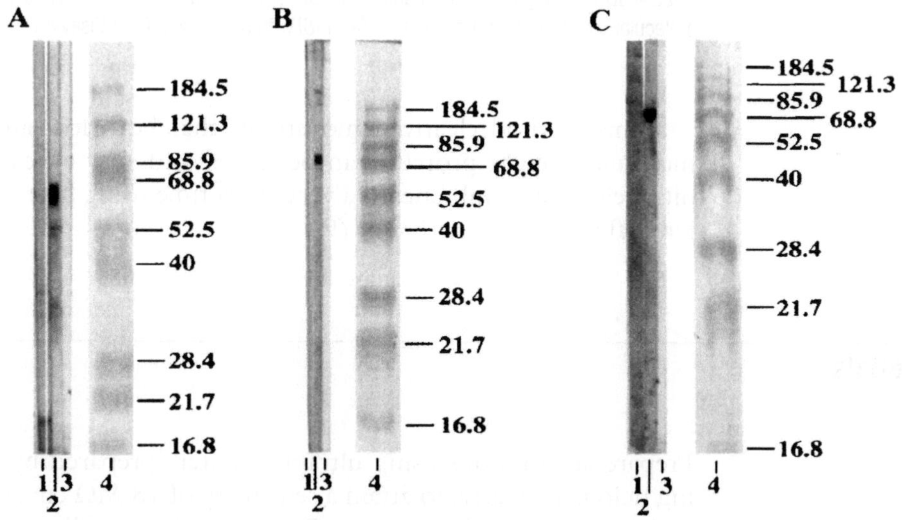

Fig. 2. Western blot transfer and immunoblotting of bovine Ro 60 and prestained molecular weight standards using 7, 10, and 12.5% gels (1.5 mm gels). (**A–C**) Proteins on a 7, 10, and 12.5% SDS-PAGE gels transferred using heat. Lane 1 – Anti-Ro 60 negative control; lane 2 – anti-Ro 60 positive control; lane 3 – conjugate control; lane 4 – prestained molecular weight standards [reproduced from *(8)* with permission from Elsevier].

detected immunologically following a conventional electrotransfer compared to that obtained following heat mediated transfer (**Fig. 4**).

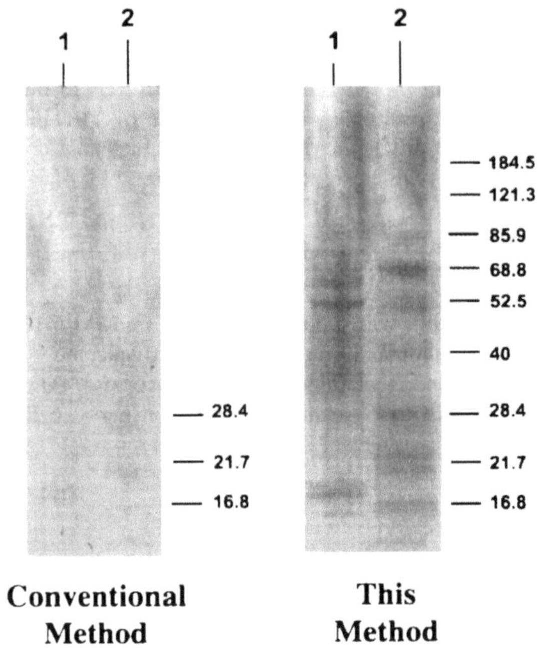

Fig. 3. Fast Green staining of proteins from a HeLa cell extract transferred to nitrocellulose membrane by a conventional method (*left*) and using the heat transfer method (*right*) from a 4–20% SDS-PAGE gradient gel. Lane 1 – HeLa cell extract and lane 2 – 10 μL of prestained molecular weight marker [reproduced from *(8)* with permission from Elsevier].

This method clearly demonstrates that both low and high molecular weight proteins can be transferred very efficiently to nitrocellulose membranes in a very short time using heated transfer buffer without methanol *(9)*.

2. Materials

Prepare all solutions using ultrapure water (prepared by purifying deionized water, to attain a sensitivity of 18 MΩ cm at 25°C) and analytical grade reagents. Prepare and store all reagents at room temperature (unless indicated otherwise). Diligently follow all waste disposal regulations when disposing waste materials. We do not add sodium azide to our reagents.

1. 10% SDS-PAGE precast gels (10-well) (ISC Bioexpress, Kaysville, UT, USA).
2. SDS lysis buffer (5×): 0.3 M Tris–HCl (pH 6.8), 10% SDS, 25% β-mercaptoethanol, 0.1% bromophenol blue, 45% glycerol. Leave one aliquot at 4°C for current use and store remaining aliquots at –20°C (*see* **Note 1**).

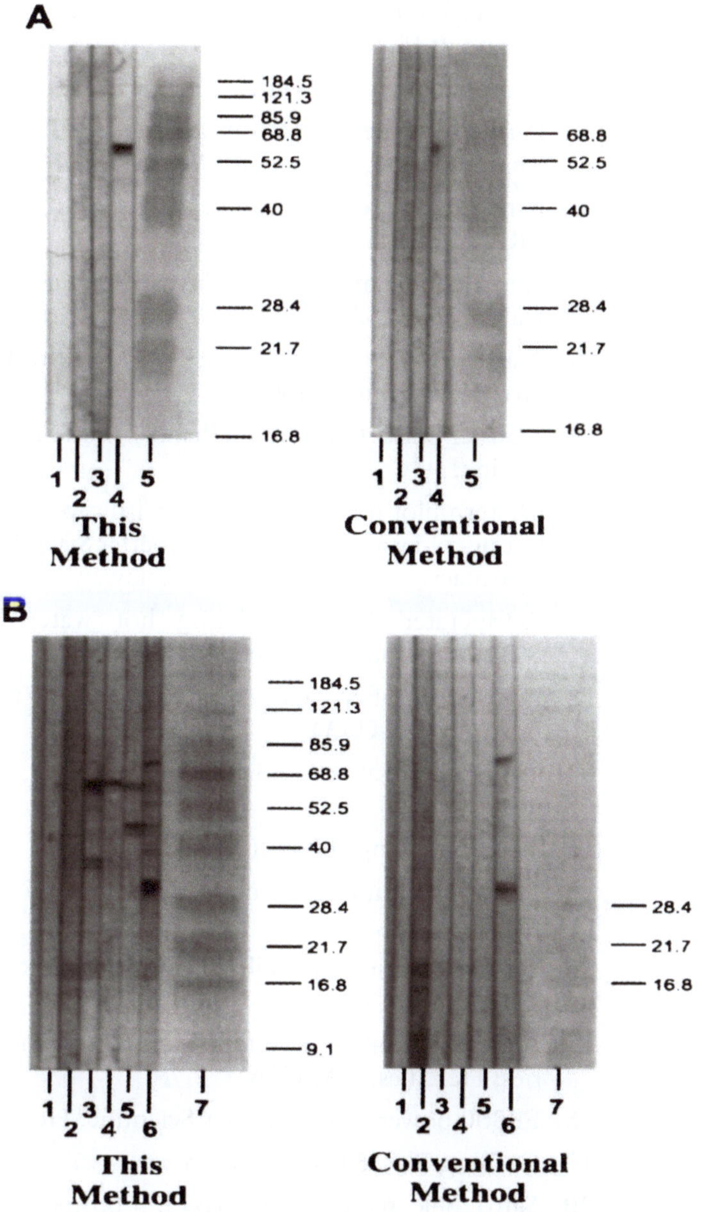

Fig. 4. Immunoblots of purified Ro 60 and proteins derived from a HeLa cell extract transferred to nitrocellulose membrane using the heat transfer method (*left*) and by a conventional method (*right*). (**A**) Purified Ro 60 autoantigen immunoblot obtained from a 12.5% SDS-PAGE gel using the heat transfer method (*left*) and by a conventional method (*right*). Lane 1 – conjugate control; lanes 2 and 3 – normal controls; lane 4 – anti-Ro 60 SLE sera; lane 5 – prestained protein molecular weight standards. (**B**) HeLa cell extract immunoblot obtained from a 4–20% gradient SDS-PAGE gel using the heat transfer method (*left*) and a conventional transfer method (*right*). Lane 1 – conjugate control; lane 2 – normal control; lane 3 – anti-Ro 60 SLE sera; lane 4 – anti-La sera; lane 5 – anti-Ro 52 sera; lane 6 – anti-Sm/nRNP sera; lane 7 – prestained protein molecular weight markers (10 μL) (reproduced from *(8)* with permission from Elsevier).

3. SDS-PAGE running buffer: 0.025 M Tris, pH 8.3, 0.192 M glycine, 0.1% SDS.
4. Phosphate buffered saline (PBS), pH 7.4.
5. Purified bovine Ro 60 autoantigen *(8, 10)* was a gift from Immunovision, Springdale, AK, USA.
6. BenchMark prestained molecular weight standards (Gibco BRL, Bethesda, MD, USA).
7. Nitrocellulose membranes (Gelman Sciences/Fisher Scientific, Dallas, TX, USA).
8. Transfer buffer (with methanol): 0.025 mM Tris, 192 mM glycine, 20% methanol.
9. Transfer buffer (without methanol): 0.025 mM Tris, 192 mM glycine.
10. Western blot transfer apparatus with capability of circulating hot or cold water at the base (Hoeffer Scientific Instruments, Pharmacia/GE Healthcare, Piscataway, NJ, USA).
11. Refrigerated bath/circulating hot water bath (Endocal RT-110, Neslab, Portsmouth, NH, USA).
12. Branson Sonifier Cell Disruptor 185 (VWR Scientific Co., Boston, MA, USA).
13. Tris buffered saline (TBS; 10×): 1.5 M NaCl, 0.1 M Tris–HCl, pH 7.4.
14. TBS containing 0.05% Tween-20 (TBST).
15. Blocking solution: 5% milk in TBS, pH 7.4. Store at 4°C (*see* **Note 2**).
16. Diluent solution: 5% milk in TBS, pH 7.4 containing 0.05% Tween. Store at 4°C (*see* **Note 2**).
17. Mini PROTEAN® 3 System Glass plates (Bio-Rad Laboratories; Hercules, CA, USA).
18. FB300 power supply (Fisher Scientific, Houston, TX, USA).
19. Corning PC-351 magnetic stirrer.
20. Nitro blue tetrazolium (NBT)/5-bromo-4-chloro-3-indolyl phosphate (BCIP): Dissolve 1 g NBT in 20 mL of 70% dimethylformamide (DMF). Dissolve 1 g BCIP in 20 mL of 100% DMF. Add 33 μL of BCIP and 66 μL of NBT to 10 mL of alkaline phosphatase buffer just before adding to membrane.

3. Methods

All procedures are carried out at room temperature unless otherwise specified.

3.1. Preparation of HeLa Cell Lysate

1. Harvest freshly cultured HeLa cells by centrifuging at 800 × *g* and wash twice with PBS.
2. Lyse cells by sonication in SDS-PAGE lysis buffer using a Branson sonicator (setting 4) and spin at 10,000 × *g* for 10 min.
3. Use an aliquot of the supernatant for SDS-PAGE.

3.2. SDS-page

1. Carry out SDS-PAGE *(11)* (*see* Chapter "Non-electrophoretic bi-directional transfer of a single SDS-PAGE gel with multiple antigens to obtain twelve immunoblots") on 7, 10, 12.5% (0.75 mm or 1.5 mm thickness) or 4–20% gradient gels.

3.3. Conventional Electrophoretic Transfer

1. Carry out conventional electrotransfer of proteins *(3)* separated on gradient and regular gels (*see* Chapter "Detection of La/SS-B by Western Blot Using Nanogold-Tagged Antibodies and Silver Enhancement") at 4°C using standard transfer buffer with methanol) for 20 min.
2. Save the nitrocellulose membrane for use with the membrane to be obtained following heat-mediated electrophoretic transfer for 20 min. Stain the polyacrylamide gel with Coomassie Brilliant Blue stain to ensure the efficiency of transfer.

3.4. Heat-Mediated Electrophoretic Transfer

1. Turn on the circulating hot water bath and set temperature at 70°C.
2. Following SDS PAGE, pry the gel plates open with the use of a spatula. The gel remains on one of the glass plates. Rinse the gel gently with deionized water and transfer carefully to a container with transfer buffer (without methanol).
3. Heat the transfer buffer (without methanol) to about 70–75°C in a 1 L glass beaker. Cover the beaker with clear plastic wrap.
4. Cut a nitrocellulose membrane to the size of the gel and immerse in transfer buffer (*see* **Note 3**).
5. Cut four sheets of Whatman 3MM filter paper to the size of the gel and transfer to the transfer buffer (*see* **Note 4**). Place two adsorbent pads also in the buffer and remove air bubbles by pressing down on it with help of fingers.
6. Place clear plastic wrap (12 in. in length) on the work-bench. Place two filter papers on top of the plastic wrap. Position the membrane on top of the filter papers. Transfer the gel to the top of the membrane in such a way that there are no air bubbles between the gel and the membrane (*see* **Note 5**). Place the remaining two filter papers on top of the gel. Place in transfer cassette. Ensure that the nitrocellulose membrane is between the gel and the anode (*see* **Note 6**).
7. Place the transfer apparatus on a magnetic stirrer and connect it to the hot water bath maintained at 70°C. Transfer the hot

transfer buffer into the apparatus (*see* **Note 7**). Place a magnetic stir bar inside the transfer apparatus to circulate the buffer.

8. Carry out the transfer at 40 V (*see* **Note 8**) for periods ranging from 10 to 20 min, depending upon the type of SDS polyacrylamide gel used (*see* **Note 9**).
9. Disconnect power supply and disassemble the sandwich. Save the membrane along with one obtained using conventional electrotransfer.
10. Stain the polyacrylamide gel with Coomassie Brilliant Blue stain to ensure the efficiency of transfer.

3.5. Immunoblotting

1. Carry out standard immunoblotting *(3)* (*see* Chapter "Detection of La/SS-B by Western Blot Using Nanogold-Tagged Antibodies and Silver Enhancement") for both sets of membranes.
2. Use human systemic lupus erythematosus (SLE) sera with autoantibodies against 60 kDa Ro, 48 kDa La, 52 kDa Ro, Sm, or nuclear ribonucleoprotein autoantigens *(8, 10)* to identify the respective antigens transferred to nitrocellulose from various gels (*see* **Notes 9,10**).

4. Notes

1. SDS precipitates at 4°C. Therefore, the lysis buffer needs to be warmed prior to use.
2. Add 100 mL of 10× TBS to a 1 L graduated cylinder and make it to about 800 mL with water. Transfer 50 g skim milk powder into the cylinder and mix stir until dissolved. Make to 1 L with water. Separate 500 mL as the blocking solution. To the remaining 500 mL add 250 μL of Tween-20 (cut end of blue tip to aspirate Tween-20 easily), dissolve, and use it as the diluent.
3. Moistening the nitrocellulose completely with the transfer buffer (without methanol) takes about 10 min. Alternately, the nitrocellulose membrane can be moistened almost instantaneously with regular transfer buffer that contains methanol. The membrane can then be rinsed with water and the transfer buffer (without methanol).
4. Preferably, the buffer used to assemble the sandwich should be at room temperature or just slightly warm. Using the hot buffer to assemble the sandwich causes the 3MM filter paper, used as a part of the gel sandwich, to disintegrate partially.
5. Hold the two top corners of the gel with each hand. Lower the bottom part of the gel first of the membrane and gently

release the gel little by little to lay the complete gel on the membrane. This will prevent trapping of bubbles in between the gel and the membrane. We use clear plastic wrap. Part of the wrap was folded over the sandwich. A 10 mL pipette was used to roll out the air bubbles from the gel membrane sandwich prior to placing in transfer cassette.

6. In the Bio-Rad transfer apparatus, we place the gel side of the transfer cassette facing the black side of the transfer cassette holder and the membrane side facing the red side.
7. Exercise caution in handling the hot contents. Use thermo gloves to hold the hot beaker and pour carefully.
8. Forty volts was close to the upper limit possible using the FB300 power supply. The current was about 500 mA at this point.
9. Transfer from a 7.5% SDS polyacrylamide gel (0.75 mm thick) can be carried out in 10 min. However, it takes longer time (20 min) for a 12.5% gel (1.5 mm thick). If a circulating water bath is not available, it is still possible to transfer using just the heated buffer, by increasing the time of transfer. The main advantage of this procedure is that it is possible to transfer high molecular weight protein, in addition to saving time.
10. Exercise universal precaution when handling human sera. Treat each serum sample as potentially dangerous.

Acknowledgement

This work was supported by NIH grant ARO1844 and Oklahoma Center for the Advancement of Science and Technology to RHS. We also express our thanks to Samantha Ganick for her excellent technical assistance.

References

1. Kurien, B.T., and Scofield, R.H. (1997) Multiple immunoblots after non-electrophoretic bidirectional transfer of a single SDS-PAGE gel with multiple antigens. *J Immunol Methods* 205, 91–94.
2. Peferoen, M., Huybrechts, R., and De Loof, A. (1982) Vacuum-blotting: a new simple and efficient transfer of proteins from sodium dodecyl sulfate-polyacrylamide gels to NC. *FEBS Lett* 145, 369–372.
3. Towbin, H., Staehelin, T., and Gordon, J. (1979) Electrophoretic transfer of proteins from polyacrylamide gels to NC sheets: procedure and applications. *Proc Natl Acad Sci USA* 76, 4350–4354.
4. Kurien, B.T., and Scofield, R.H. (2000) Association of neutropenia in systemic lupus erythematosus with anti-Ro and binding of an immunologically cross-reactive neutrophil membrane antigen. *Clin Exp Immunol* 120, 209–217.
5. Kurien, B.T., Matsumoto, H., and Scofield, R.H. (2001) Purification of tryptic peptides for mass spectrometry using polyvinylidene fluoride membrane. *Indian J Biochem Biophys* 38, 274–276.

6. Bolt, M.W., and Mahoney, P.A. (1997) High efficiency blotting of proteins of diverse sizes following sodium dodecyl sulfate-polyacrylamide gel electrophoresis. *Anal Biochem* 247, 185–192.
7. Otter, T., King, S.M., and Witman, G.B. (1987) A two-step procedure for efficient electrotransfer of both high-molecular weight (greater than 400,000) and low-molecular weight (less than 20,000) proteins. *Anal Biochem* 162, 370–377.
8. Deutscher, S.L., Harley, J.B., and Keene, J.D. (1988) Molecular analysis of the 60 kD human Ro ribonucleoprotein. *Proc Natl Acad Sci USA* 85, 9479–9483.
9. Kurien, B.T., and Scofield, R.H. (2002) Heat mediated, ultra-rapid electrophoretic transfer of high and low molecular weight proteins to NC membranes. *J Immunol Methods* 266, 127–133.
10. Ben-Chetrit, E., Gandy, B.J., Tan, E.M., and Sullivan, K.F. (1989) Isolation and characterization of a cDNA clone encoding the 60 kDa component of the human SS-A/Ro ribonucleoprotein autoantigen. *J Clin Invest* 83, 1284–1292.
11. Laemmli, U.K. (1970) Cleavage of structural proteins during assembly of the head of bacteriophage T4. *Nature* 227, 680–685.

Chapter 21

Strip Immunoblotting of Multiple Antigenic Peptides to Nitrocellulose Membrane

Biji T. Kurien

Summary

Multiple antigenic peptides (MAPs) can be efficiently separated on sodium dodecyl sulfate (SDS) polyacrylamide gel and transferred to a nitrocellulose membrane for immunoblotting. MAPs involve a hepta lysine core with end groups for anchoring multiple copies of the same synthetic peptide. MAPs are amenable to staining with Coomassie and silver on SDS polyacrylamide gels as well as by Fast Green on a blotted nitrocellulose membrane. They lend themselves to analysis on an immunoblot as they behave like low molecular weight proteins. Affinity immunoblotting for analysis of antibody clonotype distribution has also been carried out using these peptides.

Key words: Multiple antigenic peptides, SDS-PAGE, Immunoblotting, Nitrocellulose

1. Introduction

Multiple antigenic peptides (MAPs) consist of a hepta lysine backbone with end groups for anchoring multiple copies of the same synthetic peptide. Ever since MAPs were first introduced in 1988 *(1)*, they have been used by several investigators for eliciting anti-peptide antibodies *(2–7)*, antibodies to ribonucleoprotein complexes *(8)*, and in enzyme linked immunosorbant assay *(9, 10)*. The potential of MAPs as vaccines *(1, 11)* has also been investigated. In addition to these studies, MAPs have also been used in surface plasmon resonance studies, double immunodiffusion, affinity column purification *(12)*, and calcium-binding assays *(13)*.

Biji T. Kurien and R. Hal Scofield (eds.), *Methods in Molecular Biology, Protein Blotting and Detection, vol. 536*

DOI: 10.1007/978-1-59745-542-8_21

We demonstrate in this report that multiple antigenic peptides (MAPs) can be efficiently separated on sodium dodecyl sulfate (SDS) polyacrylamide gels and transferred to a nitrocellulose membrane for subsequent use in western blot. MAPs lend themselves to staining with Coomassie and silver on SDS-PAGE as well as by Fast Green on an immunoblot. Affinity immunoblotting for analysis of antibody clonotype distribution has been carried out using MAPs *(14)*.

Other investigators have used MAPs previously in SDS-PAGE *(15, 16)*. MAPs appear to migrate as smears in these studies. However, the MAPs used in this study migrated like low molecular weight proteins, as distinct bands. The difference most likely is on account of the fact that the MAPs used in our studies were purified by HPLC and shown to be homogeneous by mass spectroscopy. These data presented herein indicate that MAPs can be used in SDS-PAGE (**Fig. 1**) and immunoblot (**Fig. 2**) without any further modification.

2. Materials

Prepare all solutions using ultrapure water (prepared by purifying deionized water, to attain a sensitivity of 18 M Ω cm at 25°C) and analytical grade reagents. Prepare and store all reagents at room temperature (unless indicated otherwise). Diligently follow all waste disposal regulations when disposing waste materials. We do not use sodium azide in our reagents.

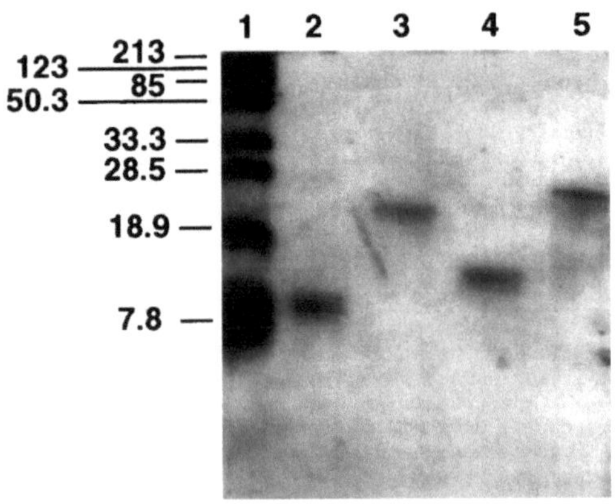

Fig. 1. MAPs from the human Ro 60 sequence analyzed on a 15% SDS-PAGE and stained with Coomassie blue. *Lanes 2, 3, 4* and *5* show the MAPs 10, 13, 19, and 21, respectively (*see* **Table 1**). *Lane 1* corresponds to prestained high range molecular weight standards. MAPs at a concentration of 5 μg/well (for Coomassie stain) and about a third of this for silver staining were used in SDS-PAGE. Immediately following electrophoresis, the gel was stained with Coomassie brilliant blue or silver according to standard procedures (reproduced from *(22)* with permission from wiley-VCH).

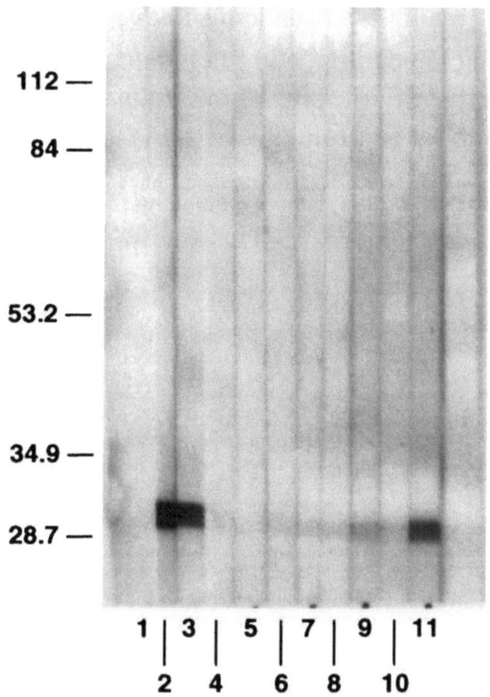

Fig. 2. Ro MAP 482–495 from the sequence of the human Ro autoantigen analyzed on an immunoblot after transfer from a 15% SDS-PAGE. *Lane 1* corresponds to low range molecular weight standards. *Lanes 2* and *3* – probed with sera from mice immunized with 479–494 60 kD Ro linear peptide binding to 482–495 Ro MAP (Table 1, MAP #21); *Lanes 4* and *5* – negative peptide mouse serum; *Lanes 6* and *7* – Freund's adjuvant only immunized mouse serum; *Lane 8* – saline only injected mouse serum; *Lane 9* – anti 52 kD Ro human serum; *Lane 10* – normal human serum; *Lane 11* – anti 60 kD Ro human serum, and *Lane 12* – conjugate control. Sera: DBA mice were immunized with a linear peptide synthesized from the 60 kD human Ro sequence (Ro 479–494). Groups of mice were immunized with an unrelated peptide (a peptide from the sequence of the Human Leucocyte Antigen-A2), with Freund's Aadjuvant only and with saline only. The peptides (100 μg) made up in 0.89% saline were mixed well with Freund's complete adjuvant and injected subcutaneously and intraperitoneally. Subsequent boosts were in Incomplete Freunds adjuvant. The mice were bled from their tail veins. Blood was centrifuged and the anti-sera used in the immunoblots. Anti-Ro 60 sera from systemic lupus erythematosus patients and normal sera were also used in immunoblots (reproduced from *(22)* with permission from Wiley-VCH).

1. Ro 60 multiple antigenic peptides (MAPs): Twenty one Ro 60 MAPs were synthesized (*see* **Note 1**) from the sequence of the Ro 60 autoantigen *(17, 18)* (Molecular Biology Resource Facility, University of Oklahoma Health Sciences Center, Oklahoma City, OK, USA) by a manual stepwise solid phase procedure. An unrelated MAP with the sequence PPPGRRPP from the Sm autoantigen *(19)* was also synthesized (*see* **Table 1**).
2. 15% SDS polyacrylamide gels (*see* **Note 2**).
3. SDS-PAGE running buffer: 0.025 M Tris-HCl, pH 8.3, 0.192 M glycine, 0.1% SDS.
4. SDS-PAGE lysis buffer (5x): 0.3 M Tris-HCl (pH 6.8), 10% SDS, 25% β-mercaptoethanol, 0.1% bromophenol blue, 45% glycerol. Leave one aliquot at 4°C for current use and store remaining aliquots at –20°C (*see* **Note 3**).

Table 1
The sequences, amino acid position, and the molecular weight (as measured by mass spectrometry) of the different Ro 60 multiple antigenic peptides constructed from the sequence of the Ro 60 protein

Amino acid sequence of 60 kD Ro MAPs used	Location on the Ro Protein	M. Wt. by mass spec. (kD)
1. TYYIKEQKLGL	45–55	11.69
2. SQEGRTTKQ	81–89	9.12
3. STKQAAFKAV	106–115	9.25
4. TFIQFKKDLKES	126–137	12.7
5. MKCGMWGRA	139–147	9.2
6. MWGRALRKAIA	143–153	11.02
7. LAVTKYKQRNGWSHK	166–180	15.37
8. LRLSHLKPS	183–191	9.25
9. VTKYITKGWKEVH	198–210	13.55
10. LYKEKALS	212–219	8.45
11. TEKLLKYL	222–229	8.9
12. EAVEKVKRTKDELE	230–243	14.23
13. HLLTNHLKSKEVWKAL	257–272	16.18
14. ALLRNLGKMTA	280–290	10.34
15. NEKLLKKARIHPFH	310–323	14.69
16. YKTGHGLRGKLKWRP	331–345	15.22
17. AAFYKTFKTV	355–364	10.25
18. VEPTGKRFL	364–372	9.21
19. MVVTRTEKDSY	401–411	11.4
20. LPMIWAQKTNTP	449–460	12.04
21. ALREYRKKMDIPAK	482–495	14.59
22. PPPGRRPP	Sm	7.63

HPLC purification and characterization by mass spectrophotometry confirmed the homogeneity and purity of these MAPs

5. BenchMark prestained high and low molecular weight standards (Gibco BRL, Bethesda, MD, USA).
6. 0.1% bromophenol blue.
7. Nitrocellulose membranes (Gelman Sciences/Fisher Scientific, Dallas, TX, USA).

8. Western blot transfer buffer: 0.025 M Tris, 0.192 M glycine, 20% methanol.
9. Western blot transfer apparatus: Mini Trans-Blot apparatus (Bio-Rad, Hercules, CA, USA).
10. Tris buffered saline (TBS) – 10× solution: 1.5 M NaCl, 0.1 M Tris-HCl, pH 7.4.
11. Blocking solution: 5% milk in TBS (*see* **Note 4**). Store at 4°C.
12. Diluent solution: 5% milk in TBST (*see* **Note 4**). Store at 4°C.
13. Membrane strip mini-incubation containers (Bio-Rad).
14. Alkaline phosphatase (AP) buffer: 100 mM Tris-HCl, pH 9.5, 100 mM NaCl, 5 mM $MgCl_2$.
15. Nitro blue tetrazolium/5-bromo-4-chloro-3-indolyl phosphate, *p*-toluidine salt (NBT/BCIP) substrate stock: Dissolve 1 g NBT in 20 mL of 70% dimethylformamide (DMF). Dissolve 1 g BCIP in 20 mL of 100% DMF (*see* **Note 5**). Store at 4°C.
16. Working substrate solution: Add 66 μL of NBT to 5 mL of AP buffer and then add 33 μL BCIP. Make up to 10 mL with AP buffer. Prepare this just prior to adding to membrane (*see* **Note 5**).
17. 0.1% Fast Green in 25% methanol and 10% acetic acid.

3. Methods

3.1. Immunoblotting

1. Perform regular 15% SDS PAGE (preparative gel or 10-well gels) (*see* **Note 6**) *(20)* (*see* chapter "Non-electrophoretic bi-directional transfer of a single SDS-PAGE gel with multiple antigens to obtain twelve immunoblots") and transfer to nitrocellulose membrane by Towbin's electrophoretic transfer method *(21)* (*see* chapter "Detection of La/SS-B by Western Blot Using Nanogold-Tagged Antibodies and Silver Enhancement"; *see* **Note 7**).
2. Stain with Fast Green to visualize the MAPs transferred to the membrane (*see* **Note 8**).
3. Trim the edges of the membrane. Wrap the Fast Green stained membrane in cling wrap and make a photocopy for record keeping purposes.
4. Excise the lane (for strip immunoblotting) with molecular weight standards from the main nitrocellulose sheet containing the transferred protein. Excise a tiny wedge from the bottom

left side of the marker lane and the main membrane sheet for orientation purposes (*see* **Note 9**).

5. Excise thin strips, about 2 mm in width, from the main nitrocellulose membrane sheet. Cut a tiny wedge from the bottom left side of the strips for orientation purposes. Place them in numbered 12-lane plastic trays with the antigen side up (*see* **Note 9**).
6. Destain strips with TBST. Decant TBST (*see* **Note 10**) once it becomes colored and replace with fresh TBST if necessary.
7. Block for 1 h with 1 mL each of blocking solution. Decant blocking.
8. Add 1 mL primary human or mouse sera at 1:100 dilution and incubate for 2 h. Decant (*see* **Note 11**).
9. Rinse strips with deionized water 2–3 times (*see* **Note 12**).
10. Wash five times with TBST, 5 min each time. Dilute anti-human or anti-mouse IgG conjugated to alkaline phosphatase 1:5,000 with diluent and add 1 mL to each strip. Incubate for 1 h. Wash five times with TBST, 5 min each time. Add 0.5 mL substrate and develop bands.
11. Align the strips (with developed bands) on card board insert (*see* **Note 13**).

4. Notes

1. Twenty one MAPs, from the Ro 60 sequence, (*see* **Table 1**) were synthesized at the Molecular Biology Resource Facility, University of Oklahoma Health Sciences Center, Oklahoma City. OK, USA. The peptide sequences were designed based on the regions on Ro 60 bound by anti-Ro 60 human sera, obtained from systemic lupus erythematosus patients, in a PIN ELISA. The MAPs were quick HPLC purified and showed a single peak on mass spectrometric analysis *(22)*. In experiments varying the amount of MAP, as little as 250 ng of MAP could be detected on an immunoblot.
2. Four to twenty percent gradient gels (Bioexpress) can also be used.
3. SDS precipitates at 4°C. Therefore, the lysis buffer needs to be warmed prior to use.
4. Add 100 mL of 10× TBS to a 1 L graduated cylinder and make it to about 800 mL with water. Transfer 50 g skim milk powder into the cylinder and stir until dissolved. Make to 1 L with water. Separate 500 mL as the blocking solution.

To the remaining 500 mL add 250 μL of Tween-20 (cut end of blue tip to aspirate Tween-20 easily), dissolve and use it as diluent.

5. Dissolve NBT in the bottle it came in. Likewise for BCIP. We started using stabilized ready-to-use BCIP/NBT substrate (Kirkegaard and Perry Laboratories, Gaithersburg, Maryland, USA) solution prior to switching to ECL plus chemiluminiscence (Amersham/GE Healthcare, Piscataway, NJ, USA).
6. Do not heat MAPs with the lysis buffer, since it makes the MAPs to migrate as doublets at about 100,000 molecular weight on SDS-PAGE. Since they contain short sequences, there is no need to denature them with heat.
7. Good contact between membrane and gel is essential for good transfer. Add extra foam if contact is insufficient.
8. Using a dilute fast green solution helps to prevent overstaining of the membrane. Dilute stock fast green solution in gel destaining solution (25% methanol, 10% acetic acid). Use TBST to destain the membrane if it is overstained with fast green.
9. Strips can be excised very nicely using a razor blade. A Mini PROTEAN® 3 System glass plate is placed at an angle on the nitrocellulose at a distance of 2 mm from the edge and the razor is used to cut the strip (pull the razor blade along the sides of the glass plate to cut).
10. The tray can be gently slanted and the entire buffer can be discarded into the sink. The membrane strips remain stuck to the plates.
11. Decant diluted human sera into container having the diluted germicidal detergent Vesphene (Steris Corporation, St. Louis, MO, USA). Let sit in this for a couple of hours and then discard into sink.
12. Rinsing the membrane strips with deionized water 2–3 times will help remove a bulk of the nonspecific antibodies and other contaminants. This will help reduce the amount of TBST used subsequently and also reduce the number of washes. This wash helps to reduce nonspecific binding of NBT/BCIP to the strip. The water, owing to its low ionic strength compared with TBST, will be able to remove contaminants much better than TBST. Water is much cheaper compared with TBST, in terms of money and labor. Other investigators have found no reduction in detection of specific signals due to washing with water *(23)*.
13. We use paper boards found in between stacks of ELISA plates in packages of ELISA plates (Costar, Cambridge, MA, USA) for this purpose.

Acknowledgement

This work was supported by NIH grant ARO1844 and Oklahoma Center for the Advancement of Science and Technology to Dr. Hal Scofield, OMRF, OKC, OK, USA.

References

1. Tam, J.P. (1988) Synthetic peptide vaccine design: synthesis and properties of high-density multiple antigenic peptide system. *Proc Natl Acad Sci USA* 85, 5409–5413.
2. Posnett, D.N., McGrath H., and Tam J.P. (1988) A novel method for producing anti-peptide antibodies. Production of site specific antibodies to the cell antigen receptor beta-chain. *J Biol Chem.* 263, 1719–1725.
3. Troalen, F., Razafindratsita A., Puisieux A., Voeltzel T., Bohuon C., Bellet D., et al. (1990) Structural probing of human lutropin using antibodies raised against synthetic peptides constructed by classical and multiple antigen peptide system approaches. *Mol Immunol.* 27, 363–368.
4. Munesinghe, D.Y., Clavijo P., Calle M.C., Nuzzenzweig R.S., and Nardin E. (1991) Immunogenicity of multiple antigen peptides (MAP) containing T and B cell epitopes of the repeat region of the *P. falciparum* circumsporazoite protein. *Eur J Immunol.* 21, 3015–3020.
5. McLean, G.W., Owsianka A.M., Subak-Sharpe J.H., and Marsden H.S. (1991) Generation of anti-peptide and anti-protein sera. Effect of peptide presentation on immunogenicity. *J Immunol Meth.* 137, 149–157.
6. Habluetzel, A., Pessi A., Bianchi E., Rotigliano G., and Esposito F. (1991) Multiple antigen peptides for specific detection of antibodies to a malaria antigen in human sera. *Immunol Lett.* 30, 75–80.
7. Francis, M.J., Hastings G.Z., Brown F., McDermed J., Lu Y.-A., and Tam J.P. (1991) Immunological evaluation of the multiple antigen peptide (MAP) system using the major immunogenic site of foot-and-mouth disease virus. *Immunology* 73, 249–254.
8. James, J.A., Gross T., Scofield R.H., and Harley J.B. (1995) Immunoglobulin epitope spreading and autoimmune disease after peptide immunization: Sm B/B'-derived PPPG-MRPP and PPGIRPP induce spliceosome autoimmunity. *J Exp Med.* 181, 453–461.
9. Tam, J.P. and Zavala F. (1989) A novel approach to increase detection sensitivity of synthetic peptides in solid-phase immunoassays. *J Immunol Meth.* 124, 53–61.
10. Briand, J.-P., Barin C., Regenmortel M.H.V.V., and Muller S. (1992) Application and limitations of the multiple antigenic peptide (MAP) system in the production and evaluation of anti-peptide and anti-protein antibodies. *J Immunol Meth.* 156, 255–265.
11. Tam, J.P. and Lu Y.-A. (1989) Vaccine engineering: Enhancement of immunogenicity of synthetic peptide vaccines related to hepatitis in chemically defined models consisting of T- and B- cell epitopes. *Proc Natl Acad Sci USA* 86, 9084–9088.
12. Scofield R.H., Kurien B.T., Zhang F., Mehta P., Kaufman K., Gross T., et al.. (1999) Protein-protein interaction of the Ro- ribonucleoprotein particle using multiple antigenic peptides. *Mol Immunol.* 36, 1093–1106.
13. Scofield R.H., Kurien B.T., Gross J.K., Reed N.B., Taylor A.K., Dominguez N., et al.. (2004) Interaction of calcium and Ro60: increase of antigenicity. *Mol Immunol.* 41, 809–816.
14. Scofield, R.H., Kurien B.T., and Reichlin M. (1997) Immunologically restricted and inhibitory anti-Ro/SSA in monozygotic twins. *Lupus* 6, 395–398.
15. McLean G.W., Cross A.M., Munns M.S., and Marsden H.S. (1992) Rapid attachment of a helper T cell epitope to branched peptides by fragment condensation to give enhanced immunogenicity. *J Immunol Methods* 155, 113–120.
16. Yao, Z.J., Kao M.C.C., Loh K.C., and Chung M.C.M. (1994) Antibody against branched epitope as an affinity ligand to separate the parent protein. *J Chromatogr A.* 679, 190–194.
17. Deutscher, S.L., Harley J.B., and Keene J.D. (1988) Molecular analysis of the 60 kD human Ro ribonucleoprotein. *Proc Natl Acad Sci USA* 85, 9479–9483.
18. Ben-Chetrit, E., Gandy B.J., Tan E.M., and Sullivan K.F. (1989) Isolation and characterization of a cDNA clone encoding the 60 kDa

component of the human SS- A/Ro ribonucleoprotein autoantigen. *J Clin Invest.* 83, 1284–1292.

19. James, J.A., Gross T., Scofield R.H., and Harley J.B. (1995) Immunoglobulin epitope spreading and autoimmune disease after peptide immunization: Sm B/B'-derived PPPG-MRPP and PPGIRPP induce spliceosome autoimmunity. *J Exp Med.* 181, 453–461.
20. Laemmli, U.K. (1970) Cleavage of structural proteins during assembly of the head of bacteriophage T4. *Nature* 227, 680–685.
21. Towbin, H., T. and Gordon J.. (1979) Electrophoretic transfer of proteins from polyacrylamide gels to nitrocellulose sheets: procedure and applications. *Proc Natl Acad Sci USA* 76, 4350–4354.
22. Kurien B.T., Jackson K., and Scofield R.H. (1998) Immunoblotting of multiple antigenic peptides. *Electrophoresis* 19, 1659–1661.
23. Wu M., Stockley P.G., and Martin W.J. 2nd. (2002) An improved Western blotting technique effectively reduces background. *Electrophoresis* 23, 2373–2376.

Chapter 22

Affinity Purification of Autoantibodies from an Antigen Strip Excised from a Nitrocellulose Protein Blot

Biji T. Kurien

Summary

A method for the affinity purification of autoantibodies from small volumes of human sera using a western blot strip containing a target antigen electrophoretically transferred from a sodium dodecyl sulfate (SDS) polyacrylamide gel is described. This method is a very useful alternative for affinity column chromatography, particularly when the antigen of interest is of low abundance. The protein mixture is resolved on a preparative SDS polyacrylamide gel and transferred to nitrocellulose membrane. A couple of strips are excised vertically from either side of the blotted membrane and immunoblotted with specific antisera to identify the target band. Then the target band is excised horizontally and used for affinity purification. We have used this procedure to affinity purify antibodies to a 70,000 molecular weight protein derived from HeLa cell extract. A sham band, excised away from the target antigen, was used as a control for sham purification of autoantibodies. The autoantibodies purified in this manner reproduced the multiple nuclear dot anti-nuclear antibody pattern obtained using crude sera from 21 patients without primary biliary cirrhosis or anti-mitochondrial antibody.

Key words: Anti-nuclear antibody, Multiple nuclear dot, Autoimmunity, Affinity purification, Nitrocellulose, Autoantibodies, Autoantigens

1. Introduction

Anti-nuclear antibodies (ANA), directed against a variety of nuclear antigens, are a common characteristic of autoimmune diseases *(1)*. Its detection depends on the evaluation of immuno-fluorescence of tissue culture cells. ANA positive sera, typically, exhibit nuclear homogeneous, nuclear speckled or nucleolar immunofluorescence patterns on HEp-2 (Human Epithelial Type 2) cells. However, in certain instances, a distinct set of fluorescent spots called multiple nuclear dot (MND) ANA occur all over the nucleus except in the

Biji T. Kurien and R. Hal Scofield (eds.), *Methods in Molecular Biology, Protein Blotting and Detection, vol. 536*

DOI: 10.1007/978-1-59745-542-8_22

nucleoli *(2, 3)*. The MND ANA are similar to the anti-centromere pattern but differs in that the discrete fluorescent dots are larger, fewer in number, and seen only in the interphase cell, whereas the anti-centromere antibodies (ACA) give a speckled or punctate staining of the chromosomes in metaphase.

We undertook a study to determine the antigen bound by sera from a group of 21 patients with MND ANA but no ACA. We found that the crude sera from these patients showed a MND ANA pattern on HEp-2 cells. These sera were also found to bind to a 70,000 molecular weight antigen on a HeLa cell extract immunoblot. This study was undertaken to see whether autoantibodies affinity purified off this 70,000 molecular weight protein could reproduce the pattern that we found using the crude sera.

Purified antibodies are essential for a number of techniques, such as immunoblots, immunoassays, or cell staining (e.g., HEp-2 cells). There are several methods for purifying antibodies, e.g., precipitation with ammonium sulfate, use of hydroxyapatite column, gel filtration, protein A/B beads, and column affinity chromatography *(4)*. Ammonium sulfate precipitation is advantageous in that it is cheap and convenient to work with large volumes. However, the antibody yields are impure, and therefore this procedure has to be coupled with other methods to obtain pure antibodies. Gel filtration is appropriate for IgM derived from all sources, since it can efficiently separate IgM from other antibody in polyclonal sera *(4)*. Sample dilution, impure antibody yields and low sample capacity are some of the disadvantages of this method. Protein A beads are useful for IgG that bind to it from various sources and results in pure antibody, providing high yield in a single step. It is disadvantageous in that it is expensive and also because it is not suitable for all species and classes *(4)*. Affinity column chromatography has been commonly used for purifying antibodies from polyclonal sera. This procedure requires pure antigen for coupling to the column, is expensive, and involves multiple steps. However, it is possible to obtain pure and specific antibody. The main disadvantage of this procedure is that it requires large amounts of pure antigen. However, when the antigen of interest is of low abundance in a mixture of proteins (like in HeLa cell extract) and antibodies to that antigen is to be characterized from limited amounts of patient sera, affinity purification using a Western blot strip containing that antigen becomes very useful. The protein mixture is electrophoresed on a SDS polyacrylamide gel, transferred to nitrocellulose membrane, and stained with Fast Green. Two thin strips, one from each side of the blotted membrane, are excised and immunoblotted with specific antisera to identify the target band. Then the target band is excised horizontally and used for affinity purification of autoantibodies from patient sera. A sham band of the nitrocellulose, approximately similar in size to the excised target band, was also

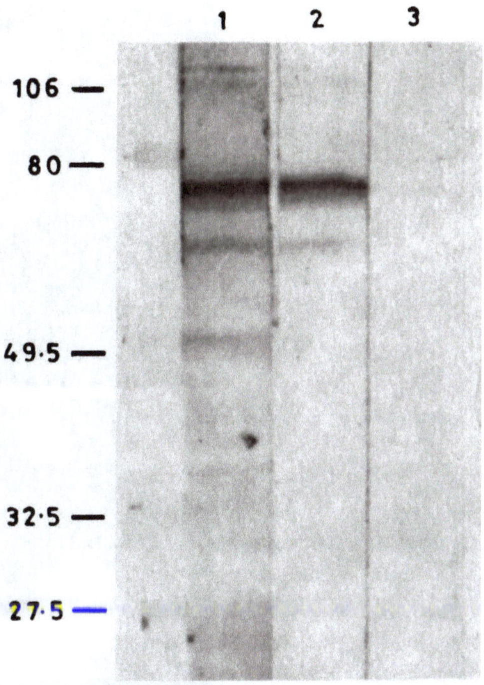

Fig. 1. Demonstration of immunoglobulin purified from the 70 kD band binding back to the 70 kD band on immunoblot. *Lane 1*: Crude MND ANA prototype human serum prior to affinity purification of anti-70 kD antibody; *Lane 2* – prototype human serum partially depleted of the anti-70 kD antibody following affinity purification (crude serum minus affinity purified antibodies); *Lane 3* – affinity purified anti-70 kD immunoglobulin (reproduced from *(5)* with permission from publisher).

cut from an area away from the 70 kD band. We have used affinity purification, using nitrocellulose membrane strips, to purify antibodies to the 70,000 molecular weight protein derived from HeLa cell extract to reproduce binding to the 70 kD protein on immunoblot (**Fig. 1**) as well as the multiple nuclear dot antinuclear antibody pattern (**Fig. 2**) detected using crude sera from 21 patients without primary biliary cirrhosis or anti-mitochondrial antibody *(5)*.

2. Materials

Prepare all solutions using ultrapure water (prepared by purifying deionized water, to attain a sensitivity of 18 M Ω cm at 25°C) and analytical grade reagents. Prepare and store all reagents at room temperature (unless indicated otherwise). Diligently follow all waste disposal regulations when disposing waste materials. We do not add sodium azide to the reagents.

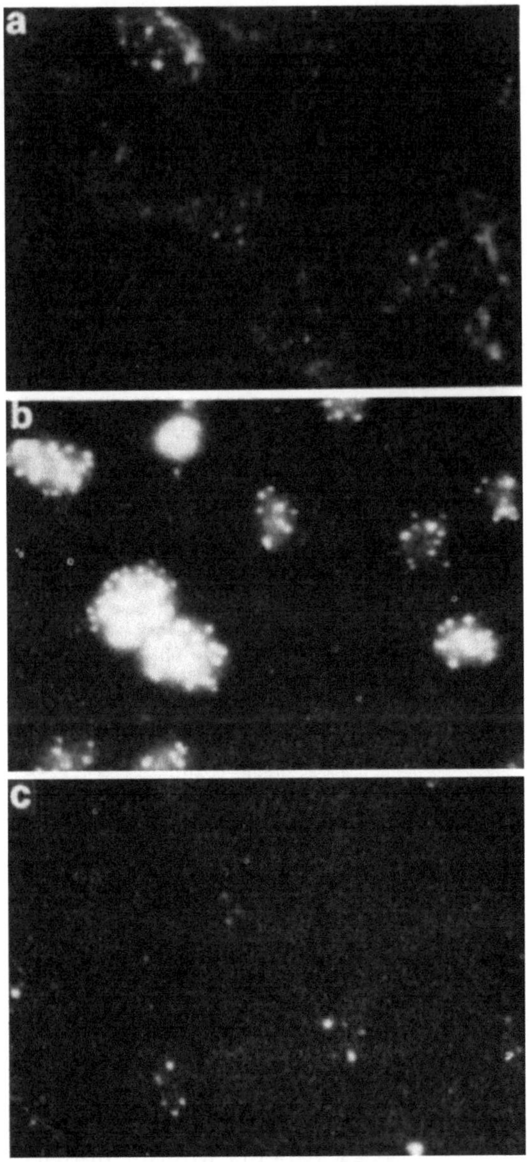

Fig. 2. Indirect immunofluourescence showing that affinity anti-70 kD antibody reproduces the pattern obtained with crude serum. *Top panel*–Purified anti-mitochondrial antibody with MND and binding to a 100 kD antigen in immunoblot; *Middle panel*– MND serum that bound to a 70 kD antigen in immunoblot; *Bottom panel*– antibody affinity purified off the 70 kD band from the MND serum. HEp-2 slides are incubated with crude or purified anti-70 kD IgG, washed with PBS and further incubated with FITC labeled anti-human IgG secondary antibody. The slides are washed with PBS and photographed (reproduced from *(5)* with permission from publisher).

1. Resolving gel buffer: 1.5 M Tris-HCl, pH 8.8. Add about 100 mL water to a 1 L graduated cylinder or a glass beaker (*see* **Note 1**). Weigh 181.7 g Tris and transfer to the cylinder. Add water to a volume of 900 mL. Mix and adjust pH with HCl (*see* **Note 2**). Make up to 1 L with water.

2. Stacking gel buffer: 0.5 M Tris-HCl, pH 6.8. Weigh 60.6 g Tris and prepare a 1 L solution as in previous step.
3. Thirty percent acrylamide/bis solution (29.2:0.8) Weigh 29.2 g of acrylamide monomer and 0.8 g bis (cross-linker) and transfer to a 100 mL graduated cylinder containing about 40 mL of water. Add a spatula of mixed-resin beads and mix for about 30 min. Make up to 100 mL with water and filter through a 0.45 μm Corning filter (*see* **Note 3**). Store at 4°C, in a bottle wrapped with aluminum foil (*see* **Note 4**).
4. Ammonium persulfate: 10% solution in water (*see* **Note 5**).
5. *N.N.N.N'*-Tetramethyl-ethylenediamine (TEMED) (Sigma Chemical Company, St. Louis, MO, USA). Store at 4°C (*see* **Note 6**).
6. SDS-PAGE running buffer: 0.025 M Tris-HCl, pH 8.3, 0.192 M glycine, 0.1% SDS (*see* **Note 7**).
7. SDS lysis buffer (5×): 0.3 M Tris-HCl (pH 6.8), 10% SDS, 25% β-mercaptoethanol, 0.1% bromophenol blue, 45% glycerol. Leave one aliquot at 4°C for current use and store remaining aliquots at −20°C (*see* **Note 8**).
8. Bromophenol blue (BPB) solution: Dissolve 0.1 g BPB in 100 mL water.
9. Phosphate buffered saline (PBS), pH 7.4. Dissolve 5 PBS tablets (Sigma) in 1 L water to obtain PBS solution, pH 7.4 at 25°C (0.01 M phosphate, 0.00027 M potassium chloride, 0.137 M sodium chloride).
10. Human lymphocytes: Purify lymphocytes from normal human peripheral blood using Lymphoprep (Fisher Scientific, Dallas, TX, USA) according to manufacturer's instruction.
11. BenchMark prestained molecular weight standards (Gibco BRL, Bethesda, MD, USA).
12. Nitrocellulose membranes (Bio-Rad, Hercules, CA, USA).
13. Western blot transfer buffer: 0.025 M Tris, 0.192 M glycine, 20% methanol (*see* **Note 9**).
14. Western blot transfer apparatus (Bio-Rad).
15. Tris buffered saline (TBS; 10×): 1.5 M NaCl, 0.1 M Tris-HCl, pH 7.4.
16. Blocking solution: 5% milk in TBS (*see* **Note 10**). Store at 4°C.
17. Diluent solution: 5% milk in TBST (*see* **Note 10**). Store at 4°C.
18. Alkaline phosphatase (AP) buffer: Weigh 6.1 g of Tris, 2.9 g sodium chloride and 0.51 g magnesium chloride-$6H_2O$ and make it to v mL with water after adjusting pH to 9.3 with HCl. Store at 4°C (*see* **Note 1**).

19. Nitro blue tetrazolium (NBT)/5-bromo-4-chloro-3-indolyl phosphate (BCIP): Add 33 μL of BCIP and 66 μL of NBT to 10 mL of AP buffer at the time of assay.
20. Glycine elution buffer (Glycine Buffered Saline-Tween-20; GBST): 0.2 M glycine, 0.15 M NaCl, 0.05% Tween-20, pH 2.7 (GBST). Store at 4°C.
21. Electrophoresis system: Mighty small vertical slab gel unit (Hoeffer Scientific) (*see* **Note 11**).
22. HEp-2 slides (INOVA, San Diego, CA, USA).
23. Centricon 30 micro-concentrators (Amicon, Danvers, MA, USA).
24. Alkaline phosphatase conjugated affinity purified goat anti-human IgG (Jackson ImmunoResearch Laboratories, West Grove, PA, USA).
25. Nitro blue tetrazolium (NBT)/5-bromo-4-chloro-3-indolyl phosphate (BCIP): Dissolve 1 g NBT in 20 mL of 70% dimethylformamide (DMF). Disolve 1 g BCIP in 20 mL of 100% DMF. Add 33 μL of BCIP and 66 μL of NBT to 10 mL of alkaline phosphatase buffer just before adding to membrane.
26. Fluorescent isothiocyanate conjugated affinity purified goat anti-human IgG (Jackson ImmunoResearch).
27. Tubes (3.5 mL) (Sarstedt 55.484, Numbrecht, Germany).
28. 0.009″ single edge razor blades–Smith Brand (Fisher Scientific).
29. Membrane strip mini-incubation container (Bio-Rad).
30. Helium.

3. Methods

All procedures are carried out at room temperature unless otherwise specified.

3.1. Preparation of Human Lymphocyte Extract

1. Lyse the lymphocytes by sonication in PBS using a Branson sonicator (setting # 4) and spin at 10,000 × g for 10 min (*see* **Note 12**). Analyze an aliquot of the supernatant by SDS-PAGE.

3.2. 10% Preparative SDS Polyacrylamide Gel

1. Mix 2.5 mL of resolving buffer, 3.33 mL of acrylamide mixture, and 4 mL water in a 50 mL conical flask. Degas with helium for 10 min. Add 100 μL of SDS, 80 μL of ammonium persulfate, and 10 μL of TEMED and cast gel within a 7.25 cm × 10 cm × 1.5 mm gel cassette. Allow space for

stacking gel and gently overlay with isobutanol or water (*see* **Note 13**).

2. Prepare the stacking gel by mixing 1.25 mL of resolving buffer, 0.67 mL of acrylamide mixture and 3 mL water in a 50 mL conical flask. Degas with helium for 10 min. Add 100 μL of SDS, 40 μL of ammonium persulfate, and 5 μL of TEMED. Insert a preparative gel comb immediately without introducing air bubbles.
3. Heat an aliquot of the lymphocyte extract at 95°C for 5 min. Do not add lysis buffer to the prestained protein standard or subject it to heat (manufacturer instructions). The heated lymphocyte extract was centrifuged at 3,000 × g for 30 s to bring down the condensate. Load the sample and the protein standard (10 μL/well) on the gel. Electrophorese at 15 mA until the sample enters the gel and then continue at 25 mA till the dye front (from the BPB dye in the samples) reached the bottom of the gel (*see* **Note 14**).

3.3. Immunoblotting

1. Transfer proteins resolved on SDS PAGE to nitrocellulose membrane by Towbin's electrophoretic transfer method *(5)* (*see* chapter "Detection of La/SS-B by Western Blot Using Nanogold-Tagged Antibodies and Silver Enhancement").
2. Stain with Fast Green to visualize proteins transferred to the membrane (*see* **Note 15**).
3. Trim the edges of the membrane. Wrap the Fast Green stained membrane in cling wrap and make a photocopy for record keeping purposes.
4. Excise the lane with molecular weight standards from the main nitrocellulose sheet containing the transferred protein. Excise a tiny wedge from the bottom left side of the marker lane and the main membrane sheet for orientation purposes (*see* **Note 16**).
5. Excise two strips, each about 2 mm in width, from either ends of the main nitrocellulose membrane sheet. Place them in a 12-lane mini incubation container.
6. Destain strips with TBST.
7. Block for 1 h with 1 mL each of blocking solution.
8. Add 1 mL primary sera at 1:100 dilution and incubate for 2 h. Wash five times with TBST, 5 min each time. Dilute anti-human IgG conjugated to alkaline phosphatase 1:5,000 with diluent and add 1 mL to each strip. Incubate for 1 h. Wash five times with TBST, 5 min each time. Add 0.5 mL substrate and develop bands.
9. Align the two strips (with developed bands) on either side of the main nitrocellulose membrane and excise the desired

antigen strip (horizontally this time). Use this strip, containing the antigen of interest, for the affinity purification of autoantibodies (*see* **Note 17**).

10. Cut a sham strip away from the target antigen strip and also similar to the size of the target antigen strip. This is used as a sham purification control.

3.4. Purification of Autoantibodies (6, 7)

1. Block the tubes for collecting purified antibodies, the tubes for incubating the membrane strips with sera (Sarstedt 55.484) and the Centriprep 30 concentrator with 1% milk/TBST solution for 30 min at 4°C (*see* **Note 18**).
2. Wash the tubes adequately with TBST to get rid of unbound milk protein.
3. Cut the nitrocellulose membrane strip with the desired antigen into smaller pieces and transfer to Sarstedt 3.5 mL "B" tube (for blocking). Incubate these pieces with 2.5 mL 5% milk/TBST for 30 min on an orbital shaker.
4. Wash well with several changes of TBST.
5. Transfer pieces from the "Blocking tube" ("B") to a serum "Incubation" ("I") tube previously blocked with 1% milk/TBST.
6. Incubate nitrocellulose pieces with 2.5 mL of a 1:50 sera dilution in 5% milk/TBST for 1 h on an orbital shaker (*see* **Note 19**).
7. At the end of the incubation, pipet off and save the diluted sera (sera diluted 1:50 in 5% milk/TBST) into another tube for use in later repeat sera incubations with the membrane pieces.
8. Transfer the membrane pieces to a TBST "Wash" tube ("W") and wash five times (10 min each time) on the orbital shaker.
9. After the wash, transfer the membrane pieces to a "Glycine Elution" tube ("E").
10. Add 2.5 mL glycine elution buffer (GBST) and shake vigorously on an orbital shaker for 2 min. Then pipette the solution to a previously blocked collection tube.
11. Repeat with 2.5 mL of GBST and pipette into collection tube.
12. Add 2.5 mL of TBST and shake for approximately 1 min and pipette this solution into the collection tube.
13. Neutralize GBST with 500 μL of 1 M Tris.
14. Transfer pieces back to "blocking" tube ("B"). Block for 30 min in 5% milk/TBST. Wash with TBST.
15. Repeat steps 4–10 for two more elution cycles. However, care need to be taken to use a 1:100 dilution of sera in 5% milk/TBST saved in step 7.
16. Upon completion of 3 elution cycles, transfer nitrocellulose pieces back to "blocking" tube ("B"). Block in 5% milk/TBST

for 30 min. The nitrocellulose can be stored at 4°C for 1–2 weeks if further elutions are needed.

17. Concentrate the eluted antibody solution to required volume (*see* **Note 20**). Antibody is now ready for immunoblotting, (*see* **Note 21**) carried out as in **Subheading 3.3** or ANA testing.
18. Arrange the immunoblotted strips on card board inserts (*see* **Note 22**).

4. Notes

1. Having water at the bottom of the cylinder helps to dissolve the Tris much faster, allowing the magnetic stirrer to go to work immediately. If using a glass beaker, the Tris can also be dissolved faster if the water is warmed to about 37°C. However, the downside is that care should be taken to bring the solution to room temperature before adjusting the pH.
2. Concentrated HCl (12N) can be used at first to narrow the gap from the starting pH to to the required pH. From then on it would be better to use a series of HCl (e.g., 6N and 1N) with lower ionic strengths to avoid a sudden drop in pH below the required pH.
3. Wear a mask when weighing acrylamide. To avoid exposing acrylamide to coworkers, cover the weigh boat containing the weighed acrylamide with another weigh boat (similar size to the original weigh boat containing the weighed acrylamide) when transporting it to the fume hood. Transfer the weighed acrylamide to the cylinder inside the fume hood and mix on a mixer inside the hood. Unpolymerized acrylamide is a neurotoxin, and care should be exercised to avoid skin contact.
4. The acrylamide solution can be stored at 4°C for one month. The acrylamide mixture, buffer, and water can be prepared in large batches, frozen in aliquots (for greater day-to-day reproducibility) and used indefinitely (*see* ref. 4). Remove the required amount, bring to room temperature, and add the other ingredients for polymerization.
5. We find it is best to prepare this fresh each time.
6. We find that storing at 4°C reduces its pungent smell.
7. Simple method of preparing running buffer: Prepare 10× native buffer (0.25 M Tris, 1.92 M glycine). Weigh 30.3 g Tris and 144 g glycine, mix and make it to 1 L with water. Dilute 100 mL of 10× native buffer to 990 mL with water and add 10 mL of 10% SDS. Care should be taken to add SDS solution last, since it makes bubbles.

8. SDS precipitates at 4°C. Therefore, the lysis buffer needs to be warmed prior to use.
9. Dilute 100 mL of 10× native buffer to 800 mL with water and add 200 mL of methanol. Avoid adding methanol directly to the 10× buffer, since it precipitates its ingredients. Even in such a scenario the precipitate can be redissolved by the addition of 800 mL water.
10. Add 100 mL of 10× TBS to a 1 L graduated cylinder and make it to about 800 mL with water. Transfer 50 g skim milk powder into the cylinder and stir until dissolved. Make to 1 L with water. Separate 500 mL as the blocking solution. To the remaining 500 mL add 250 μL of Tween-20 (cut end of blue tip to aspirate Tween easily), dissolve, and use it as diluent.
11. Hoeffer Scientific was taken over by Pharmacia LB and now by GE Healthcare Piscataway, NJ, USA.
12. Cells were sonicated four times, 15 s each time. The tube with the lysate is chilled on ice in between sonication steps to prevent the contents from heating. Do this in a fume hood with an ear protector.
13. The gel cassette was sealed at the base using 1% agarose. Overlay the resolving gel with water for gels having acrylamide concentration lower than 8% and use isobutanol (or isobutanol saturated with water) for gels of 10% or greater (*see* ref. 4). This overlay prevents contact with atmospheric oxygen, (which inhibits acrylamide polymerization) in addition to helping to level the resolving gel solution.
14. Centrifuging the samples prior to the run helps remove insoluble debris, which could produce streaks in the protein lanes (revealed when stained with Coomassie blue). Add a drop of 0.1% BPB to the upper chamber buffer. This helps to form a much stronger dye front during the electrophoretic run.
15. Using a dilute Fast Green solution help to prevent overstaining of the membrane. Dilute stock Fast Green solution in gel destaining solution (25% methanol, 10% acetic acid). Use TBST to destain the membrane if it is overstained by Fast Green.
16. Strips can be excised very nicely using a razor blade. A Mini PROTEAN® 3 System glass plate is placed at an angle on the nitrocellulose at a distance of 2 mm from the edge and the razor is used to cut the strip (pull the razor blade along the sides of the glass plate to cut).
17. The Fast Green stained proteins bands in the main nitrocellulose membrane should be still visible. After aligning the two strips on either side of the membrane, the developed band on the strips will line up with one of the bands on the

membrane. This will permit the easy excision of the desired band from the main membrane.

18. This will prevent antibody loss as a consequence of adherence to the tubes.
19. Exercise universal serum handling precaution when handling human sera. Treat each serum sample as potentially hazardous.
20. It is possible to concentrate up to 500 μL with Centri-Prep 30 concentrator. If there is a need to reduce the volume further, use a 10,000 or 30,000 molecular weight cut off microcentrifuge tube.
21. If assaying for antibody purity (of the purified antibody) on western blot, best results are obtained by incubating on nitrocellulose strips overnight at room temperature on the orbital shaker. This allows better binding of low titer antibody recovery and also washes should be longer (10 min instead of 5 min) and more frequent (5 vs. normally 4 washes).
22. We use paper boards placed in between stacks of ELISA plates in packages of ELISA plates (Costar, Cambridge, MA, USA) for this purpose.

Acknowledgement

This work was supported by NIH grant ARO1844 and Oklahoma Center for the Advancement of Science and Technology to Dr. Hal Scofield, OMRF, OKC, OK, USA.

References

1. Kurien, B.T. and Scofield, R.H. (2006) Autoantibody determination in the diagnosis of systemic lupus erythematosus. *Scand J Immunol.* 64, 227–235. Review.
2. Hansen, B.U., Eriksson, S., and Lindgren, S. (1991) High prevalence of autoimmune liver disease in patients with multiple nuclear dot, anti-centromere, and mitotic spindle antibodies. *Scand J Gastroenterol.* 26, 707–713.
3. Bernstein, R.M., Neuberger, J.M., Bunn, C.C., Callender, M.E., Hughes, G.R., and Williams, R. (1984) Diversity of autoantibodies in primary biliary cirrhosis and chronic active hepatitis. *Clin Exp Immunol.* 55, 553–560.
4. Electrophoresis. Appendix I. In: Harlow, E., and Lane, D., Ed. Antibodies. A laboratory manual. Cold Spring Harbor Laboratory Press, New York, 1988, 283–318.
5. Kurien, B.T., and Scofield, R.H. (1998) Multiple nuclear dot antinuclear antibody in patients without primary biliary cirrhosis. *Indian J Biochem Biophys.* 35, 41–47.
6. Towbin, H., Staehelin, T., and Gordon, J. (1979) Electrophoretic transfer of proteins from polyacrylamide gels to NC sheets: procedure and applications. *Proc Natl Acad Sci USA* 76, 4350–4354.
7. Olmsted, J.B. (1981) Affinity purification of antibodies from diazotized paper blots of heterogeneous protein samples. *J Biol Chem.* 256, 11955–11957.

Chapter 23

Double-Blotting: A Solution to the Problem of Nonspecific Binding of Secondary Antibodies in Immunoblotting Procedures

Françoise Lasne

Summary

Nonspecific interactions between blotted proteins and unrelated secondary antibodies generate false positives in immunoblotting techniques. Some procedures have been developed to reduce this adsorption but they may work in specific applications and be ineffective in other ones. "Double-blotting" has been developed to overcome this problem. It consists of interpolating a second blotting step between the usual probings of the blot membrane with the primary antibody and the secondary antibodies. This step, by isolating the primary antibody from the interfering proteins, guarantees the specificity of the probing with the secondary antibody. This method has been developed for the study of erythropoietin in concentrated urine since a strong nonspecific binding of biotinylated secondary antibodies to some urinary proteins is observed using classical immunoblotting protocols. However, its concept makes it usable in other applications that come up against this kind of problem. This method is expected to be especially useful for investigating proteins that are present in minute amounts in complex biological media.

Key words: Double-blotting, Immunoblotting, False positives, Secondary antibodies, Erythropoietin

1. Introduction

The development of a test for anti-doping control of erythropoietin (EPO), a hormone used in endurance sport to stimulate the red blood cell production, has been a long and exacting task. The method is based on differentiation of natural and recombinant (used in case of doping) hormones in urine by their isoelectric profiles *(1)*. For this, urine is first submitted to ultrafiltration to concentrate EPO in retentates that are then subjected to isoelectric focusing. Immunoblotting of EPO is then performed

Biji T. Kurien and R. Hal Scofield (eds.), *Methods in Molecular Biology, Protein Blotting and Detection, vol. 536*

DOI: 10.1007/978-1-59745-542-8_23

using primary monoclonal anti-human EPO antibodies and secondary biotinylated goat anti-mouse IgG antibodies *(2)*. The major drawback of the ultrafiltration step is the resulting very high protein content of the retentates that are then subjected to the next step of isoelectric focusing. This is particularly true for the urine samples taken at the end of a competition, because of proteinuria induced by physical exercise. This results in retentates with huge total protein contents (about 50 g/L for samples taken at rest and up to 200 g/L for samples taken after a physical exercise) for an EPO concentration generally no more than 4 μg/L. Such a situation is a real challenge for the classical immunoblotting procedures. Indeed, a strong nonspecific binding of secondary antibodies to some proteins present in the retentates was observed, completely masking the detection of EPO. All attempts to prevent or reduce this nonspecific binding were ineffective when working directly on the blotting membrane. The problem was solved by isolating the primary antibody from the interfering proteins on a second membrane that was then probed by the secondary antibody without any risk of nonspecific binding *(3, 4)*. Gershoni has emphasized the difficulty of resolving such a problem *(5)*. For this, after it has been probed by the primary antibody, the membrane with the blotted proteins is assembled with a second blank membrane and submitted to a second blotting under acidic conditions. The primary antibody molecules are thus desorbed from their corresponding antigen and transferred onto the second membrane, whereas the antigen and the interfering proteins remain bound to the first one. The second membrane can then be probed by the secondary antibodies without the risk of nonspecific binding.

2. Materials

Since double blotting (DB) takes place after probing of the blotting membrane (B membrane) with a primary antibody and before probing with the secondary one (**Fig. 1**), the reagents and materials for these steps (blotting membrane, blocking and washing buffers, primary and secondary antibody solutions, possibly amplification system, development reagents) will not be indicated here, being specific for the application in which DB is introduced (*see* **Note 1**).

Only the materials used for the DB step itself will be detailed.

1. Semi-dry transfer unit.
2. Roller.
3. Immobilon P (Millipore, Molsheim, France) polyvinylidene fluoride membranes (PVDF), 0.45 μm pore size.

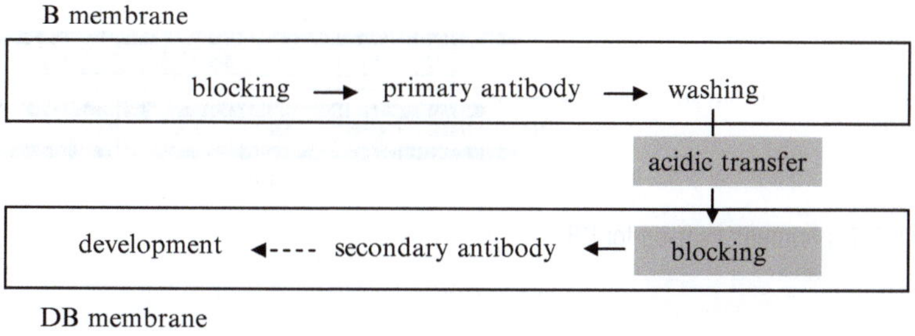

Fig. 1. Positioning of the double-blotting process into an immunoblotting procedure. The additional steps specifically related to DB are grey coloured.

4. Durapore (Millipore) hydrophylic PVDF membranes, 0.65 µm pore size
5. Filter paper sheets: electrode paper Novablot (GE Healthcare, Saclay, France).
6. 0.7% (v/v) acetic acid solution.
7. Phosphate buffered saline (PBS) pH 7.4.
8. Sealing film, parafilm.

3. Methods

1. Proceed to the usual blotting, blocking, primary antibody probing and washing steps of your B membrane, according to your application. DB is performed after the last wash of your B membrane (*see* **Fig. 1** and **Note 2**).
2. Cut two stacks of 9 filter paper sheets, a Durapore (intermediate membrane) and an Immobilon P (DB membrane) membrane to the dimensions of the blotting membrane.
3. Condition them in 0.7% acetic acid: just immerse the Durapore in the acidic solution for at least 10 min, prewet the Immobilon P membrane in methanol for 3 s, rinse in water for 2 min before equilibration in acidic solution for 10 min. The stacks of filter paper are moistened in acetic acid solution by capillary action.
4. During the same time, perform a rinsing of the B membrane in two changes of PBS.
5. Layer the B membrane of your application onto a first stack of filter paper with the blotted proteins facing up and cover it with the intermediate and DB membranes successively. Quickly put the second stack of filter paper onto the DB membrane to prevent the membranes from drying (*see* **Note 3**).

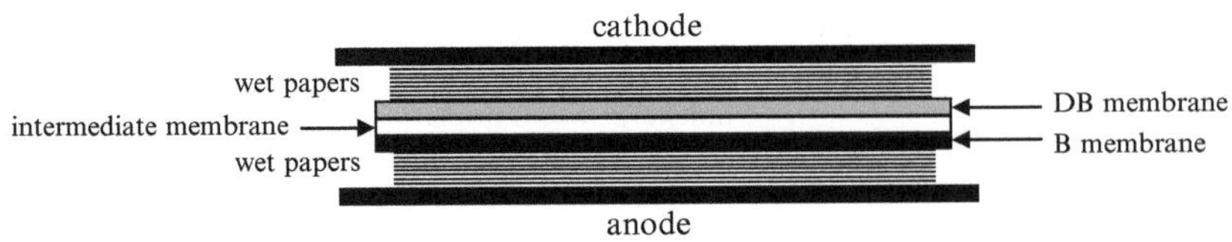

Fig. 2. Experimental set-up for DB.

6. Position this sandwich on the anode plate of the semi-dry electrophoretic blotting instrument so that the B and the DB membranes will face the anode and the cathode, respectively (**Fig. 2**).
7. Place a sealing film onto the sandwich and carefully press out the air bubbles with the roller. Remove the film.
8. Place the cathode plate on the sandwich and connect the blotting instrument to the power supply.
9. Apply a constant intensity of 0.8 mA per cm^2 for 10 min (*see* **Notes 4–6**).
10. Disconnect the blotting instrument and disassemble the membranes (*see* **Note 3**).
11. Keep the B membrane in PBS at 4°C (*see* **Note 7**).
12. Rinse the DB membrane quickly in two changes of PBS (*see* **Note 3**).
13. Proceed to the blocking of the DB membrane according to your usual procedure
14. Proceed to the usual steps of your application from the probing with the secondary antibody to the final development, on the DB membrane.

4. Notes

1. Reagents for detection of erythropoietin. The method is illustrated by its application to immunodetection of erythropoietin. The specific reagents used in the case of EPO are: (a) Primary antibodies: monoclonal mouse anti-human EPO (clone AE7A5 from R&D, Abingdon, England), (b) Secondary antibodies: biotinylated goat anti-mouse IgG (H + L) (31800 from Pierce, Rockford, USA), (c) Amplifying systems: Streptavidin/biotinylated peroxidase complexes (GO14–61 from Biospa, Milano, Italy), (d) Development system: Chemiluminescence (SuperSignal West Femto from Pierce), (e) Blocking buffer: 5% (w/v) non-fat milk in PBS buffer.

2. DB has been developed using PVDF as blotting membrane and has not been tested with other types of membranes.
3. Be aware that the DB membrane is very sensitive to drying when not saturated. This, while being not visually perceptible during the handling of the membrane, will produce a high background in the final image. It is thus extremely important to quickly perform the steps in which the membrane is handled without any liquid contact.
4. The acidic pH of this step induces dissociation of the primary antibody molecules from their corresponding antigen. The released antibodies, being positively charged due to the acidic pH, migrate toward the cathode, passing through the intermediate membrane and thus are transferred onto the DB membrane. Since the acidity does not affect hydrophobic interactions with PVDF, the antigen and the unrelated proteins (interfering proteins and proteins used for blocking the B membrane) are retained on the B membrane (**Fig. 3**). It should be noted that the result is actually an "image" of the probed antigen since it is only the antibody on the second membrane and not the antigen that gives rise to the final signal. However, this image is quite representative of the probed antigen (**Fig. 4**).
5. The interposition of the intermediate membrane between the B and the DB was very useful in our application (EPO analysis), which uses nonfat milk as a blocking agent, some "holes" in the final image obtained with the DB membrane being sometimes observed when this intermediate membrane was omitted. Although the explanation for this is purely hypothetical (local releases of clumps of casein precipitated by the acidic pH), the interpolated microporous Durapore membrane worked as a barrier definitely remedying this problem. It is possible that this membrane is not necessary in other applications but its use is strongly advised in case of "holes" in the final image.
6. The use of an electric field speeds up the transfer of the primary antibody from the blotting to the DB membrane. However, a simple contact between the membranes (passive transfer) without applying an electric field for a prolonged time (30

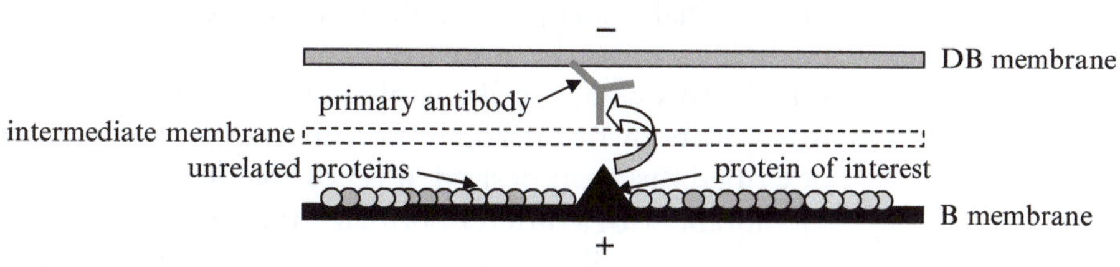

Fig. 3. Principle of double-blotting.

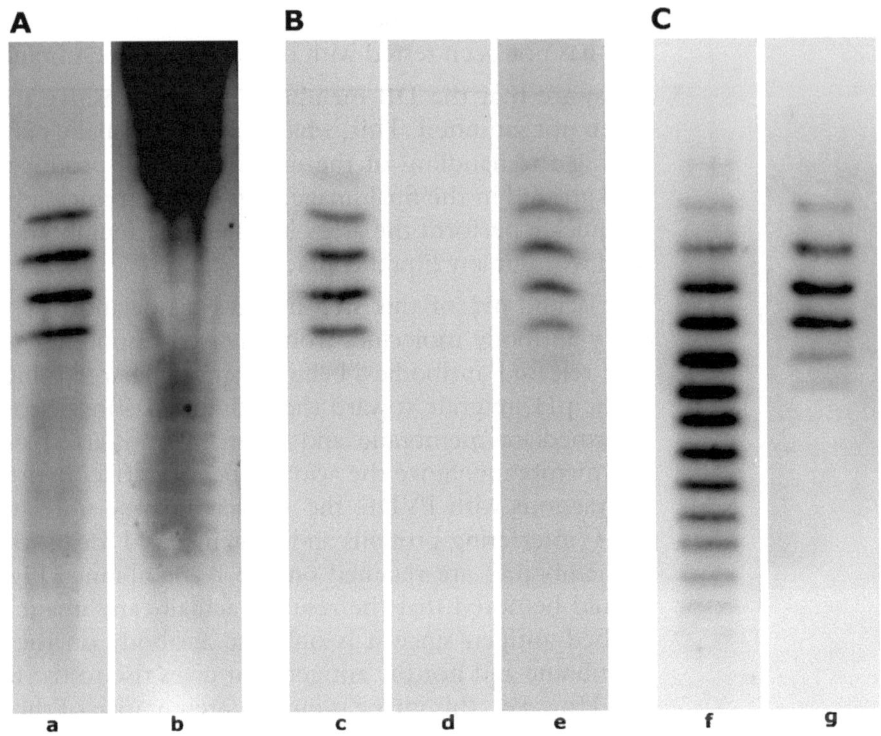

Fig. 4. Isoelectric patterns of EPO (**A**) images obtained without DB of (**a**) pure recombinant human EPO (rHuEPO) (Epoetin α), (**b**) a retentate obtained from ultrafiltration of a urine sample (chosen because it is devoid of endogenous EPO); (**B**) images obtained with DB of (**c**) pure rHuEPO (Epoetin α), (**d**) the same retentate as in b, (**e**) rHuEPO introduced into this same retentate; (**C**) images obtained with DB of (**f**) a retentate from urine containing natural endogenous EPO, (**g**) a retentate from urine containing excreted rHuEPO. In the case of pure rHuEPO, comparison of the images obtained without (**a**) and with (**c**) DB shows that no significant change in the isoelectric pattern is induced by the DB process. In the case of a retentate, due to the binding of the secondary antibody on the urinary proteins, a very strong EPO non-specific signal of is observed with the classical immunoblotting process (**b**). DB totally eliminates this nonspecific signal (**d**) and enables to specifically detect the introduced rHuEPO (**e**). An illustration of the use of DB in anti-doping control analysis is given in (**C**), natural endogenous urinary EPO (**f**) and administered rHuEPO excreted in urine (**g**) are differentiated by their isoelectric patterns.

min) has been tested in dot blot experiments and proved to be usable too (data not shown).

7. Storage of the B membrane (on which the blotted proteins are retained) in PBS may be useful. If some problem is observed on the final image, e.g. high background (*see* **Note 3**), it is possible to reuse the stored B membrane to perform a second DB. In this case, the B membrane is reincubated in the primary antibody of the application and DB is then performed as described earlier. If desired, it is possible to probe the stored B membrane with a different primary antibody prior to performing a second DB. This enables obtaining an image of another antigen from the same sample.

Acknowledgements

The DB process has been patented *(2 786 273)* by "Hospices Civils de Lyon", and by "Laboratoire National de Dépistage du Dopage"with "Hospices Civils de Lyon" as PCT/ FR01/01331.

References

1. Lasne, F., and de Ceaurriz, J. (2000) Recombinant erythropoietin in urine. *Nature* **405**, 635.
2. Lasne, F., Martin, L., Crepin, N., and de Ceaurriz, J. (2002) Detection of isoelectric profiles of erythropoietin in urine: differentiation of natual and administered recombinant hormones. *Anal Biochem.* **311**, 119–126.
3. Lasne, F. (2001) Double-blotting: a solution to the problem of non-specific binding of secondary antibodies in immunoblotting procedures. *J Immunol Meth* 253, 125–131.
4. Lasne, F. (2003) Double-blotting: a solution to the problem of non-specific binding of secondary antibodies in immunoblotting procedures (protocol). *J Immunol Meth* 276, 223–226.
5. Gershoni, J.M. (1988) Protein blotting: a manual, in Methods of biochemical analysis (Glick D., ed.), Vol. 33, Wiley, New-York, pp. 1–58.

Chapter 24

Efficient Electroblotting of Very Large Proteins Using a Vertical Agarose Electrophoresis System

Marion L. Greaser and Chad M. Warren

Summary

Very large proteins (subunit sizes >200 kDa) are difficult to electrophoretically separate, and they are also challenging to analyze by western blotting because of their incomplete transfer out of polyacrylamide gels. An SDS vertical agarose gel system has been developed that has vastly improved resolving power for very large proteins. The large pores of the agarose also allow full transfer of proteins as large as titin (Mr =3,000–3,700 kDa) onto blots. Inclusion of a reducing agent in the upper reservoir buffer and transfer buffer has been found to be a key technical procedure in blotting large proteins.

Key words: SeaKem agarose, Titin, DATD, Large protein blotting

1. Introduction

Very large molecular weight proteins are difficult to separate by electrophoresis because of their poor penetration into gels using the Laemmli SDS (sodium dodecyl sulfate) polyacrylamide system *(1)*. Protein migration in SDS gels has been found to be linear with the log of the molecular weight *(2)*, so the larger the protein, the more poorly it is resolved from other big proteins. Many workers have attempted to solve this problem using very low concentration acrylamide gels *(3)*, acrylamide mixed with agarose *(4)*, or acrylamide gradients *(5)* to better separate large proteins. Low concentration acrylamide gels are difficult to use because of their mechanical fragility and distortion during handling; these problems are magnified when blotting is attempted. An additional difficulty in blotting very large proteins is their poor transfer to the membrane. Inclusion of 2-mercaptoethanol

Biji T. Kurien and R. Hal Scofield (eds.), *Methods in Molecular Biology, Protein Blotting and Detection, vol. 536*

DOI: 10.1007/978-1-59745-542-8_24

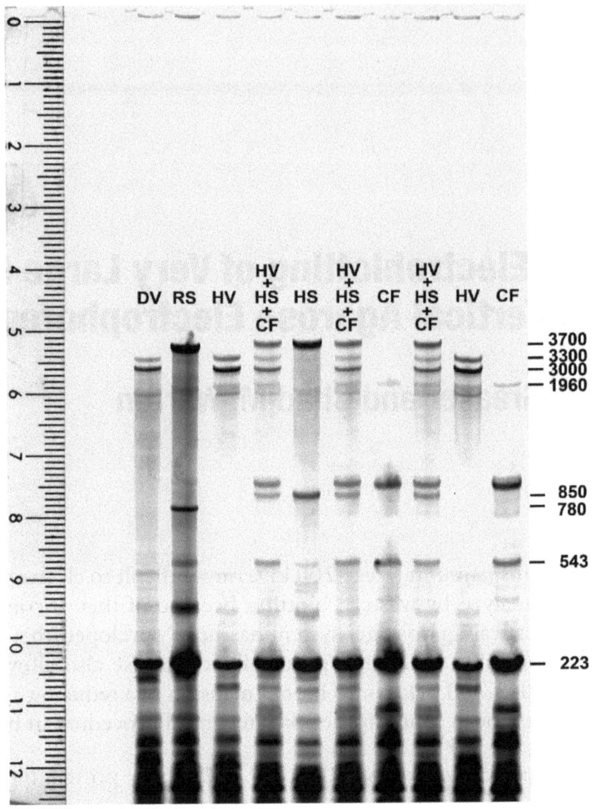

Fig. 1. SDS 1% agarose gel stained with silver. A centimeter ruler is shown on the *left* and the sizes of the various protein bands in kDa are listed on the *right*. Abbreviations: *DV* dog ventricle; *RS* rat soleus; *HV* human ventricle; *HS* human soleus; *CF* crayfish claw muscle. Human soleus titin is 3,700 kDa and human ventricle has two titin bands of 3,300 and 3,000 kDa. The bands at 780 and 850 kDa are rat and human nebulin, respectively. The myosin heavy chain is 223 kDa. Blotting proteins this size from acrylamide gels usually results in incomplete transfer, but full transfer can be achieved with agarose *(7)*.

in the transfer buffer improves transfer efficiency, but acrylamide gels stained after transfer typically still contain most of the giant muscle protein titin *(6)*.

A new electrophoresis system using SDS agarose for protein electrophoresis and blotting has been described *(7)*. An example showing the resolution for several muscle samples containing large proteins is shown in **Fig. 1**. Migration distance shows a linear relationship with the log of the molecular weight *(7)*. This system allows quantitative transfer of proteins from the gel with much higher reproducibility than can be achieved with methods using low percentage acrylamide.

2. Materials

2.1. Apparatus

1. SE 600 Slab Gel Unit with 16 × 18 cm glass plates (Hoefer) or a similar commercial gel unit (*see* **Note 1**).
2. 65°C oven.
3. 3 A constant current power supply.
4. Circulating cooler.
5. TE62 Tank Blotting Unit, Hoefer (GE Health Services, Piscataway, NJ, USA).
6. DATD (N-N′ diallyltartardiamide) (Bio-Rad, Hercules, CA, USA).
7. CAPS (*N*-cyclohexyl-3-aminopropanesulfonic acid) (Sigma Chemical Company, MO, USA).
8. *NNNN′*-tetramethyl-ethylenediamine (TEMED).
9. Acrylamide.
10. 0.45 µm filter (such as a Millex-HA, Millipore Corporation, USA).

2.2. Stock Solutions

1. 38.5% acrylamide for gel for plug: Weigh 37.5 g of acrylamide and 1 g DATD into a beaker, add about 50 mL of water, stir till dissolved, dilute to 100 mL. Filter through a 0.45 µm filter (such as a Millex-HA, Millipore Corporation). Store in a brown bottle in the cold room (4°C). Danger! Avoid skin contact.
2. Reservoir and agarose gel buffer concentrate (5×): 0.25 M Tris – 1.92 M glycine – 0.5% SDS. Store at room temperature (RT).
3. Ammonium persulfate: Prepare a 100 mg/mL solution in water; store frozen in 0.5 mL aliquots (stable indefinitely at -20°C).
4. Sample buffer: 8 M urea, 2 M thiourea, 0.05 M Tris-HCl (pH 6.8), 75 mM DTT, 3% SDS, 0.05% bromophenol blue (adapted from **ref.** *8*). (Dissolve urea and thiourea and treat with mixed bed resin to remove ionic constituents; then add remaining ingredients. Store at -20°C).
5. 50% v/v glycerol.
6. Transfer buffer: 20 mM Tris, 150 mM glycine, 20% v/v methanol *(7, 8)* or 10 mM CAPS, pH 11 *(9)*. For high molecular weight proteins, add SDS and 2-mercaptoethanol to 0.1% and 10 mM, respectively, to the transfer buffer.

3. Methods

3.1. Gel Preparation

1. Volumes listed will provide enough solution for two 16 × 18 cm gels with 1.5 mm spacers. One is used for staining either with Coomassie Blue or with silver with a special procedure for agarose *(7)*, the other for blotting.
2. Clean plates and spacers with soap, rinse with distilled water and finally with ethanol.
3. Assemble gel plates. Place plate on clean bench top. Place spacers hanging half the way off each side of plate. Place second plate on top. Stand up plates and place one side into the clamp. Align spacer with side of plates and clamp and push spacer down so that bottom is flush with the glass plates (top buffer will leak if spacers are not flush with plates).
4. Pour acrylamide plugs in bottom of gel plate assembly (*see* **Note 2**): In a 15 mL plastic beaker add: 1.924 mL deionized water, 1.7 mL 50% glycerol, 2.12 mL 3 M Tris (pH 9.3), 2.72 mL acrylamide (40%), 24 μL of 10% ammonium persulfate, and 13 μL TEMED (*see* **Note 2**). Mix by pipeting a few times. Immediately add 2.5 mL to each gel assembly. Add a small amount of water on top of each plug to level the upper surface and provide an oxygen barrier. Allow gel to polymerize for 20–30 min. Drain off water layer by inverting gel plate assembly on a paper towel.
5. Place assembly, 20 lane sample combs, and 60 mL plastic syringe in a 65°C oven for 10 min (*see* **Note 3**).
6. Weigh 0.8 g of SeaKem Gold agarose (Lonza Group Ltd) (*see* **Note 4**) into a 600 mL beaker (*see* **Note 5**). To a 100 mL graduated cylinder add 48 mL of 50% v/v glycerol (*see* **Note 6**), 16 mL of 5× electrophoresis buffer, and bring up the volume to 80 mL with deionized water. Place parafilm over top of the graduated cylinder, mix by inverting a few times, and pour solution into the 600 mL beaker containing the agarose. Place saran wrap over top of beaker and poke a few holes in the saran wrap. Weigh beaker with contents. Place beaker in a microwave oven along with a separate beaker of deionized water. Heat for a total of 2 min (stop every 30 s to swirl – protect hand with an insulated glove) (*see* **Note 7**).
7. Allow agarose to cool for a few min at RT. Reweigh, and add sufficient heated deionized water to replace that lost by evaporation.
8. Draw up about 40 mL of agarose in the prewarmed 60 mL Luer-Lock syringe and pour each gel slowly until it just overflows the top of the plates. Try to avoid formation of bubbles (if bubbles present, bring them to the top of the gel and pinch

them with the sample comb). Insert sample combs and allow unit to cool at RT for about 45 min (*see* **Note 8**).

3.2. Electrophoresis Setup and Sample Loading

1. Add 4 L of buffer to lower chamber (3,200 mL deionized water plus 800 mL of 5× electrophoresis buffer). Start cooling unit and stir bar (gels run at 6°C).
2. Prepare 600 mL upper chamber electrophoresis buffer (same concentration as lower chamber buffer). Add 2-mercaptoethanol (final concentration of 10 mM). Buffer should be poured into top chamber after samples are loaded and assembly placed in unit.
3. Take combs out of gels by bending them back and forth to detach from gel and slowly pulling them up. Pour a small amount of upper chamber buffer into a 15 mL beaker and pipette buffer into first and last wells (the rest will fill over). Add buffer to remove any trapped bubbles. Insert pipette tip to deposit sample in bottom of the sample well. Skip the first and last lanes (*see* **Note 9**).

3.3. Running Gels

1. Once samples are loaded, put upper chamber on the assembly. Pour upper chamber buffer into upper chamber from corners (do not pour buffer directly over wells). Place lid on unit, and connect to power supply. Turn electrophoresis unit on and run at 30 mA (2 gels) for 3 h.

3.4. Staining and Western Blotting

1. After tracking dye reaches the bottom of the acrylamide plug, turn off the power and disassemble the plates. Cut off sample wells and acrylamide plug and discard. Soak the remaining agarose gel in 10 mM CAPS (pH 11), 0.1% SDS, and 10 mM 2-mercaptoethanol for 30 min with gentle shaking.
2. The gel is then placed on top of either a sheet of PVDF (polyvinylidene difluoride) or nitrocellulose, assembled into the transfer unit, and the protein electrophoretically transferred using 40 V constant voltage for 2–3 h (*see* **Note 10**).
3. Blotted proteins can then be treated using conventional procedures with either colorimetric (horseradish peroxidase or alkaline phosphatase substrates) or ECL (enhanced chemiluminescence) methods.

4. Notes

1. The agarose gel procedure works equally well with small format gels (i.e., 8 × 10 cm).

2. The acrylamide plug is used to prevent the agarose from slipping out of the vertical gel plate assembly. Use of DATD as the cross-linker provides a stickier bond of the acrylamide to the glass plates than if a conventional bisacrylamide cross-linker is used. Plugs can be poured a day before making the gel (place tape or parafilm over the top of the plates to prevent drying and store in cold room).
3. Preheating the glass plate assembly, well comb, and syringe prevents premature agarose gelling when the solution touches the colder surfaces. In addition, the plates are less likely to crack during pouring if they are closer to the temperature of the hot agarose.
4. The supplier for SeaKem Gold agarose has changed twice since 2003. Biowhittaker was succeeded by Cambrex who was followed by Lonza Group Ltd, Muenchensteinerstrasse 38, CH-4002 Basel, Switzerland.
5. It is essential to use SeaKem Gold agarose for optimal migration of high molecular weight proteins. This type has large pore size and excellent mechanical stability. Other types of agarose may be used, but the protein mobility will be significantly reduced.
6. Glycerol is included in the mixture to increase the solution viscosity inside the gel and thus sharpen the protein bands.
7. Periodic swirling during the heating step eliminates non-hydrated agarose granules in the final gel.
8. Sample combs should extend no longer than 1 cm into agarose; otherwise they may be difficult to remove. Gels can be used right away or stored overnight in a cold room.
9. Conventional sample buffers may not be dense enough for the sample to stay at the bottom of the well. If necessary add additional glycerol (up to 30% v/v final concentration) to increase sample density.
10. The disulfide bond formation of large proteins during electrophoresis also retards their migration out of the gel onto blots during transfer. Thus inclusion of 2-mercaptoethanol in the transfer buffer improves efficiency of transfer of high molecular weight proteins. The use of the agarose electrophoresis system with inclusion of 2-mercaptoethanol in the transfer buffer results in complete transfer of all high molecular weight proteins out of the gel, including titin (Mr 3,000–3,700 kDa subunit size) *(7)*. Alternatively, protein can be alkalated to prevent disulfide bond formation during the transfer process *(10)*.

Acknowledgements

This work was supported by the College of Agricultural and Life Sciences, University of Wisconsin-Madison, Purdue University College of Agriculture, and from grants (MLG- NIH HL77196 and Hatch NC1131.

References

1. Laemmli, U.K. (1970) Cleavage of structural proteins during the assembly of the head of bacteriophage T4. *Nature* **227**, 680–685.
2. Weber, K., and Osborn, M. (1969) The reliability of molecular weight determinations by dodecyl sulfate-polyacrylamide gel electrophoresis. *J Biol Chem.* **244**, 4406–4412.
3. Granzier, H., and Wang, K. (1993) Gel electrophoresis of giant proteins: solubilization and silver-staining of titin and nebulin from single muscle fiber segments. *Electrophoresis* **14**, 56–64.
4. Tatsumi, R., and Hattori, A. (1995) Detection of giant myofibrillar proteins connectin and nebulin by electrophoresis in 2% polyacrylamide slab gels strengthened with agarose. *Anal Biochem.* **224**, 28–31.
5. Cazorla, O., Freiburg, A., Helmes, M., Centner, T., McNabb, M., Wu, Y., et al. (2000) Differential expression of cardiac titin isoforms and modulation of cellular stiffness. *Circ Res.* **86**, 59–67.
6. Fritz, J.D., Swartz, D.R., and Greaser, M.L. (1989) Factors affecting polyacrylamide gel electrophoresis and electroblotting of high-molecular-weight myofibrillar proteins. *Anal Biochem.* 180, 205–210.
7. Warren, C.M., Krzesinski, P.R., and Greaser, M.L. (2003) Vertical agarose gel electrophoresis and electroblotting of high-molecular-weight proteins. *Electrophoresis* 24, 1695–1702.
8. Yates, L.D., and Greaser, M.L. (1983) Quantitative determination of myosin and actin in rabbit skeletal muscle. *J Mol Biol.* 168, 123–141.
9. Matsudaira, P. (1987) Sequence from picomole quantities of proteins electroblotted onto polyvinylidene difluoride membranes. *J Biol Chem.* 262, 10035–10038.
10. Sechi, S., and Chait, B.T. (1998) Modification of cysteine residues by alkylation. A tool in peptide mapping and protein identification. *Anal Chem.* 70, 5150–5158.

Chapter 25

Semi-Dry Protein Transfer and Immunodetection of P-Selectin Using an Antibody to its C-terminal Tag

Padmaja Mehta

Summary

P-selectin is a multidomain glycoprotein expressed on activated endothelial cells. We previously expressed a recombinant form of P-selectin containing only its N terminal lectin and EGF domains in CHO-K1 cells and showed that these two domains are sufficient to mediate ligand binding. We have now expressed the same construct in CHO-Lec1 cells that make truncated glycans. The uniform glycosylation in these cells should make it easier to crystallize this protein.

Key words: SDS-PAGE, Western blotting, Chromogenic detection, Immobilon-P, P-selectin, P-selectin Glycoprotein ligand-1 (PSGL-1)

1. Introduction

Selectins are multidomain glycoproteins that play an important role both in inflammation and tissue injury *(1)*. P-selectin is expressed on activated endothelial cells and platelets, E-selectin is expressed on activated endothelial cells, and L-selectin is expressed on leukocytes. The tetrasaccharide sialyl Lewis x forms the core structure recognized by all selectins *(2)*. In addition, P-selectin and L-selectin require sulfation of one or more N-terminal tyrosine residues in their cognate ligand P-selectin glycoprotein ligand-1 (PSGL-1). L-selectin also requires sulfation of one or more GlcNAc residues in its ligands on HEV mucins on lymph nodes *(3)*.

All selectins contain an N-terminal lectin domain, an EGF domain, a variable number of consensus repeats, a transmembrane

Biji T. Kurien and R. Hal Scofield (eds.), *Methods in Molecular Biology, Protein Blotting and Detection, vol. 536*

DOI: 10.1007/978-1-59745-542-8_25

domain, and a cytoplasmic tail. Using a recombinant form of soluble P-selectin containing only the lectin and EGF domains expressed in CHO-K1 cells, we have shown that the lectin and EGF domains of P-selectin are sufficient to bind its ligand, PSGL-1 *(4)*.

We also generated a recombinant, soluble, monomeric form of P-selectin in the CHO-Lec1 cell line for further structure–function studies. The recombinant protein was purified by affinity chromatography over an HPC4 antibody column *(5)*, and its purity checked by sodium dodecyl sulfate polyacrylamide gel electrophoresis (SDS-PAGE) followed by silver-staining and western blotting.

2. Materials

2.1. P-selectin lec-EGF

1. P-selectin lec-EGF was purified by HPC4 affinity chromatography from conditioned medium of transfected CHO-Lec1 cells.

2.2. SDS-Polyacrylamide Gel Electrophoresis (SDS-PAGE)

1. 30% acrylamide/bis solution (37.5:1 with 2.6% C) (*see* **Note 1**).
2. Separating gel buffer (3×); 1 M Tris-HCl, pH 8.8. Store at 4°C.
3. Stacking gel buffer (5×); 1 M Tris-HCl, pH 6.8. Store at 4°C.
4. Gel loading buffer: 62.5 mM Tris-HCl, pH 6.8, 5% β-mercaptoethanol, 10% glycerol, 2% SDS. A 3× version of the gel-loading buffer was also made by reducing the amount of water used to make up the final volume (*see* **Note 2**).
5. 10% SDS solution. (Exercise care when using this chemical in the dry form.) Store at room temperature (RT).
6. *N,N,N,N'* tetramethyl-ethylenediamine (TEMED): Make a 10% solution in water. This should be made fresh every time to ensure good polymerization of gels.
7. Ammmonium persulfate (APS): Make a 10% solution in water. This should also be made fresh every time to ensure good polymerization (*see* **Note 3**).
8. Running buffer for SDS-PAGE: 25 mM Tris, 190 mM glycine, 0.2% (w/v) SDS. This buffer was made as a 10× solution and diluted with water before use. Store at RT.
9. β-mercaptoethanol (Bio-Rad, Hercules, CA, USA).
10. Rainbow Molecular Weight Markers (Bio-Rad).

2.3. Western Blotting

1. Setup buffer: 25 mM Tris, 190 mM glycine, 20% (v/v) methanol. Do not adjust the pH of the Tris solution.
2. Immobilon-P membrane (Millipore, Massachusetts, USA).
3. Methanol.
4. Thick filter paper.

2.4. Immunostaining

1. Equilibration buffer 25 mM Tris, 190 mM glycine, 20% methanol. Do not adjust the pH of the Tris solution.
2. 20 mM Tris-HCl pH 7.5, 100 mM NaCl (TBS), 1 mM $CaCl_2$ (*see* **Note 4**).
3. Wash buffer: TBS with 1% Tween-20 (*see* **Note 5**).
4. Blocking solution: Nonfat dry milk, 5% in TBS (*see* **Note 6**).
5. Primary antibody: Diluted to 5 μg/mL in blocking solution.
6. Secondary antibody (biotinylated): 5 μg/mL in TBS containing 0.1% BSA.
7. Vectastain kit (ABC) (peroxidase conjugate).
8. Chromogenic substrate: 4-chloro-1-naphthol.

3. Methods

Over the years, several antibodies have been raised against P-selectin *(6)*. Many of these antibodies have been shown to have epitopes that lie mainly within the lectin domain. Most of these antibodies recognize three-dimensional (conformational) epitopes, and hence will stain only gels that have been run without treating the samples with β-mercaptoethanol (i.e., under nonreducing conditions). This will be applicable to other proteins as well, hence when staining any protein, the antibody used for the Western blot should be selected carefully. The antibody used for staining here is an antibody to its C-terminal epitope tag, and will stain under reducing conditions as well.

3.1. P-Selectin lec-EGF (Sample for Western Blotting)

P-selectin lec-EGF was purified by HPC4 affinity chromatography from conditioned medium of transfected CHO-Lec1 cells as previously described *(4)*.

3.2. SDS-Polyacrylamide Gel Electrophoresis (SDS-PAGE)

A minigel apparatus, made by Atto (Tokyo, Japan) was used. The proteins were electrophoresed in buffer containing 2-mercaptoethanol essentially as described by Laemmli *(7)*, using a discontinuous SDS-polyacrylamide gel system with a 12.5% separating gel (*see* chapter "Non-electrophoretic bi-directional transfer of a single SDS-PAGE gel with multiple antigens to obtain twelve immunoblots").

3.3. Western Blotting (Semi-Dry Transfer)

This technique is used to electrophoretically transfer the samples run on SDS-PAGE, to a membrane support, like Immobilon-P, which is a PVDF membrane *(8)*. The directions below are specifically applicable to the Bio-Rad Trans-Blot SD Semi-dry Transfer cell. The gel equilibration steps and the assembly of the transfer cassette will be similar even when using other types of transfer apparatus, including a transfer tank system. However, the volume of buffer required and the length of run will differ.

1. After the dye front in the separating gel reaches the base of the gel, turn the power off and disconnect the gel unit. Pour out the running buffer, remove the gel cassette from the unit and disassemble. Very carefully, pry the two plates open, allowing the gel to remain on the straight plate. The straight plate is laid down on the bench with the gel face up. Cut out the stacking gel and make a small mark in the left corner at the base of the separating gel near the dye front.
2. Place the separating gel in a small flat trough (the clean cover of a tip box is a good option) containing western blotting set-up buffer. Allow the gel to equilibrate in this buffer on a rocker for 5–10 min.
3. Fill two more flat troughs with the set-up buffer, and one trough each with water and methanol.
4. Cut a piece of Immobilon-P the same size as the separating gel (*see* **Note** 7). Place it in the methanol container for one min, drip dry, and place it in the trough of water for 2 min. Place this container on the rocker and ensure that the membrane is completely submerged in the water. After 2 min, transfer the membrane to a fresh trough containing set-up buffer.
5. Take two pieces of extra thick filter paper and wet them in a fresh trough containing set-up buffer. Place one wet filter paper on the transfer apparatus surface. Layer this with the piece of Immobilon that is soaked in buffer. Place the equilibrated gel over the Immobilon and complete the assembly by placing the second piece of moist filter paper over the sandwich.
6. Place the lid of the transfer apparatus on this sandwich and connect the electrodes to the power supply. Run the transfer at 15 V for 20 min with the current limit at 350 mA per gel. (Increase the time if transferring two gels, but do not increase the voltage beyond 25 V.)

3.4. Immunostaining

1. After the transfer is complete, remove the lid of the transfer apparatus carefully. Separate the gel sandwich and remove the Immobilon membrane (*see* **Note 8**).
2. Place the Immobilon membrane in the blocking solution in a flat trough and keep this on a rocker at RT for 1 h (*see* **Note 9**).

3. Dilute the primary antibody (HPC4, an anti Protein C antibody with its epitope fused to the C-terminus of P-selectin) to a concentration of 5 μg/mL into blocking solution (add one drop of normal horse serum) and incubate the proteins on the membrane with the primary antibody for 1 h at RT (this incubation can also be performed at 4°C overnight).
4. Wash the membrane with four changes of transfer wash buffer over 20 min, on a rocker.
5. Transfer the membrane to a 5 μg/mL solution of the biotinylated secondary antibody, made in dilution buffer (add 2 drops antibody for every 20 mL buffer). Incubate the membrane with the secondary antibody at RT for 45 min, on a rocker.
6. Wash the membrane with four changes of transfer wash buffer over 20 min, on a rocker.
7. In the meantime, prepare the Vectastain ABC reagent by adding 4 drops of reagent A and 4 drops of reagent B to 20 mL of the dilution buffer. Mix well and let it stand at RT for 30 min.
8. Transfer the membrane to the Vectastain ABC reagent prepared above and incubate on a rocker for 30 min at RT.
9. Wash the membrane with three changes of transfer wash buffer over 15 min, on a rocker. Wash once with TBS.
10. In the meantime prepare the peroxidase substrate solution containing 0.5 mg/mL of 4-chloro-1-naphthol as follows: dissolve 30 mg of 4-chloro-1-naphthol in 10 mL of chilled methanol. To this, add 50 mL TBS and 30 μL of 30% hydrogen peroxide (*see* **Note 10**).
11. Incubate the membrane with the substrate solution at RT. Color development should occur within 5–10 min (*see* **Note 11**).
12. After sufficient color has developed and the bands are clearly visible as shown in **Fig. 1**, wash the membrane twice in deionized water and air dry for storage. After the membrane is dry, store it wrapped in saran-wrap.

4. Notes

1. This is a neurotoxin, so great care should be taken while handling this chemical. Use gloves and do not breathe fumes.
2. When running samples under nonreducing conditions, the β-mercaptoethanol should be omitted from the buffer.

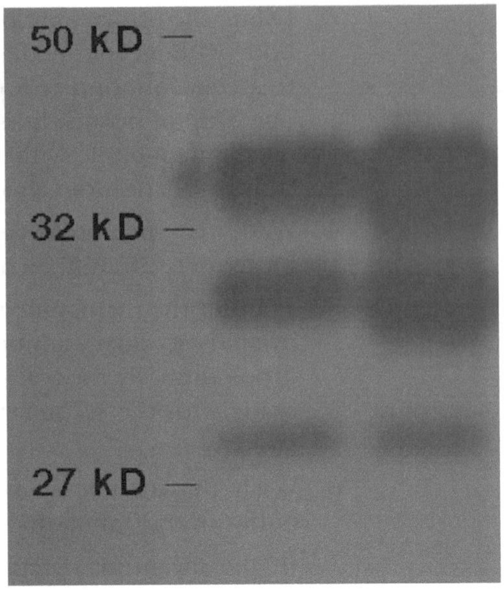

Fig. 1. Electrophoresis of P-selectin lec-EGF expressed in CHO-Lec1-cells. After SDS-PAGE, the proteins were transferred to Immobilon-P and stained with the HPC4 antibody. P-lec-EGF has three potential N-glycosylation sites and the three bands seen on the western blot represent three different glycosylated forms of the protein.

3. For the sake of convenience, this can be made in a larger volume and then stored in single use (200 μL) aliquots at –20°C.
4. The calcium chloride is not required for the technique itself and its addition is optional. We add it for the stability of our protein and its interaction with the primary antibody.
5. We have observed that a wash buffer containing Triton-X-100 instead of Tween-20 is not as effective.
6. Nonfat dry milk contains carbohydrates. When using detection reagents that stain carbohydrates (such as lectins and certain antibodies), a different blocking solution (such as 3% BSA) should be used, or else you will get a high background.
7. Always use gloves when handling the gel and the Immobilon membrane.
8. After the transfer, the gel can be stained with a protein stain to assess the quality of the transfer so that it can be optimized for subsequent runs. If a large amount of higher molecular weight bands are still visible on the gel, this indicates that the proteins should be transferred for a longer time.
9. The amount of blocking or antibody solution used should be sufficient to keep the membrane completely submerged even when the container is on the rocker.
10. The conjugated enzyme will vary depending on the ABC kit used. For peroxidase conjugates, other substrates that

may be used include TMB and DAB-Ni. The entire immunostaining protocol will remain the same, only the substrate preparation and incubation times will vary.

11. It is recommended that for color development, a fresh container be used for incubation with substrate.

Acknowledgement

The author thanks Dr. Charles Esmon for the HPC4 antibody.

References

1. McEver, R.P. (2001) Adhesive interactions of leukocytes, platelets and the vessel wall during hemostasis and inflammation. *Thromb Hemost.* 86, 746–756.
2. McEver, R.P. (2002) Selectins: Lectins that initiate cell adhesion under flow. *Curr Opin Cell Biol.* 14, 581–586.
3. McEver, R.P. (2005) A sulfated address for lymphocyte homing. *Nat Immunol.* 6, 1067–1069.
4. Mehta, P., Patel, K.D., Laue, T., Erickson, H.P., and McEver, R.P. (1997) Soluble monomeric P-selectin containing only the lectin and epidermal growth factor domains binds to P-selectin Glycoprotein Ligand-1 on leukocytes. *Blood* 90, 2381–2389.
5. Stearns, D.J., Kurosawa, S., Sims, P.J., Esmon, N.L., and Esmon, C.T. (1988) The interaction of a Ca^{2+}-dependent monoclonal antibody with the protein C activation peptide region. Evidence for obligatory Ca^{2+} binding to both antigen and antibody. *J Biol Chem.* 263, 826–832.
6. Geng, J.G., Bevilacqua, M.P., Moore, K.L., McIntyre, T.M., Prescott, S.M., Kim, J.M.et al. , (1990) Rapid neutrophil adhesion to activated endothelium mediated by GMP-140. *Nature* 343, 757–760.
7. Laemmli, U.K. (1970) Cleavage of structural proteins during the assembly of the head of bacteriophage T4. *Nature* 227, 680–685.
8. Towbin, H., Staehelin, T., and Gordon, J. (1979) Electrophoretic transfer of proteins from polyacrylamide gels to nitrocellulose sheets: procedure and some applications. *Proc Natl Acad Sci USA* 76, 4350–4354.

Chapter 26

The Multiple Antigen Blot Assay: A Simple, Versatile and Multipurpose Immunoenzymatic Technique

Oscar Noya, Sandra Losada, Marilyan Toledo, and Belkisyolé Alarcón de Noya

Summary

This technique is based on the sensitization of different antigens in a single nitrocellulose strip, which react when exposed to an immune serum and thereafter with the appropriate peroxidase conjugate and the corresponding substrate. Signals in those reactive spots are recorded as black squares in a negative photographic film, using a chemiluminiscent substrate or as blue spots when a precipitable colorimetric substrate is used. This technique allows the simultaneous demonstration of antigenicity of different antigens (peptides, recombinant molecules, and crude preparations), with a high sensitivity and specificity. Its major value is based on its versatility, since it is possible to rapidly evaluate and to compare various antigenic preparations and to use it for diagnosis of different infectious, allergic and autoimmune diseases, at a low cost.

Key words: Multiple, Blot, Antigen, MABA, Immunochromatographic, Enzyme-immunoassay, Immunodiagnosis, Luminescent, Colorimetric

1. Introduction

Three diagnostic techniques comply with the concept of simultaneous multidiagnosis: the antigen/antibody microarrays *(1)*, the multiple laser detection beads (Multiplex ®) *(2)*, and the multiple antigen blot assay (MABA) *(3)*. The first two have been used for the detection of antigens, antibodies, or nucleic acids, while the third one only for antibodies, so far.

The occurrence of simultaneous infections is not an infrequent finding worldwide; therefore, it would be ideal for the

Biji T. Kurien and R. Hal Scofield (eds.), *Methods in Molecular Biology, Protein Blotting and Detection, vol. 536*

DOI: 10.1007/978-1-59745-542-8_26

health personnel to be aware about the patient's current or past history of contact with the most important pathogens, as well as his immune status, to have a more integral knowledge of each patient for a more rationale medical practice. With this purpose, MABA can be implemented and adapted for particular populations and institutions, such as blood banks, pediatric diseases, immunocompromised patients, vaccine evaluation, etc., being a major line of research of our laboratory. It is also a practical, rapid, reproducible, sensitive, and cheap technique for the identification and evaluation of antigens (antigenicity) and sera. This can be done with crude, chemically synthesized and recombinant antigens *(4–8)*. Allergic conditions detecting specific IgE have been successfully evaluated with this technique *(9)*. Another advantage is that a single MABA that allows the evaluation of at least 28 different sera against 28 different antigens is equivalent to eight conventional ELISA plates. The cost of evaluation of each antigen per patient is 0.078$ by MABA vs. 0.114$ by a conventional ELISA.

2. Materials

1. Carbonate-bicarbonate buffer, pH 9.6: 1.59 g sodium carbonate (Na_2CO_3) (0.015 M); 2.93 g sodium bicarbonate ($NaHCO_3$) (0.03 M). Adjust pH to 9.6 and with distilled water complete to a final volume of 1,000 mL. Store at 2–8°C.
2. Phosphate buffered saline (PBS)-0.05% Tween-20 (PBST), pH 7.5: Solution **A** (0.2 M NaH_2PO_4): 24.0 g NaH_2PO_4 (anhydrous) in 1,000 mL distilled water; solution **B** (0.2 M Na_2HPO_4): 5.68 g Na_2HPO_4 (anhydrous) in 200 mL distilled water;13 mL solution **A**+ 87 mL solution **B**+ 8.76 g NaCl + up to 800 mL distilled water. Adjust pH to 7.5 with solutions **A** or **B**. Then, complete with distilled water up to 1,000 mL and add 0.5 mL of Tween-20. Once prepared, keep at 4°C.
3. Blocking solution (5% non-fat milk in PBST): 5 g nonfat milk in 100 mL PBST. Prepare freshly.
4. SuperSignal® (West Pico Chemiluminescent Substrate, Pierce, USA, Cat. # 34077). The two solutions should be mixed 1:1. Prepare freshly.
5. Acetate buffer 50 mM: Solution **A**: acetic acid 200 mM: 2 mL acetic acid + 172 mL distilled water; Solution **B**: 5.44 g sodium acetate in 200 mL distilled water. The buffer may be stored at 2–8°C for up to 1 month;110 mL Solution **A**+ 15 mL Solution **B**+ distilled water up to a volume of 500 mL.

6. TMB Membrane Peroxidase Substrate (KPL; Gaithersburg MD, USA, Cat. # 50-77-03): This substrate contains 3,3′, 5,5′ - tetramethylbenzidine in acidic buffer. The concentration of hydrogen peroxide (H_2O_2) is 0.02%. It is a very sensitive chromogenic substrate for peroxidase detection. It develops a deep blue color at the reaction sites labeled with horseradish peroxidase-labeled conjugates. No reagent preparation is needed.
7. Sodium metaperiodate-SMP (Sigma Chemical Company, St. Louis, MO, USA, Cat. #S-1878) (1 mM) in acetate buffer: 4.28 mg SMP in 20 mL acetate buffer. Prepare freshly.
8. 50 mM sodium borohydride ($NaBH_4$) (Sigma Chemical Company, St. Louis, MO, USA, S-9125) in PBS: 0.189 g $NaBH_4$ in 100 mL PBS. Prepare freshly.
9. Anti-human IgG (Fc specific) peroxidase antibody produced in goat (Sigma A-0170). For continuous use, store at 2–8°C up to 1 month. For extended storage, the solution may be frozen in working aliquots at −20°C, since repeated freezing and thawing is not recommended. Dilute to 1:30,000 in blocking solution. Prepare freshly.

   In some cases, the quality of the water is critical. In those situations it is necessary to use Milli Q water.

3. Methods

3.1. Luminescent MABA (3)

1. A rectangle of nitrocellulose (NC) sheet (BIORAD® 0.45 µm Cat. 162-0115) 7.5 × 10 cm is cut (**Fig. 1**) and labeled with an indicator line in one of the borders for reference, done with a fine ball point-water resistant pen (**Fig. 2**).
2. The paper is soaked in distilled water for 5 min (**Fig. 3**).
3. The wet paper is aligned in the upper part of Miniblotter® 28SL (Immunetics Inc, Cambridge, MA, USA), with the reference line parallel to channel 1 (**Fig. 4**). Then, a plastic cushion is placed onto the nitrocellulose sheet (**Fig. 5**).
4. The Miniblotter® is closed with the lower piece and tightened with the two provided screws (**Fig. 6**). Avoid membrane damage by excessive tightening. Remaining water in the parallel channels is removed by aspiration with an Eppendorf-type pipette (**Fig. 7**). Next steps should be carried out in less than half and hour, to prevent drying and possible cracking of the paper.

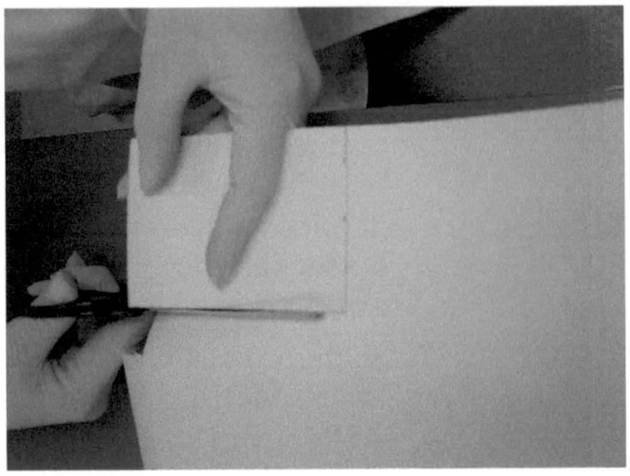

Fig. 1. Cut a rectangle of NC sheet 7.5 × 10 cm.

Fig. 2. Reference line in one of the borders.

Fig. 3. NC paper is soaked in distilled water for 5 min.

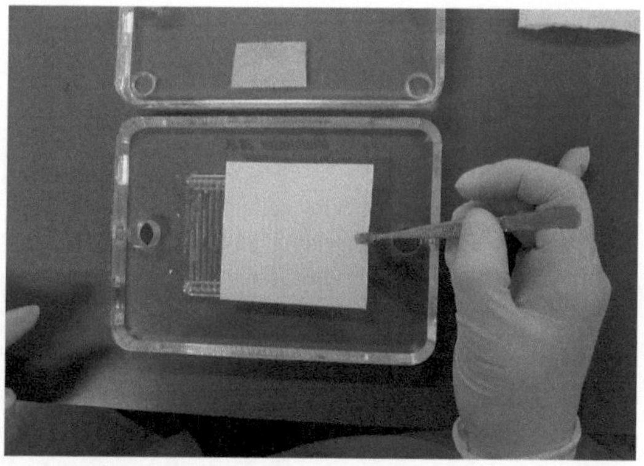

Fig. 4. NC is aligned in the upper part of Miniblotter® with the reference line parallel to channel 1.

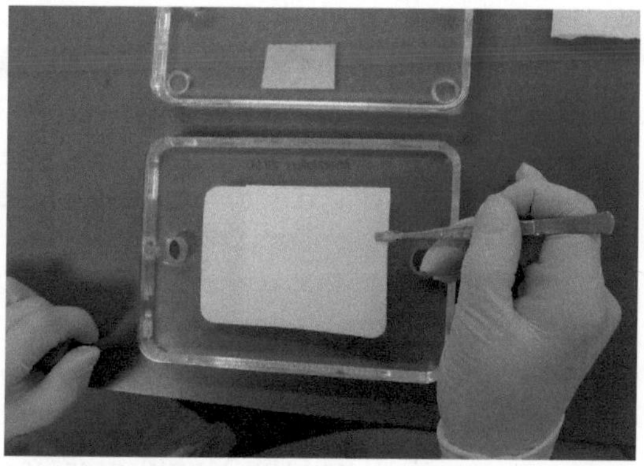

Fig. 5. A plastic cushion is placed onto the NC.

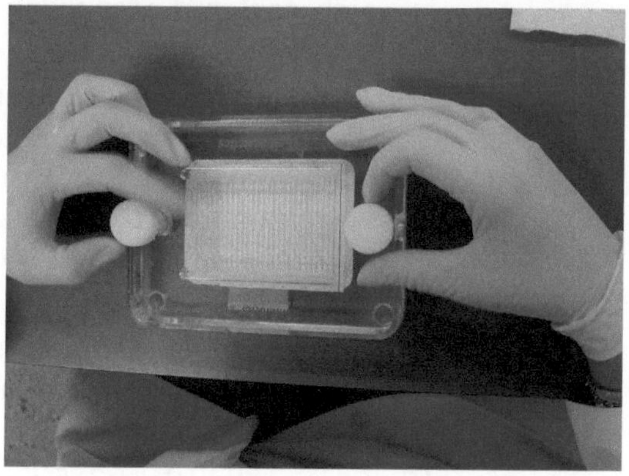

Fig. 6. The Miniblotter® is closed with the lower piece and tightened with the two provided screws.

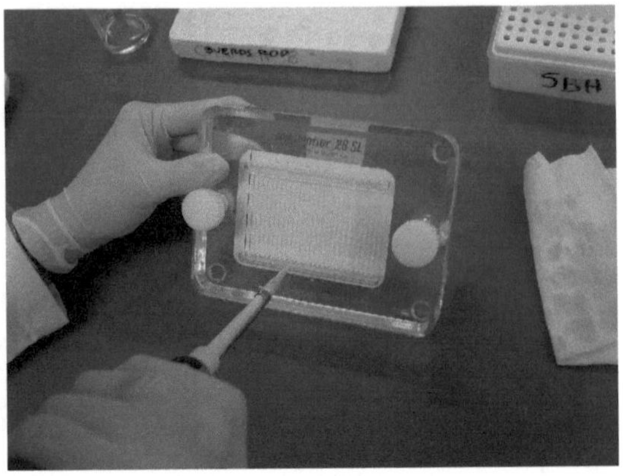

Fig. 7. Remaining water in the channels is removed by aspiration with an Eppendorf-type pipette.

5. The different antigen preparations (crude, synthetic peptides, and recombinant molecules) up to a maximum of 28 can be analyzed in each NC sheet, at a concentration between 10–50 μg/mL in a carbonate-bicarbonate buffer, pH 9.6.
6. To sensitize the NC sheet, 60 mL of each antigen are introduced in each groove of the Miniblotter® (**Fig. 8**), avoiding the formation of bubbles, and incubated for 60 min on an orbital or horizontal shaker, at room temperature (RT) (**Fig. 9**).
7. The antigen solutions are removed by washing with 40 mL PBST, using a washing bottle (**Fig. 10**) or the manifold provided with the equipment (**Fig. 11**).
8. Then the Miniblotter® is opened and the NC sheet is washed three times with 10 mL PBST on a shaker for 10 min each.
9. Blocking is achieved by immersing the NC in 5% nonfat milk in PBST for 2 h on a shaker, at RT (**Fig. 12**).
10. Once blocked, the NC sheet could be processed immediately or stored wrapped in filter paper and inside a plastic bag at –20°C.
11. Numbered 2 mm wide strips are cut perpendicular to the channels and the reference line, with a scalpel (**Fig. 13**). Therefore, each strip would contain a row of square spots corresponding to a maximum of 28 different antigens. Shorter strips with less spot rows can be designed.
12. Strips are immersed individually in the troughs of an incubation tray, in serum diluted (1:100 or 1:200) in blocking solution (**Figs. 14–16**) and incubated for 90 min at RT on a shaker (**Fig. 17**).
13. Strips are washed 3–5 times depending on the background of each system, for 10 min each, with PBST (**Fig. 18**), and then incubated for 90 min with the corresponding

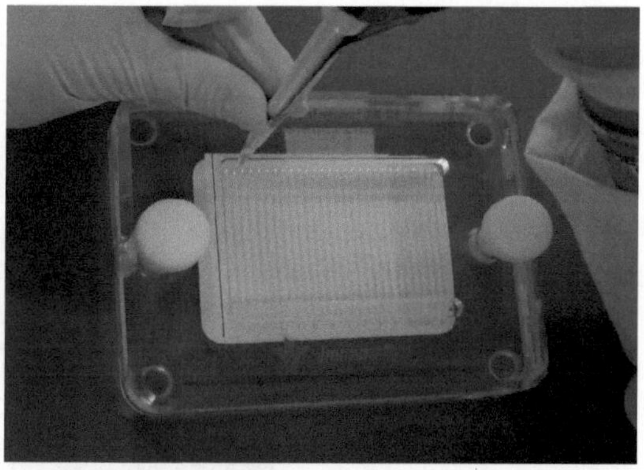

Fig. 8. Sixty microliter of each antigen are introduced in each groove.

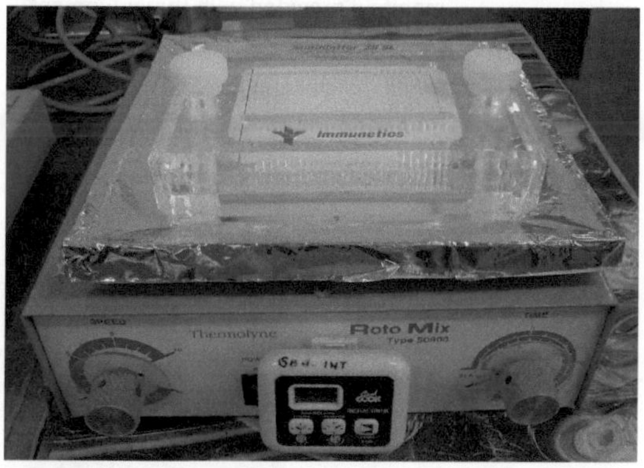

Fig. 9. Incubation for 60 min on an orbital or horizontal shaker.

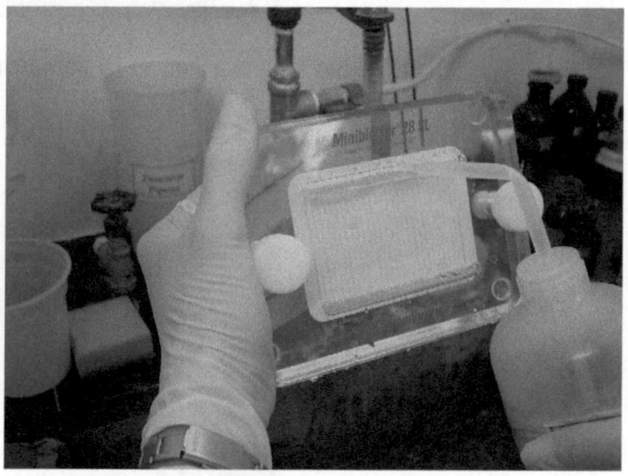

Fig. 10. Antigen excess is removed by washing with 40 mL PBST using a washing bottle.

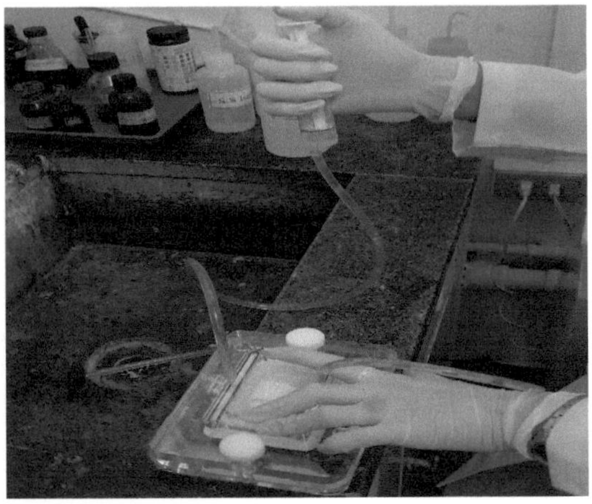

Fig. 11. Antigen excess can also be removed by washing with 40 mL PBST using the manifold provided with the equipment.

Fig. 12. Blocking is achieved by immersing the NC sheet in 5% nonfat milk in PBST for 2 h on a shaker.

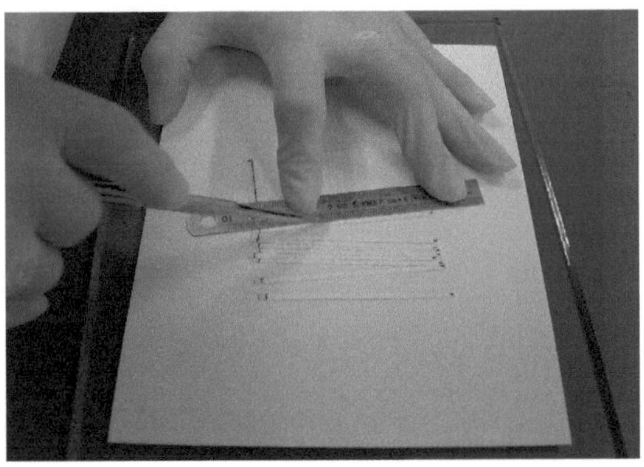

Fig. 13. Numbered 2 mm wide strips are cut perpendicular to the channels and the reference line, with a scalpel.

Fig. 14. Add 600 μL of blocking solution.

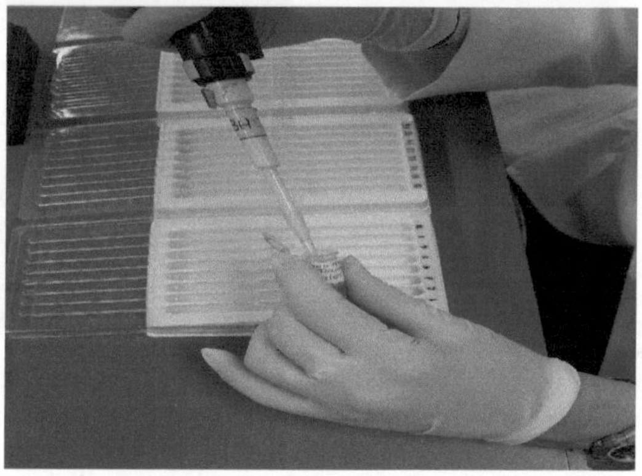

Fig. 15. Three to six microliter of sera (dilution of 1:200 or 1:100, respectively) in blocking solution.

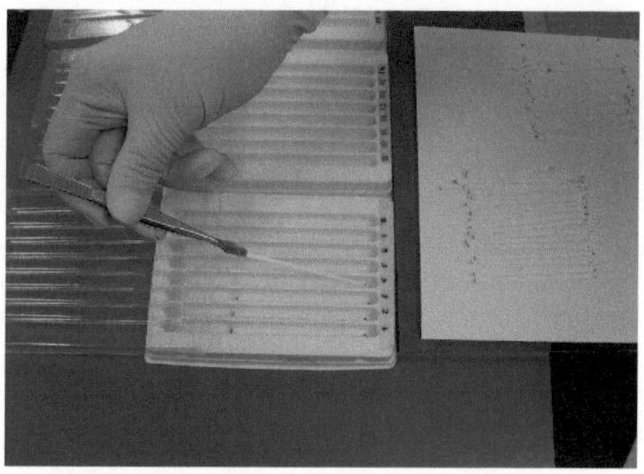

Fig. 16. Strips are immersed individually in the troughs of an incubation tray.

Fig. 17. Incubation for 90 min at RT on a shaker.

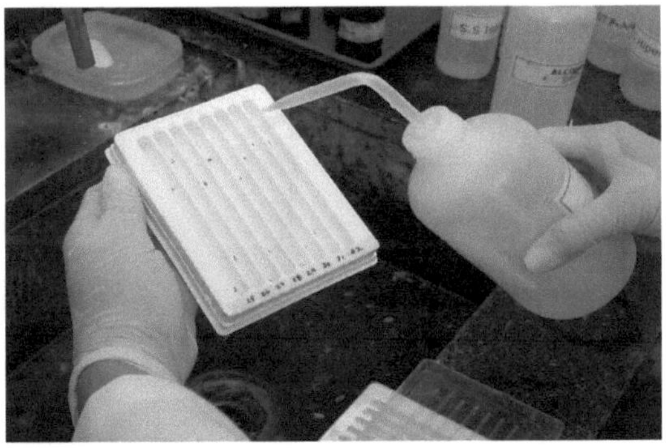

Fig. 18. Strips are washed 3–5 times for 10 min each, with PBST on a shaker.

anti-serum horseradish peroxidase labeled conjugate (**Fig. 19**), diluted (1:1,000–1:50,000) in blocking solution at RT, on a shaker.

14. Three washes of strips are carried out for 10 min each with PBST, as in **Fig. 18**.
15. All strips are put together in a single tray. In case of using a chemiluminescent substrate, strips are immersed in approximately 1 mL of SuperSignal® solution for at least 1 min (**Fig. 20**) and arranged in parallel onto a glass (**Fig. 21**) and covered with a plastic wrap (**Figs. 22** and **23**).
16. Finally, the strips are exposed to Hyperfilm® (Amersham, UK, Cat. # RPN3103K), under darkness and developed photographically (**Fig. 24**). Routinely, two different exposure times are carried out: 5 and 20 s, depending on the intensity of the signal for each condition. For each antigen and antibody systems, all steps must be standardized. If it is available a luminescent detection system (ChemiDoc ™, BIO-RAD Laboratories,

Fig. 19. Incubation for 90 min with the corresponding anti-serum horseradish peroxidase-labeled conjugate.

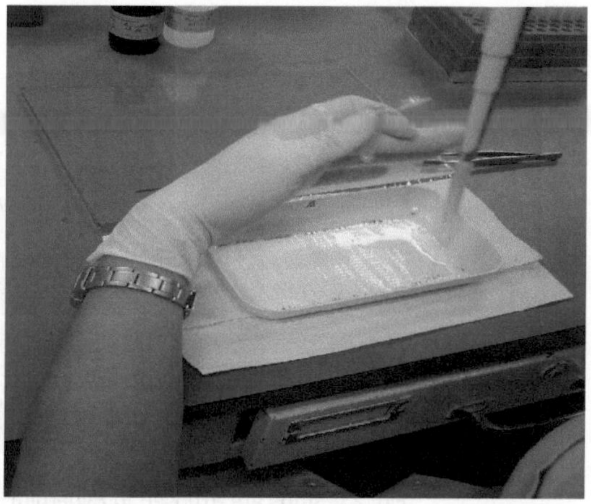

Fig. 20. Strips are immersed in approximately in 1 mL SuperSignal® for at least 1 min.

Fig. 21. NC strips are arranged in parallel onto a glass.

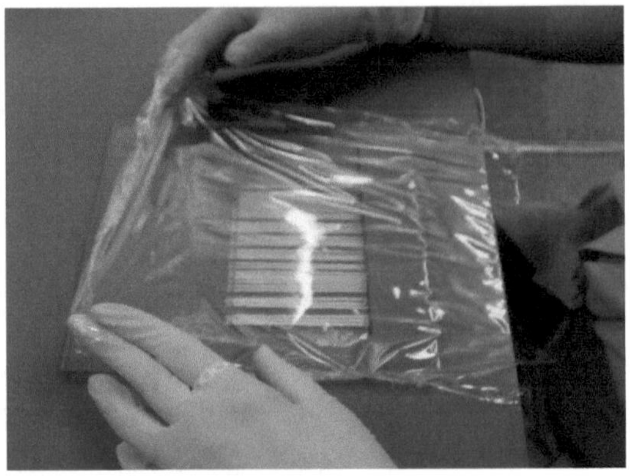

Fig. 22. Strips are covered with a plastic wrap.

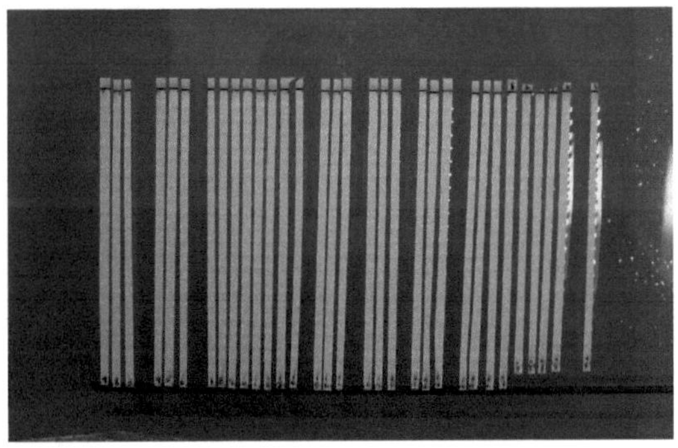

Fig. 23. Strips are exposed to Hyperfilm® under darkness.

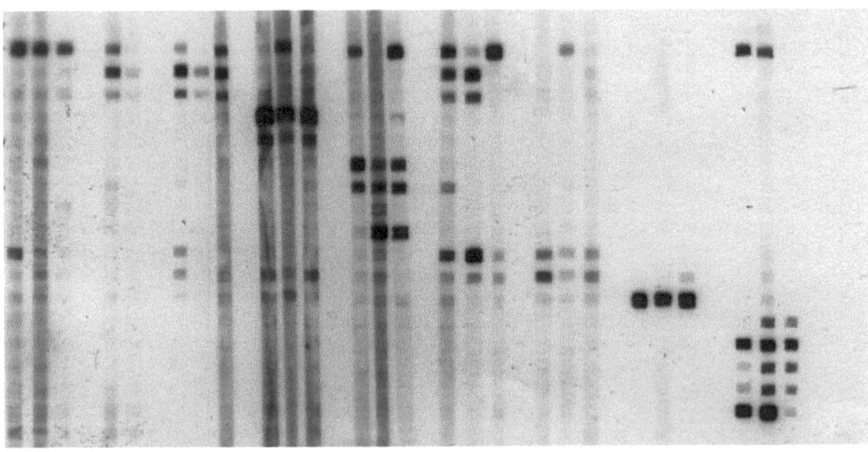

Fig. 24. Representative MABA of antigenicity of crude extracts and synthetic peptides from different infectious agents recognized by patient's sera developed with chemiluminiscent substrate.

Inc. CA, USA), the equipment directly registers electronically the luminescent signal.

17. In case of nonspecific reactivities due to cross-reacting crude antigens, treatment of the antigen preparation with sodium metaperiodate is recommended (*see* **Note 1**).

3.2. Colorimetric MABA

(*See* **Subheading 3.1, steps 1–12** are the same for luminescent and colorimetric MABA)

1. To develop with a precipitable colorimetric substrate, the peroxidase-labeled conjugate should be diluted in PBS/0.3%Tween-20 without milk and incubated for 90 min at RT with continuous gentle agitation on a shaker.
2. Wash the strips four times for 10 min each with PBS-0.3%-Tween-20.
3. Remove the PBS and incubate in the TMB® membrane peroxidase substrate solution previously warmed at RT for 5–15 min, or until desired color is achieved.
4. Stop reaction by immersing the NC in distilled water for 20–30 s. The reaction should be stopped before background color becomes too intense (**Fig. 25**). For storage, dry the membrane, seal with clear plastic and store in the dark to minimize fading.

4. Notes

1. MABA-SMP: Certain crude antigens have limitations because of false-positive results due to immunological cross-reactions that are frequent within certain organisms such as helminths.

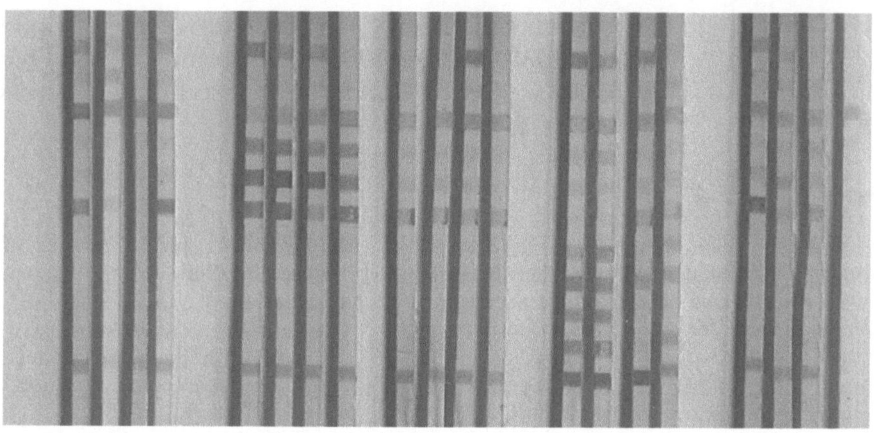

Fig. 25. Representative MABA of antigenicity of crude extracts and synthetic peptides from different infectious agents recognized by patient's sera developed with colorimetric substrate.

Most of these molecules correspond to glycosylated epitopes *(10)*. To minimize this serological cross-reactivity, oxidation with sodium metaperiodate (SMP) has been used to elucidate the role of carbohydrate portion of antigenic glycoproteins in the reactivity by serology *(11, 12)*. After treatment of crude antigens with SMP, the hydroxyl groups of sugars are oxidized to aldehydes under acidic pH. In the presence of a reducing agent such as sodium borohydride ($NaBH_4$), aldehydes become alcohols, being not recognized by the cross-reactive anti-carbohydrate epitope antibodies *(11, 12)*. The method is as follows: (a)NC membrane is sensitized and processed as previously described (*see* **Subheading 3.1, steps 1–10**); (b) NC sheet is washed under agitation 1 min in PBST on a shaker; (c) Membrane is washed under agitation 3 min in acetate buffer (50 mM) pH 4.5; (d) For SMP treatment it is necessary to standardize the SMP concentration (1–3 mM) in acetate buffer for 60 min, in the dark, at RT and prepare it just before use; (e) A brief washing is made with 50 mM acetate buffer for 3 min; (f) Add 50 mM sodium borohydride ($NaBH_4$) in PBS and incubate for 30 min at RT, with continuous gentle agitation on a shaker. The reagent should be prepared just before use. It should be bubbling and if the reagent is excessively hydrated and does not bubble, treatment will not work appropriately; (g) Wash 3 min with PBS under agitation; (h)Continue the procedure as it has been described previously (*see* **Subheading 3.1, steps 11–16**).

Acknowledgement

This work has been partially supported by grants N° G-2005000387 and F-2005000122 from FONACIT-Venezuela. We thank Lic. Cecilia Colmenares for technical assistance.

References

1. Schena, M., Heller, R.A., Theriault, T.P., Konrad, K., Lachenmeier, E., and Davis, R.W. (1998) Microarrays: biotechnology´s discovery platform for functional genomics. *Trends Biotechnol.* 16, 301–306.
2. Dunbar, S.A., Vander Zee, C., Oliver, K., Karem, K., and Jacobson J. (2003) Quantitative, multiplexed detection of bacterial pathogens: DNA and protein applications of the Luminex LabMAP™ system. *J Microbiol Methods.* 53, 245–252.
3. Noya, O., and Alarcón de Noya, B. (1998) The multiple antigen blot assay (MABA): a simple immunoenzymatic technique for simultaneous screening of multiple antigens. *Immunol Lett.* 63, 53–56.
4. Noya, O., Patarroyo, M.E., Guzmán, F., and Alarcón de Noya, B. (2003) Immunodiag-

nosis of parasitic diseases with synthetic peptides. *Curr Protein Pept Sci.* 4, 299–308.

5. Noya,O., Alarcón de Noya,B., Guzmán, F., and Bermúdez, H. (2003) Immunogenicity of Sm32 synthetic peptides derived form the *Schistosoma mansoni* adult worm. *Immunol Lett.* 88, 211–219.
6. Chacón,N., Losada,S., Bermúdez, H., Cesari, I.M., Hoebeke, J., and Noya, O. (2003) Immunogenicity of polymerizable synthetic peptides derived from a vaccine candidate against schistosomiasis: the asparaginyl endopeptidase (Sm32). *Immunol Lett.* 88, 199–210.
7. Wide, A., Capaldo, J., Zerpa, N., Pabón,R., Noda,A., Noya,B., et al. (2006) Sharing of antigens between *Plasmodium falciparum* and *Anopheles albimanus. Rev Inst Med Trop S Paulo.* 48, 327–332.
8. Losada, S., Chacón, N., Colmenares, C., Bermúdez, H., Lorenzo, A., Pointier, J.P., et al. (2005) *Schistosoma*: Cross-reactivity and antigenic community among different species. *Exp Parasitol.* 111, 182–190.
9. Puccio, F.A., Lynch, N.R., Noya, O., Noda, A., Hagel, I., López, E., et al. (2004) Importance of including *Blomia tropicalis* in the routine diagnosis of Venezuelan patients with persistent allergic symptoms. *Allergy* 59, 753–757.
10. Alarcón de Noya, B., Colmenares, C., Lanz, H., Caracciolo, M.A., Losada, S., and Noya,O. (2000) *Schistosoma mansoni*: immunodiagnosis is improved by sodium metaperiodate which reduces cross-reactivity due to glycosylated epitopes of soluble egg antigen. *Exp Parasitol.* 95, 106–112.
11. van Lenten, L., and Ashwell, G. (1971) Studies on the chemical and enzymatic modification of glycoproteins. *J Biol Chem.* 246, 1889–1894.
12. Woodward, M., Young, W., and Bloodgood, R.A. (1985) Detection of monoclonal antibodies specific for carbohydrate epitopes using periodate oxidation. *J Immunol Meth.* 78, 143–153.

Chapter 27

Blotting from Immobilized pH Gradient Gels: Application to Total Cell Lysates

Harry Towbin

Summary

Isoelectric focusing as used in the first dimension of two-dimensional gel electrophoresis separates protein isoforms such as those due to phosphorylation and acetylation. The immunoblotting method described here reveals this diversity by a one-dimensional separation. Using commercially available immobilized pH gradient plates or strips, the resolved proteins are transferred to PVDF membranes by diffusion and are probed with protein specific antibody. The system is useful for monitoring changes of banding patterns and permits parallel processing of samples. Since the effect of posttranslational modifications on the isoelectric point can be predicted, inferring the number and extent of modifications is possible.

Key words: Isoelectric focusing, Posttranslational modifications, Phosphorylation, Immunoblotting

1. Introduction

When looked at closely, proteins both intracellular and secreted, display a bewildering number of modifications. Because of the importance of posttranslational modifications for regulating biological activities of proteins, there is a corresponding effort aiming at detecting and monitoring their occurrence. An increasing number of antibodies are becoming available that allow specific detection of modifications on selected proteins, mostly well characterized components of signaling cascades. Such antibodies are, however, often not available for other proteins. The method described in this chapter can be used to detect and track protein modifications that lead to alterations of their charge and consequently of a change of the isoelectric point. Such charge isoforms can be separated by isoelectric focusing and are evident from

Biji T. Kurien and R. Hal Scofield (eds.), *Methods in Molecular Biology, Protein Blotting and Detection, vol. 536*

DOI: 10.1007/978-1-59745-542-8_27

two-dimensional separations where trains of spots are seen that are frequently due to different phosphorylation states. Especially high resolution is achieved by IEF on immobilized pH gradient gels. Although SDS-PAGE is sometimes capable of resolving modified proteins, the extent or even direction of the changed migration is difficult to predict.

The IEF blotting method described here was developed for monitoring alterations occurring on intracellular proteins that require denaturing conditions for extraction and separation *(1)*. The method allows sensitive detection and estimation of the relative abundance of isoforms with antibodies. The latter is a quantity that is not easily measured by modification specific reagents. Quantification is facilitated by running samples side by side under the same conditions. In principle, similar information can be obtained from two-dimensional separations by running a series of westerns. Results from such experiments are, however, difficult to quantify because of unavoidable technical variability. The separation method used here is very similar to the first dimension of two-dimensional separation systems *(2, 3)*. It uses commercially available precast gels and allows direct comparison with two-dimensional separations. Transfer from the plastic supported gels is achieved by diffusion blotting.

2. Materials

1. IPG plates (immobilized pH gradient gels) Immobiline Dry-Plates (GE Healthcare or other manufacturers) or immobilized pH gradient strips (*see* **Note 1**). Various pH ranges are available.
2. R-Buffer *(4)*: 4% CHAPS, 7 M urea, 2 M thiourea, 10 mg/mL dithiothreitol, and 1% carrier ampholytes (pH 3–10) (Pharmalytes; Amersham Pharmacia Biotech, Uppsala, Sweden), 0.02 mg/mL bromophenol blue. Store at −80°C.
3. Reswelling buffer: R-buffer without carrier ampholytes and without bromophenol blue. Store at −80°C.
4. TCA fixative: 12% trichloroacetic acid, 3.5% sulfosalicylic acid.
5. Multiphor II Electrophoresis System (GE Healthcare).
6. Sample applicator for IEF (*see* **Note 2**).
7. Transfer buffer: 4 M guanidinium chloride, 0.05 M Tris-HCl pH 8, 1 mg/mL DTT (added shortly before use).
8. Filter paper e.g. Whatman 1Chr, Cat. No. 3001917.
9. Polyvinylidene difluoride (PVDF) membrane.

3. Methods

3.1. Preparation of Extracts

1. Collect cells by centrifugation, wash once with PBS, suspend the pellet, transfer into a tared microcentrifuge tube, centrifuge once more and remove as much of the supernate as possible (*see* **Note 3**). Estimate the volume by weighing and add R-buffer corresponding to five volumes of the cell pellet, suspend and leave on ice for about 1 h. Centrifuge at 1,000 g for 15 min in the cold. Store the sample at –80°C. Protein concentration may be determined by a Coomassie binding assay *(5)* (*see* **Note 4**). Protein concentrations are in the range of 3–10 mg/mL.

3.2. Isoelectric Focusing on IEF Immobiline Plates

1. Reswelling polyacrylamide sheets in R-buffer devoid of ampholines (*see* **Note 5**): The procedure is otherwise according to instructions of the manufacturer. Cut required size if only part of a sheet is used, and mark the polarity of the pH-gradient by cutting off a corner from the sheet. Place the volume of buffer required for the sheet in the reswelling tray and place the sheet face down on the liquid. Avoid trapping air bubbles and spilling of buffer over the top side of the sheet. Leave overnight at room temperature in a box containing wet paper towels for achieving a water-saturated atmosphere. Instead of Immobiline plates, it is also possible to use individual IEF strips (*see* **Note 1**).
2. Place about 2 mL of kerosene on the cooling stage of the Multiphor apparatus and position the IPG plate paying attention to match the IPG plate to the required polarity of the electrodes. Remove excess buffer with a filter paper (provided for that purpose by the manufacturer). Position electrode filter paper strips and sample applicator on the gel surface (about 2 cm from the anode). The position may vary according to the proteins of interest.
3. Using a micropipette apply the samples (1 µL or up to 4 µL for the larger sample comb), which will be held by capillary force underneath the indentations (*see* **Note 6**).
4. Running conditions: 15°C, from 0 to 300 V in 30 min, then to 900 V in 9 h, keeping at 900 V for 6 h, raising to 3,500 V within 1 h, keep at 3,500 V for a final 4 h. The entire run takes 20 h with an accumulated Volt-hour product of ~27,000 Vh. We avoided focusing at the highest voltages without attention. Adhere to the safety instructions of the IEF manufacturer. Depending on the pH range and type of protein, different focusing conditions may be needed.

3.3. Transfer to PVDF Membrane

1. Incubate the gel with TCA fixative for 1 h. Wash three times with water, 10 min each (*see* **Note 7**).

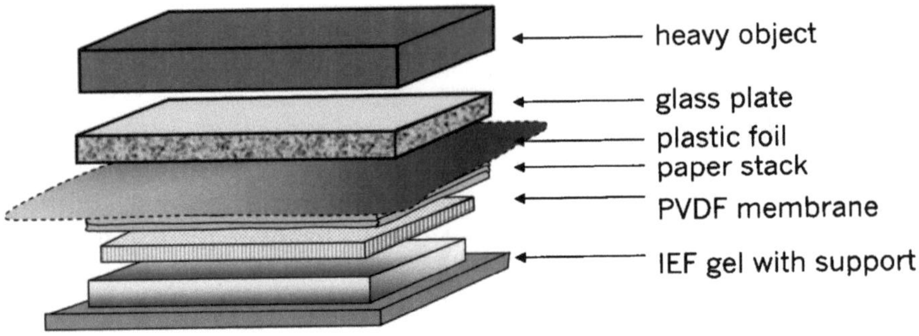

Fig. 1. Setup used for diffusion transfer from plastic backed Immobiline DryPlates.

2. Cut a piece of PVDF to the size of the gel, moisten it by soaking in ethanol followed by washing in water.
3. Prepare 10 pieces of filter paper (e.g., Whatman 1Chr, Cat. No. 3001917), 1 cm larger than the membrane, and soak in transfer buffer (about 35 mL is required for half a gel, ~13×13 cm^2).
4. Assemble the sandwich by placing the gel on a glass plate, followed by the PVDF membrane, and a stack of filter papers soaked in the transfer buffer. Overall thickness of the paper stack will be about 2 mm. Wrap the assembly with a plastic foil to prevent drying out and cover with glass plate (**Fig. 1**). Ensure good contact by compressing with a heavy object and leave overnight at room temperature.
5. Wash the membrane with water to remove guanidinium chloride.
6. At this stage, the membrane is processed like any western blot.

3.4. Visualization of Total Protein Pattern (Optional)

To assess any disturbances or differences between samples, we find it useful to visualize and photograph the total protein pattern on the PVDF membrane using stains that are compatible with subsequent immunostaining (e.g., SyproRuby). Also, if needed, IEF standards may be added in adjacent lanes and thus be visualized on the membrane.

3.5. Interpretation

It is useful to judge whether an assumed modification can account for the observed shift of the isoelectric point of the protein under investigation. Thus, for proteins with known amino acid composition, the isoelectric point can be predicted *(6)*. The effect of phosphorylation can also be calculated *(7)*. For acetylation on lysine groups, it is possible to simply omit one lysine from the sequence before computation of the pI. Similarly, other modifications may be incorporated by considering their effect on the charge of the protein. **Figure 2** shows an example of an application of the IEF blotting method.

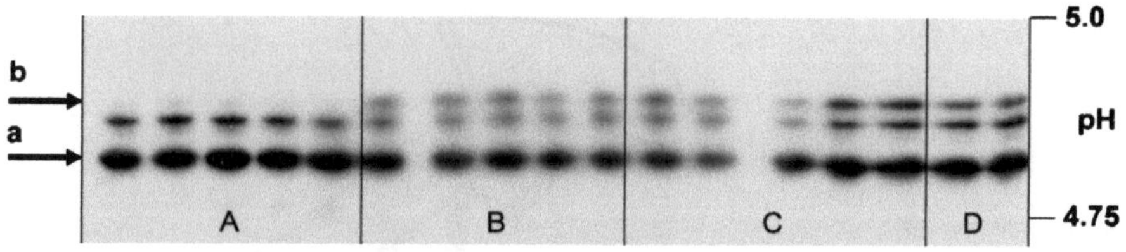

Fig. 2. Example of an IEF blotting experiment with extracts from rat peripheral blood white blood cells. The protein extract loaded (4 µL) corresponds to cells obtained from 80 µL of blood. The separation was carried out on Immobiline plates pH range 4.5–5.4. Transfer to PVDF membranes was achieved by diffusion blotting. An antibody against 14-3-3gamma protein detects two bands in untreated animals and three bands in animals that had been treated with bengamide. This compound inhibits methionine aminopeptidase (8). As a consequence, acetylation of the penultimate N-terminal aminoacid is also inhibited and a slightly more basic protein accumulates. The position of normal acetylated protein is marked as "a," the unprocessed form as "b." The nature of the intermediate form is not known. A: untreated; B, C, and D: animals treated with increasing doses of bengamide.

4. Notes

1. Instead of Immobiline DryPlates, individual strips may also be used. Here, the procedures for the first dimension of 2D electrophoresis are closely adhered to. Thus, reswelling of the gel strips proceeds in the presence of the sample, thereby allowing application of higher sample volumes. After electrophoresis, the strips are processed exactly as described for the plates. For the blotting step, the strips can be arranged side by side. A large choice of strips covering wide and narrow pH ranges is commercially available. For strips, an electrophoretic transfer method, which requires removal of plastic backings, has been described (9).
2. The applicator comb may be built by cutting a disposable applicator intended for the Phast Gel™ system (PhastGel™ Sample Applicators, 8× of 1 µL or 6× of 4 µL, Cat. No. 18-1618-01 or 18-0012-29, respectively). The comb needs to be cut horizontally to fit to the limited height available on the Multiphor II. A rectangular piece of the plastic material is glued to one end of the comb with a drop of dichloromethane (this solvent is toxic and irritant, work under a fume hood) (*see* **Fig. 3**).
3. It is important to keep the salt concentration as low as possible as high conductivity leads to disturbances during running.
4. We find it convenient to carry the assay out in 96-well microtiter plate by adding 0.5 and 1 µL of the sample to 10 µL water and then adding 150 µL prediluted Bradford reagent to the wells. This amount of R-buffer is tolerated in this assay.
5. We found that ampholines at the relatively high concentrations used for focusing as well as CHAPS have a tendency to bind to the blotting membrane. Hydrophilic proteins may poorly bind in the presence of these components.

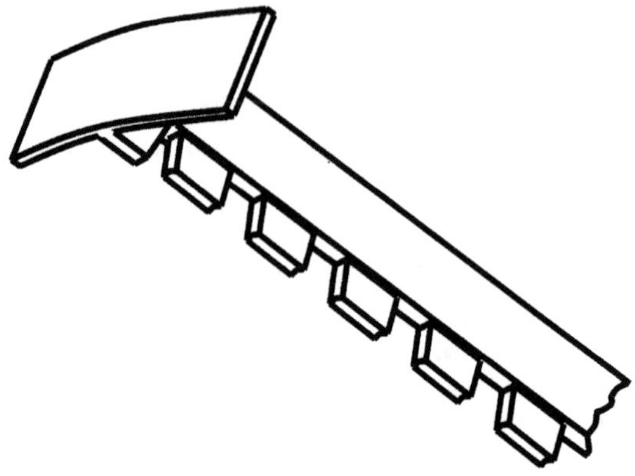

Fig. 3. T-shaped sample applicator.

6. Other methods of sample application may also be useful, such as cup loading or paper pieces.
7. This step serves to wash out CHAPS, which interferes with adsorption of many proteins. Large or hydrophobic proteins may not require the fixation and CHAPS washout step, one example being tubulins *(1)*. If IEF blotting is to be carried out repeatedly, it is worthwhile to test simplified transfer procedures.

References

1. Towbin, H., Ozbey, O., and Zingel, O. (2001) An immunoblotting method for high-resolution isoelectric focusing of protein isoforms on immo-bilized *p*H gradients. *Electrophoresis* 22, 1887–1893.
2. Görg, A., Weiss, W., and Dunn, M.J. (2004) Current two-dimensional electrophoresis technology for proteomics. *Proteomics* 4, 3665–3685.
3. Hoving, S., Voshol, H., and van Oostrum, J. (2005) Using Ultra-Zoom Gels for High-Resolution Two-Dimensional Polyacrylamide Gel Electrophoresis, The Proteomics Protocols Handbook, Humana Press Inc.,U.S., John M. Walker, ed. pp. 151–166
4. Rabilloud, T. (1998) Use of thiourea to increase the solubility of membrane proteins in two-dimensional electrophoresis. *Electrophoresis* 19, 758–760.
5. Bradford, M. (1976) A rapid and sensitive method for the quantitation of microgram quantities of protein utilizing the principle of protein-dye binding. *Anal Biochem* 72, 248–254.
6. Bjellqvist, B., Hughes, G.J., Pasquali, Ch., Paquet, N., Ravier, F., Sanchez, J.-Ch., et al (1993) The focusing positions of polypeptides in immobilized *p*H gradients can be predicted from their amino acid sequences. *Electrophoresis* 14, 1023–1031.
7. Obenauer, J.C., Cantley, L.C., and Yaffe, M.B. (2003) Scansite 2.0: Proteome-wide prediction of cell signaling interactions using short sequence motifs. *Nucleic Acids Res.* 31, 3635–3641.
8. Towbin, H., Bair, K.W., DeCaprio, J.A., Eck, M.J., Kim, S., Kinder, F.R., *et al* (2003) Proteomics-based target identification: bengamides as a new class of methionine aminopeptidase inhibitors. *J Biol Chem.* 278, 52964–52971.
9. Poland, J., Bohme, A., Schubert, K., and Sinha, P. (2002) Revisiting electroblotting of immobilized pH gradient gels: a new protocol for studying post-translational modification of proteins. *Electrophoresis* 23, 4067–4071.

Chapter 28

Immunoprecipitation-Western Blot for Proteins of Low Abundance

Edward P. Trieu, Joanne K. Gross, and Ira N. Targoff

Summary

Combining the procedures of immunoprecipitation and immunoblotting can help to overcome some of the limitations of each separate procedure. Immunoblotting can identify immunoprecipitated proteins more specifically and with higher sensitivity than nonspecific protein stains or autoradiography. Immunoprecipitation can enrich proteins of interest to improve sensitivity for detection when compared with immunoblotting of whole cell extracts. Recently, immunoprecipitation-blotting helped to characterize a new autoantibody, anti-p155, and to test for the presence of the autoantibody in patient sera to study its clinical associations. The procedure for immunoprecipitation-blotting, with specific reference to this autoantibody test ("reverse" immunoprecpitation-blotting), is reported here in detail.

Key words: Immunoprecipitation, Immunoblot (western blot), Autoantibody, Autoantigen, Dimethyl pimelimidate

1. Introduction

Immunoprecipitation has been a useful method for purification of specific proteins and protein complexes from cell extracts or other mixtures, and has been used extensively to characterize autoantigens and autoantibodies in autoimmune diseases. Usually the analyses of the immunoprecipitated material is by gel electrophoresis, but sometimes, more specificity is required. For autoantibodies, although the band pattern by electrophoresis is often enough to determine identity between the autoantibodies in different sera, this band pattern may not be specific, particularly when only a single protein band is immunoprecipitated.

Biji T. Kurien and R. Hal Scofield (eds.), *Methods in Molecular Biology, Protein Blotting and Detection, vol. 536*

DOI: 10.1007/978-1-59745-542-8_28

Combining immunoprecipitation with immunoblotting can often solve these problems. It can provide greater specificity for protein identification, and, by using different sera for immunoprecipitation and blotting, can help determine immunologic identity. Immunoblotting can also help identify the antigenic protein within an immunoprecpitated multi-protein complex. The high sensitivity of immunoblots can also enhance the sensitivity of detection of immunoprecipitated proteins, particularly when present in complex mixtures or crude extracts. In turn, immunoblotting of immunoprecipitates can solve some of the problems that arise with immunoblotting of crude extracts. It can often enhance detection and recognition of low-abundance protein. For autoantibodies, it can improve the sensitivity and specificity of autoantibody recognition when compared with immunoblotting of whole extract.

The method we describe here facilitates the detection of low-abundance proteins, and increases the ability to detect antibodies to these proteins in a specific manner. For autoantibody testing, immunoprecipitation is used to enrich the low-abundance proteins, which can then be electrophoresed, transferred, and used to test other sera for autoantibodies to these proteins. Some autoantibodies do not recognize denatured antigens, a limitation that can be overcome by reversing this approach. For reverse immunoprecipitation-blotting, immunoprecipitates are prepared using the test sera, and blotted with known positive reference sera. The use of this method for detection of a recently described autoantibody found in patients with dermatomyositis, anti-p155, is used as an example. Immunoprecipitation-western blots could be used to detect p155 autoantibody even in sera that could not recognize the denatured form of p155 antigen after electrophoresis and western blotting (*see* **Figs. 1** and **2**) *(1)*. Thus, immunoprecipitation-western blot is a basic tool to provide useful data about a new antibody that other methods could not.

This procedure consisted of five different stages:

1. *Preparation of antibody-protein A gel:* The antibody is allowed to bind to protein A by incubating the antibody with protein A gel. After incubation, the column is washed with buffer to wash away the unbound proteins. The antibody is cross-linked to the protein A gel with a cross-linking solution to avoid IgG contaminating the eluate *(2)*. Following cross-linking, block the remaining unbound sites with an amine group from a chemical such as Tris or Glycine. Next, a mild eluting agent removes the unbound antibodies that remain. Then, equilibrate the column with antigen-binding buffer.

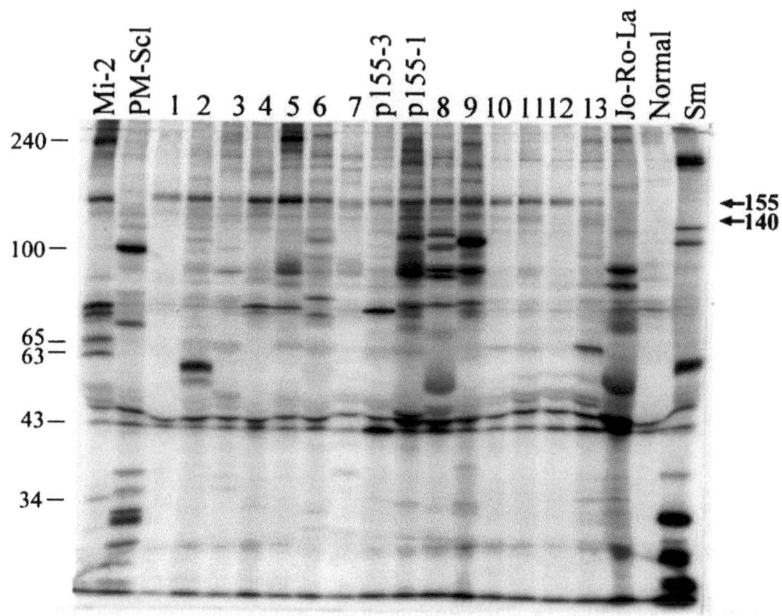

Fig. 1. Before proceeding to the immunoprecipitation-western blot, 13 anti-p155 autoantibody-positive sera (*lanes 1–13*) were screened by an immunoprecipitation of S35-labeled Hela cell extract. This is an autoradiograph of the polyacrylamide gel electrophoresis of the immunoprecipitates. *Lanes 1–13* were sera that precipitated a similar appearing protein to that of the prototype sera, p155-1 and p155-3; however, several sera show additional coexistent autoantibodies in the same region. To identify positively the p155 band, the sera are next screened with the immunoprecipitation western blot. Control autoantibodies used in the other lanes were Mi-2, PM-Scl, Jo-Ro-La, and Sm. Normal is serum from a normal subject. The numbers on the left side of the gel are molecular weight markers in kDa (reproduced from *(1)* with permission from John Wiley & Sons).

2. *Immunoprecipitation:* Incubate the cell extract with the antibody-protein A gel. The antibody immunoprecipitates the antigen then the washes from the buffer remove the unbound proteins. After washing, eluted protein from the beads with SDS sample buffer is the purified antigen.
3. *Gel electrophoresis of immunoprecipitate:* Run the purified antigen on SDS PAGE to resolve the proteins according to size by the Laemmli method.
4. *Western blot:* The antigen separated via discontinuous SDS PAGE is electrophoretically transferred onto a nitrocellulose membrane *(3)*.
5. *Antibody detection:* Cut the nitrocellulose paper with purified antigen into strips to develop with different sera for the detection of antibody.

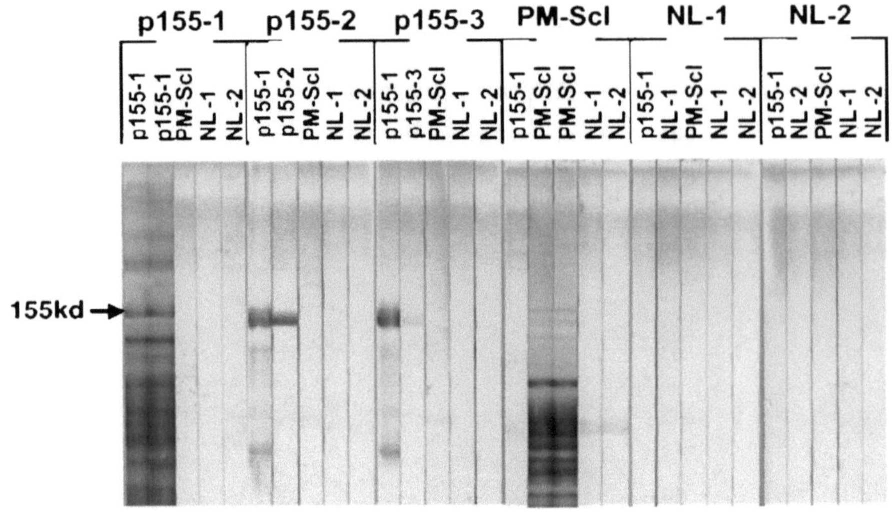

Fig. 2. Results from a typical immunoprecipitation-western blot. Immunoprecipitates were prepared from the six sera listed across the top, separated via electrophoresis and transferred to nitrocellulose. The sera as indicated on the figure blotted each of the 30 lanes. In this experiment, the immunoprecipitates were the immunoprecipitation of test serums and the sera in the blotting phase were the reference sera. The result shows that the three anti-p155-positive sera, p155-1, p155-2, and p155-3, recognized the native form of p155 antigen, precipitated it and p155-1 reference serum detected it on the blot. In the antibody development panel, p155-1 blots the first strip in each set and it stained the p155 band with all three p155-positive sera. The second strip in each set, stained with the serum used for immunoprecipitation. P155-1 and p155-2 recognized the denatured form of the antigen but p155-3 did not *(1)*. Thus, this "reverse" immunoprecipitation-blotting procedure allowed the detection of the anti-p155 antibody in serum p155-3, whereas direct immunoprecipitation-western blotting (immunoprecipitation using reference serum and blotting using test serum) would not have detected this antibody. Thus, the serum used in the blotting phase must be able to recognize the denatured form of the antigen. The control sera PM-Scl, NL-1, and NL-2 were all negative for p155 antigen (reproduced from *(1)* with permission from John Wiley & Sons).

2. Materials

2.1. Preparation of Antibody-Protein A Gel

1. Protein A sepharose or agarose (GE Healthcare, Sweden).
2. Micro Bio-Spin Chromatography Columns, caps, and tip enclosures (Bio-Rad, Hercules, CA, USA).
3. Sera or monoclonal antibodies: Mouse IgG subclass 2a, 2b, and 3 can bind to protein A but they are weak. Protein G can bind to all mouse IgG sub class 1–3.
4. Antibody-Binding Buffer (ABB): 0.2 M triethanolamine, pH 8.9, 0.5 M NaCl. Store at 4°C.
5. Cross-linking solution: 30 mg/mL of dimethyl pimelimidate (DMP, Sigma Chemical Co, St. Louis, MO, USA) in antibody binding buffer. Store DMP at 4°C or lower since it is unstable and hygroscopic. Prepare the solution immediately before use.
6. Quenching buffer: 0.1 M Tris, pH 8.5, 0.5 M NaCl. Store at 4°C.

7. Unbound antibody elution solution: 1.0 M acetic acid, 0.15 M NaCl. Store at RT.
8. Immunoprecipitation buffer (IPP): 10 mM Tris, pH 8.0, 0.5 M NaCl, 0.1% NP-40 or Igepal CA-630 (Sigma). Store at 4°C.
9. 8-Channel Nunc-Immuno* washer (Fisher Scientific) (*see* **Note 1**).
10. Sonicator (Ultra Sonic Processor).

2.2. Immunoprecipitation

1. K562 cell culture pellets in log phase stored frozen at −80°C in 15 mL conical tubes.
2. SDS sample buffer: 10 mM Tris, pH 6.8, 10% glycerol, 2% SDS, 0.01% bromphenol blue, 1% mercaptoethanol. Store frozen at −20°C (*see* **Note 2**).
3. 4× stacking buffer: 0.5 M Tris-HCl, pH 6.8. Stored frozen at −20°C in 50 mL aliquots.

2.3. Gel Electrophoresis of Immunoprecipitate

1. Protean II xi gel apparatus from Bio-Rad, 16 cm glass plates, 25-well combs, and 1.5 mm spacers.

2.4. Western Blot

1. Trans-Blot cell (Bio-Rad).
2. A small power supply with a current up to 450 mA.
3. Refrigerated circulator (Fisher Scientific, Dallas, TX, USA), or alternatively a refrigerator with an electrical outlet inside.
4. Filter paper (Whatman 3MM).
5. Nitrocellulose sheets with 0.2 μm spore size. (Fisher; Bio-Rad).
6. Pre-cut cellophane sheet (Bio-Rad).
7. Protein transfer buffer: 25 mM Tris, 192 mM glycine, 20% methanol, (pH 8.3). Methanol is very dangerous. Read MSDS before using it.

2.5. Antibody Development

1. Mini incubation trays (Bio-Rad).
2. Antibody development panel:
   a. Sera to be used for immunoprecipitation, including "test" or positive sera and "control" sera.
   b. Sera to be used for blotting, including "test" or "positive" sera and "control" sera.
   c. Control sera can include those with antibodies to other antigens and normal sera.
3. Nitro Blue Tetrazolium, (NBT) (Sigma): 50 mg/mL in 70% dimethylformamide (DMF). Light sensitive. To avoid loss of

activity prepare in a small amount. Store at 4°C to –20°C in a dark bottle.

4. 5-5-Bromo-4-Chloro-3-Indolyl Phosphate, (BCIP), (Sigma): 25 mg/mL in DMF. Light and temperature sensitive. Store at –20°C in a dark bottle.
5. Alkaline phosphatase (AP) buffer: 100 mM Tris-HCl, pH 9.5, 100 mM NaCl and 5 mM $MgCl_2$ ($MgCl_2$ is optional and tends to precipitate).
6. AP substrate solution: Add 15 μL from each of NBT and BCIP for every 5 mL AP buffer. Use within 1 h and discard any unused solution.
7. High-precision stainless steel forceps. (Sigma; Fisher Scientific).
8. Scalpel handles No. 3 and Scalpel blade No. 10 (Sigma).
9. Nalgene reusable plastic utility box, dimension: 71/2 × 6 × 11/2 inches, (Fisher).
10. Colloidal gold staining solution (Bio-Rad).
11. Rolling ruler, 12 in., part No. 961-812 (Staedtler).
12. 10× TBST: 0.1 M Tris-HCl, pH 7.4, 1.5 M NaCl, 0.5% Tween-20.
13. Instant Nonfat Dry Milk: To rehydrate the powder completely, warm up the mixture of milk powder and TBST to 37°C in a warm water bath for 10–20 min before use or leave it at 4°C overnight.
14. Fast Green FCF (Fisher Scientific): 0.2% in 20% methanol, 10% acetic acid.
15. White tulle from fabric or craft stores or Spectra/Mesh woven nylon filters (Spectrum Laboratories, Inc, Rancho Dominguez, CA, USA).

3. Methods

3.1. Preparation of Antibody-Protein A Gel

3.1.1. Monoclonal Antibody or Sera for a Gel Protocol

1. For column with monoclonal antibody: Measure the amount of antibody available to add to protein A gel. To get an optimal amount of monoclonal antibody, first use 100–200 μg of mouse monoclonal IgG of subclass 2a, 2b or 3 per column. Usually one milliliter of protein A gel can bind 20–40 mg of IgG.
2. For column with human serum: The amount of each serum used for preparation of column is 25–50 μL based on estimated 15–20 mg of IgG per milliliter of human

serum of which specific antibody of interest is very little. For one gel, we recommend preparing 12 protein A columns simultaneously. The control sera usually include at least one that is positive for the antibody of interest and one that is normal or negative, and often sera with other antibodies that are positive.

3.1.2. Equilibration of Protein A-Sepharose or Agarose

1. Reconstitution of protein A gel may be necessary. The gel can be reconstituted with ABB buffer before use. For 12 columns add 0.3 g protein A gel to 10 mL of ABB. Usually 1 mL of preswelled gel consists of 0.2 g dry beads.
2. Cap, vortex, and mix on a rotator for 5 min.
3. Centrifuge at low speed about 500 × g without brake.
4. Wash beads one time with ABB to remove preservatives that prevent antibody binding.
5. Aspirate the buffer to gel level. Add exactly the amount of ABB to equal the amount of gel to make a 1:1 ratio.

3.1.3. Preparation of Antibody-Protein A-Sepharose or Agarose

1. To each Micro Bio-Spin Column add 50 or 100 μL of protein A gel from step 5 above (*see* **Note 3**).
2. Add 25 or 50 μL of each serum into the columns (*see* **Note 4**).
3. Add 25 or 50 μL of ABB to the columns for a better suspension.
4. Cap and place on a rotator for 20 min at RT.
5. After 20 min, the binding to protein A gel is complete. To bring all the beads to the bottom of the column, add 0.9 mL of ABB to each of the column, cap, invert to mix and centrifuge. After centrifuging, remove the lids and snap off the column tips. If no manifold is available, put the columns into a 2 mL 96 deep well plate or put the columns into tubes (13 × 100 mm) in a rack to keep the columns upright for washing. Wash the beads thrice with ABB (*see* **Note 5**). Remove and put back the tip enclosures on the columns after the washing.
6. Add 100 μL of fresh cross-linking solution.
7. Shake the columns horizontally to mix. Let them sit for 15 min with momentary shaking.
8. Add 200 μL of blocking solution.
9. Cap and place on a rotator for 10 min at RT.
10. Wash the columns with IPP buffer twice and add 300–400 μL of elution buffer.
11. Wash columns with IPP buffer thrice or more to remove any trace of acetic acid and the columns are ready for the next step.

3.2. Immunoprecipitation

1. Estimate the volume of K562 frozen cell pellet to make the cell extract. The volume of dry cell pellet necessary per column is roughly equal from one to one and a half volume of serum loaded on the protein A column (e.g., for 12 columns which were loaded with 50 μL sera one would need 0.6–0.9 mL of pelleted cells (*see* **Note 6**).
2. Take the cell pellet out of the freezer and add PMSF for a final concentration of 1 mM. Allow it to thaw in a beaker of water at RT. Vortex to loosen the pellet/s and to speed thawing.
3. Add six volumes of thawed cell pellet with lysis buffer (0.3% NP-40 and 1 mM PMSF in IPP buffer) and vortex to mix.
4. Sonicate the cells in a beaker with ice and water at power of 30W 6 times with 20 s sonication and 10 s rests between each sonication.
5. Transfer the cell extract into an ultracentrifuge tube or microcentrifuge tubes. Centrifuge at 10,000 × g for 20 min. Pour cell extract into a new tube and discard the cell debris pellet.
6. Remove tip enclosures from columns to let IPP buffer come down to gel level.
7. Put the tip enclosures back on the columns, and then add six column volumes of cell extract. If 25 μL of serum is used then add 150–200 μL of cell extract; if 50 μL of serum is used then add 300–400 μL of cell extract to each column.
8. Put columns on a rotator and incubate them for 3 h at 4°C.
9. When incubation is complete, wash the beads 10× with ice-cold IPP buffer.
10. Wash each column with 1 mL of ice cold 50 mM Tris, pH 6.8 (from 1/10 dilution of stacking buffer).
11. Remove the micro-columns from the manifold and put into 2 mL microfuge tubes. Spin briefly to remove remaining water.
12. Transfer columns to new-labeled 1.5 mL tubes.
13. To each column, add a volume of SDS sample buffer equal to the original volume of serum used to make the column. If 25 μL of serum is used, add 25 μL of sample buffer. If 50 μL of serum is used, then add 50 μL of sample buffer.
14. Centrifuge briefly, about 20 s.
15. Add into each column a second volume of SDS sample buffer (the same volume as the first) to remove remaining proteins still in the beads.
16. Cap tubes, heat denature samples at 95°C for 5 min to inactivate proteinases (*see* **Note 7**). Store samples at –20 to –80°C until gel electrophoresis.

3.3. Gel Electrophoresis of Immunoprecipitate

1. Prepare a discontinuous 7% polyacrylamide SDS gel, 3.3% C with 25-well comb, 1.5 mm thick spacer. The procedure is based on the Laemmli method *(4)* but with some modifications: The pH of running gel buffer was 8.7 for better separation of proteins because the gel runs a little bit slower. The stacking gel was 3.9% with pH 6.8. Cover the top with Parafilm and leave it overnight at RT.
2. The next day remove the comb. Combine two wells by removing one gel finger (well wall) in between every two wells to make the lanes wider (*see* **Note 8**). To remove a gel finger, insert two strips of plastic (*see* **Note 9**) on either side of a gel finger, close the end to cut and pull it up. Leave the last well for the molecular weight markers.
3. Heat-denature the samples again at 95°C for 5 min to remove disulfide bonds that may form during freezing (optional).
4. Load the samples and molecular weight markers into the gel (*see* **Note 10**).

   Check manufacturer's instruction for the best current setting for the gel apparatus. In general, with Protean II xi, start the current at 26 mA per gel and adjust it up to 32 mA per gel when the blue dye enters the running gel. Turn off current when the blue dye comes down to 2–3 cm from the bottom of gel.

3.4. Western Blot

1. Before gel electrophoresis is complete, make up the transfer buffer. Degas to remove bubbles that may have formed from reaction of methanol with water.
2. Remove clamps from gel assembly. Separate the two glass plates. Leave gel on one glass plate. Cut off the stacking gel. Mark the gel for orientation, for example, by cutting off a small piece of corner of the molecular weight marker lane without cutting into the sample lanes.
3. Cut a piece of nitrocellulose paper and two pieces of 3MM filter paper. The nitrocellulose paper should be 5–10 mm bigger than the size of the gel.
4. Pour some transfer buffer into another container and put one of the 3MM filter paper into the tray. Push it down to the bottom of the container. Put nitrocellulose paper into the transfer buffer. Start from one corner of membrane onto the surface of the solution. Let the nitrocellulose membrane absorb buffer and then gradually lay the whole piece down. Orient nitrocellulose paper onto the center of the first filter paper (*see* **Note 11**).
5. Invert the glass plate and drop the gel into tray. Put the gel with molecular weight markers side (a corner was marked with a cut) to the left. Now the gel is on the middle of the nitrocellulose paper with 3MM filter paper at the bottom. Put a second 3MM filter paper on top of gel with a cut on one corner to recognize the

gel side. Remove any bubbles below the second filter paper. Align the filter papers on top of each other.

6. Place the whole gel sandwich onto a glass plate with the nitrocellulose paper side up.
7. Using a roller or a long tube, lightly roll out any excess water or bubbles between the gel and nitrocellulose paper. Repeat a few times (*see* **Note 12**).
8. Transfer gel sandwich with membrane side toward the anode (clear or red). Assemble with Scotch Brite pad supports on both sides of the Trans-Blot cassette.
9. Close the cassette and put it into a slot of Trans-Blot cell that holds the cassette (*see* **Note 13**).
10. Pour transfer buffer into Trans-Blot cell to cover the gel inside the cassette. Put the cover with the leads on and connect them to power supply.
11. Set the current at 200 mA.
12. Transfer the protein overnight at 4°C about 16–22 h in a cold room or using refrigerated circulator (*see* **Note 14**). Transfer at 4°C in a refrigerator is much more heat transfer effective than using a refrigerated circulator.

3.5. Antibody Development

1. Turn off the power when the protein transfer is complete. Remove the leads that connect to power supply. Take out protein transfer cassette and put it onto a tray.
2. Remove membrane from gel and filter papers with fine point forceps.
3. Stain with Fast Green solution.
4. Destain with ddH_2O.
5. When destaining is complete, air-dry the membrane on a large Kim wipe until completely dry. The drying of the membrane could reduce the background *(4)*.
6. Wet the membrane with ddH_2O.
7. Put the membrane on a clean glass plate, use a sharp scalpel on a handle and trim off excess nitrocellulose paper with a ruler.
8. It may be necessary to use background rather than the protein of interest itself to identify the lanes, since the latter may not be visible.
9. Set up to cut the membrane into strips: Lay the membrane on a clean glass plate. Make sure the membrane is always wet. Put a rolling ruler on the membrane with a cutting strip template underneath the glass plate. The line guide to cut the strips should be visible through both top and bottom

of membrane. The strip template should be wide enough within or inside the protein lane.

10. After cutting the nitrocellulose membrane into strips and putting into the wells of the mini incubation trays, add ddH_2O into the wells. Put the lane with molecular weight markers into a separate container with ddH_2O.
11. Set up washing devices: Prepare 1× TBST and pour it into a bottle. Use one of the ports on the 8-channel multi-well plate washer/dispenser manifold to dispense the TBST buffer and the second port to aspirate the buffer with vacuum. Adjust the amount of washing buffer from a syringe that connects to a port with a three-way Luer stop cock to switch from dispensing to loading of buffer or vice versa. Manifold tips will contact the membrane strips and leave marks on them. To avoid the marks of needles on the strips, cover the tips of manifold with some silicone tubing size 18 GA extended out from the needle about 1–2 mm.
12. Destain Fast green on membrane strips with 1 mL of 2 mM NaOH. Destain the molecular weight marker lane with 10–20 mL. Put them on the shaker for a few min.
13. Block the molecular weight markers strip with 0.3% Tween-20 for 1 h at RT. Wash with ddH_2O and add enough colloidal gold staining solution to cover the strip. Put on shaker for a few h or overnight at 4°C. When staining is complete keep it in ddH_2O until antibody development is finished. Colloidal gold stain is as sensitive as silver stain and it does not shrink the membrane *(5)*. Reuse the colloidal gold staining solution until it is exhausted.
14. When most of Fast Green comes off from membrane strips, aspirate 2 mM NaOH. Remaining small amount of Fast Green will come off at the end of development. Add into each well 1 mL of blot blocking solution at 37°C.
15. Block the membranes with 5% instant nonfat dry milk in TBST on a shaker for 1–3 h at 22–37°C or at 4°C overnight *(6)*.
16. Add primary antibodies for blotting (*see* **Subheading 2.5, item 2**). Dilute the serum 1/100 in blocking solution by adding 5 µL of each antibody panel serum into each well of mini incubation tray that contain 0.5 mL of 1% instant nonfat dry milk in TBST with membrane strip *(7)*. An eight multichannel pipettor will shorten the time of adding antibody panel sera except the serum used to prepare column.
17. Incubate first antibody on shaker at 4°C overnight or at least 8 h.
18. When first antibody incubation time is over, put mini incubation trays on a tray large enough for one set of 12 mini incubation trays or less. To aspirate the buffer, raise one side of tray with an object.

19. Wash the membrane strips with TBST thrice (5–10 min each on shaker at RT to allow antibodies inside of membrane to equilibrate with the buffer outside).
20. Add second antibody in 5% milk blocking solution.
21. Incubate second antibody on shaker at 4°C overnight or at least 8 h.
22. Wash membrane strips with TBST thrice (5–10 min for each wash on a shaker at RT).
23. Set up to develop the strips with substrate solution (*see* **Note 15**). Place on bench top in this order:
    a. A piece of steel or stainless steel sheet larger than the membrane size.
    b. A piece of saran wrap large enough to cover the inside of Nalgene Reusable Plastic Utility Box.
    c. A piece of nylon filter or white tulle about the same size with a piece of steel or stainless steel (*see* **Fig. 3**). Wet the nylon filter or white tulle with TBST.
24. Aspirate the wash buffer inside the incubation trays.
25. Transfer the strips on the incubation trays with fine tip forceps onto the wetted nylon filter or white tulle according to their original order of lane number. Use fingers to move the strips next to each other to make room for the next strips.

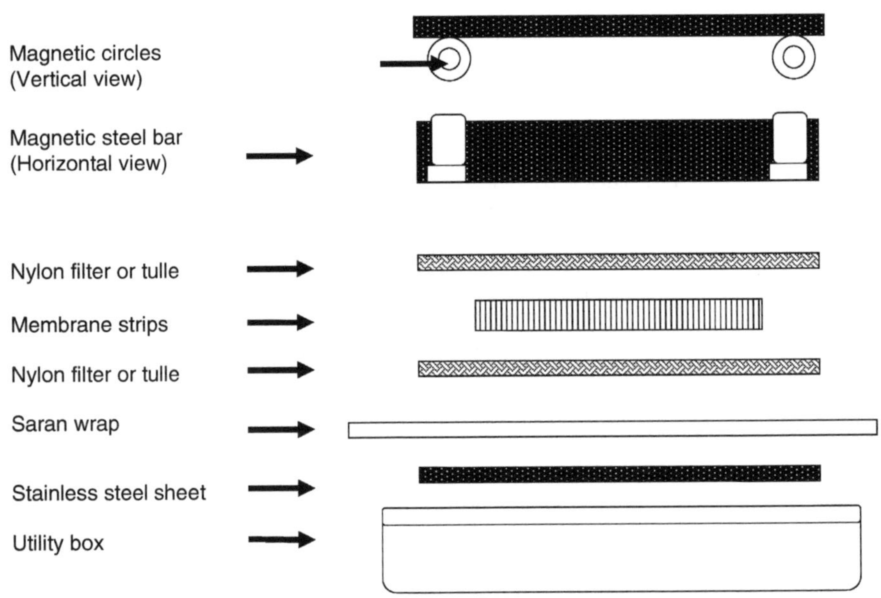

Fig. 3. The membrane strips sandwich used in substrate development.

26. Using two rulers, align the top and bottom of strips to line them straight.
27. Cover the membranes with a second nylon filter or white tulle.
28. Put on each top and bottom edge of membrane a stainless steel bar or plastic bar with two pieces of super magnet circles attached on either side (*see* **Fig. 4**).
29. Put the whole set up of membrane sandwich between two pieces of nylon filter or white tulles on stainless steel sheet into a Nalgene Reusable Plastic Utility Box.
30. Prepare 50 ml of the alkaline phosphatase substrate solution and pour it into the utility box that contains the membrane strips.
31. Color changes rapidly. Stop the reaction with dH_2O before it becomes too dark. After development is complete, scan the membrane strips for analysis.
32. Often things turn out wrong unexpectedly (*see* **Note 16** for trouble shooting western blot results).

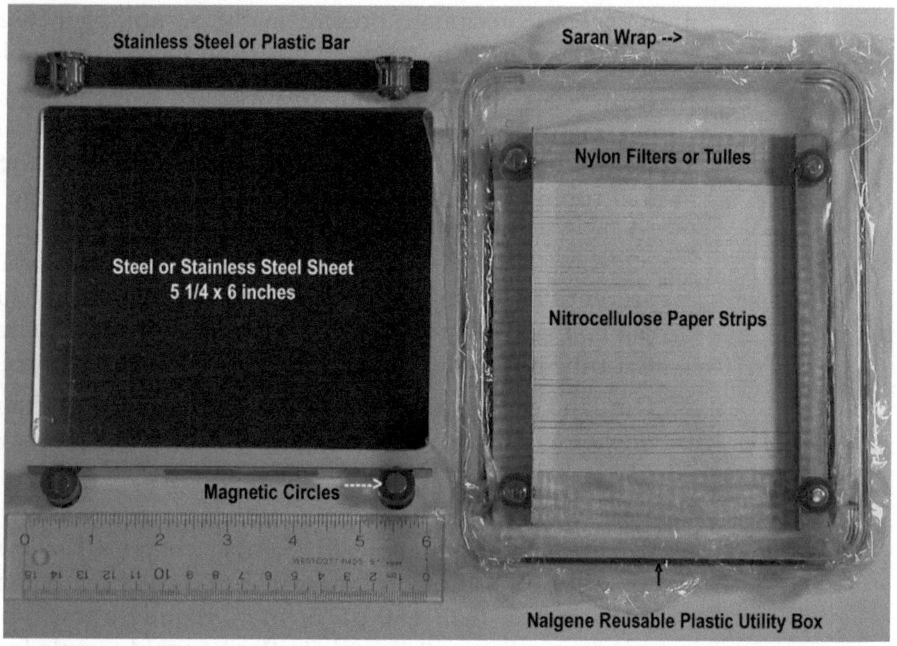

Fig. 4. A homemade device secures the membrane strips together and facilitates the development of all strips simultaneously in one tray: (1). Make two strong magnetic circles (The Magnetic Source) and permanently attach by glue or screw on either side of a piece of plastic or metal bar. (2) Cut a piece of stainless steel sheet 1-mm thick about the size of the inside development tray (5¼ × 6 inches). Round off the corners and smooth out the rough edges to avoid cuts. (3) Two pieces of nylon filter or white tulle (5 × 6 inches) aid in holding the membrane strips together and in order, while letting the substrate solution go through during development.

4. Notes

1. Lock the tips of an 8-channel Nunc-immuno* Washer to fit the pattern of columns on a vacuum manifold. The ports of dispenser manifolds come in different gauges. Sometimes it will be possible to use the rubber cover of a Vacutainer needle to plug some of the ports. Other times it may be necessary to insert thin walled plastic tubing (18 GA) into the rubber covers for a tight fit.
2. SDS sample buffer usually contains bromophenol blue (BPB) to aid sample visualization when loading and running the gel. A small amount of this dye in the sample buffer will not affect protein migration during gel electrophoresis, but it may affect it if there is too much. The amount of BPB can be reduced from 0.05% to 0.01% for better gel electrophoresis separation.
3. Protein A-sepharose or agarose beads equilibrated 1:1 with another buffer, such as immunoprecipitation buffer, is an option since there are three washes of antibody-binding buffer that would remove any buffers that might interfere with the cross-linking reaction. Allow the wash buffer to come down to the gel level before adding new buffer.
4. If there is precipitate present in the serum, centrifuge it at 10,000 × g for 5 min. Avoid pipetting the pellet or lipid.
5. If there are many columns to wash, use a vacuum manifold in conjunction with a dispenser manifold. When a column is attached to a manifold, the water comes down into the manifold tubing by gravity. The weight of water inside the tubing creates a vacuum; therefore, if the tubing is longer, the vacuum is greater. Conveniently, capillary action inside the tubing will prevent the column from running dry. A homemade manifold can be made from any source, such as any polyethylene micro test tube or 4 mm syringe filter that can attach to a column.
6. Our lab keeps a stock of frozen cell pellets at –70°C and routinely uses pliers to pool frozen cell pellets by squeezing from one end of a 15 mL tube.
7. Microwaving the samples for 90 s is a quick way of heat denaturing the samples *(8)*. Sample volumes less than 100 µL will not boil or explode while microwaving. Sample volumes greater than 150 µL or more require a test run in the microwave first with sample buffer to avoid sample loss. This technique is not recommended for samples having high salt content, concentrated samples from a large sample volume or samples containing a lot of glycerol because when microwaved they can be unstable. Tubes should be uncapped.
8. Cutting off the gel fingers allows for more versatility by creating a wider lane to accomodate a larger sample.

9. Tools to make wider wells: Cut out two thin strips of polyethylene plastic, the same thickness as the spacer. Make one strip with a narrow sharp hook to cut the finger and the other strip with no hook for support. The tools should be thin, strong and long enough to hold the gel fingers while inserting them into the upper gel chamber. To make the new well, cut the gel finger at the bottom of the well and remove it.
10. Prepare one lane of unstained molecular weight marker: Colloidal gold staining is very sensitive and requires very little molecular weight marker. One lane of a 25-well comb requires about 2–2.5 μg of unstained molecular weight markers. Reconstitute the markers at 0.1 μg/μL and add 20–25 μL of this to 20–25 μL of sample buffer to bring the volume of markers equal to the volume of the sample wells.
11. The nitrocellulose paper wets by capillary action inside the nitrocellulose membrane. After long-term storage, nitrocellulose paper will become brittle, will no longer be wettable, and should be replaced with a new one.
12. If there are two or more gels, separate them with cellophane membrane. The cellophane membrane prevents proteins from depositing on the membrane belonging to the other gel.
13. Orient the membrane toward the positive electrode (red color) and the gel toward the negative electrode (black color). If the membrane is not in the path of protein migration, the protein will not deposit onto the membrane.
14. Protein transfer buffer will be partially exhausted after 16–22 h. Most power supplies are not equipped to detect this change, but some for semi-dry transfer systems can detect the conductivity of the protein transfer buffer during electrophoresis. To regain conductivity, add 10–20 mM NaOH to the buffer after the first use; it will last several times until discarded *(9)*. To avoid unwanted artifacts depositing onto the membrane from a small amount of protein leaking through the membrane in the old buffer, add cellophane membrane in front of the gel.
15. Development of membrane strips with substrate solution on mini incubation trays has a disadvantage in that if there are many incubation trays to develop at the same time, the reaction of substrate solution may happen too rapidly. This may lead to undesirable high background. A new method to develop all of the membrane strips simultaneously is to put all of the membrane strips for one gel side by side on one tray (*see* **Fig. 4**).
16. Troubleshooting western blot results: (a)High background, no signal: Bad cell pellet. Use a new cell pellet at the log phase; (b) High background, with signals: the concentration of secondary antibody is too high is too high, causing high background even if extensively washed. Dilute the conjugate

further from 1:10,000 to 1:30,000. (c) Net like appearance after substrate development: Tulle or nylon filter may cause this appearance. When developing strips with substrate, place tray on a shaker at slow speed or lightly agitate by hand. The substrate solution makes the top tulle or nylon filter float a little above the membrane strips, but the strips will still be secure. After stopping development with ddH_2O keep the strips in ddH_2O in the tray for at least 10 min to prevent further development; (d) A strong shadow of IgG and artifacts: Insufficient blocking of membrane causes these artifacts. Block the membrane strips longer and at higher temperature with 5% instant nonfat dry milk in TBST at 42°C for 3–4 h. Instant nonfat dry milk may not dissolve completely if it has been stored dry for too long; (e)The image of the protein band turns light when stained by strong antibodies but dark when stained by normal serum: The pH of the blocking solution may be lower because the milk in solution has soured; (f) High background within the lane: Cell extract or the antigen source, such as recombinant protein, is too concentrated or needs further centrifugation; (g) Bubbles appear on the membrane: Gas in the transfer buffer released during the protein transfer or failure to carefully roll out the water between the membrane and the gel. Degas the transfer buffer before use or prepare it a few hours ahead; (h) High molecular weight proteins are still in the gel after transfer: The higher molecular weight proteins are harder to transfer and require a longer electrophoresis time. Add 0.01% SDS to the transfer buffer to facilitate the transfer of the largest proteins by solubilizing them *(10)*. Use no more than 0.01% of SDS because it has the reverse effect.

References

1. Targoff, I. N., Mamyrova, G., Trieu, E. P., Perurena, O., Koneru, B., O'Hanlon, T. P., et al. (2006) A novel autoantibody to a 155-kd protein is associated with dermatomyositis. *Arthritis Rheum* 54, 3682–3689.
2. Schrieider, C., Newman, R.A., Sutherland, D.R., Asser, U., and Greaves, M.F. (1982) A one-step purification of membrane proteins using a high efficiency immunomatrix. *J Biol Chem.* 257, 10766–10769.
3. Towbin, H., Staehelin, T., and Gordon, J. (1979) Electrophoretic transfer of protein from polyacrylamide gels to nitrocellulose sheets: procedure and some applications. *Proc Natl Acad Sci USA* 76, 4350–4354.
4. Laemmli, U.K. (1970) Cleavage of structural proteins during the assembly of the head of bacteriophage T4. *Nature* 227, 680–685.
5. Moeremans, M., Daneels, G., and Mey, J. De. (1985) Sensitive colloidal metal (gold or silver) staining of protein blots on nitrocellulose membranes. *Anal Biochem.* 145, 315–321.
6. Johnson, D. A., Gautsch, J. W. J., Sportsman, R., and Elder, J. H. (1984) Improved technique utilizing nonfat dry milk for analysis of proteins and nucleic acids transferred to nitrocellulose. *Gene Anal Techn* 1, 3–8.
7. Sheng, S., and Schuster, S. M. (1992) Simple modifications of a protein immunoblotting protocol reduce nonspecific background. *BioTechniques* 13, 704–708.
8. Horscroft, N. J. and Roy, P. (1997) Thermal denaturation of proteins for SDS-PAGE analysis by microwave irradiation. *BioTechniques* 22, 224–226.

9. Brody, J. R., and Kern, S. E. (2004) Sodium boric acid: a Tris-free, cooler conductive medium for DNA electrophoresis. *BioTechniques* 36, 214–216.

10. Wang, K., Fanger, B. O., Guyer, C. A., and Staros, J. V. (1989) Electrophoretic transfer of high-molecular-weight proteins for Immunostaining. *Meth Enzymol.* 172, 687–696

Chapter 29

Native Electrophoresis and Western Blot Analysis: Method and Applications

Ioannis N. Manoussopoulos and Mina Tsagris

Summary

Native electrophoresis and western blot analysis (NEWeB) has been developed for the study of plant virus characteristics, among others, virus particle–protein interactions, electrophorotype formation, and strain separation. The method is based on the property of electrophoretic mobility of virus particles (VP) and proteins and combines the analytical capacity of electrophoresis with the specificity of western blot. One of its advantages is that it deals with entire VP that can be studied in cause and effect or in time-interval experiments. Some of the most interesting approaches include VP structural studies, VP interaction with host or viral proteins, and also the characterization of VP–protein complexes. In this protocol, NEWeB is used to demonstrate the interaction of *Plum pox virus* particles with the helper component, a virus encoded protein. It is expected that the method could be used in analogous studies of other viruses or large protein complexes, where similar principles apply.

Key words: Native electrophoresis, NEWeB, Virus particle–protein interaction, Helper component, HCPro, Coat protein, *Plum pox virus*, *Potyvirus*

1. Introduction

Most viruses consist of particles with helical (rod or filamentous) or icosahedral (roughly spherical) architecture. Virus particles, in their simplest form, are made up of an infectious nucleic acid (the genetic material) encapsidated in a protective coat (the capsid). The capsid is constructed by the spatial arrangement of multiple copies of one or more types (depending on the virus) of virus-encoded protein subunits. VPs are thus large entities, having masses in the MD range compared with single proteins that have molecular masses usually in the kilodalton range.

Biji T. Kurien and R. Hal Scofield (eds.), *Methods in Molecular Biology, Protein Blotting and Detection, vol. 536*

DOI: 10.1007/978-1-59745-542-8_29

One important physicochemical property of VPs is their net charge, resulting from the ionization of the side groups of the basic and acidic amino acids in the surface of their capsid proteins *(1)*. Thus, at low pH, the amino groups of lysine and arginine and the nitrogen atoms in the imidazole ring of histidine are charged, whereas at high pH the carboxyl groups of aspartic and glutamic acids are charged *(2)*. At a specific pH, known as the isoelectric point (pI), characteristic for each protein, negative and positive charges are balanced, and net charge equals zero. Virus particles or proteins will carry a negative net charge when the pH of the solution is above their pI, while they will be positively-charged when the pH is below it. When in an electric field such negatively or positively-charged VPs or proteins will move toward the anode or the cathode, respectively, whereas uncharged particles will be immobile. This property is known as the "electrophoretic mobility" (EMb) and the rate of migration of the virus particles or proteins is mainly dependent on both their size and net charge *(2)*. As a result, the EMb of free capsid or other proteins is different from that of large VPs. Typically, such proteins will move much faster than VPs, covering longer distances in the gel during electrophoresis.

NEWeB exploits the EMb of VPs and proteins to allow the study of several properties of viruses. Its main advantage is that it deals with whole VPs that can be treated and detected as such, enabling a number of approaches and studies that cannot be designed otherwise.

Some of the most interesting NEWeB approaches are outlined below:

1. Study of electrophorotype formation without the need of virus purification (electrophorotypes are virus particle populations of the same virus strain with different EMb).
2. Characterization of protein–VP interaction and stability. Proteins could be either of viral (structural or other proteins depending on the virus genus) or host origin. Unknown host proteins interacting with VPs could be further explored by proteomic approaches.
3. Time interval or "cause and effect" experiments, as for example a follow up of the infection kinetics and the formation of viral genome-protein complexes or the effect of different factors on VP, respectively.

In this protocol some of the above approaches are demonstrated using a Greek stain of *Plum pox virus* (PPV- Lar) *(3)*.

Plum pox virus belongs to genus *Potyvirus* having members with flexuous filamentous particles of about 680–900 nm length and 12 nm width. The genetic material is a ssRNA of positive polarity, coding for at least eight mature proteins, most having multifunctional properties. The capsid consists of about 2,000 protein subunits of the virus encoded coat protein (CP) molecule.

Members of the *Potyvirus* genus are transmitted by aphids in a mechanism involving the interaction of VPs with another virus encoded protein, the helper component (HC) *(4–8)*. Although much progress has been made on the subject, the exact mechanism of transmission has not yet been elucidated. Using NEWeB, it was shown that particles of PPV-Lar were in association with HC in extracts of diseased plants *(6, 8)*, implying a VP-HC interaction already in the plant cell (**Fig. 1**). Using the same method, it was found that a small population of VPs associated with HC had survived ultracentrifugation at high salt concentrations *(8)* suggesting a strong interaction between VP and HC in these populations. In support to these observations, "subpopulations" of VPs associated with HC have been recently discovered for two Potyviruses, namely *Potato virus Y* (PVY) and *Potato virus A (9)*. Furthermore, it was shown by NEWeB that VP-HC complexes from plant extracts were stable under low urea concentrations (**Fig. 2**).

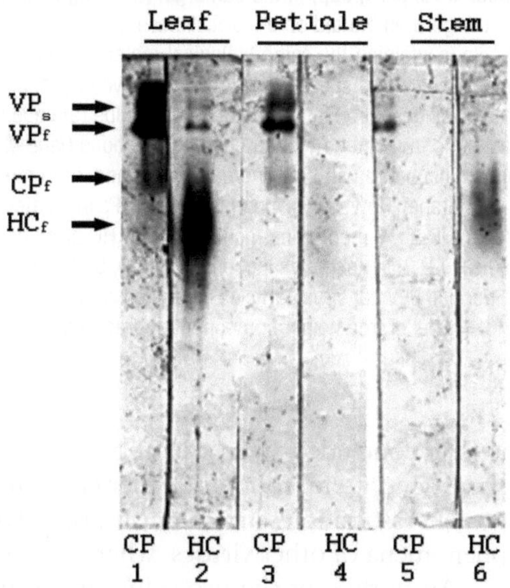

Fig. 1. Electrophorotype formation and VP-HC complex distribution in different tissues of a PPV-Lar infected *Nicotiana benthamiana* plant. Sap from a leaf, its petiole, and a piece of the stem was extracted separately in TBE (*p*H 8.3) about 15 days post-infection. After centrifugation, 100 μL of the supernatant was mixed in equal volumes with loading buffer. Two groups of three wells each in the same gel were loaded with 25 μL of each sample. Samples were loaded and electrophoresed in a 0.8% agarose gel, which was blotted on a nitrocellulose membrane. After transferring, the membrane was cut into two identical sheets (bearing the same biological material). The one sheet was probed with coat protein specific antibodies (CP) and the other with antibodies specific for PPV helper component (HC), as described in the protocol. Both membranes were incubated with secondary antibodies conjugated to alkaline phosphatase. After NBT/BCIP incubation and color development, the membranes were cut into strips corresponding to the above lanes for comparison. Two electropphorotypes (VP_s and VP_f) were detected in the leaf the petiole and the stem (traces) (lanes 1, 3, and 5). Particles of both electrophorotypes were in association with HC in the leaf (compare lanes 1 and 2), but not in the other tissues (compare lanes 3 with 4 and 5 with 6). Free HC aggregates (HC_f) were detected in the leaf and the stem but not in the petiole (lanes 2, 4, and 6), whereas free coat protein aggregates (CP_f) were detected in the leaf and the petiole but not in the stem (lanes 1, 3 and 5).

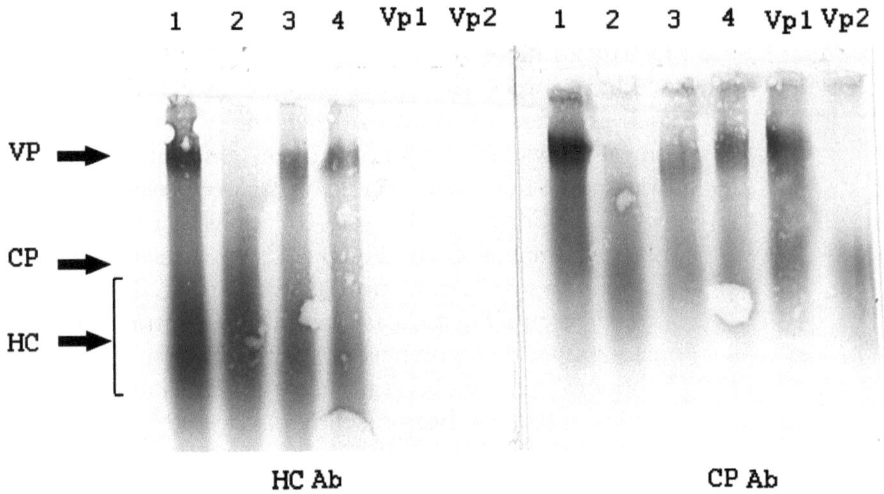

Fig. 2. Incubation of PPV-Lar with different concentrations of urea. Leaves from a PPV-Lar *N. benthamiana* infected plant were extracted in TBE (pH 8.3) and sap was incubated after low speed centrifugation in aliquots with 0, 1.25, 2.5, and 5 M (final concentration) of urea for 15 min at room temperature. Samples were mixed with 10% sucrose, loaded in two identical groups of wells (four wells per group) in the same gel (0.6% agarose-20% acrylamide-bis) and run in TBE pH 8.3. After electrophoresis, the gel was blotted on a nitrocellulose membrane overnight, in the same buffer. The membranes were cut into two sheets (containing the same biological material) and each sheet was probed separately with virus (CP Ab) and helper component (HC Ab) specific antibodies, followed by incubation with secondary antibodies conjugated with alkaline phosphatase, as described in the protocol. After substrate addition (NBT/BCIP) and color development the membranes were washed in tap water. At low urea concentrations (*lane 4*) all complexes were stable. VPs detected with the CP Abs (CP Ab) and incubated with 2.5 M urea (*lane 3*) appeared more unstable than those detected with HC (compare *lane 3* in each membrane). However, all complexes and VPs incubated with 5 M urea were dissociated (compare *lane 2* in each membrane). HC Ab = membrane incubated with HC primary antibodies; CP Ab = membrane incubated with CP primary antibodies; 1,2,3,4 = samples incubated with 0, 5, 2.5, and 1.25 M urea, respectively; VP = level of virus particles; VP1, VP2 = purified PPV Lar virus particles incubated with 0 and 5 M urea, respectively; CP_f = level of free coat protein aggregates; HC_f = level of free helper component aggregates.

As concluded from the above synopsis, NEWeB could be a simple and useful tool for further characterization of *potyvirus* VP-protein associations and could also be helpful in the study of similar phenomena of other viruses, where similar principles apply.

An outline of the procedure is given below followed by the detailed protocol:

VPs are extracted from infected tissue in a buffer adjusted to the appropriate pH, loaded in agarose or mixed agarose-acrylamide gels, electrophoresed under native conditions and blotted onto nitrocellulose or other suitable membranes.

When VP–protein complexes are studied, two or more identical gels are prepared, each blotted to a different membrane. However, two or more replicas (groups) of a few wells in the same gel, loaded with the same samples, will be adequate for most studies.

After blotting, each membrane is treated with the antibodies (Ab) of interest, and following probing with secondary Abs, substrate incubation and color development, membranes are compared against each other for signal positioning. If a VP–protein complex exists, a signal coming from the (host or viral) protein under study will be spotted at the same level with that of VPs detected in the second membrane (protein shift) by the coat protein specific antibody (detecting VPs), suggesting a VP–protein interaction (**Figs. 1** and **2**).

2. Materials

2.1. Virus Isolates

Virus particles can be used either from purified preparations diluted to the pH range in extraction buffer, or detected directly in extracts of diseased plants.

2.2. Solutions

2.2.1. Electrophoresis and Blotting

1. Extraction and electrophoresis buffer: 0.2 M sodium phosphate pH 7.5, or TBE pH 8.3 (89 mM Tris base, 89 mM boric acid, 2.5 mM EDTA).
2. Loading buffer: 10% sucrose in extraction buffer. Sucrose is added to increase the density of the sample thus enabling smooth precipitation into the well during loading. It is inert and will not influence the integrity of the complexes.
3. Tracing solution (1×): 0.25% bromophenol blue, 0.25% xylene cyanol FF, 40% (w/v) sucrose in water. Weigh 0.25 g bromophenol blue, 0.25 g xylene cyanol FF, and 40 g sucrose and dissolve thoroughly in 100 mL dH_2O by continuous steering. Store at 4°C.
4. Acrylamide solution: 19% acrylamide 1% *N,N'*-methylene-bis-acrylamide (bis).Weigh 19 g acrylamide and 1 g bis and make up to 100 mL with distilled water (dH_2O) (*see* **Note 1**). Adjust a heated magnetic stirrer to about 37°C and allow the solution to mix by continuous stirring in a fume cupboard. Store at 4°C in dark bottles.
5. Ammonium persulfate (10%): Weigh 1 g ammonium persulfate and make up to 10 mL with dH_2O. Mix in a magnetic stirrer. Make this solution always fresh.
6. Transfer buffer: Same as extraction buffer.

2.2.2. Immunodetection

1. Tris buffered saline (TBS)-Tween (TBST): 10 mM Tris, 0.9% NaCl, 0.05% Tween-20, pH 7.4. Add 1.21 g Tris-HCl in 800 mL dH_2O and adjust the pH to 7.4 with HCl. Add 9 g NaCl and 0.5 mL Tween-20 and make up to 1 L with dH_2O.
2. TBS/T/Milk (TBS/T/M): Dissolve 10 g skimmed milk powder in 150 mL TBS/T. Stir in a magnetic stirrer for 10 min. After homogenization make up to 200 mL with TBS/T.

3. TBST/M/Sap (TBST/M/S): Extract 5 g of leaves from a healthy plant, if possible the same species as that of the diseased one, in 50 mL TBST/M. Centrifuge for 10 min at 10,000 rpm in a bench centrifuge to get rid of debris and collect the supernatant.
4. Alkaline phosphatase buffer (100 mM Tris-HCl, 100 mM NaCl, 5 mM $MgCl_2$ pH 9.5): Weigh 2.4 g Tris, 1.16 g NaCl, and 0.1 g $MgCl_2$. Dissolve in 100 mL dH_2O and adjust the pH to 9.5 with HCl. Make up to 200 mL with dH_2O.
5. Substrate solution:
   a) NBT (nitro blue tetrazolium). Dissolve 1 g NBT in 20 mL of 70% dimethylformamide.
   b) BCIP (5-bromo-4-chloro-3-indolyl phosphate): Disolve 1 g BCIP in 20 mL of 100% dimethylformamide. Store at 4°C in dark vials.

3. Methods

3.1. Extraction

The pH of the extraction and electrophoresis buffer should be selected so that the VPs and the proteins of interest have the desired charge. In the case of PPV both VP and HC were negatively-charged at pH above 7.3. However, when both, positively and negatively proteins and VPs are examined in the same gel, a modified version called two-directional NEWeB (td-NEWeB) should be applied. In td-NEWeB, gels are prepared as described in the protocol but the comb is placed in the middle of the tray. After electrophoresis, each part of the gel upwards and downwards the loading points (wells) is blotted according to the polarity of the particles or molecules it contains (**Fig. 3**). Thus, the blotted half of the gel containing the negative particles will be facing the anode during transfer, while the half blotted part containing the positive particles will be facing the cathode (**Fig. 3**). The blotting membranes should be cut to the exact size of each gel part to be blotted. The pieces of the Whatman paper should be cut to the exact size of the whole gel side, and applied accordingly along with the pads as described in the protocol.

1. Extract tissue with a pestle and mortar in extraction buffer (1 g in 5 mL).
2. Centrifuge for 10 min at 10,000 rpm in a bench centrifuge to separate cell debris and use the supernatant.

3.2. Gel Preparation

3.2.1. Agarose (0.8%)

1. Weigh 0.8 g agarose and dissolve in 100 mL extraction buffer (*see* **Note 2**) in a conical flask.
2. Heat in a microwave oven until the agarose dissolves giving a clear solution. Cool the solution to about 55°C.

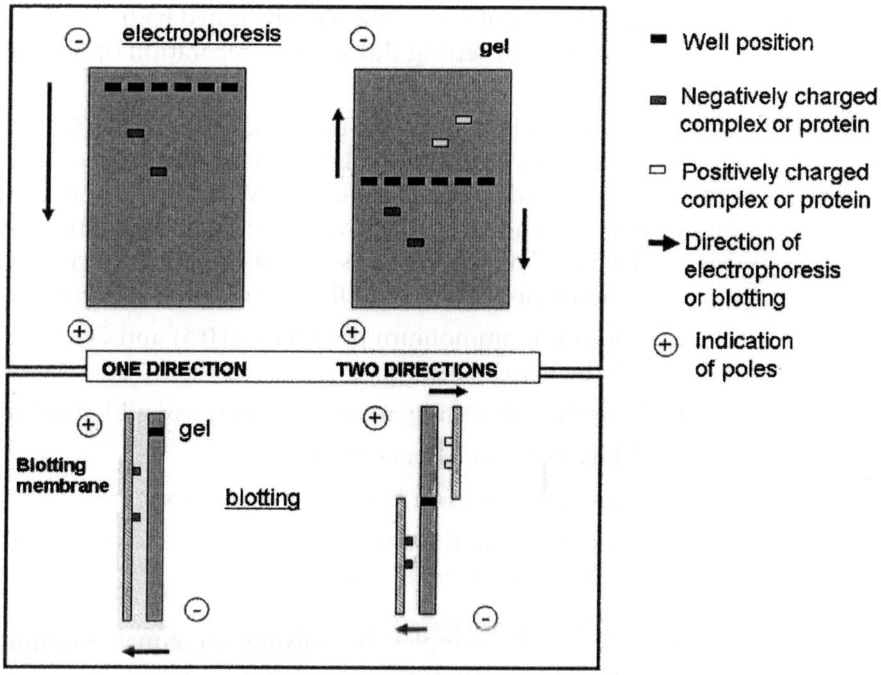

Fig. 3. One and two-directional NEWEB: In one-directional NEWeB (*left*) samples are loaded near the top of the gel and complexes and proteins with the same polarity are separated moving according to their net charge toward the opposite pole. After electrophoresis, the gel is blotted and treated as described in the protocol. In two-directional (td) NEWeB (*right*) samples are loaded in the middle of the gel and VPs and proteins of different polarities are separated, according to their net charge in the two gel directions moving to the opposite poles. After electrophoresis, the parts of the gel containing molecules and particles of opposite polarity are blotted accordingly on nitrocellulose membranes cut to the exact size of each gel part to be blotted. Gels are placed in the sandwich so that the blotted part containing positive molecules faces the cathode and the blotted part containing negative molecules faces the anode. Following transfer the membranes are treated as described in the protocol for signal development and analysis. The gel is shown in *gray* and the membranes in a downward diagonal line pattern. *Black arrows* indicate the direction of particle migration during electrophoresis or blotting.

3. Pour the solution in an appropriately assembled gel tray.
4. Fix the comb at the appropriate position, either on top or in the middle of the tray according to the expected particle charge and the desired migration direction.
5. After setting, place the gel in the electrophoresis unit and cover with electrophoresis buffer.

3.2.2. Mixed Agarose–Acrylamide (0.6% in Agarose, 0.76% in Acrylamide, 100 mL Total Volume

Composite agarose/acrylamide gels have improved mechanical characteristics (strength and elasticity). However, the type of gel to be used should be selected by experimentation. Yet, other acrylamide/agarose combinations can also be used. In our hands, both types worked equally well for at least PPV and PVY. Agarose gels were preferred because of their simplicity in preparation.

1. Switch on a water bath and adjust temperature to about 60°C.
2. In a conical flask, add 0.6 g powder agarose and 60 mL extraction buffer. Heat in a microwave until the agarose dissolves giving a clear solution.

3. Put the conical flask in the already heated bath. This will prevent agarose from setting during the preparation of the acrylamide solution.
4. In another flask mix 24 mL dH_2O, 6 mL TBE (10×), and 4 mL acrylamide-bis solution (19% acrylamide, 1% bis), and place it in the already warmed water bath.When the acrylamide solution is warm enough, mix it with the agarose solution by gentle shaking. The acrylamide solution should be warm enough for avoiding premature solidification of agarose during mixing.
5. Add 6 mL ammonium persulfate (10%) and 50 μL TEMED. Mix by gentle shaking.
6. Pour the solution in an appropriately assembled gel tray.
7. Adjust the combs as needed.
8. Allow about half an hour for the gel to set.
9. Place the gel in the electrophoresis unit and cover it with electrophoresis buffer.

3.3. Sample Loading

1. Prepare the samples by mixing an equal amount of the extracted preparations (e.g., 100 μL) with an equal volume of loading buffer. Mix thoroughly and give a quick centrifugation in a bench centrifuge.
2. Load in sequence 20–25 μL of each sample in each well. Samples are usually colorless or slightly greenish. Fixing a red tape below the electrophoresis tray at the level of the wells will increase contrast, making them visible and facilitate loading. When a VP-complex is examined, load one or more identical groups of wells, in the same gel. Let one or two wells empty between the groups.
3. Add 20 μL tracker solution in the outside wells. Dyes of the tracker solution may interfere with electrophoretic behavior; therefore, it is better to load them separately.
4. Run at 35–45 V until bromophenol blue reaches the end of the gel (about 5 h).

3.4. Blotting

1. Mark the gel by cutting the corner indicating the starting point of loading.
2. Wearing gloves cut four pieces of 3MM Whatman paper, in the exact dimensions of the gel and soak them in transfer buffer. Cut a piece of nitrocellulose membrane at the exact size of the gel. Allow the piece of nitrocellulose membrane to wet by capillary action in transfer buffer.
3. The following steps are described for one-directional NeWEB. For the two-directional NeWEB, another set up of the blotting array is needed (**Fig. 3**). Slide the gel from the electrophoresis tray onto the supporting plate of the blotting sandwich, the

one that will be facing the anode during transferring. Lay the supporting plate with the gel on an open tray containing some amount of transfer buffer. Put the two pieces of the wetted paper on the gel so that they will fit exactly. Roll out any air bubbles with a glass pipette and add a porous pad over the papers. Cover with the second supporting plate.

4. Holding the two supporting plates firmly, turn the sandwich up-side-down. Remove the upper supporting plate to expose the side of the gel to be blotted.
5. Align the moistened nitrocellulose membrane on the gel and mark, with a pen, the corner corresponding to that of the marked gel. Mark gently, with a suitable pen, the positions of the wells in which the samples have been loaded. Squeeze out air bubbles by gently rolling a glass pipette on the membrane. Cover with two pieces of wet paper and add a porous pad.
6. Fix the sandwich with rubber bands, and place it in the tank with the blotted part facing the anode. Fill the tank with transfer buffer.
7. Run at 15–40 V 120 mA overnight keeping the buffer cooled (*see* **Note 3**).

3.5. Immunodetection

1. Switch off the blotting apparatus.
2. Disassemble the sandwich and wearing gloves carefully remove the membrane.
3. Incubate the membrane in 15 mL TBS/T/M by gentle shaking for 1 h.
4. Wash the membranes by shaking in TBS/T three times for five min each.
5. If complexes are studied, cut the membrane appropriately so that two (or more) identical sheets are obtained (each will be treated with different primary Abs). Mark each membrane according to the Ab it will be treated.
6. Incubate the membranes separately with Abs specific for each of the antigens examined, diluted in TBS/T/M/S at predetermined ratios defined by experimentation. This is a necessary step for blocking (inactivating) nonspecific Abs that may have been raised against host proteins copurified with the virus particles used for Ab production. However, when monoclonal or Ab raised against recombinant proteins are used, incubation could be done in TBS/T/M. In the above examples shown in **Figs. 1** and **2** one membrane was incubated in CP particle specific Abs and the other in HC specific ones.
7. Repeat washing step 4.

8. Incubate each membrane separately for 1 h in appropriately conjugated secondary Abs suspended in TBS/T/M at a predetermined titration dilution depending on manufacturer.
9. Repeat washing step 4.
10. For each membrane mix 66 μL NBT in 10 mL alkaline phosphatase buffer and then add 33 μL BCIP. Incubate each membrane for appropriate time (usually 5–60 min) by gentle shaking (*see* **Note 4**).
11. After satisfactory signal development, immerse the membranes in tap water for 20 min and let them dry between two sheets of dry Whatman paper. Take pictures of the membranes and compare the signals, examining VP and protein positions as revealed in each membrane.

4. Notes

1. Acrylamide is a potent neurotoxin. Weighing and mixing should be done in a fume cupboard, whereas a lab coat, gloves, and a mask must be worn during handling. Even after should be handled with care because of the possibility of containing small amounts of unpolymerized acrylamide.
2. This amount has been calculated for gels with approximate dimensions of 16-cm in length, 12-cm in width, and 0.5-cm in height. Smaller gels can be prepared by reducing the ingredients proportionally.
3. Some blotting apparatuses have incorporated buffer cooling systems. Blotting can also be performed in a cold room, but precautions should be taken that the power supply apparatus be compatible for operation in such environments.
4. NBT/BCIP and alkaline phosphatase conjugated antibody has good sensitivity. However, secondary Abs conjugated to fluorophores (e.g. Qdots etc) could also be tested for further increasing sensitivity. In that case, membranes appropriate for fluorescence detection should be used (e.g., Immobilon FL (Millipore)), whereas a documentation system will also be needed for best signal detection.

Acknowledgements

We thank Prof. E. Maiss for providing PPV CP and HC specific antibodies and M. Bastaki for help during her rotation. This work was partly supported by the Greek Ministry of Development (Project EPET II, 453) and by the INTAS 989 program of the EC.

References

1. Levy A.J., Fraenkel-Conrat H., & Owens A.R. (1994) Virology (Prentice Hall International Limited, London UK).
2. Gibbs A., & Harrison B.D. (1976) Plant Virology: The Principles (London, Edward Arnold).
3. Tsagris M., Liu., B., Tzortzakaki, S., & Tabler, M. (1993) Construction of a full length cDNA clone of an aphid transmissible PPV strain from Greece p66–21. IX International Congress of Virology.
4. Govier, D. A., & Kassanis, B. (1974) Evidence that a component other than the virus particle is needed for aphid transmission of potato virus Y. *Virology* 57:285–286.
5. Govier, D. A., & Kassanis, B. A. (1974) Virus-induced component of plant sap needed when aphids acquire potato virus Y from purified preparations. *Virology* 61:420–426.
6. Manoussopoulos, I. N., Maiss, E., & Tsagris, E. (1996). Detection of virus particle-helper component complexes in *Nicotiana benthamiana* plants infected with plum pox virus. PW37-8. Jerusalem, Israel, Xth International Congress of virology.
7. Blanc, S., Lopez-Moya, J. J., Wang, R. Y., Garcia-Lampasona, S., Thornbury, D. W., & Pirone, T. P. (1997) A specific interaction between coat protein and helper component correlates with aphid transmission of a potyvirus. *Virology* 231:141–147.
8. Manoussopoulos, I. N., Maiss, E., & Tsagris, M. (2000) Native electrophoresis and Western blot analysis (NEWeB): a method for characterization of different forms of potyvirus particles and similar nucleoprotein complexes in extracts of infected plant tissues. *J Gen Vir* 81:2295–2298.
9. Torrance, L., Andreev, I. A., Gabrenaite-Verhovskaya, R., Cowan, G., Makinen, K., & Taliansky, M. E. (2006) An unusual structure at one end of potato potyvirus particles. *J Mol Biol* 357:1–8.

Chapter 30

Grid-Immunoblotting

Olga Yeliosof and R. Hal Scofield

Summary

Grid-immunoblotting is a fast, simple, and efficient method for simultaneously testing multiple allergens utilizing small amount of antibody.

Key words: Allergens, Antibodies, Nitrocellulose

1. Introduction

Grid-immunoblotting is a technique developed by Reese et al. *(1)* that could alleviate many problems often encountered in allergy diagnosis and research. This method is particularly valuable for it requires only about 150–200 μL of the serum sample compared with 10 times more serum required for enzyme-linked immunosorbant assay (ELISA). Furthermore, this technique permits simultaneous testing of up to 20 different antibodies against 20 different allergens. In clinical or laboratory settings, the amount of sera drawn from allergic subjects is often limited. This limitation really becomes a barrier when blood is drawn from allergic children or patients with complicated preexisting conditions who are unable to give sufficient amounts of blood.

The grid-immunoblotting procedure consists of three basic steps. First, the protein is immobilized on a carrier nitrocellulose membrane. This membrane is placed onto a multichannel manifold (Surf-blot apparatus) and various proteins are applied to the channels. After a 1-h incubation period, the membrane is washed and blocked. Second, the blot is incubated with the primary

Biji T. Kurien and R. Hal Scofield (eds.), *Methods in Molecular Biology, Protein Blotting and Detection, vol. 536*

DOI: 10.1007/978-1-59745-542-8_30

antibody. Finally, specific binding is detected using a detection system *(1, 2)*.

2. Materials

2.1. Protein Immobilization and Blocking

1. 12.5 cm × 12.5 cm reinforced nitrocellulose membrane (Optitran BA-S 85, pore size 0.45 μm, Schleicher & Schuell – Perkin Elmer Life Sciences, Waltham, MA, USA).
2. Cyanogen bromide (CNBr) used to activate the nitrocellulose membrane and improve protein binding.
3. Tris-buffered saline (TBS): 100 mM Tris–HCl, 100 mM NaCl, 2.5 mM $MgCl_2$, pH 7.4.
4. Surf-blot apparatus (Idea Scientific, Minneapolis, MN, USA).
5. ELISA carbonate/bicarbonate coating buffer: 60 mM Na_2CO_3, 140 mM $NaHCO_3$, pH 9.6.
6. Allergen solutions (200 μL/channel).
7. 0.01% Pyronin Y solution (200 μL).
8. Washing buffer (TBST): TBS supplemented with 0.05% Tween-20.
9. Blocking solution: 200 mL of TBST containing 1% (w/v) nonfat dry milk powder (Carnation, Nestle Food Company, Glendale, CA, USA).

2.2. Blot Incubation with Primary Antibody

1. Antibodies specific to the allergens.
2. Detection antibody (*see* **Note 1**).

2.3. Detection of Specific Binding with a Detection System (see Note 4)

1. Alkaline phosphatase (AP) buffer: 100 mM Tris–HCl, 100 mM NaCl, 5 mM $MgCl_2$, pH 9.5.
2. Substrate/chromogen mixture for alkaline phosphatase at 37°C [450 μM 5-bromo-4-chloro-indonyl-phosphate disodium salt (BCIP; Sigma, St. Louis, MO, USA) and 400 μM nitroblue tetrazolium chloride (NBT; Sigma) solubilized in AP buffer].

3. Methods

3.1. Protein Immobilization and Blocking

1. Activate reinforced nitrocellulose membranes with CNBr according to the technique used by Demeulemester et al. *(3)* (can store this membrane at 4°C for extended periods of time).

2. Soak the membranes for 1 min in Tris-buffered saline.
3. Place the soaked membranes onto the Surf-blot apparatus and assemble the apparatus according to the manufacturer's instructions (*see* **Note 2**).
4. Dilute the allergens in standard ELISA carbonate/bicarbonate coating buffer with a concentration ranging from 100 to 1,000 μg/mL.
5. Pipette 200 μL of allergen solution into each channel of the assembled Surf-blot apparatus except for the last channel. Coat the last channel with 200 μL of a 0.01% Pyronin Y solution instead of the allergen in order to avoid confusion about the orientation of the membrane. This will result in a permanently stained pink lane on the blot.
6. Incubate the Surf-blot apparatus for 1 h at room temperature (RT) with end-over-end rotation (*see* **Note 3**).
7. Perform three washing steps (5 min each) with 2 mL per channel of TBST wash buffer.
8. Disassemble the apparatus using a regular 1 mL pipette and block the unoccupied binding sites on the membrane with 200 mL of TBST containing 1% (w/v) nonfat dry milk for 1 h at RT.
9. Rinse the membranes twice for 10 min in 200 mL of the washing buffer TBST and air-dry until use.

3.2. Blot Incubation with Primary Antibody

1. For detection of specific antibody reactivity, soak the membrane in TBST for 1 min.
2. Arrange the membranes in such a way that the antigen-coated lanes on the membrane are perpendicular to the incubation channels of the manifold, thus permitting each antibody to interact with each of the 20 antigens.
3. Remove the excess buffer and pipette 150–200 μL of diluted (1:100) antibody-containing solution into the lanes.
4. Incubate the blot for 1 h with end-over-end rotation.
5. Perform another washing procedure (3×) (*see* **Subheading 3.1**, **step 7**) with TBST and incubate the membrane in 100 mL of TBST containing 1% nonfat dry milk powder and detection antibody for 1 h (*see* **Note 3**).

3.3. Detection of Specific Binding with a Detection System

1. Use calorimetric detection method to make the antibody binding visible (*see* **Note 4**).

3.3.1. Calorimetric Detection

1. Wash the membranes with 100 mL of TBST and 50 mL TBS-AP buffer for 5 min each.
2. Incubate the blots in substrate/chromogen mixture for alkaline phosphatase at 37°C till spots appear.
3. Stop color development by washing with TBST.

4. Notes

1. For the detection of mouse antibodies, use 1:20,000 diluted alkaline phosphatase conjugated goat anti-mouse IgG + IgM antibody (Jackson ImmunoResearch, West Groove, PA). For the detection of IgE antibodies, incubate the blot with 1:1,000 diluted alkaline phosphatase conjugated monoclonal anti-human IgE (Southern Biotechnology Associates, Birmingham, AL, USA).
2. Aspirate any TBS remaining in the channels before the membrane is coated with allergen.
3. Make sure that the antigen evenly coats the membrane and that no bubbles are trapped in the channels.
4. As an alternative to colorimetric detection, one can use a chemiluminescence substrate for alkaline phosphatase. The following materials [*see* (a)–(c)] and methods [*see* (d)–(g)] need to be used for this procedure.
   (a) Washing assay buffer (100 mM diethanolamine/HCl, 1.0 mM $MgCl_2$, pH 10).
   (b) 1:20 Diluted nitroblock chemiluminescense enhancer (Tropix, Bedford, MA, USA).
   (c) 250 μM CSPD (disodium 3-(4-methoxy-spiro{dioxetane-3,2 -(5 chloro) tricyclo [3.3.1.1.^{3,7}] decan}-4-yl)phenyl phosphate; Tropix).
   (d) Wash the blots with freshly prepared assay buffer (100 mM diethanolamine/HCl, 1 mM $MgCl_2$, pH 10).
   (e) Incubate the blots in the 1:20 diluted nitroblock chemiluminescense enhancer for 5 min.
   (f) Incubate the blots in 250 μM CSPD (disodium 3-(4-methoxy-spiro{dioxetane-3,2′(5′ chloro) tricyclo [3.3.1.1.^{3,7}] decan}-4-yl)phenyl phosphate; Tropix) for 5 min and drain any excessive liquid.
   (g) Expose the blots to autoradiography film for 15, 30, 60, and 120 s sealing the blots between transparencies.

References

1. Reese, G., Schmechel, D., Ayuso, R., and Lehrer S.B. (2001) Grid-immunoblotting: a fast and simple technique to test multiple allergens with small amounts of antibody for cross-reactivity. *J Chromatogr B* 756, 151–156.
2. Kurien, B.T. and Scofield, R.H. (2003) Protein blotting: a review. *J Immunol Methods* 274, 1–15.
3. Demeulemester, C., Peltre, G., Laurente, M., Panheleux, D., David, B. (1987) Cyanogen bromide-activated nitrocellulose membranes: a new tool for immunoprint techniques. *Electrophoresis* 8, 71–73.

Chapter 31

Sequential Use of Immunoblots for Characterization of Autoantibody Specificities

Holger Bartsch and Michael Bachmann

Summary

Sera of patients with systemic autoimmune diseases frequently contain autoantibodies to nuclear autoantigens. Immunoblotting of recombinant and native autoantigens is a commonly used technique for the identification and characterization of autoantibody specificities. Here, we describe an easy procedure that facilitates the comparison of antibody specificities by reusing the same immunoblot at least three times in order to detect an abundantly expressed autoantigen in total cellular extracts.

Key words: Systemic autoimmunity, Autoantibodies, Immunoblotting

1. Introduction

Most if not all patients suffering from systemic autoimmune diseases such as systemic lupus erythematosus (SLE) produce high levels of autoantibodies often exceeding concentrations of 1 mg/mL. Among the best characterized autoantibodies, which are of diagnostic value, are antibodies to dsDNA, Sm, RNP, Ro/SS-A, La/SS-B and P *(1)*. To (a) characterize the epitope specificities of autoantibodies or monoclonal antibodies directed to the autoantigens or, (b) identify cross reactivities among species or with other protein antigens, immunoblotting experiments are frequently performed. Antigen preparation and repeated transfer is time consuming and, especially in case of a series of recombinantly expressed deletion mutants, cost-prohibitive. Because of protein concentration variations quantitative comparisons may become difficult. Thus, it can become desirable to reuse the same immunoblot as fre-

Biji T. Kurien and R. Hal Scofield (eds.), *Methods in Molecular Biology, Protein Blotting and Detection, vol. 536*

DOI: 10.1007/978-1-59745-542-8_31

quently as possible. Here, we describe the conditions that allowed us to reuse the same immunoblot at least three times to detect the abundantly expressed nuclear autoantigen La/SS-B, including in total cellular extracts.

2. Materials

2.1. Cell Culture and Lysis

1. Dulbecco's Modified Eagle's Medium (DMEM) (Invitrogen, Karlsruhe, Germany) supplemented with 10% fetal bovine serum (FBS, Biochrom, Berlin, Germany).
2. The following cell lines have been used: Human Maba cell line (**Fig. 1**, lane 1, h); rat PC12 cell line (**Fig. 1**, lane 2, r); rat glioblastoma cell line (**Fig. 1**, lane 3, r); green African monkey

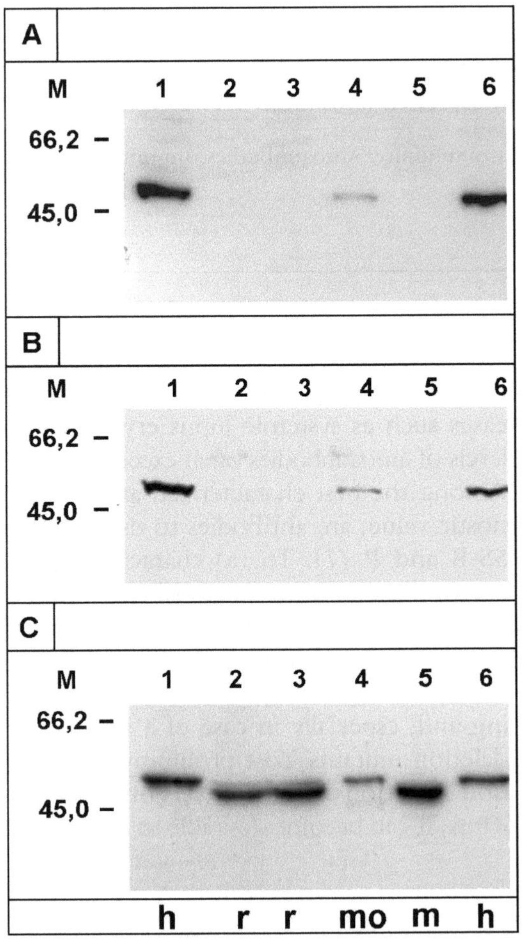

Fig. 1. Total extracts from various cell lines were prepared and analyzed by immunoblotting against three different antibodies (**A–C**) directed to the nuclear autoantigen La/SS-B. The cell lines were either of human (h), rat (r), monkey (mo), or mouse (m) origin. The blots show that only the antibody used in C cross reacts with human, monkey, and murine La protein forms while the antibodies used in A and B are directed to human La only.

kidney cell line CV1 (**Fig. 1**, lane 4, mo); mouse 3T3 cell line (**Fig. 1**, lane 5, m); and human Raji cell line (**Fig. 1**, lane 6, h).

3. Solution of trypsin (0.25%) and ethylenediamine tetraacetic acid (EDTA) (1 mM) (Invitrogen, Karlsruhe, Germany).
4. Modified Laemmli *(2)* buffer for cell lysis and gel loading: 200 mM dithiothreitol (DTT), 4% (w/v) SDS, 20% (v/v) glycerol, 0.2% (w/v) bromophenol blue, 100 mM Tris–HCl (pH 6.8). Store in aliquots at –20°C (*see* **Note 1**).
5. Teflon cell scrapers (Greiner Bio One, Frickenhausen, Germany).
6. Prestained molecular weight marker: Page Ruler (Fermentas, St.Leon-Roth, Germany).

2.2. Western Blotting for Detection of La/SS-B

Immunoblotting was performed according to Matsudaira *(3)* using the following modifications:

1. Transfer buffer: Roti-Blot A and Roti-Blot K (Carl-Roth, Karlsruhe, Germany).
2. PVDF membrane: Hybond-P (Amersham/GE-HealthCare Life Science, Munich, Germany) (*see* **Note 2**).
3. 20% Methanol for equilibration of PVDF membrane.
4. 3 MM chromatography paper from Whatman (Maidstone, UK).
5. Tris-buffered saline with Tween (TBST) 10×: 1.5 M NaCl, 0.5 M Tris–HCl, pH 8, 1% Tween-20. Dilute 100 mL with 900 mL water for use.
6. Blocking buffer: 5% (w/v) Blocking reagent (Roche, Mannheim, Germany) in TBST.
7. Antibody incubation buffer: TBST supplemented with 1% (w/v) fraction V bovine serum albumin (BSA).
8. Secondary antibody: Anti-mouse IgG conjugated to peroxidase (Sigma-Aldrich, Taufkirchen, Germany) (*see* **Note 3**).
9. Enhanced chemiluminescent (ECL Plus) reagents from GE-HealthCare Life Science (Munich, Germany).
10. ChemiDoc XRS system for analysis (BioRad, Munich, Germany).

2.3. Reprobing

1. Stripping buffer: 0.2 M glycine, 1.5 M NaCl, pH 2.3.
2. TBST.

3. Methods

For the identification of cross-reactivities of different monoclonal antibodies directed against the autoantigen La/SS-B among species or with other protein antigens, immunoblotting experiments were performed. Antigen preparation and repeated transfer is time consuming and cost-prohibitive. Also, different sample preparations might differ slightly, making it difficult to compare the results obtained. Therefore, we have reused the same blot at least three times for different antibodies (**Fig. 1**).

3.1. Preparation of Samples for Detection of La/SS-B by Western Blotting

1. The cells used herein are passaged when approaching confluence with trypsin/EDTA to provide new maintenance cultures on 100 mm tissue dishes and experimental cultures on 35 mm dishes. One 35 mm dish is required for one experiment (*see* **Note 4**).
2. A labeled microcentrifuge tube for each sample with a hole poked in the cap using a 26-gauge syringe needle; a heat-block at 100°C, and cell lysis buffer preheated to 100°C are held ready.
3. To each culture 100 μL of heated cell lysis buffer is added and the material is scraped into the appropriate labeled tube (*see* **Note 5**).
4. The tubes are closed and then boiled for a further 10 min. After cooling to room temperature, they are ready for separation by SDS-PAGE.

3.2. SDS-PAGE

1. Carry out SDS-PAGE (1-mm thick, 10% minigels) essentially according to Laemmli *(2)*.

3.3. Western Blotting for Detection of La/SS-B

1. After the separation by SDS-PAGE, the samples are transferred to PVDF membranes electrophoretically. These directions assume the use of a semidry system provided by Bio-Rad (TransBlot™ Semi-Dry Transfer Cell, Bio-Rad, Munich, Germany). Three trays with methanol, transfer buffer Roti-Blot A, and transfer buffer Roti-Blot K, respectively, are prepared with a size slightly bigger than the dimension of the gel.
2. Cut the PVDF membrane and eight pieces of Whatman 3 MM paper to the size of the separating gel.
3. The PVDF membrane is submerged in 100% methanol for 10 min and then reequilibrated in transfer buffer Roti-Blot A (10 min). Cut one edge of the membrane for orientation.
4. Four sheets of 3 MM are moistened in transfer buffer A and transferred to the anode plate of the blotting device. On top of this stack, the reequilibrated PVDF membrane is added.
5. The gel unit is disconnected from the power supply and disassembled. The stacking gel is removed and discarded and one corner (corresponding to the membrane) is cut from the separating gel to allow its orientation to be tracked. The separating gel is then laid on top of the PVDF membrane.

6. Another four sheets of 3 MM paper are wetted in transfer buffer Roti-Blot K and carefully laid on top of the gel, ensuring that no bubbles are trapped in the resulting sandwich. You can remove air bubbles by carefully rolling a glass pipette on top of the stack.
7. The lid is put on top of the stack and the power supply is activated. Transfers can be accomplished at 0.8 mA/cm^2 blot size for 1 h.
8. Once the transfer is complete the stack is carefully disassembled. The 3 MM paper and the gel can then be discarded. If you want, you can stain the gel in Coomassie Blue staining reagent for 1 h and destain thereafter in order to check for completeness of transfer of the separated proteins. On the PVDF membrane the colored molecular weight markers should be clear.
9. The membrane is then incubated in 10 mL blocking buffer overnight at 4°C. Alternatively, blocking is done for 1.5 h at room temperature (RT) on a rocking platform.
10. Discard the blocking buffer. Add the appropriate hybridoma supernatant and incubate for 1 h at RT on a rocking platform.
11. The hybridoma supernatant is removed (*see* **Note 6**) and the membrane is washed four times for 5 min each with 10 mL TBST.
12. The secondary antibody is freshly prepared for each experiment as 1:80,000-fold dilution in antibody incubation buffer and added to the membrane for 1 h at RT on a rocking platform.
13. Discard the secondary antibody and wash the membrane five times for 5 min each with 10 mL TBST.
14. During the final wash, for each blot warm 1 mL solution A and 25 μL solution B of the ECL Plus solutions to RT. Once the final wash is removed from the blot, the ECL Plus reagents are mixed together and then immediately added to the blot, which is then incubated for 5 min at RT. Take care that the membrane is always covered with liquid.
15. The blot is removed from the ECL Plus reagents, dried and analyzed in a ChemiDoc XRS system (BioRad, Munich, Germany) or equivalent system.

3.4. Stripping of Western Blot and Reprobing

1. Stop the development of the blot and wash the blot briefly in TBST.
2. Strip the blot by incubating in stripping buffer for 5 min at RT on a rocking platform.
3. Continue with blocking again (**Subheading 3.3**, step 9) for 1 h at RT or overnight at 4°C and proceed with the following steps for probing with different antibodies (*see* **Note 7**).

4. Notes

1. Unless stated otherwise, all solutions should be prepared in water that has a conductivity of 0.056 μS/cm and total organic content of less than 5 ppb. This standard is referred to as "water" in this text.
2. *The usage of PVDF membrane is mandatory.* The use of nitrocellulose membrane is not possible with the method described herein.
3. We have found this antibody to be excellent for western blotting, but as numerous competitive reagents are available from other commercial sources, they might work equally well.
4. This protocol can be adapted for many other cell culture systems, e.g., also for cells that grow in suspension, such as Jurkat T cells.
5. The cell material will be very viscous at this stage owing to release of DNA and thus hard to pipette. It is easiest to use the pipette tip to transfer the sample to the tube by dragging and pushing rather than by drawing the sample up and down.
6. After the incubation the hybridoma supernatant can be kept at –20°C and reused up to three times.
7. The reprobing could be done at least three times and you can try more. Nevertheless, when you observe high background in the first usage of your blot, most likely, this would not be abrogated by the stripping procedure. The viscosity of the material will decrease during subsequent boiling, allowing accurate loading of the gel.

Acknowledgements

Special thanks for excellent technical support to K. Joanne and Tim Gross (Oklahoma Medical Research Foundation, OK, USA).

References

1. Tan, A.N. (1989) Antinuclear antibodies: diagnostic markers for autoimmune diseases and probes for cell biology. *Adv Immunol.* 44, 93–151.
2. Laemmli, U.K. (1970) Cleavage of structural proteins during the assembly of the head of bacteriophage T4. *Nature* 227, 680–685.
3. Matsudaira P. (1987) Sequence from picomole quantities of proteins electroblotted onto polyvinylidene difluoride membranes. *J Biol Chem.* 262, 10035–10038.

Chapter 32

Renaturation of Recombinant Ro 60 Autoantigen by Calcium Ions on PVDF Membrane

Biji T. Kurien and Michael Bachmann

Summary

Calcium is pivotally involved in many biochemical processes in different cell types. This divalent cation mediates its function by interacting with specific calcium binding proteins that serve as calcium sensors and regulatory proteins. In our earlier studies we found that calcium was involved in the protein–protein interaction observed between Ro 60 multiple antigenic peptides (MAPs) and Ro 60 autoantigen. Since calcium was found to bind Ro 60 MAPs we hypothesized that it would renature human recombinant Ro 60 on a protein blot. As hypothesized antibodies to Ro 60 bound significantly higher to the recombinant Ro antigen that was incubated with calcium compared with that incubated without calcium on a polyvinyline fluoride (PVDF) blot. Since the immmunological epitopes of Ro 60 are mainly conformational, we believe that calcium induced a more native tertiary structure in recombinant Ro 60 autoantigen following blotting to a PVDF membrane.

Key words: Calcium, Protein–protein interaction, Recombinant Ro 60 autoantigen, Systemic lupus erythematosus, PVDF

1. Introduction

Calcium (Ca^{2+}) is involved in many biochemical processes in different cell types, from the beginning of new life and egg fertilization to the end of life and cell death *(1)*. Intracellular free calcium ions are important as a second messenger for external stimuli and as a key regulator of numerous cellular processes. Cytosolic free calcium ($[Ca^{2+}]_i$) is an important second messenger during stimulation in a wide variety of cells, including polymorphonuclear leukocytes (PMNs). Its mobilization in PMNs is altered in various diseases such as atherosclerosis and ageing *(2)*. This divalent cation carries out most of its functions by interacting with

Biji T. Kurien and R. Hal Scofield (eds.), *Methods in Molecular Biology, Protein Blotting and Detection, vol. 536*

DOI: 10.1007/978-1-59745-542-8_32

binding proteins that serve as calcium sensors and regulatory proteins *(3)*. Some of the well-known calcium binding proteins are calmodulin, calreticulin, albumin, and the La autoantigen. Calmodulin undergoes a series of conformational changes upon binding calcium and exerts its action on the cell cycle mainly, but not exclusively, through activation of protein phosphorylation and dephosphorylation *(1)*.

We found that calcium was involved in mediating the intramolecular protein–protein interaction occurring between Ro 60 multiple antigenic peptides (MAPs) (constructed from the sequence of the Ro 60 autoantigen) and the Ro 60 autoantigen itself *(4)*. This 60,000 molecular weight antigen is part of the Ro ribonucleoprotein particle and is noncovalently associated with at least one of four short uridine-rich human cytoplasmic RNAs (the hY RNAs) *(5, 6)*. Anti-Ro 60 is found in up to 40% of patients with systemic lupus erythematosus (SLE) and is associated with subacute cutaneous lupus, photosensitive skin rash, deficiency of early complement components, renal disease, neonatal lupus, lymphopenia, and neutropenia *(7, 8)*.

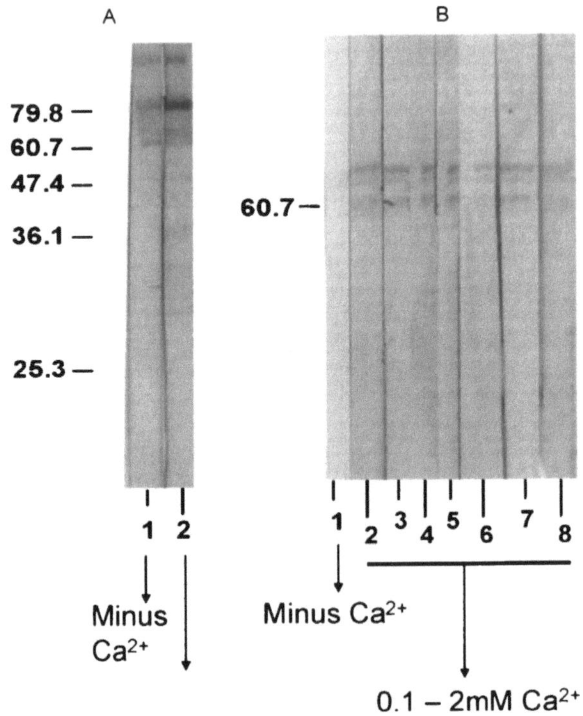

Fig. 1. Renaturation of Ro 60 autoantigen by calcium on PVDF immunoblot. (**A**) Calcium was added following immunoblotting and the membranes were blocked with 2% BSA. Regular immunoblotting was followed from this step onward. Lane 1 – no calcium; lane 2 – 2 mM calcium (final concentration). (**B**) Lane 1 – no calcium; lanes 2–8 – plus calcium – 0.1, 0.25, 0.5, 1.0, and 2 mM, respectively (final concentration) (reproduced from *(4)* with permission from Elsevier).

Renaturation of proteins on a blot is an essential step for appropriate recognition of blotted proteins *(9)*. Since calcium was found to bring about the interaction of Ro 60 MAPs and Ro 60 autoantigen we hypothesized that calcium might increase the antigenicity of recombinant Ro 60 autoantigen following blotting to a PVDF membrane by inducing a more native tertiary structure. We found that antibodies to Ro 60 bound significantly higher to the recombinant Ro antigen that was incubated with calcium compared with that incubated without calcium on a polyvinylidine fluoride (PVDF) blot (**Fig. 1**). This showed that calcium was able to renature recombinant Ro 60 antigen on the blot.

2. Materials

Prepare all solutions using ultrapure water (prepared by purifying deionized water, to attain a sensitivity of 18 MΩ cm at 25°C) and analytical grade reagents. Prepare and store all reagents at room temperature (unless indicated otherwise). Diligently follow all waste disposal regulations when disposing waste materials. We do not add sodium azide to our reagents.

1. 10% SDS-PAGE precast gels (10 well) (ISC Bioexpress, Kaysville, UT, USA).
2. SDS lysis buffer (5×): 0.3 M Tris–HCl (pH 6.8), 10% SDS, 25% β-mercaptoethanol, 0.1% bromophenol blue, 45% glycerol. Leave one aliquot at 4°C for current use and store remaining aliquots at –20°C (*see* **Note 1**).
3. SDS-PAGE running buffer: 0.025 M Tris–HCl, pH 8.3, 0.192 M glycine, 0.1% SDS (*see* **Note 2**).
4. Phosphate buffered saline (PBS), pH 7.4: Dissolve five PBS tablets (Sigma Chemical Company, St. Louis, MO, USA) in 1 L water to obtain PBS solution, pH 7.4 at 25°C (0.01 M phosphate, 0.00027 M potassium chloride, 0.137 M sodium chloride).
5. Purified recombinant Ro 60 *(4)* was a gift from Dr. Michael Bachmann, Dresden, Germany.
6. BenchMark prestained molecular weight standards (Gibco BRL, Bethesda, MD, USA).
7. Polyvinylidene difluoride membrane (Millipore, Bedford, MA, USA).
8. Western blot transfer buffer: 0.025 M Tris, 0.192 M glycine, 20% methanol (*see* **Note 3**).
9. Western blot transfer apparatus (Hoeffer Scientific Instruments, GE Healthcare, Piscataway, NJ, USA).

10. Tris buffered saline (TBS; 10×): 1.5 M NaCl, 0.1 M Tris–HCl, pH 7.4.
11. Blocking solution: 2% bovine serum albumin in PBS, pH 7.4. Store at 4°C (*see* **Note 4**).
12. Diluent solution: 2% bovine serum albumin in PBS, pH 7.4 containing 0.05% Tween-20. Store at 4°C (*see* **Note 4**).
13. Normal and anti-Ro 60 human sera (Clinical Immunology Laboratory, Oklahoma Medical Research Foundation, Oklahoma City, OK, USA).
14. Nitro blue tetrazolium (NBT)/5-bromo-4-chloro-3-indolyl phosphate (BCIP): Dissolve 1 g NBT in 20 mL of 70% dimethylformamide (DMF). Disolve 1 g BCIP in 20 mL of 100% DMF. Add 33 μL of BCIP and 66 μL of NBT to 10 mL of alkaline phosphatase buffer just before adding to membrane.
15. Alkaline phosphatase (AP) buffer: Weigh 6.1 g of Tris, 2.9 g sodium chloride, and 0.51 g magnesium chloride·$6H_2O$ and make it to 500 mL with water after adjusting pH to 9.3 with HCl. Store at 4°C
16. 0.009-in. Single edge razor blades – Smith Brand (Fisher Scientific, Dallas, TX, USA).

3. Methods

All procedures are carried out at room temperature unless otherwise specified.

3.1. SDS-PAGE

1. Heat an aliquot of the recombinant Ro 60 at 95°C for 5 min in SDS lysis buffer. Do not add lysis buffer to the prestained protein standard or subject it to heat. Centrifuge the heated sample at 3,000 × *g* for 30 s to bring down the condensate. Load the sample (in required number of wells) and the protein standard (10 μL/well) on the gel. Electrophorese at 15 mA until the sample has entered the gel and then continue at 25 mA till the dye front (from the BPB dye in the samples) reached the bottom of the gel (*see* **Note 5**).

3.2. Western Blotting

1. Following electrophoresis, pry the gel plates open with the use of a spatula. The gel remains on one of the glass plates. Rinse the gel gently with deionized water and transfer carefully to a container with western blot transfer buffer.
2. Cut a PVDF membrane to the size of the gel and immerse in methanol. Rinse twice in distilled water and once with transfer buffer.

3. Cut four sheets of Whatman 3M filter paper to the size of the gel and transfer to the transfer buffer. Place two adsorbent pads also in the buffer (*see* **Note 6**).
4. Place plastic wrap (about 12 in.) on the workbench. Place two filter papers on top of the plastic wrap. Position the PVDF membrane on top of the filter papers. Transfer the gel to the top of the membrane in such a way that there are no air bubbles between the gel and the membrane. Place the remaining two filter papers on top of the gel. Place in a transfer cassette (*see* **Note 7**). Make doubly sure that the PVDF membrane is between the gel and the anode (*see* **Note 8**). Transfer was carried out overnight (~18 h) at 30 V (constant voltage).
5. Disconnect power supply and disassemble the sandwich. Keep the membrane (*see* **Note 9**) moist at all times.
6. Stain the polyacrylamide gel with Coomassie Brilliant Blue stain or by silver to determine the efficiency of transfer.

3.3. Renaturation of Recombinant Ro 60 on Membrane

1. Excise the lane with molecular weight standards from the main PVDF sheet containing the transferred Ro 60 protein lanes. Cut a tiny wedge from the bottom left side of the marker lane and the main membrane sheet for orientation purposes (*see* **Note 10**).
2. Align the comb across the top of the PVDF membrane and mark the areas for cutting. Cut strips (about 2–3 mm in width) containing the Ro antigen (*see* **Note 10**).
3. Incubate one set of strips with just water and another set of strips with calcium chloride (0.1–2 mM final concentration) (*see* **Note 11**) for a minimum of 1 h.
4. Block the strips with blocking solution for 1 h.
5. Dilute human anti-Ro 60 sera and control sera 1:1,000 with BSA diluent and add 1 mL to appropriate strips. Incubate for 2 h on an orbital shaker.
6. Wash 5× times, 5 min each time with TBST (*see* **Note 12**).
7. Add 1 mL of anti-human IgG alkaline phosphatase conjugate (1:5,000 dilution, diluted in diluent) to each strip and incubate for 1 h.
8. Wash as in **step 6**.
9. Add 500 μL of nitroblue tetrazolium/5-bromo-4-chloro-3-indolyl phosphate substrate and let bands develop (*see* **Note 13**).
10. Wash with TBST for 5 min. Let the strips dry and arrange in order on paper board inserts (*see* **Note 14**).

4. Notes

1. SDS precipitates at 4°C. Therefore, the lysis buffer needs to be warmed prior to use.
2. Simple method of preparing running buffer: Prepare 10× native buffer (0.25 M Tris, 1.92 M glycine). Weigh 30.3 g Tris and 144 g glycine, mix, and make it to 1 L with water. Dilute 100 mL of 10× native buffer to 990 mL with water and add 10 mL of 10% SDS. Care should be taken to add SDS solution last, since it makes bubbles.
3. Dilute 100 mL of 10× native buffer to 800 mL with water and add 200 mL of 10% SDS. Avoid adding methanol directly to the 10× buffer, since it precipitates its ingredients. Even in such a scenario the precipitate can be redissolved by the addition of 800 mL water.
4. Do not use milk for blocking or as a diluent since it contains calcium and will interfere with the experiment.
5. Ensure that all wells have the same amount of SDS lysis buffer to ensure that all samples migrate at the same rate. Centrifuging the samples prior to the run helps remove insoluble debris, which could produce streaks in the protein lanes (revealed when stained with Coomassie Blue). Add a drop of 0.1% BPB to the upper chamber buffer. This helps to form a much stronger dye front during the electrophoretic run.
6. Take care to exclude air bubbles from the support pads. Press the pads with fingers while the pads are immersed in the buffer (to get rid of air bubbles). The buffer should be at 4°C.
7. Hold the two top corners of the gel with each hand. Lower the bottom part of the gel first on the membrane and gently release the gel little by little to lay the complete gel on the membrane. This will prevent trapping of bubbles in between the gel and the membrane. We use Saran plastic wrap. Part of the Saran wrap was folded over the sandwich. A 10 mL pipette was used to roll out the air bubbles from the gel membrane sandwich prior to placing in a transfer cassette.
8. In the Bio-Rad transfer apparatus, we place the gel side of the transfer cassette facing the black side of the transfer cassette holder and the membrane side facing the red side.
9. Ensure that the membrane is always moist. If for some reason, the membrane strips dry out, immerse the membrane in methanol and wash with buffer.

10. The strips can be excised very nicely using a razor blade. A Mini PROTEAN® 3 System glass plate is placed at an angle on the nitrocellulose at a distance of 2 mm from the edge and the razor is used to cut the strip (pull the razor blade along the sides of the glass plate to cut).
11. Prepare stock solutions of calcium and dilute to obtain the desired final concentration.
12. Rinsing the strips with deionized water 2–3 times will help remove a bulk of the nonspecific antibodies and help reduce the amount of TBST used subsequently and also reduce the number of washes. This wash helps to reduce unspecific binding of NBT/BCIP to the strip. The water, owing to its low ionic strength compared with TBST, will be able to remove contaminants much better than TBST.
13. NBT/BCIP comes as a ready-to-use stabilized mixture.
14. We use paper boards found in between stacks of ELISA plates in packages of ELISA plates (Costar, Cambridge, MA, USA) for this purpose.

Acknowledgement

This work was supported by NIH grant ARO1844 and Oklahoma Center for the Advancement of Science and Technology to Dr. Hal Scofield, OMRF, OKC, OK, USA. We also express our thanks to Natalie B. Reed and Alexandra Taylor for their excellent technical assistance.

References

1. Santella, L., Ercolano, E., and Nusco, G.A. (2005) The cell cycle: a new entry in the field of Ca^{2+} signaling. *Cell Mol Life Sci.* 62, 2405–2413. Review.
2. Karpati, I., Seres, I., Matyus, J., Ben, T., Paragh, G., Varga, Z., and Kakuk, G. (2001) Which parameters affect cytosolic free calcium in polymorphonuclear leukocytes of haemodialysis patients? *Nephrol Dial Transplant.* 16, 1409–1415.
3. Koledova, V.V. and Khalil, R.A. (2006) Ca^{2+}, calmodulin, and cyclins in vascular smooth muscle cell cycle. *Circ Res.* 98, 1240–1243. Review.
4. Scofield, R.H., Kurien, B.T., Gross, J.K., Reed, N.B., Taylor, A.K., Dominguez, et al. (2004) Interaction of calcium and Ro60: increase of antigenicity. *Mol Immunol.* 41, 809–816.
5. Farris, A.D., Gross, J.K., Hanas, J.S., and Harley, J.B. (1996) Genes for murine Y1 and Y3 Ro RNAs have class 3 RNA polymerase III promoter structures and are unlinked on mouse chromosome 6. *Gene* 174, 35–42.
6. O'Brien, C.A. and Wolin, S.L. (1994) A possible role for the 60-kDa Ro autoantigen in a discard pathway for defective 5S rRNA precursors. *Genes Dev.* 8, 2891–2903.
7. Harley, J.B., Scofield, R.H., and Reichlin, M. (1992) Anti-Ro in Sjögren's syndrome and systemic lupus erythematosus. *Rheum Dis Clin N Am.* 18, 337–358.
8. Kurien, B.T., Newland, J.G., Paczkowski, C., Moore, K.L., and Scofield, R.H. (2000).

Anti-Ro mediates neutropenia in systemic lupus erythematosus by binding a 60,000 MW antigen on the surface of neutrophils. *Clin Exp Immunol.* 120, 209–217.

9. Muro, M.D., Fernández, C., and Moneo, I. (1996) Renaturation of blotted allergens increases the sensitivity of specific IgE detection. *J Invest Allergol Clin Immunol.* 6, 166–171.

Chapter 33

Intermittent Microwave Irradiation Facilitates Antigen–Antibody Reaction in Western Blot Analysis

Yu-Ting Liu and Shinya Toyokuni

Summary

We established a shortened protocol for western blot analysis using intermittent microwave irradiation. With this method, the procedure is completed within 1 h after applying the primary antibody, and thus greatly saves time. This procedure appears to be applicable to any antibody based on our experience of several years.

Key words: Intermittent microwave irradiation, Western blot analysis

1. Introduction

Specific reaction between antigen and antibody can be visualized by various methods such as western blot analysis and immunohistochemistry. In the March 2002 issue of *Pathology International*, Li et al. presented the first report describing the use of intermittent microwave irradiation in western blot analysis *(1)*. We have independently developed a more rapid western blot protocol, which we would like to introduce here.

Microwave radiation, an energy wave with a frequency of approximately 1–300 GHz, has been used not only for radar, cooking procedure, radio communication but also for pathological methods as an antigen retrieval method in formaldehyde-fixed paraffin-embedded specimens *(2, 3)*. Recently, intermittent microwave irradiation has been also applied to fluorescent in situ hybridization technique *(4)*.

Biji T. Kurien and R. Hal Scofield (eds.), *Methods in Molecular Biology, Protein Blotting and Detection, vol. 536*

DOI: 10.1007/978-1-59745-542-8_33

Western blot analysis is an indispensable biochemical method for protein analysis *(5, 6)*, and is used in most laboratories dealing with wet biological samples. However, the procedure takes half a day to 2 days after the blotting process in the currently used protocols *(7)*. Microwave irradiation oscillates the exposed molecules. In the present protocol, we applied intermittent microwave irradiation to western blot analysis.

2. Materials

2.1. SDS-Polyacrylamide Gel Electrophoresis (SDS-PAGE)

1. Stacking buffer (4×): 0.5 M Tris–HCl, pH 6.8, 0.4% SDS. Store at room temperature (RT).
2. Separating buffer (4×): 1.5 M Tris–HCl, pH 8.8, 0.4% SDS. Store at RT.
3. 44% Acrylamide/1.2% Bis solution (Nacalai, Kyoto, Japan). Store at 4°C.
4. *N,N,N,N'*-Tetramethyl-ethylenediamine (TEMED) (Wako, Osaka, Japan). Store at 4°C.
5. Ammonium persulfate: Prepare 10% solution in water (Nacalai, Kyoto, Japan). Store at –20°C.
6. Running buffer (5×): 125 mM Tris, 960 mM glycine, 17.3 mM SDS. Store at RT.
7. Precision Plus protein dual-color standard marker (Bio-Rad, Hercules, CA, USA). Store at –20°C.

2.2. Western Blotting for Actin

1. Polyvinylidene difluoride (PVDF) membranes (Immobilon, Millipore, Tokyo, Japan).
2. Blockace (Yukijirushi Co. Ltd., Sapporo, Japan).
3. Wash buffer: 0.1% Tween-20, 100 mM Tris-buffered saline (TBST, pH 7.4).
4. Moisture chamber (Incubation chamber 10DO, Cosmo Bio, Tokyo, Japan).
5. Parafilm "M" (American National Can, Menasha, WI, USA).
6. Intermittent microwave irradiation machine (Microwave processor, MI-77, Azumaya, Tokyo, Japan; **Fig. 1**).
7. Primary antibody: Anti-actin monoclonal antibody (Chemicon International Inc., Temecula, CA, USA), diluted to 1:2,000.
8. Secondary antibody: Horseradish peroxidase-conjugated rabbit anti-mouse IgG (Dako Japan Co. Ltd., Kyoto, Japan), diluted 1:2,000.
9. ECL detection reagents (Amersham Pharmacia Biotech, Tokyo, Japan).

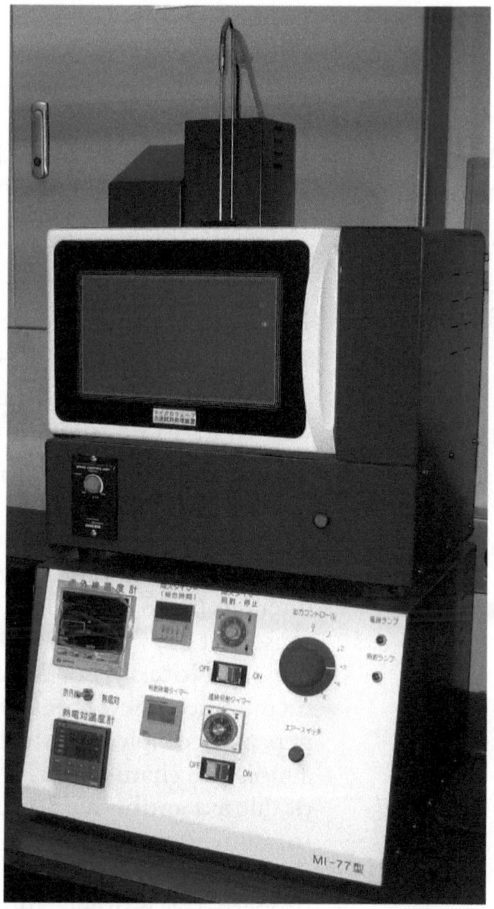

Fig. 1. Microwave processor, MI-77.

2.3. Tissue Homogenization

1. RIPA (radioimmunoprecipitation) buffer: 20 mM Tris–HCl, 0.1% SDS, 1% Triton X-100, 1% sodium deoxycholate, pH 7.4.

3. Methods

3.1. Preparation of Samples

1. Three 6-week-old specific pathogen-free male Wistar rats (Shizuoka Laboratory Animal Center, Shizuoka, Japan) were purchased, and two of them received a single intraperitoneal injection of 15 mg iron/kg of ferric nitrilotriacetate (Fe-NTA) freshly prepared as previously described *(8)*.
2. The rats were killed 6 and 24 h after Fe-NTA administration, respectively. One animal served as an untreated control.
3. After euthanasia of the animal, kidneys were immediately excised and homogenized in RIPA buffer.

4. The supernatant collected after centrifugation at 15,000 × *g* for 15 min was used after the determination of protein concentration with BCA protein reagents (Pierce, Rockford, IL, USA).

3.2. SDS-PAGE and Membrane Transfer

Western blotting was carried out after 10% SDS-PAGE under reducing conditions (4 μg of protein in each lane) and semi-dry blotting on PVDF membranes with a standard method as described in other chapters.

3.3. Western Blotting for Actin

1. The PVDF membrane after transfer was incubated in Blockace solution for 1 h at RT on a rocking platform.
2. The blocking solution was discarded and the membrane was washed in TTBS solution three times for 5 min each at RT.
3. They were incubated with primary antibody (anti-actin monoclonal antibody, diluted to 1:2,000). During the antibody incubation period, intermittent microwave irradiation was applied. The conditions were as follows: 4-s irradiation at 4-s intervals (2.45 GHz; 150, 250, or 350 W) with a temperature sensor set at 37°C and the rotation of the table at 6 rounds per min (*see* **Note 1**). For the antigen–antibody reaction on PVDF membrane, a new sheet of parafilm "M" was pressed with fingers and completely attached to the smooth plastic surface of a moisture chamber. A PVDF membrane was placed over each of diluted antibody (50 μL antibody/cm^2 membrane) on the parafilm with the protein surface downward (**Fig. 2**). After the reaction, the antibody can be reused several times if stored properly (*see* **Notes 2** and **3**).
4. Horseradish peroxidase-conjugated rabbit anti-mouse IgG (diluted 1:2,000) was applied in the same way after washing in TTBS (*see* **Notes 2** and **3**).

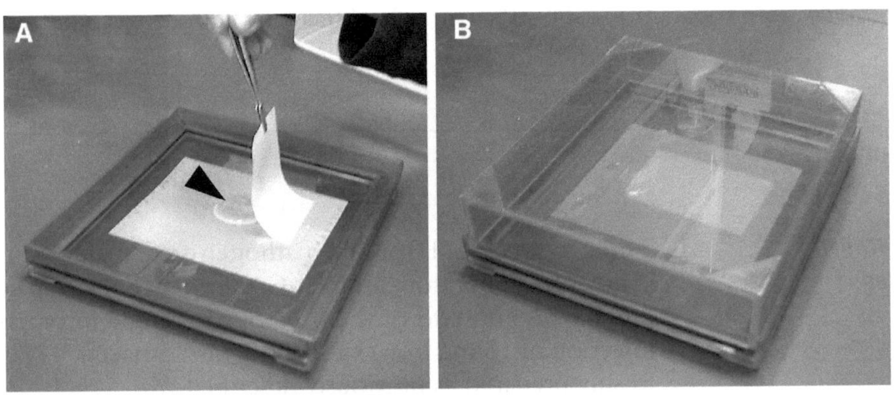

Fig. 2. Procedures for microwave irradiation. **(A)** Blotted membrane was placed over an aliquot of diluted antibody on parafilm "M"; *arrow*, antibody. The antibody was distributed homogeneously by surface tension. A moisture chamber was used upside down. **(B)** The lid was replaced, and the membrane was ready for microwave irradiation. See **Subheadings 2** and **3** for details (reproduced from *(9)* with permission).

5. Finally, ECL detection reagents were used according to the manufacturer's instruction for visualization of the signals on the film. The exposure time of the membrane to the film was 1 min (*see* **Notes 2** and **3**).

4. Notes

1. We advise to avoid using microwave machines used for cooking because their power is usually too strong (600–1,200 W) for this purpose.
2. After membrane transfer and blocking, our former routine procedure consisted of 1-h incubation with primary antibody at RT, three 5-min washes in TBST, 1-h incubation with secondary antibody at RT, and three 5-min washes with TBST. When the incubation periods with antibody were shortened to 15 and 5 min, respectively, the signals were greatly decreased (**Fig. 3**). However, if the incubation was performed under intermittent microwave irradiation with the same shortened incubation periods, signals with enough intensity were observed. While a microwave intensity of 150 and 250 W worked well, 350-W irradiation induced either rather decreased signals (**Fig. 3**) or unspecific bands (data not shown). When the incubation periods were shortened to 5 and 1 min, respectively, the signals were too weak even with microwave irradiation (data not shown). Furthermore, with the use of microwave irradiation, we found that two 5-min washes with ample TTBS were sufficient for a clean background. This is probably because nonspecific attachment of antibodies to the membrane is minimized with shorter incubations.
3. Standard incubation periods under this intermittent microwave irradiation are as follows: 15 min for primary antibody, 2 × 5 min washes, 5 min for secondary antibody, 2 × 5 min washes.

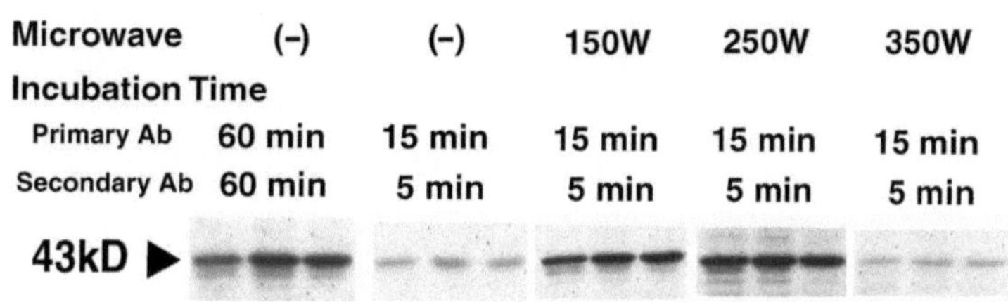

Fig. 3. Intermittent microwave irradiation facilitates antigen–antibody reaction in western blot analysis. Lysates of three different rat kidney samples were subjected to 10% sodium dodecyl sulfate-polyacrylamide gel electrophoresis, and transferred to polyvinylidene difluoride membrane. Western blot analyses for actin were performed. Samples from left to right; untreated control kidney, kidney obtained 6 and 24 h after a single ferric nitrilotriacetate administration. Refer to text for details (reproduced from *(9)* with permission).

Acknowledgements

This work was supported in part by a Grant-in-Aid from the Ministry of Education, Science, Sports and Culture of Japan, a Grant-in-Aid for Cancer Research from the Ministry of Health, Labour and Welfare of Japan, and the Eiko Yasuhara Memorial Fund.

References

1. Li, W., Murai, Y., Okada, E., Matsui, K., Hayashi, S., Horie, M., et al. (2002) Modified and simplified western blotting protocol: use of intermittent microwave irradiation (IMWI) and 5% skim milk to improve binding specificity. *Pathol Int.* 52, 234–238.
2. Leong, A. and Milios, J. (1986) Rapid immunoperoxidase staining of lymphocyte antigens using microwave irradiation. *J Pathol.* 148, 183–187.
3. Shi, S., Cote, R., and Taylor, C. (1997) Antigen retrieval immunohistochemistry: past, present, and future. *J Histochem Cytochem.* 45, 327–343.
4. Kitayama, Y., Igarashi, H., and Sugimura, H. (2000) Initial intermittent microwave irradiation for fluorescence *in situ* hybridization analysis in paraffin-embedded tissue sections of gastrointestinal neoplasia. *Lab Invest.* 80, 779–781.
5. Towbin, H., Staehelin, T., and Gordon, J. (1979) Electrophoretic transfer of proteins from polyacrylamide gels to nitrocellulose sheets: procedure and some applications. *Proc Natl Acad Sci USA* 76, 4350–4354.
6. Burnette, W. (1981) "Western blotting": electrophoretic transfer of proteins from sodium dodecyl sulfate-polyacrylamide gels to unmodified nitrocellulose and radiographic detection with antibody and radioiodinated protein A. *Anal Biochem.* 112, 195–203.
7. Harlow, E. and Lane, D. (1988) Antibodies: A Laboratory Manual. Cold Spring Harbor, NY: Cold Spring Harbor Laboratory.
8. Toyokuni, S., Uchida, K., Okamoto, K., Hattori-Nakakuki, Y., Hiai, H., and Stadtman, E.R. (1994) Formation of 4-hydroxy-2-nonenal-modified proteins in the renal proximal tubules of rats treated with a renal carcinogen, ferric nitrilotriacetate. *Proc Natl Acad Sci USA* 91, 2616–2620.
9. Toyokuni, S., Kawaguchi, W., Akatsuka, S., Hiroyasu, M., and Hiai, H. (2003) Intermittent microwave irradiation facilitates antigen–antibody reaction in western blot analysis. *Pathol Int.* 53, 259–261.

Chapter 34

Detection of Protein–Protein Interactions by Far-Western Blotting

Kazuya Machida and Bruce J. Mayer

Summary

Far-western blotting is a convenient method to characterize protein–protein interactions, in which protein samples of interest are immobilized on a membrane and then probed with a nonantibody protein. In contrast to western blotting, which uses specific antibodies to detect target proteins, far-western blotting detects proteins on the basis of the presence or the absence of binding sites for the protein probe. When specific modular protein binding domains are used as probes, this approach allows characterization of protein–protein interactions involved in biological processes such as signal transduction, including interactions regulated by posttranslational modification. We here describe a rapid and simple protocol for far-western blotting, in which GST-tagged Src homology 2 (SH2) domains are used to probe cellular proteins in a phosphorylation-dependent manner.

Key words: Protein–protein interaction, Far-western blotting, GST fusion protein, Affinity purification, SH2 domain, Tyrosine phosphorylation, Reverse-phase assay

1. Introduction

Far-western blotting is a method of characterizing protein–protein interactions, in which protein samples of interest are separated by gel electrophoresis, immobilized on a membrane, and then probed with a nonantibody protein *(1)*. The term "Far-western" was derived from western blotting, a similar method in which membranes are probed directly with specific antibodies, and is also referred to as a West-western or blot overlay assay *(2–4)*. Nonantibody proteins have been also used as a means to screen phage-based expression libraries *(2, 5–7)*.

Biji T. Kurien and R. Hal Scofield (eds.), *Methods in Molecular Biology, Protein Blotting and Detection, vol. 536*

DOI: 10.1007/978-1-59745-542-8_34

Far-western blotting is very different from other commonly used methods to detect and characterize protein–protein interactions, and therefore complements these other approaches. Because the probe protein directly binds to denatured/separated proteins immobilized on a membrane, far-western blotting detects only direct interactions; by contrast, most non-far-western protein binding assays, such as immunoprecipitation and pull-down assays, may detect either direct association (two proteins make contact directly) or indirect association (two proteins do not make contact, and another molecule in the ternary complex mediates the association). Thus the far-western assay is often used to confirm direct interaction following immunoprecipitation or pull down assays.

The ability of far-western blotting to detect direct interactions is offset by limitations in the types of protein–protein interactions that can be detected. Because target proteins in a cell lysate are usually denatured in the process of gel electrophoresis, it may be difficult or impossible to detect interactions that require the native, folded conformation of the target protein. For this reason, far-western blotting has been particularly useful in characterizing the binding partners of modular protein binding domains that bind to short, linear peptide motifs. It is now apparent that many signaling proteins interact with their partners via such modular binding domain–peptide interactions, thus the far-western approach is quite useful for analysis of signaling networks. However, these differences highlight the importance of taking care to use multiple approaches to assess specific protein–protein interactions.

In far-western blotting, either a whole protein or a fragment of a protein containing a suspected binding interface is used to probe interaction partners immobilized on a membrane. The interaction is visualized by direct labeling of the probe or by its subsequent detection with antibodies (**Fig. 1a**). There are a number of considerations in selecting the specific probe. First, ease of growth and purification of the probe must be considered. For the sake of cost and convenience, expressing probe proteins in bacteria is advantageous. However, only relatively small proteins (less than ~100 kDa) tend to remain soluble when grown in bacteria, so in general a fragment of a protein containing only the known or suspected binding domain will be easier to work with than the full-length protein. Second, it is useful to fuse the probe protein or domain to a tag sequence for ease of purification and detection. We routinely make probe proteins as glutathione S-transferase (GST) fusions, which have the dual advantage of allowing purification of proteins on glutathione columns, and allowing detection of bound probe with glutathione conjugates or with anti-GST antibodies. A further advantage of GST fusions is that GST exists as a stable dimer in solution. As in the case of antibodies, a dimeric probe binds with much greater avidity

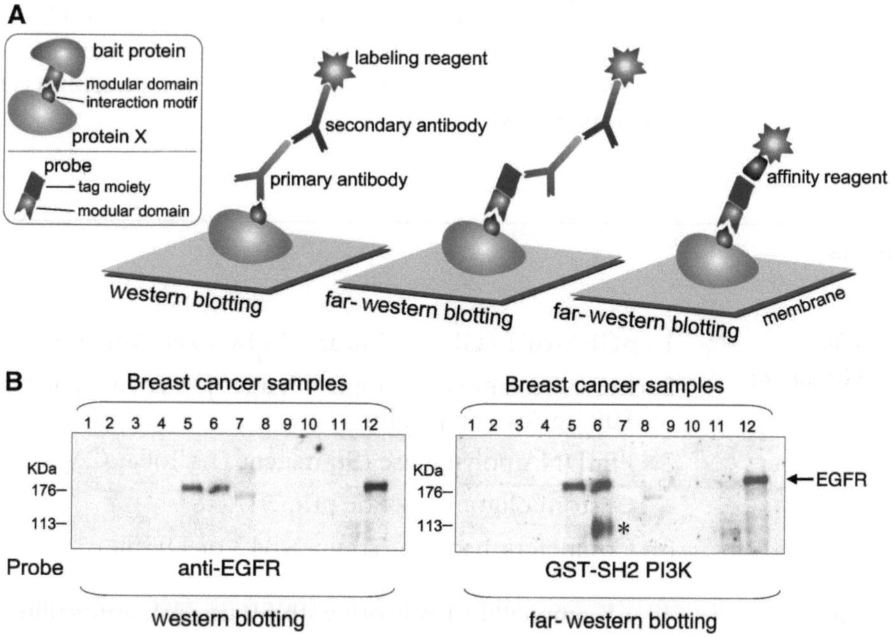

Fig. 1. Comparison of western and far-western blotting. (**A**) Schematic representation of both methods. Box, bait protein and protein X interact via modular interaction domain and interaction motif. *Left*: western blotting uses antibodies raised against a target protein X to detect its presence. *Middle* and *right*: far-western blotting uses a protein probe containing a modular interaction domain or a short interaction motif, fused to a tag moiety (e.g., GST), to detect protein X based on presence of its binding sites. Specific interaction of far-western blotting is visualized by antibody-based detection (*middle*) or by direct labeling using isotope or an enzyme-conjugated affinity reagent, e.g., glutathione–HRP (*right*). (**B**) Application of both methods. Breast cancer patient-derived samples (tumor 1–12) were immobilized on a membrane, and then identical membranes were probed with anti-EGFR antibody (*left*, western blotting) or GST-PI3K SH2 domain fusion (*right*, far-western blotting). Whereas both results look similar, interpretation of the results is somewhat different; the western result indicates a presence of EGFR in sample lane 5, 6, and 12, while the far-western result indicates a presence of PI3K SH2 domain recognition sites on a protein which is likely to be EGFR (*arrow*). In addition, the far-western probe also detected an uncharacterized PI3K SH2-binding protein (*asterisk*) that can potentially distinguish tumor 6 from 5 and 12, not otherwise detected by western blotting.

compared to a monomer to targets containing multiple binding sites, such as a membrane surface bearing many molecules of a target protein.

In contrast to western blotting where a target protein is usually known in advance, far-western blotting can detect proteins on the basis of presence or absence of binding sites without any previous knowledge about their identities (**Fig. 1a, b**). From the intensity of bands observed on a far-western blot of a complex mixture of proteins, one can gain insight into both the number and the relative affinity of binding partners for the probe in that sample. Furthermore, since some protein binding domains recognize their targets only after specific posttranscriptional modifications, far-western blotting can be used to assess the modification status of multiple proteins in a sample *(8–10)*. In this chapter, we will present a specific example of the utility of far-western blotting methods, in which GST-tagged Src homology 2 (SH2) domains, which bind specifically to tyrosine-phosphorylated target proteins, are used to probe the state of tyrosine phosphorylation of cellular proteins.

2. Materials

2.1. Subcloning of GST-SH2 Construct

1. pGEX-6P1 (GE Healthcare, Milwaukee, WI, USA).
2. Luria-Bertani (LB)-ampicillin agar plate: LB agar plate with 100 μg/mL ampicillin.
3. Pfu DNA polymerase (Stratagene, La Jolla, CA, USA).
4. Custom oligonucleotide primers.
5. Competent bacteria (strain NB42 or DH5α).

2.2. Evaluation of GST-SH2 Clones

1. LB-ampicillin: LB broth with 50 μg/mL ampicillin.
2. Isopropyl-β-d-thiogalactoside (IPTG).
3. Bacteria Triton X-lysis buffer (BXB): Phosphate buffered saline (PBS) with 100 mM ethylene diamine tetraacetic acid (EDTA), 1% Triton X-100; add phenylmethyl sulphonyl fluoride (PMSF) to 1 mM, aprotinin (Sigma Chemical Co., St. Louis, MO, cat. no. A6279) to 1% v/v (3 trypsin international units (TIU)/mL), dithiothreitol (DTT) to 1 mM just before use.
4. Sonicator with microtip probe (e.g., Branson Sonifier 450 or equivalent).
5. 5X sample buffer: 0.3 M Tris–HCl pH 6.8, 10% sodium dodecyl sulfate (SDS), 25% β-mercaptoethanol, 0.1 mg/mL bromophenol blue, 45% glycerol.
6. Glutathione Sepharose 4B (GE Healthcare).
7. 12% SDS-polyacrylamide gel electrophoresis (PAGE) mini gel (*see* **Subheading 2.4**).
8. Control lysates (*see* **Subheading 2.4**).
9. Antiphosphotyrosine antibody (Cell Signaling Technology, Beverly, MA, PY100).
10. Coomassie Blue solution: 40% methanol, 10% acetic acid, 0.25% Coomassie Blue R-250.
11. Fix solution: 20% methanol, 10% acetic acid.
12. Bacteria stock solution: 50% glycerol (autoclaved).
13. Cryogenic tubes.

2.3. Large-Scale Preparation of GST-SH2 Probe

1. Tris–NaCl–EDTA (TNE) buffer: 50 mM Tris–HCl pH 7.4, 150 mM NaCl, 10 mM EDTA; add aprotinin to 1% (3 TIU/mL), PMSF to 1 mM just before use.
2. Chromatography column (Bio-Rad Laboratories, Hercules, CA, poly-prep, 0.8 × 4 cm).
3. Elution buffer: 20 mM reduced glutathione, 100 mM Tris–HCl, pH 8.0; add aprotinin to 1% (3 TIU/mL), PMSF to 1 mM just before use.
4. Sephadex G-25 PD-10 column (GE Healthcare).
5. Dialysis membrane tubing (Spectra/Por, Laguna Hills, CA, molecular weight cut-off 3500).
6. PBS with 10% glycerol.
7. Bio-Rad Bradford dye reagent (Bio-Rad Laboratories, cat. no. #500–0006).
8. Ultrafiltration membrane (Amicon, Danvers, MA, YM10).

2.4. Far-Western Blotting

1. Kinase lysis buffer (KLB): 150 mM NaCl, 25 mM Tris–HCl pH 7.4, 5 mM EDTA, 1% Triton X-100, 10% glycerol, 0.1% sodium pyrophosphate, 10 mM β-glycerophosphate, 10 mM sodium fluoride (NaF); add aprotinin to 1% (3 TIU/mL), PMSF to 1 mM, pervanadate (50 mM orthovanadate, 4% hydrogen peroxide) to 50 μM just before use.
2. Sodium orthovanadate: dissolve powdered sodium ortho-vanadate (final concentration will be 50 mM, but leave some extra volume for multiple rounds of pH adjustment); adjust with NaOH to pH of 10 (solution will turn bright yellow); boil in microwave until colorless, then stir until cooled to room temperature; adjust pH once again to 10, and repeat boiling; continue boiling and adjusting pH as above until pH stays at 10 after boiling (usually three rounds total); adjust volume for 50 mM, filter, and store at room temperature (RT).
3. Pervanadate solution: mix 16-μL concentrated hydrogen peroxide and 100 μL 50 mM sodium orthovanadate; incubate at RT for 30 min (not stable, needs to be freshly prepared before use).
4. Positive control lysates: equal amounts of KLB lysate from pervanadate-treated NIH 3T3, HepG2, A431, and MR20 cells were combined.
5. Negative control lysate: lysates of each cell line were prepared in the absence of vanadate, combined, and treated with tyrosine phosphatase PTP-1B for 1 h at RT.
6. 12% SDS-PAGE gel (1.0 mm thickness, for three mini-gels).

Stock solutions	Resolving gel	Stacking gel
Acrylamide/bis-acrylamide	20 mL	2.2 mL
(30%/0.8%, w/v)		
1 M Tris–HCl (pH 8.8)	18.6 mL	–
1 M Tris–HCl (pH 6.8)	–	2.1 mL
Distilled water	10.5 mL	12.2 mL
10% SDS	500 μL	167 μL
10% Ammonium persulfate	500 μL	125 μL
TEMED	16.7 μL	16.7 μL

7. Electrophoresis buffer: 25 mM Tris, 192 mM glycine, 0.1% SDS.
8. Nitrocellulose membrane (Schleicher & Schuell, Keene, NH, 0.2-μm pore, BA83).
9. Transfer buffer: 10 mM 3-(cyclohexylamino)-1-propanesulfonic acid (CAPS), pH 11, 20% methanol, kept at 4°C.
10. Tris buffered saline-Tween 20 (TBST): 25 mM Tris–HCl, pH 8, 150 mM NaCl, and 0.05% v/v Tween-20.
11. Blocking solution: 10% fat-free milk in TBST, 1 mM EDTA, 1 mM sodium orthovanadate.
12. Labeling reagent: Glutathione–HRP conjugate (Sigma, G6400, production discontinued) or anti-GST–HRP conjugate (Sigma, A7340).
13. Chemiluminescence kit:

Product	Maker	Sensitivity	Background
RPN2106	GE Healthcare	Medium	Low
NEL103	PerkinElmer, Boston, MA	High	Medium
#34079	Pierce, Rockford, IL	Very high	High
RPN2132	GE Healthcare	Very high	High

We routinely use NEL103.

14. Imaging: X-ray film (Kodak, Rochester, NY, cat. no. #1651454); CCD detection system (Kodak, Image Station 2000R).

2.5. Stripping and Reprobing

1. Acidic stripping buffer: 100 mM Glycine–HCl pH 2.0.
2. SDS-ME stripping buffer: 2% SDS, 100 mM β-Mercaptoethanol, 62.5 mM Tris–HCl pH 6.8.

3. Methods

As for western blotting, protein samples are separated by SDS-PAGE and transferred to a nitrocellulose or polyvinylidene fluoride (PVDF) membrane. The blocked membrane is then incubated with a probe followed by appropriate wash and bound probes are visualized. Generally any protein samples compatible with western blotting can be used including whole cell lysates, purified proteins, and native or denatured samples. The far-western method described here is a rapid and simple protocol in which the membrane is prepared without denature–renature procedures, the probe is labeled directly, and probing is performed in one step *(8)*. This protocol has been optimized to detect in vitro interaction between modular binding domain probes and immobilized proteins containing short peptide motifs (*see* **Note 1**). Of note, alternative protocols are available including another example in which proteins containing modular domains on a membrane are probed with labeled binding motifs *(4, 11, 12)*. Below, we present a specific protocol for generating GST-SH2 domain probes and using them to probe tyrosine-phosphorylated whole cell lysates. Of course these procedures can be adapted for any modular protein binding domain and its binding partners with minor modifications.

For all far-western blotting methods, detection of specific signal is strongly dependent on the quality of the probe protein. Insolubility, aggregation, or denaturation of the protein tends to cause nonspecific background, and even the native probe may bind nonspecifically to abundant proteins in the sample. Thus it is important to (1) confirm purified probe is soluble, folded, and not significantly degraded; (2) evaluate activity of the probe and optimize binding conditions if needed; and (3) always include appropriate positive and negative controls for each experiment to ensure any positive signal is indeed specific. To address these considerations, in the following section we will describe a detailed protocol for generation and evaluation of GST-SH2 fusion probes. Appropriate controls should be prepared considering the intended physiological activity of the probe; as an example, several control blots for specific SH2–phosphotyrosine interaction are presented in **Fig. 2b**. At minimum, GST alone, or more ideally GST fused to the domain of interest bearing a mutation known or suspected to abolish specific binding activity, should be used as a negative control probe.

3.1. Subcloning of GST-SH2 Construct

1. Retrieve cDNA and protein sequences of a SH2 domain-containing protein of interest, e.g., at NCBI Entrez Gene (*see* **Note 2**).
2. Find location of the SH2 domain using the protein sequence at ScanProsite.

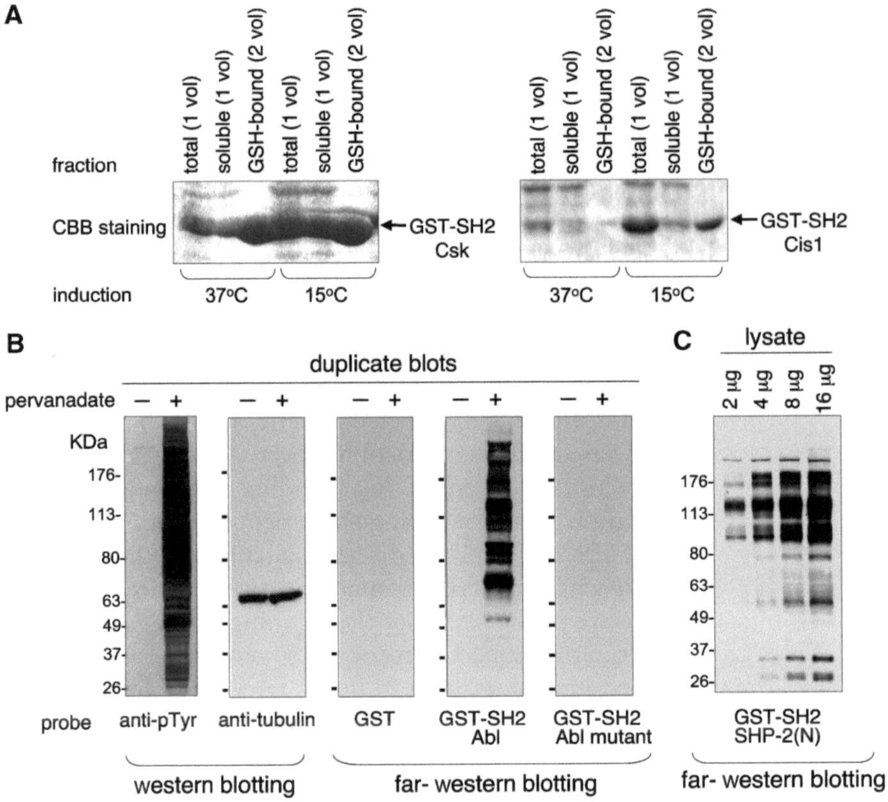

Fig. 2. Generation and evaluation of GST-tagged probes. (**A**) Testing solubility of GST fusion proteins. GST fusions of Csk and Cis1 SH2 domains were expressed in *E. coli* at regular (37°C, 1 h) and low (15°C, overnight) culture temperatures. Protein fractions from affinity purification are visualized by Coomassie Brilliant Blue (CBB) staining: whole cell fraction (total); Triton-soluble fraction (soluble); and fraction bound to glutathione-agarose beads (GSH-bound). The beads fraction represents twice the relative amount of original culture as total and soluble fractions. Cis-SH2 was much less soluble than Csk-SH2, but its solubility was improved by expression at low temperature. (**B**) Evaluation of GST-SH2 probes. Identical blots of pervanadate-treated lysate (*plus*) and POV-untreated, phosphatase-treated lysate (*minus*), were prepared. To validate specific interaction of a probe with target proteins, different probes were used for following purposes: antiphosphotyrosine antibody (anti-pTyr), indicator of tyrosine phosphorylation; anti-tubulin antibody, loading control; GST, negative control probe; GST-SH2 domain of Abl, experimental probe; and GST-SH2 Abl mutant, loss-of-function mutant probe. (**C**) Optimization of protein loading. A membrane with various amounts of POV-treated lysate was probed with a GST fusion with N-terminal SH2 domain of SHP-2. High affinity ligand proteins for the SH2 domain are selectively detected with a 2 μg protein per lane, while more proteins are detected with increased protein loading.

3. Find the nucleotide sequence corresponding to the SH2 domain using the sequence editor program (*see* **Note 3**).
4. Find academic or industry source for corresponding cDNA (*see* **Note 4**), otherwise clone the cDNA by RT-PCR method.
5. Design primers for PCR (*see* **Note 5**).
6. Amplify the SH2 fragment by PCR using the oligonucleotide primers and the cDNA template.

7. Digest the PCR product, and then purify the fragment from an agarose gel.
8. Insert the purified SH2 fragment into a pGEX vector digested with appropriate restriction enzymes.
9. Transform competent bacteria and grow overnight on LB-ampicillin agar plate (*see* **Note 6**).

3.2. Evaluation of GST-SH2 Clones

Protein expression and solubility of GST-SH2 clones can be tested quickly in small-scale bacterial cultures (*see* **Note 7**), and the activity of the probe can be tested at the same time by a control pull-down assay.

1. Inoculate 0.4 mL dense liquid culture of bacteria into 1.6 mL fresh prewarmed LB-ampicillin.
2. Shake at 37°C for 1 h.
3. Add IPTG to 0.1 mM and shake at 37°C for 1 h to induce protein expression.
4. Transfer 1.5 mL bacteria to microcentrifuge tubes and spin at 10,000 × *g* at 4°C for 2 min in microfuge.
5. Remove supernatant and resuspend bacterial pellet in 0.4 mL BXB.
6. Vortex to resuspend, then sonicate briefly (e.g., 2–5 s at relatively low power) on ice to break cells, let sit on ice, then repeat. Try to avoid foaming; if foaming occurs, let rest on ice for a few minutes to allow foam to dissipate.
7. Remove 10 μL of total lysate, add 2.5 μL 5× sample buffer for gel (total cell fraction, **Fig. 2a**).
8. Spin rest of lysate for 5 min in microfuge at 10,000 × *g* at 4°C, transfer supernatant to a new tube.
9. Take 10 μL of the cleared lysate for gel as above (soluble fraction, **Fig. 2a**).
10. Take 100 μL of cleared lysate and add to 10 μL glutathione-agarose bead slurry (cut end off pipet tip with razor blade to more accurately pipet beads).
11. Rotate at 4°C for 30 min.
12. Spin out briefly, and wash beads 3× with 1 ml cold BXB.
13. Resuspend the bead pellet with BXB, take 10 μl, and add 2.5 μl 5× sample buffer for gel (GSH-bound fraction, **Fig. 2a**) (*see* **Note 8**).
14. Boil all samples and run on 12% SDS gel.
15. When gel is done, stain for 15 min with Coomassie blue solution, destain with fix solution, and then dry (**Fig. 2a**) (*see* **Note 9**).

Steps 16–19: Evaluate pTyr binding activity by GST pull-down assay (optional)

16. Incubate remaining beads (GSH-bound fractions for pGEX-SH2 clones and a control pGEX clone, if any) with 10 μg cell lysates (*see* **Note 10**).
17. Rotate for 1 h at 4°C, and wash three times with BXB.
18. Boil all samples and run on 12% SDS gel (*see* **Note 11**).
19. Perform western blotting with anti-phosphotyrosine antibody (*see* **Note 12**).
20. Subject positive clones to DNA sequencing and store bacteria in 25% (v/v) sterile glycerol at –70°C.

3.3. Large-Scale Preparation of GST-SH2 Probe

GST-SH2 probe is purified following the standard protocol for preparation of GST fusion proteins using pGEX series bacterial expression vectors *(13)* (*see* **Note 13**).

1. Inoculate frozen stock culture of a verified GST-SH2 clone in 50-mL LB-ampicillin overnight.
2. Inoculate 50 mL of dense overnight culture to 1 L LB-ampicillin. Shake at 37°C for 2 h (*see* **Note 14**).
3. Add IPTG to 0.1 mM. Shake at 37°C for 3 h (*see* **Note 15**).
4. Centrifuge at 5,000 × *g* at 4°C for 10 min.
5. Resuspend pellet with in 5–20 mL ice-cold BXB, transfer to 50-mL tube, and sonicate on ice until cells are broken (*see* **Note 16**).
6. Centrifuge at 5,000 × *g* at 4°C for 10 min to remove debris (*see* **Note 17**).
7. Add glutathione-agarose to supernatant: 3 mL bead slurry per liter original culture.
8. Rotate at 4°C about 1 h (up to 2 h).
9. Wash beads with TNE buffer: spin out beads at low speed, remove supernatant, and resuspend beads in fresh buffer. Repeat 3–5 times.
10. To elute GST-SH2, pour beads into small disposable column, elute with approximately three bead volumes of elution buffer (*see* **Note 18**).
11. Change buffer by gel filtration on a Sephadex G-25 PD-10 column according to the supplier's instructions. Briefly, equilibrate column with approximately 25 mL PBS-10% glycerol. Discard the flow-through. Add sample followed by buffer up to a total volume of 2.5 mL. Discard the flow-through. Elute with 3.5 mL buffer (collect seven 0.5 mL fractions of the eluate in separate tubes (*see* **Note 19**).

12. Estimate relative protein concentration with Bio-Rad Bradford dye reagent, combine top three fractions into one tube (*see* **Note 20**).
13. Determine protein concentration by Bradford assay, take 500 μg of protein and dilute to 0.1 μg/μL with PBS-10% glycerol, and aliquot 50 μL diluted probe into chilled microcentrifuge tubes. Store the aliquots and the undiluted probes at –70°C (*see* **Note 21**).
14. Evaluate the purification fractions by 12% SDS gel as in **Subheading 3.2** (*see* **Note 22**).

3.4. Far-Western Blotting

1. Separate proteins on SDS-polyacrylamide gels and transfer to nitrocellulose or PVDF membranes following general western blotting protocol (*see* **Note 23**).
2. Block membranes in blocking solution for about 1 h at room temperature or at 4°C overnight (*see* **Notes 24** and **25**).
3. To label probe, thaw the stored probe on ice and add 5 μL GSH–HRP conjugate (0.1 μg/μl) to 50 μl of diluted probe (0.1 μg/μl) (*see* **Note 26**).
4. Incubate on ice for about 1 h. Labeled probes can be stored at 4°C (*see* **Note 27**).
5. Dilute labeled probe to optimal concentration with blocking buffer and apply to the blocked membrane (*see* **Note 28**).
6. Let probe bind for 1–2 h at RT, then wash with multiple changes TBST for 20 min.
7. Visualize signal by enhanced chemiluminescence according to manufacturer's instruction (*see* **Note 29**).
8. Take appropriate exposure using X-ray films or an image analyzer e.g., Kodak Image Station system (*see* **Note 30**).

3.5. Stripping and Reprobing

Generally, fresh membranes are best for far-western blotting; stripping and reprobing of the membrane may result in significant signal loss and increased nonspecific background. Nevertheless recycling membranes might be beneficial if sample is limited, or if precise comparison of specific bands is needed within the same membrane.

1. After initial probing, keep membrane wet; it can be stored wet, wrapped in plastic wrap, at 4°C (up to a week) or –20°C (for longer period).
2. Rinse the membrane twice with TBST.
3. Immerse membrane in stripping buffer at RT for 20 min under gentle rocking agitation (*see* **Note 31**).
4. Wash membrane in large volume of TBST at RT for 45 min with frequent shaking.
5. Proceed to blocking and reprobing.

4. Notes

1. This approach, termed as reverse-phase assay, maximizes signal to noise when the concentration of analyte is low (e.g., tyrosine-phosphorylated proteins in whole-cell lysates), because efficient binding to the immobilized analyte can be driven by high concentrations of the SH2 domain probe in solution *(9)*.
2. Useful information about modular interaction domains is available at Web-based databases: NCBI Entrez Gene (http://www.ncbi.nlm.nih.gov/sites/entrez?db = gene); ScanProsite (http://us.expasy.org/cgi-bin/scanprosite); The Human Protein Reference Database (http://www.hprd.org/index_html); UniGene (http://www.ncbi.nlm.nih.gov/sites/entrez?db= unigene); ATCC (http://www.atcc.org/catalog/all/allIndex.cfm); The SH2 domain information sites of the Nash Lab (http://proteoscape.uchicago.edu/sh2/), the Pawson Lab (http://pawsonlab.mshri.on.ca/index.php?option=com_content&task=view&id=30&Itemid=63), Li Lab (http://lilab.uwo.ca/SMALI.htm), and Mayer Lab (http://155.37.79.139/fmi/iwp/cgi?-db=Mayer%20Lab%20SH2%20Database&-loadframes).
3. Several sequence editing programs are available, e.g., CLC Free Workbench (http://www.clcbio.com/index.php?id=28).
4. Large collections of full-length or partial cDNAs are now commercially available at reasonable prices, e.g., Image Consortium (http://image.llnl.gov/). To find appropriate cDNA clone, go back and forth between NCBI Entrez Gene site and Unigene sites; referring to the SH2 nucleotides, find Image clones that contain intact SH2 domain. Check availability of the Image clone at ATCC site and locate an ATCC ordering number.
5. To ensure maximal binding activity and solubility, we usually include 5–10 amino acids N- and C-terminal to the domain boundary for an SH2 domain probe. In some cases, yield and solubility of the fusion protein are greatly affected by the precise borders of the construct. Structural studies, if any, are useful when planning the borders for any modular domain. PCR primers should have approximately 24 nucleotides of homology to the template; the 5′ end should have five C's (this clamps the end, allowing efficient digestion with restriction nucleases) followed by a restriction site to be used for cloning in-frame into the pGEX expression vector, followed by the region with homology to template.
6. We routinely use *E. coli* NB42, which lacks the two major proteases; this may help increase yields by limiting degradation *(14)*. Similar results can be obtained with typical laboratory strains such as DH5α, though yields may be somewhat lower and bacterial growth is slower. We do not routinely

perform restriction endonuclease mapping or DNA sequencing of plasmids at this step, instead we immediately test protein expression of clones as below.

7. Solubility is one of the major determining factors of probe activity in far-western blotting. In our hands about half of the GST-SH2 domains are relatively insoluble, and highly insoluble domains give poor yields and tend to lack detectable binding activity *(9)*. Multiple strategies to improve solubility have been reported *(15–18)*. We have observed that, for about two-thirds of insoluble GST-SH2 proteins, solubility could be significantly improved when protein is expressed in bacteria at lower temperature (e.g., 15°C vs. 37°C, **Fig. 2a**).
8. If solubility of the GST fusion is unknown, volume used to resuspend beads should be adjusted to load 2–10-fold more of the bead-bound fraction relative to total cell fractions on gel to visualize the band (e.g., twofold, **Fig. 2a**).
9. A GST-SH2 band of about 40 kDa in size should be easily visible as in **Fig. 2a**, although the degree of protein expression and solubility may vary depending on the constructs (*see* Csk SH2 vs. Cis SH2 in **Fig. 2a**). If protein degradation is observed on gel, care must be taken in the large-scale purification.
10. Use positive and negative control lysates for pull-down; as an active SH2 domain should bind to tyrosine-phosphorylated proteins, we use pooled lysates of pervanadate-treated cell lines as a positive control (pervanadate inhibits endogenous protein tyrosine phosphatases, thus strongly enhancing tyrosine phosphorylation in vivo). Corresponding pooled lysates lacking tyrosine phosphorylation (prepared in the absence of phosphatase inhibitors and then treated with phosphatase in vitro) serve as a negative control.
11. The gel lanes should contain: positive control lysate 5 μg; negative control lysate 5 μg; pull-down positive control with GST-bound beads; pull-down negative control with GST-bound beads; pull-down positive control with GST-SH2-bound beads; pull-down negative control with GST-SH2-bound beads.
12. In the pull-down result, a functional GST-SH2 protein should have increased anti-pTyr signal relative to GST control. If GST-SH2 is of correct size (and sequence) but the pull-down result is not clearly positive, activity of the probe could be re-evaluated by far-western assay (*see* **Subheading 3.4** and **Fig. 2b**).
13. We routinely obtain about 5–10 mg fusion protein per liter of bacteria using the protocol described here (less if the protein is less soluble).
14. Usually an OD of 0.4–0.6 is optimal at this step.

15. If the protein is highly insoluble, consider protein expression at lower temperature: cool down culture with ice before adding IPTG, then add IPTG to 0.1 mM and shake at 30°C for 4 h or 15°C about 16 h.
16. Sonication time 2–3 min total with microtip at power setting 3–4; solution should become slightly darker and less turbid-looking. Avoid foaming and overheating the lysate (let solution rest on ice if it warms up detectably).
17. Supernatant does not have to be absolutely clear at this step.
18. If the elution yield is found to be suboptimal, the following conditions may help: shorter incubation time of lysate with glutathione beads (~30 min); higher pH (up to 9.6) and glutathione concentration (up to 50 mM) in elution buffer; iterative batch elution (2–3 × 30 min).
19. Alternatively, the eluate can be dialyzed overnight against several large volumes of PBS-10% glycerol.
20. To quickly check relative protein concentration, add 2 μL of each PD-10 elution fraction to 50 μL diluted Bradford dye and vortex. Pool the top three fractions with brightest blue color. If color change is not obvious (this occurs when protein concentration is less than 0.2 μg/μL), take third to fifth fractions and proceed to ultrafiltration for concentration (below).
21. If protein concentration is low (<0.5 μg/μL or so), sample can be concentrated by ultrafiltration (YM-10, Amicon) according to manufacturer's instructions.
22. It is important to take aliquots at every step of purification. Take the same fraction of the total at each step (e.g., 1/5,000 of total prep) to monitor percent recoveries at each step. Retain the pellet after sonication and the supernatant after bead binding at –70°C in case the majority of the fusion protein is there. If final product is highly degraded, we recommended re-evaluating parameters of purification such as prep scale, temperature of induction, protease inhibitors, EDTA concentration, sonication strength, etc.
23. Amount of protein loading on gel is important for an optimal result, and thus should be carefully considered (**Fig. 2c**). We usually load 10–50 μg of whole-cell protein lysate per lane for far-western blotting with GST-SH2 domains. We routinely use a nitrocellulose membrane with 0.2-μm pore size. We find the advantages of nitrocellulose compared to PVDF are ease of use (no pretreatment is needed) and better signal-to-noise ratio (our observation).
24. Composition of blocking solution is important to maximize signal to noise. Strong blocking buffer lowers background

but may decrease or eliminate specific signal. We obtained best results with 10% non-fat milk-based buffer for GST-SH2 domain probes. For other probes, optimal conditions should be determined empirically; blocking buffer containing 5–10% non-fat milk (strongest), 1–5% bovine serum albumin (BSA) (moderate), or 1–5% ovalbumin (moderate) can be tested.

25. Transferred membrane can be stored at −20°C for later use: rinse with TBST once and wrap in plastic wrap. Care should be taken to avoid damaging the membrane, which causes undesirable background signal.
26. GSH–HRP (1 μg/μl in PBS-10% glycerol) can be stored at −70°C. Recently Sigma has discontinued production of this conjugate. If GSH conjugate is unavailable, we have found that anti-GST–HRP conjugate can be used in the same way, with some decrease in signal strength. To label GST fusion probe, add 0.2 μL (~3 μg) antibody to 50 μL (5 μg) of diluted probe (0.1 μg/μL). This direct labeling method presumably promotes oligomerization of probe and HRP, increasing the avidity of binding to the target and the enhancing signal. In addition, this one-step probing and washing procedure is time saving and may avoid dissociation of probe during washes if probe binding is weak.
27. Purification of the labeled probe is not needed. Labeled probe can be stored at 4°C with protection from light up to a month.
28. Optimal concentration (maximum signal and low nonspecific background) should be determined empirically in pilot experiments. For GST-SH2 domains, optimal concentration ranges from 0.01 to 0.15 μM (most typically 0.05 μM) *(9)*. Probing can be done either in a sealed nylon bag or in an open tray with a sufficient volume of solution (1–10 mL/10 cm^2 membrane) with occasional agitation. Do not allow the membrane to dry at any point of the probing procedure through detection.
29. Select chemiluminescence kit with appropriate sensitivity (*see* **Subheading 2.4**). It is important to cover whole blot area evenly during incubation, and to drain off excess chemiluminescence reagent well (retained solution causes high background).
30. Like western blotting, multiple parameters of the assay influence the level of signal and nonspecific background, such as blocking buffer, probe concentration, incubation time and temperature, washing stringency, and detection system. Thus optimal conditions for a given GST-SH2 or other GST fusion probe should be customized empirically. In general, weak signal can be enhanced by decreased concentration of

non-fat dry milk in blocking buffer, increased probe concentration, milder washing, more sensitive chemiluminescence kit, and longer exposure times.

31. If stripping is not sufficient with this method, try a more stringent condition using SDS-ME stripping buffer: briefly, immerse membrane at 50°C for 30 min under vigorous agitation, and then wash with TBST for 60 min with frequent shaking.

Acknowledgements

We are grateful to P. Nollau for his effort for the early development of the method, C. Thompson for technical assistance, K. Claffey for breast cancer samples, and A. Das for a bacterial culture system. This work was partially supported by grants from Breast Cancer Alliance and Connecticut Breast Health Initiative (to K.M.) and NIH grant CA107785 (to B.J.M.).

References

1. Edmondson, D. G., and Dent, S. Y. (2001) Identification of protein interactions by far western analysis, in: *Current Protocols in Protein Science* (Coligen, J. E., Ed.), Vol. 2, pp. 1–10, Wiley, Hoboken, NJ.
2. Hoeffler, J. P., Lustbader, J. W., and Chen, C. Y. (1991) Identification of multiple nuclear factors that interact with cyclic adenosine 3′,5′-monophosphate response element-binding protein and activating transcription factor-2 by protein–protein interactions. *Mol Endocrinol* 5, 256–266.
3. Akiyama, T., Ohuchi, T., Sumida, S., Xu, S. Q., and Toyoshima, K. (1992) Phosphorylation of the anti-oncogene products and control of the cell cycle. *Tohoku J Exp Med* 168, 153–157.
4. Hall, R. A. (2004) Studying protein–protein interactions via blot overlay or Far western blot. *Methods Mol Biol* 261, 167–174.
5. Cicchetti, P., Mayer, B. J., Thiel, G., and Baltimore, D. (1992) Identification of a protein that binds to the SH3 region of Abl and is similar to Bcr and GAP-rho. *Science* 257, 803–806.
6. Macgregor, P. F., Abate, C., and Curran, T. (1990) Direct cloning of leucine zipper proteins: Jun binds cooperatively to the CRE with CRE-BP1. *Oncogene* 5, 451–458.
7. Blackwood, E. M., and Eisenman, R. N. (1991) Max: a helix-loop-helix zipper protein that forms a sequence-specific DNA-binding complex with Myc. *Science* 251, 1211–1217.
8. Nollau, P., and Mayer, B. J. (2001) Profiling the global tyrosine phosphorylation state by Src homology 2 domain binding. *Proc Natl Acad Sci USA* 98, 13531–13536.
9. Machida, K., Thompson, C. M., Dierck, K., Jablonowski, K., Karkkainen, S., Liu, B., Zhang, H., Nash, P. D., Newman, D. K., Nollau, P., Pawson, T., Renkema, G. H., Saksela, K., Schiller, M. R., Shin, D. G., and Mayer, B. J. (2007) High-throughput phosphotyrosine profiling using SH2 domains. *Mol Cell* 26, 899–915.
10. Moorhead, G., and MacKintosh, C. (2004) Affinity methods for phosphorylation-dependent interactions. *Methods Mol Biol* 261, 469–478.

11. Burnham, M. R., DeBerry, R., and Bouton, A. H. (2001) Detection of phosphorylation-dependent interactions by far-western gel overlay. *Methods Mol Biol* 124, 209–220.
12. Einarson, M. B., and Orlinick, J. R. (2002) Identification of protein–protein interactions with glutathione-s-transferase fusion proteins, in: *Protein–Protein Interactions* (Golemis, E., Ed.), pp. 37–57, Cold Spring Harbor Press, Cold Spring Harbor, NY.
13. Smith, D. B., and Johnson, K. S. (1988) Single-step purification of polypeptides expressed in Escherichia coli as fusions with glutathione S-transferase. *Gene* 67, 31–40.
14. Mayer, B. J., Jackson, P. K., Van Etten, R. A., and Baltimore, D. (1992) Point mutations in the abl SH2 domain coordinately impair phosphotyrosine binding in vitro and transforming activity in vivo. *Mol Cell Biol* 12, 609–618.
15. Babon, J. J., McManus, E. J., Yao, S., DeSouza, D. P., Mielke, L. A., Sprigg, N. S., Willson, T. A., Hilton, D. J., Nicola, N. A., Baca, M., Nicholson, S. E., and Norton, R. S. (2006) The structure of SOCS3 reveals the basis of the extended SH2 domain function and identifies an unstructured insertion that regulates stability. *Mol Cell* 22, 205–216.
16. Lamla, T., Hoerer, S., and Bauer, M. M. (2006) Screening for soluble expression constructs using cell-free protein synthesis. *Int J Biol Macromol* 39, 111–121.
17. Welsh, M., Mares, J., Karlsson, T., Lavergne, C., Breant, B., and Claesson-Welsh, L. (1994) Shb is a ubiquitously expressed Src homology 2 protein. *Oncogene* 9, 19–27.
18. Jones, R. B., Gordus, A., Krall, J. A., and MacBeath, G. (2006) A quantitative protein interaction network for the ErbB receptors using protein microarrays. *Nature* 439, 168–174.

Chapter 35

On-Membrane Tryptic Digestion of Proteins for Mass Spectrometry Analysis

Jose L. Luque-Garcia and Thomas A. Neubert

Summary

Identification of proteins and characterization of posttranslational modifications are crucial steps for many biological, biochemical, and biomedical studies, and mass spectrometry has become the method of choice for these analyses. Here we describe two methods for the on-membrane digestion of proteins electroblotted onto nitrocellulose membranes prior to analysis by mass spectrometry. These on-membrane methods take approximately half the time of in-gel digestion and provide better digestion efficiency, due to the better accessibility of the protease to the proteins adsorbed onto the nitrocellulose, and better protein sequence coverage, especially for membrane proteins where large and hydrophobic peptides are commonly present.

Key words: On-membrane digestion, Mass spectrometry, Nitrocellulose membranes, Tryptic digestion, Protein identification, Protein coverage, Membrane proteins

1. Introduction

Identification of gel-separated proteins is commonly followed by in-gel enzymatic digestion of the proteins immobilized in the gel followed by analysis of the digested peptides by mass spectrometry *(1)*. However, the effectiveness of in-gel digestion can be compromised by (a) the limited accessibility of the enzyme or the chemical cleavage to the proteins embedded into the gel and (b) the low recovery of large and/or hydrophobic digested peptides from the gel *(2, 3)*.

Different alternatives based on the electroblotting of the proteins in a gel onto nitrocellulose or poly(vinyldifluoride) (PVDF) membranes followed by on-membrane digestion have been developed

Biji T. Kurien and R. Hal Scofield (eds.), *Methods in Molecular Biology, Protein Blotting and Detection, vol. 536*

DOI: 10.1007/978-1-59745-542-8_35

over the years *(4, 5)*. One of the advantages of electroblotting the proteins is that after the transfer, the proteins are adsorbed onto the membrane and therefore, they are more available to the proteases, thus allowing more efficient digestions *(2)* and solving one of the limitations associated with in-gel digestion.

The peptides generated by on-membrane digestion can be analyzed by MS using several different approaches (a) They can be directly analyzed on the membrane by MALDI-MS *(6–8)*. In this case the matrix solution used to dissolve the nitrocellulose-bound peptides is crucial in order for the peptides to be incorporated properly into the matrix crystals. (b) The peptides can be extracted from the membrane *(9, 10)*. This approach, as for in-gel digestion, has an extraction step that may result in low recovery of large and/or hydrophobic peptides due to the strong protein-membrane binding *(11)*. (c) They can be dissolved together with the membrane. In this approach, nitrocellulose membranes are used exclusively. It is based on the dissolution of the nitrocellulose in an organic solvent and thus, of the peptides adsorbed onto it. This last approach overcomes the limitations associated with in-gel digestion mentioned before because it bypasses the low-yield step of peptide extraction. This approach was first described by Liang et al. *(12)* using MALDI matrix solution prepared in 100% acetone to dissolve the nitrocellulose membrane followed by MS analysis of the digested peptides. However, the 100% acetone used to dissolve the nitrocellulose can lead to partial peptide precipitation resulting in loss of peptides and thus, decreased sensitivity.

Here we describe two methods *(2, 13)* in which most of the steps involved in the procedure are identical (those steps have been described just once in the text); the methods only vary in the way in which the peptides generated by on-membrane digestion are recovered for MS analysis. The "direct dissolution method" is an improvement of the method described by Liang et al. *(12)* and it uses a mixture of acetonitrile:methanol instead of 100% acetone to prepare the MALDI matrix solution, therefore minimizing protein loss most likely due to precipitation. On the other hand, in the "nitrocellulose-free method," the nitrocellulose is removed from the sample before MS analysis allowing the possibility of using (LC)-ESI-based mass spectrometers and increasing the sensitivity.

The methods described here are suitable for either soluble or membrane proteins although they are especially recommended for the study of the latter, which can be very problematic due to the presence of hydrophobic domains. It is also noteworthy that the on-membrane methods take approximately half the time of conventional in-gel digestion, and can provide more efficient tryptic digestions, thus producing fewer missed cleavage peptides.

2. Materials

2.1. SDS-PAGE and Electroblotting

1. Separating buffer (10×): 1.5 M Tris–HCl, pH 8.8, 1% SDS. Store at room temperature (RT) (*see* **Note 1**).
2. Stacking buffer (10×): 0.5 M Tris–HCl, pH 6.8, 1% SDS. Store at RT.
3. Thirty percent acrylamide/bis solution (at a 29:1 acrylamide monomer:cross-linker ratio) (Bio-Rad, Hercules, CA). Unpolymerized acrylamide is a neurotoxin so care should be taken to prevent skin contact. Store at 2–8°C.
4. *N,N,N,N'*-Tetramethyl-ethylenediamine (TEMED). Store at RT in a desiccator.
5. Ammonium persulfate: prepare 10% solution in water and immediately freeze in single use aliquots at –20°C.
6. Isopropanol. Store at RT.
7. Laemmli sample buffer (Bio-Rad, Hercules, CA, USA). Store at RT.
8. Running buffer (5×): 125 mM Tris, 960 mM glycine, 0.5% SDS, pH 8.3. Store at RT.
9. Kaleidoscope prestained molecular weight markers (Bio-Rad, Hercules, CA, USA). Store at –20°C.
10. 100% Triton-free nitrocellulose membranes (pore size, 0.2 μm) (Bio-Rad, Hercules, CA, USA) (*see* **Note 2**).
11. 3MM Whatman chromatography paper (Whatman International, UK).
12. Transfer buffer: 25 mM Tris, 190 mM glycine, 20% methanol, 0.05% SDS, pH 8.3. Store at RT.

2.2. Membrane Staining and Destaining

2.2.1. Ponceau-S Staining

1. Ponceau-S solution: 5% acetic acid, 0.2% Ponceau-S (w/v) (Sigma, St. Louis, MO). Store at RT.

2.2.2. MemCode™ Staining

1. MemCode™ reversible protein stain kit for nitrocellulose membranes (Pierce, Rockford, IL). Store at RT.

2.3. Blocking of the Membrane Before Digestion

1. PVP-40 solution: 100 mM acetic acid, 0.5% poly-(vinylpyrrolidone) (PVP-40) (w/v) (Sigma, St. Louis, MO, USA).

2.4. On-Membrane Digestion

1. Digestion buffer: 50 mM ammonium bicarbonate, pH 7.8. Use freshly prepared buffer.
2. Trypsin Gold, Mass Spectrometry Grade (Promega, Madison, WI). Store the lyophilized powder at –20°C. Reconstitute in

50 mM ammonium bicarbonate (pH 7.8) and store in single use aliquots at −20°C (*see* **Note 3**).

2.5. Recovery of Peptides

1. HPLC-grade acetone, acetonitrile, and methanol (Fisher Scientific, Morris Plains, NJ, USA).

2.6. Mass Spectrometry Analysis

2.6.1. MALDI-TOF

1. α-cyano-4-hydroxycinnamic acid (α-CHCA) (Sigma). Store at −20°C (*see* **Note 4**).
2. Trifluoroacetic acid (TFA) (Sigma). Store at RT

2.6.2. (LC)-ESI-MS/MS

1. Sample solution: 2% acetonitrile, 0.1% formic acid

3. Methods

As with any sample preparation method for MS analysis, special attention must be paid to avoid loss or contamination of the samples during processing. Please be very careful not to use bare hands, loose hair, dirty glass- and plastic-ware, and always wear gloves (powder free and rinsed with water and ethanol before use) to minimize contamination by keratins, or these proteins will overwhelm low level protein samples and preclude successful analysis of the proteins of interest.

3.1. SDS-PAGE and Electroblotting

1. Prepare a 1-mm thick 10% gel (8 × 7.3 cm) by mixing 1 mL of separating buffer with 1 mL acrylamide/bis solution, 1.94 mL water, 15 μL ammonium persulfate solution, and 5 μL TEMED. Pour the gel, leaving space for the stacking gel, and overlay with isopropanol. The gel should be polymerized in about 30 min.
2. Pour off the isopropanol and rinse the top of the gel with water.
3. Prepare the stacking gel by mixing 0.5 mL of the stacking buffer with 0.2 mL acrylamide/bis solution, 1.272 mL water, 7.5 μL ammonium persulfate, and 2.5 μL TEMED. Use about 0.5 mL of this to quickly rinse the top of the gel and then pour the stacking solution and insert the comb. The gel should polymerize overnight to eliminate cross-linking of unpolymerized acrylamide to proteins during electrophoresis (*see* **Note 5**).
4. Prepare the running buffer by diluting 200 mL of the 5× running buffer with 800 mL of water.
5. Mix the sample with the Laemmli sample buffer 1:1, vortex, centrifuge and incubate at 95°C for 5 min to denature the proteins.
6. Add the running buffer to the chambers of the gel unit and carefully remove the comb. Load the sample in a well. Include one well for prestained molecular weight markers.

7. Complete the assembly of the gel unit and connect to a power supply. Run the gel at 50 V until all the sample enters the gel (around 20 min) and then increase voltage to 80–100 V (maintain below 20 mA) until the dye front reaches the bottom of the gel.
8. Prepare a tray of water large enough to lay out a transfer cassette with its two pieces of foam. Cut a sheet of nitrocellulose slightly larger than the size of the separating gel.
9. Wet a sheet of Whatman chromatography paper of approximately the same size as the nitrocellulose membrane and place it on top of one of the wet foam sheets. Then submerge the nitrocellulose sheet on top of the Whatman paper.
10. Disassemble the gel unit. Remove the stacking gel and discard. After briefly rinsing the separating gel with water, place it on top of the nitrocellulose membrane.
11. Wet an additional sheet of Whatman paper and carefully place it on top of the gel, ensuring no bubbles are trapped in the resulting sandwich. Place the second wet foam sheet on top and close the transfer cassette.
12. Place the cassette into the transfer tank such that the nitrocellulose membrane is between the gel and the anode. It is very important to ensure this orientation to avoid the loss of the proteins into the transfer buffer instead of being electrotransferred to the nitrocellulose membrane.
13. Place the transfer tank into an ice bath, put on the lid and activate the power supply. Carry out the transfer at 100 V for 1 h.
14. Once the transfer is complete, disassemble the casette, discard the gel and the Whatman sheets. Transfer the nitrocellulose membrane to a clean tray and rinse briefly with water. The colored molecular weight markers should be clearly visible on the membrane.

3.2. Membrane Staining and Destaining

Either of two different protein stains, both of them reversible and MS-compatible, can be used. While Ponceau-S is less expensive, it has a limit of detection of ~250 ng of protein on the membrane. MemCode™ is more sensitive, with a limit of detection of ~25 ng of protein on the membrane. Thus, the staining procedure should be selected according to the concentration of the protein/s of interest.

3.2.1. Ponceau-S Staining

1. Add 10 mL of Ponceau-S solution to the membrane. Agitate at RT for 1–2 min on a rotary platform shaker or until the stained proteins appear as red bands (*see* **Note 6**).
2. Remove the Ponceau-S solution and wash the membrane with water several times until the background staining is completely removed.

3. Excise the band/s of interest with a new clean scalpel or razor blade and transfer to clean 1.5-mL eppendorf tubes.

Wash several times with 1 mL of water for 15 min with agitation (1,000 rpm) on a thermomixer rotating platform at RT until the stain is completely removed. (If not digested immediately, membrane samples should be stored at –20°C in water.)

3.2.2. MemCode™ Staining

1. Rinse the nitrocellulose membrane with water and quickly decant (*see* **Note 6**).
2. Add ~25 mL of MemCode™ reversible protein stain to the membrane. Agitate for 30 s on a platform shaker at RT. Stained proteins appear as turquoise-blue bands.
3. To remove the background, add ~25 mL MemCode™ destaining reagent to the nitrocellulose membrane and quickly decant. Repeat this step two additional times.
4. Add ~25 mL of the MemCode™ destain reagent to the membrane and gently agitate for 5 min on a rotary platform shaker at RT.
5. Decant the destaining solution and rinse the membrane four times by adding water to the tray and quickly decanting.
6. Wash the membrane with water for 5 min with agitation.
7. Excise the band/s of interest with a new clean scalpel or razor blade and transfer to 1.5-mL eppendorf tubes. Add 1 mL of MemCode™ stain eraser to the membrane and agitate for 2 min on a thermomixer at 1,000 rpm and room temperature (2 min of agitation with stain eraser is optimal for most proteins but it may be extended to 5 min).
8. Rinse the membrane four times by adding 1 mL of water and quickly decanting.
9. Wash the membrane with water for 5 min with agitation using a thermomixer at RT and 1,000 rpm . (If not digested immediately, samples should be stored at –20°C in water.)

3.3. Blocking of the Membrane Before Digestion

1. Remove the water and add 1 mL of PVP-40 solution to block nonspecific protein (trypsin) binding sites on the nitrocellulose. Incubate at 37°C for 30 min with gentle agitation using a thermomixer at 300 rpm .
2. Wash the membrane 6–10 times with 1 mL of water to remove excess PVP-40.
3. Transfer the nitrocellulose membranes to a 200-μL eppendorf tube (PCR style).

3.4. On-Membrane Digestion

1. Add 20 μL of trypsin at 12.5 ng/μL in 50 mM ammonium bicarbonate buffer and add enough digestion buffer to completely cover the membrane bands (~20 μL).

2. Incubate at 37°C for 5 h or overnight with gentle agitation using a thermomixer at 300 rpm. Cap the tubes tightly and cover with parafilm to avoid evaporation.
3. After completing the digestion, dry the samples under vacuum.

3.5. Recovery of Peptides

There are two different methods for recovering the peptides for MS analysis. The first one (direct dissolution method) is based on the direct dissolution of the nitrocellulose membrane, and thus of the peptides adsorbed onto it, in the MALDI matrix solution. This method can be used only for MALDI-based mass spectrometers. The second method (nitrocellulose-free method) is based on the removal of the nitrocellulose before MS analysis. This second method takes longer but it can be used for either MALDI- or (LC)-ESI-based mass spectrometers.

3.5.1. Direct Dissolution Method

1. Prepare the MALDI matrix solution consisting of 10 mg of α-CHCA, 300 μL of methanol, 700 μL of acetonitrile and 10 μL of TFA.
2. Add approximately 10 μL of MALDI matrix solution per mm^2 of nitrocellulose (40 μL for a typical 4 mm^2 protein band).
3. Vortex and sonicate for 10 min.
4. Spot 1 μL of the final solution onto the MALDI plate for MS analysis.

3.5.2. Nitrocellulose-Free Method

1. Add 90 μL of acetone per 4 mm^2 of nitrocellulose. Vortex and incubate for 30 min at RT to allow complete dissolution of the nitrocellulose and precipitation of the tryptic peptides adsorbed onto it.
2. Centrifuge 10 min at 14,000 × *g* and carefully remove the acetone containing the dissolved nitrocellulose. Air-dry the precipitated peptides.
3. Resuspend the peptides. For MALDI-MS analysis, add 20 μL of MALDI matrix solution (10 mg/mL of α-CHCA prepared in 50:50 acetonitrile:water and 1% TFA). For LC-MS/MS analysis, add 20 μL of sample solution (2% acetonitrile, 0.1% formic acid).
4. Sonicate 10 min.

3.6. Mass Spectrometry Analysis

3.6.1. MALDI-TOF

1. Calibrate the instrument in reflectron mode using molecular weight standard peptides such as angiotensin I (average mass, 1,296.5 Da) and corticotropin-like intermediate lobe peptide (ACTH clip 18–39, average mass, 2,465.7).
2. Analyze the samples of interest. Typically, 100–200 laser shots are summed into each mass spectrum. Mass spectra should be

acquired in an interactive (manual) mode that allows modifying the laser energy and other key acquisition parameters as well as the visual selection of the sample area targeted to obtain a good signal-to-noise ratio (*see* **Note 7**). Examples of results obtained using this method as compared to in-gel digestion are shown in **Figs.1** and **2**.

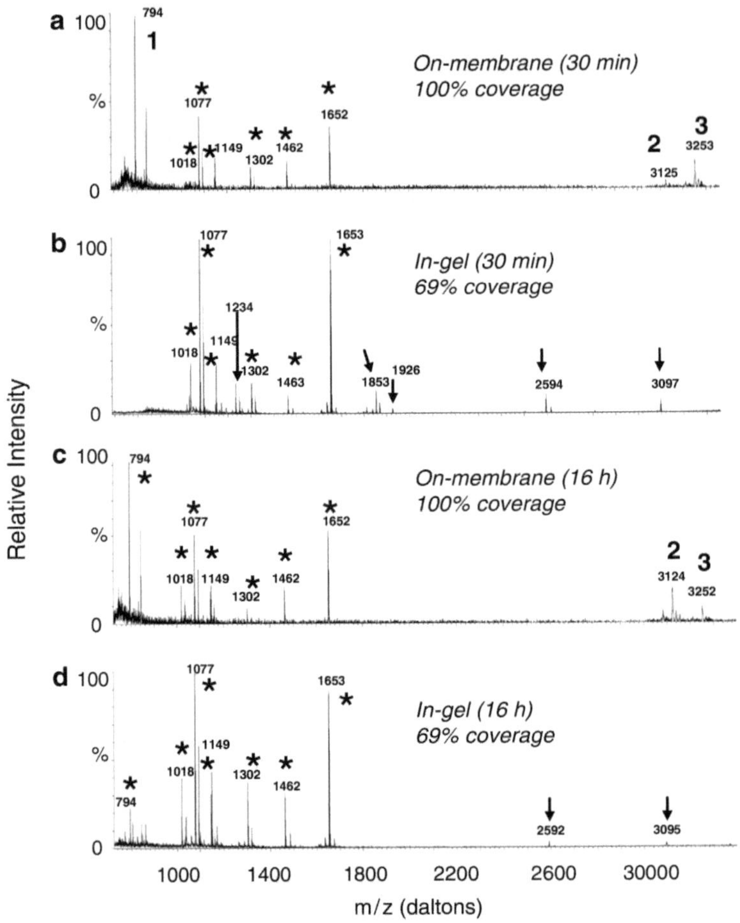

e ELVSVVDSGSGFTVTRLSAYQVTNLAPGTKYYISY
LVTK**GASTESSR**[1]EIPMSTFPRRKAESIGLAMAR**TG**
GMVVITVLLSVAMFLLVLGLIIALALGAR[2] **K**[3]

Fig. 1. Comparison of MALDI MS spectra obtained from 10 pmol of uroplakin II (UPII) after (**a**) 30 min on-membrane digestion, (**b**) 30 min in-gel digestion, (**c**) 16 h on-membrane digestion, (**d**) 16 h in-gel digestion, (**e**) amino acid sequence of bovine mature UPII. The *underlined* amino acids correspond to the transmembrane domain of the protein. In the spectra, the *stars* indicate UPII peptides detected after both in-gel and on-membrane digestion, the *arrows* indicate missed cleavage peptides that appear only after in-gel digestion, and the *numbers* indicate peptides from the UPII sequence shown in (**e**) detected only after on-membrane digestion (reproduced from *(2)* with permission from the American Chemical Society).

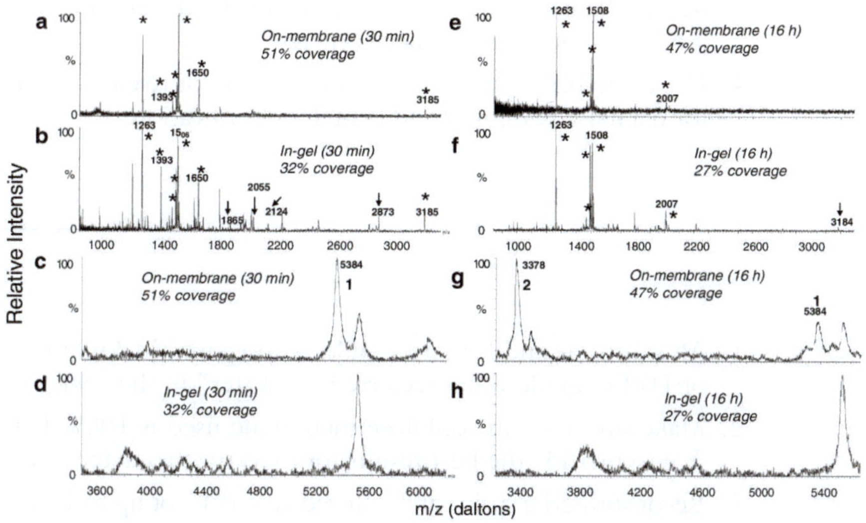

i MPPLWVVLALGCLRLGSGVNLQPQLASVTFATNNPTLTTVALEKPLCMFDSSAALHGTYEVYLYVLVDSASFRNASVQDST KTPLSSTFQQTQGGRTGPYKAAAFDLTPCSDSPSLDAVRDVSRASEILNAYLIRVGTNGTCLLDPNFQGLCNPPLSAATEYR FKYVLVNMSSGLVQDQTLWSDPIRTDRLTLYSAIDTWPGRR**SGGMIVITSILGSLPFFLLIGFAGAIVLSLVDR**[2] **GDADGATS HDSQITQEAVPK**[1] SLGTSEPSYTSVNRGPSLDRAEVYASKLQD

Fig. 2. Comparison of MALDI MS spectra obtained from 3 pmol of uroplakin III (UPIII) after (**a**, **c**) 30 min on-membrane digestion, (**b**, **d**) 30 min in-gel digestion, (**e**, **g**) 16 h on-membrane digestion, (**f**, **h**) 16 h in-gel digestion, (**i**) amino acid sequence of bovine UPIII. The *underlined* amino acids correspond to the transmembrane domain of the protein. In the spectra, the *stars* indicate UPII peptides detected after both in-gel and on-membrane digestion, the *arrows* indicate missed cleavage peptides that appear only after in-gel digestion, and the *numbers* indicate peptides from the UPII sequence shown in (**i**) detected only after on-membrane digestion. Spectra shown in (**a**), (**b**), (**e**), and (**f**) were collected in reflectron mode and spectra shown in (**c**), (**d**), (**g**), and (**h**) in linear mode (reproduced from *(2)* with permission from the American Chemical Society).

3.6.2. (LC)-ESI-MS/MS

1. The sample is transferred to a sample vial compatible with the autosampler of the liquid chromatograph, or injected manually into the HPLC. Some remaining nitrocellulose may precipitate in the presence of water in the final solution. Transfer the sample carefully to the sample vial avoiding the transfer of the nitrocellulose which otherwise could cause clogging of the LC system (*see* **Note 8**).
2. The following instructions assume the use of a CapLC (Waters) HPLC system coupled directly to a Q-TOF micro (Micromass) mass spectrometer or similar instruments. Load the peptides onto a 0.3 × 1-mm C18 nano-precolumn and wash for 5 min with 2% acetonitrile in 0.1% formic acid at a flow rate of 20 μL/min.
3. After washing, reverse the flow through the precolumn and elute the peptides with a gradient of 2–90% acetonitrile in 0.1% formic acid. The gradient is delivered over 120 min at a flow rate of 200 nL/min, using a 15:1 precolumn flow split, through a 75 μm × 15 cm fused silica capillary C18 HPLC column to a fused silica distal end-coated tip nanoelectrospray

needle. The electrospray needle can vary depending on the ion source of the mass spectrometer.

4. Use a database search engine such as Mascot (Matrix Science) for database searching and protein identification.

4. Notes

1. All solutions and buffers should be prepared in MilliQ water or HPLC-grade water (referred as "water" in this text).
2. Make sure the nitrocellulose membrane used is 100% Triton-free to provide the best results for mass spectrometry analysis.
3. Reconstituted trypsin can be stored at –20°C for up to 1 month. For long-term storage, freeze reconstituted trypsin at –70°C. Before use, thaw the reconstituted trypsin at room temperature, placing on ice immediately after thawing. To maintain sufficient enzymatic activity, limit the number of free-thaw cycles to 5.
4. If the dry matrix has a mustard-yellow color instead of bright yellow color, it may contain impurities. To recrystallize (purify), dissolve the CHCA matrix in warm ethanol (exact temperature is not important). Filter and add two volumes of deionized water. Let stand overnight in an explosion proof-refrigerator. Filter and wash the precipitate with cold water.
5. Unpolymerized acrylamide cross-linked to proteins may interfere with tryptic digestion. Thus, gels should be prepared the day before the experiment is carried out. Gels can be stored at 4°C for several days wrapped in a sturdy lab tissue paper (such as Kimwipes) wetted with running buffer and sealed with plastic wrap to prevent drying.
6. For all steps, use sufficient volumes to completely immerse the membrane, and agitate at a moderate speed on a rotary platform shaker. Do not allow the membrane to become dry during the procedure.
7. Usually, the large amount of dissolved nitrocellulose in the sample creates a heterogeneous crystal. Best signal is usually obtained in those areas with the least nitrocellulose. Those with high levels of nitrocellulose require greater laser energy to provide a good signal. Also, after selecting an area, the first shots may provide a low signal-to-noise ratio that will increase with the number of shots.
8. To ensure that remaining traces of nitrocellulose do not reach the analytical column, we strongly recommend the use of a precolumn between the injection valve and the analytical column in the LC system.

Acknowledgements

We acknowledge support from NIH grants P30 NS050276 and P30 CA016087 to T.A.N., and the Ramón y Cajal Program (Ministry of Science and Technology, Spain) for support to J.L.L.-G.

References

1. Shevchenko, A., Wilm, M., Vorm, O., and Mann, M. (1996) Mass spectrometric sequencing of proteins silver-stained polyacrylamide gels. *Anal Chem* 68, 850–858.
2. Luque-Garcia, J. L., Zhou, G., Sun, T. T., Neuber, T. A. (2000) Use of nitrocellulose membranes for protein characterization by matrix-assisted laser desorption/ionization mass spectrometry. *Anal Chem* 7, 5102–5108.
3. Jonsson, A. P., Aissouni, Y., Palmberg, C., Percipalle, P., Nordling, E., Daneholt, B., Jornvall, H., Bergman, T. (2001) Recovery of gel-separated proteins for in-solution digestion and mass spectrometry. *Anal Chem* 73, 5370–5377.
4. Bai, J., Qian, M. G., Liu, Y., Liang, X., and Lubman, D. M. (1995) Peptide mapping by CNBr degradation on a nitrocellulose membrane with analysis by matrix-assisted laser desorption/ionization mass spectrometry. *Anal Chem* 67, 1705–1710.
5. Dukan, S., Turlin, E., Biville, F., Bolbach, G., Touati, D., Tabet, J. C., et al (1998) Coupling 2D SDS-PAGE with CNBr cleavage and MALDI-TOF-MS: a strategy applied to the identification of proteins induced by a hypochlorous acid stress in Escherichia coli. *Anal Chem* 70, 4433–4440.
6. Schleuder, D., Hillenkamp, F., and Strupat, K. (1999) IR-MALDI-mass analysis of electroblotted proteins directly from the membrane: comparison of different membranes, application to on-membrane digestion, and protein identification by database searching. *Anal Chem* 71, 3238–3247.
7. Strupat, K., Karas, M., Hillenkamp, F., Eckerson, C., and Lottspelch, F. (1994) Matrix-assisted laser desorption ionization mass spectrometry of proteins electroblotted after polyacrylamide gel electrophoresis. *Anal Chem* 66, 464–470.
8. Vestling, M. M., and Fenselau, C. (1994) Poly(vinylidene difluoride) membranes as the interface between laser desorption mass spectrometry, gel electrophoresis, and in situ proteolysis. *Anal Chem* 66, 471–477.
9. van Oostveen, I., Ducret, A., and Aebersold, R. (1997) Colloidal silver staining of electroblotted proteins for high sensitivity peptide mapping by liquid chromatography-electrospray ionization tandem mass spectrometry. *Anal Biochem* 247, 310–318.
10. Bunai, K., Nozaki, M., Hamano, M., Ogane, S., Inoue, T., Nemoto, T., et al. (2003) Proteomic analysis of acrylamide gel separated proteins immobilized on polyvinylidene difluoride membranes following proteolytic digestion in the presence of 80% acetonitrile. *Proteomics* 3, 1738–1749.
11. Jorgensen, C. S., Jagd, M., Sorensen, B. K., McGuire, J., Barkholt, V., Hojrup, G. (2004) Efficacy and compatibility with mass spectrometry of methods for elution of proteins from sodium dodecyl sulfate-polyacrylamide gels and polyvinyldifluoride membranes. *Anal Biochem* 330, 87–97.
12. Liang, X., Bai, J., Liu, Y. H., and Lubman, D. M. (1996) Characterization of SDS--PAGE-separated proteins by matrix-assisted laser desorption/ionization mass spectrometry. *Anal Chem* 68, 1012–1018.
13. Luque-Garcia, J. L., Zhou, G., Spellman, D. S., Sun, T. T., and Neubert, T. A. (2008) Analysis of electroblotted proteins by mass spectrometry: Protein identification after western blotting. *Mol Cell Proteomics* 7, 308–314.

Chapter 36

Western Blot Analysis of Adhesive Interactions under Fluid Shear Conditions: The Blot Rolling Assay

Robert Sackstein and Robert Fuhlbrigge

Summary

Western blotting has proven to be an important technique in analysis of receptor-ligand interactions (i.e., by ligand blotting) and for identifying molecules mediating cell attachment (i.e., by cell blotting). Conventional ligand blotting and cell blotting methods employ nondynamic (static) incubation conditions, whereby molecules or cells of interest are placed in suspension and overlaid on membranes. However, many cell–cell and cell–matrix adhesive interactions occur under fluid shear conditions, and shear stress itself mediates and/or facilitates the engagement of these physiologically appropriate receptors and ligands. Notably, shear forces critically influence the adhesion of circulating cells and platelets to vessel walls in physiologic cell migration and hemostasis, as well as in inflammatory and thrombotic disorders, cancer metastasis, and atherosclerosis. Use of nondynamic blotting conditions to analyze such interactions can introduce bias, overtly missing relevant effectors and/or exaggerating the relative role(s) of nonphysiologic adhesion molecules. To address this shortfall, we have developed a new technique for identifying binding interactions under fluid shear conditions, the "blot rolling assay." Using this method, molecules in a complex mixture are resolved by gel electrophoresis, transferred to a membrane that is rendered semi-transparent, and the membrane is then incorporated into a parallel-plate flow chamber apparatus. Under controlled flow conditions, cells or particles bearing adhesion proteins of interest are then introduced into the chamber and interactions with individual immobilized molecules (bands) can be visualized in real-time. The substrate molecule(s) supporting adhesion under fluid shear can then be identified by staining with specific antibodies or by excising the relevant band(s) and performing mass spectrometry or microsequencing of the isolated material. This method thus allows for the identification, within a complex mixture and without prior isolation or purification, of both known and previously uncharacterized adhesion molecules operational under dynamic conditions.

Key words: Adhesion molecules, Shear stress, Shear conditions, Fluid shear, Flow conditions, Parallel plate flow chamber, Western blot, Selectins, Selectin ligands, Cell blotting, Blot rolling assay, Gel electrophoresis

Biji T. Kurien and R. Hal Scofield (eds.), *Methods in Molecular Biology, Protein Blotting and Detection, vol. 536*

DOI: 10.1007/978-1-59745-542-8_36

1. Introduction

Adhesive interactions of cells with other cells and/or extracellular matrix under fluid shear conditions are critical to a variety of physiologic and pathobiologic processes, including hemostasis, leukocyte trafficking, tumor metastasis, and atherosclerosis. Although various techniques for analyzing cell adhesion have been described, most involve binding assays under nondynamic (static) conditions. Importantly, certain adhesive receptor–ligand interactions occur preferentially if not solely under physiologic shear stress (e.g., binding of L-selectin to its ligands) or depend on low affinity and rapidly reversible interactions to serve their functions (e.g., rolling of leukocytes via selectins). Under static binding assay conditions, effectors mediating these types of interactions may be overtly neglected or overshadowed by molecules specialized to form more stable adhesions.

Methods for in vitro study of adhesion under dynamic conditions have been described, including the Stamper–Woodruff *(1)* and the parallel plate flow chamber *(2)* assays. However, these methods are limited in that they require the availability of reagents (e.g., antibodies) that can specifically interfere with the activity of relevant receptors and/or ligands or require isolated substrate materials that can be immobilized on the chamber surface. Thus, the applicability of these methods to examine the structure or function of previously unrecognized/uncharacterized ligands, especially within a complex mixture, is highly constrained.

To address these issues, we developed a method for direct real-time observation of adhesive interactions between cells or particles in flow and proteins separated by SDS-PAGE and immobilized on membranes *(3)*. This technique, which we have termed the "blot rolling assay," has allowed for the identification of new glycoprotein ligands *(4–6)*, as well as a glycolipid ligand *(7)* for selectins. This method allows real-time assessment and measurement of interaction parameters (e.g., rolling vs. firm attachment, specificity, and reversibility with inhibitors) in both physiologic and nonphysiologic shear conditions, thus permitting a unique user interface for the observation of adhesive events on membrane-immobilized materials. Ligands under investigation can be immobilized directly or segregated by gel electrophoresis (e.g., SDS-PAGE, isoelectric focusing) or other methods prior to transfer to the membrane. This method also provides for real-time manipulation of interaction conditions including wall shear stress, ion requirements, temperature, the influence of metabolic inhibitors, and the presence of activating agents or inhibitors of cell function. Blot-immobilized substrates can be used repeatedly, allowing in situ manipulation of the substrate under continuous direct visualization or direct comparison of different conditions or different cell populations in shear flow. The capacity to observe sequential experimental and control conditions on a single substrate, and to observe physiologic behaviors and responses to manipulations in real-time, provides particularly powerful advantages of this method over conventional static binding assays. Similarly, the ability to observe interaction with individual components of a complex mixture without requiring prior purification or knowledge of the nature of the components provides advantages over conventional flow-based assays. In this chapter, we will describe a detailed protocol for performing blot rolling assays of membrane proteins. A video recording of a representative experiment will be presented to illustrate the utility of this new analytic technique.

2. Materials

The required components for performing blot rolling assays include an inverted microscope and a parallel plate flow chamber apparatus. **Figure 1** shows the setup utilized in our laboratories. The parallel plate device we use has internal chamber diameters of 2 cm × 0.5 cm × 0.025 cm. A commercial product with similar properties is available from GlycoTech Corp., Rockville, MD, USA (Circular flow chamber, Product #31–001). Flow is adjusted

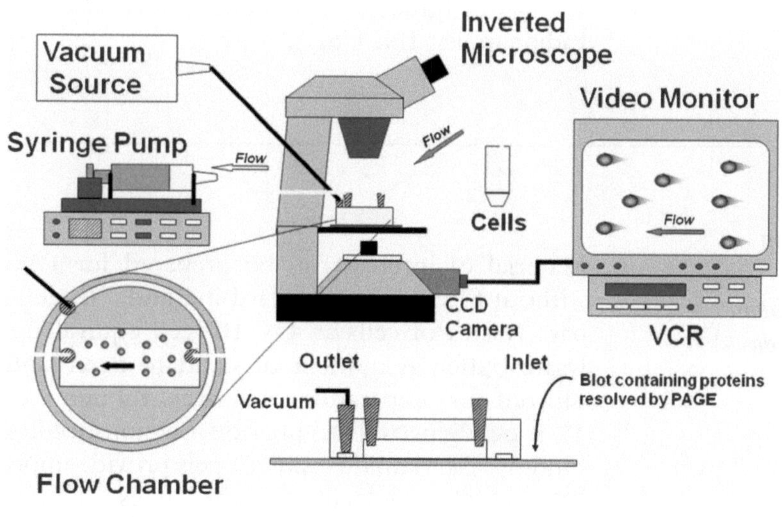

Fig. 1. Diagram of blot rolling assay assembly. The components of the blot rolling assay are shown, including the inverted microscope, flow chamber, CCD camera, syringe pump, vacuum source, video monitor, and blot membrane (resting on a Petri dish located on the microscope stage, with the flow chamber positioned on top of it).

by negative pressure regulated by a high precision, programmable syringe pump (Harvard Apparatus). Wall shear stress (dyne/cm^2) is calculated according to the formula: Wall Shear Stress (τ) = $6\mu Q/wh^2$, where w is the chamber width in cm, h the chamber height in cm, Q the volumetric flow rate in mL/s, and μ is the viscosity in poise, or dyne × s/cm^2; for water at 25°C, $\mu = 0.009$, which is adjusted for contribution of 10% glycerol (*see* below), according to the equation $\mu = 1.37 \times$ medium ($\mu = 0.0123$ at 25°C) (see below for calculations based on our chamber dimensions). Interaction events can be captured on videotape by using a standard CCD camera and video recording assembly (*see* **Fig. 1**), for later off-line analysis.

1. PAGE gel system.
2. equi-Blot™ PVDF Membrane (Bio-Rad, Hercules, CA, USA, #162–0184).
3. Fetal bovine serum.
4. Phosphate buffered saline/0.1% Tween-20 (PBST).
5. Prestained molecular weight markers.
6. Western Blue stabilized substrate for alkaline phosphatase (Promega #S3841).
7. Blot rolling medium (H/H Medium): Hank's Balanced Salt Solution (Gibco #14170–112) with 10 mM HEPES, pH 7.4 (Cellgro #25–060). For "binding medium", add 2 mM $CaCl_2$ and 10% glycerol to H/H medium.
8. Monoclonal or polyclonal antibody reagents for identifying any epitope(s) of interest (*see* **Note 1**).
9. Acrylamide gradient gels (Criterion 4–20%; Bio-Rad).
10. 1% *n*-octylglucoside (Roche Molecular Biochemicals, Indianapolis, IN, USA).

3. Methods

3.1. Preparation of Western Blots/Visualization of Protein Bands

1. Material of interest can be prepared for PAGE (with or without SDS) using standard methods. In general, we prepare lysates of cells at 4 × 10^8 cell equivalents per mL of lysate buffer. A number of standard detergents have been utilized for preparation of lysates, though we have found 1% *n*-octylglucoside to provide low nonspecific background binding. Acrylamide gradient gels provide improved resolution in higher molecular weight ranges. The amount of protein required will vary with the material under study, but we have found 25–100 µg of protein per lane to be sufficient for most of our studies. Both reducing and nonreducing gel

conditions have been utilized with good results. Omission of detergent from the gel and/or the nondenaturing gel electrophoresis systems can be utilized to optimize structural integrity of resolved bands. In all gels, it is helpful to include prestained molecular weight standards in the lanes adjacent to the material of interest to assist in alignment and localization of regions supporting adherence (*see* **Notes 2** and **3**).

2. When gel electrophoresis is complete, the component proteins are transferred to PVDF membrane using standard transfer methods; although other membrane materials may work as well, we have preferred BioRad Sequi-Blot™, which has 0.2-μm pore size and minimal surface irregularities, thus reducing turbulence and promoting laminar flow over the membrane (*see* **Note 4**).
3. After transfer, the blot is placed protein side facing up (i.e., the side of the membrane that was facing the gel) in a blocking solution (50% FBS in PBST and 100% newborn calf serum have been used with good results) to reduce nonspecific interactions and incubated with gentle agitation (on a rocker platform or rotating platform) for 1 h at room temperature. The blot is then washed twice in PBST for 5 min each with gentle agitation.
4. Immunostaining, if desired, can be performed using standard procedures. Membranes are typically stained with primary antibody for 1 h under gentle agitation, washed 2× in PBS/0.1% Tween 20 for 5 min and then incubated with secondary alkaline phosphatase-conjugated antibody in TBS (Tris buffered saline) for 1 h under gentle agitation. The blot is again washed 2× in PBST for 5 min, then rinsed in PBS followed by TBS (to remove residual PBST), and then developed with alkaline phosphatase substrate (e.g., Western Blue™). After developing blots to an appropriate signal-to-background perspective, reactions are stopped with PBS and the blot is washed twice with PBS. Blots to be used for flow experiments may be stored in buffer at 4°C until use. Blots will maintain binding fidelity for several days, though best results are obtained with freshly prepared materials (*see* **Note 5**).
5. The membranes are again blocked by incubation in 50% newborn calf serum or FBS in binding media (H/H with Ca^{2+}/10% glycerol) for at least 1 h at 4°C and equilibrated in binding media prior to use in the flow assay.

3.2. Preparation of Selectin-Expressing Cells

To assay blots for selectin ligand activity, we typically employ lymphocytes expressing L-selectin or Chinese hamster ovary (CHO) cells stably transfected to express P-selectin or E-selectin.

1. Human peripheral blood lymphocytes (PBL) can be prepared from whole blood by Ficoll density gradient separation. For analysis of L-selectin function on lymphocytes, monocytes may be removed by plastic adherence or other methods to reduce nonspecific binding (e.g., magnetic bead separation). Murine spleen and lymph node T cells and rat thoracic duct lymphocytes have also been utilized in the same fashion. E-selectin and P-selectin transfected cell lines are typically split the day prior to use to achieve log phase growth and are harvested by replacing the media with H/H 5 mM EDTA. The cells can then be released from the flask by manual agitation or scraping. We do not use trypsin to release the cells as this may alter the function of some surface adhesion molecules.
2. After counting, the cells are suspended at 10× the concentration desired for use in the flow chamber (e.g., 10–20 × 10^6/mL in H/H without Ca^{2+} or Mg^{2+} and without glycerol) and stored on ice. Clumping of adherent cell populations (i.e., CHO cells) may develop after prolonged storage on ice. This can be remedied by washing again in H/H 5 mM EDTA before use (*see* **Note 6**).
3. Controls to confirm specificity for selectin binding to blots include use of mock-transfected CHO cells and/or preincubation of cells with function-blocking anti-selectin mAb.

3.3. Inverted Microscope Setup

1. Setup will vary with type of microscope. In general, we do not use any filters.
2. **Figure 2** shows the placement of the parallel plate chamber on the blot. Begin visualization at low power to align the chamber and identify the area of interest. Many observations are best performed at relatively low magnification (e.g., 10× objective) to allow observation of larger areas and numbers of cells. Observation under increased magnification can be performed to observe individual cell characteristics. Adjust the shutter arm so that the light narrows specifically over the area of interest on the blot.

3.4. Blot Rolling Assay

1. All media should contain 10% glycerol. Equilibration in dilute glycerol alters the opacity of the PVDF membrane sufficiently to allow transmission of light and the direct visualization of cells interacting with the surface of the blot by standard light microscopy. Assessment of some functions (e.g., integrin activation) requires maintaining the chamber and media at 37°C. This can be achieved with a stage warming device and placement of supply media in water baths adjacent to the apparatus (*see* **Note 7**).
2. The chamber should be mounted onto the blot so that the flow channel is aligned over the lane of interest. The flow

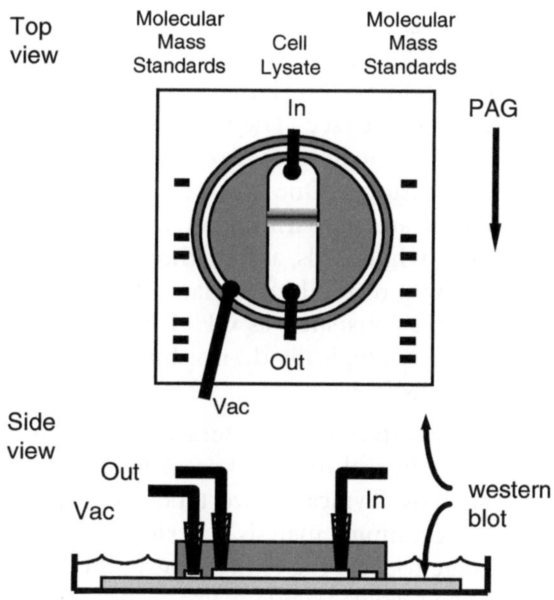

Fig. 2. Blot rolling apparatus. Western blots stained for components of interest are equilibrated in binding media containing 10% glycerol and placed in a Petri dish. A parallel plate flow chamber mounted over the area of interest is held in place by low continuous vacuum pressure (Vac). The inlet (In) is connected to a buffer reservoir and the outlet (Out) to a high precision syringe pump. The apparatus is mounted on the stage of an inverted phase-contrast microscope equipped for videomicroscopy. Molecular mass(es) of interacting protein(s) resolved in the polyacrylamide gel (PAG) are estimated by alignment with colored mass standards in adjacent lanes (reproduced with permission from *(3)*, Copyright 2002, The American Association of Immunologists, Inc.).

chamber is secured by application of low vacuum pressure in the same fashion as is used for attachment to glass or plastic surfaces. We have not found it necessary to perform any special maneuvers to maintain an adequate seal on Sequi-Blot PVDF membrane (*see* **Note 8**).

3. The flow field is rinsed with media introduced through the 20-mL syringe, attached to the three-way stopcock, to remove bubbles and prepare the blot surface for cell input (*see* **Note 9**).
4. Immediately before use, aliquots of cells (e.g., PBL or transfected CHO cells) are diluted tenfold into binding medium (H/H w/Ca^{2+} with 10% glycerol; final concentration is typically 1–2 × 10^6 cells/mL) at room temperature (or 37°C as indicated). This minimizes exposure to glycerol and divalent cations that may affect cell viability and/or promote the formation of cell aggregates (*see* **Note 10**).
5. The cell input line is flushed with assay medium and placed into the tube containing cells. Flow is regulated by function of a downstream syringe pump. Use of a programmable pump allows for automation of the assay technique and more reproducible results. We typically initiate flow at a high rate of flow (2 mL/min or ~7.5 dyne/cm^2) to bring cells

through the sample tubing and into the chamber. Upon arrival of cells in the field of view, the flow rate is reduced (typically to 0.5–1 dyne/cm^2) to allow cells to interact with the blot surface. The flow rate can be adjusted up or down to increase or decrease, respectively, the stringency of binding interactions. One can program step-wise increases in flow (shear stress) without interruption. Maintaining continuous flow while cells are in the chamber minimizes nonspecific binding to the blot surface. This is critical with adherent cell populations such as CHO cells or monocytes, though lymphocytes typically do not bind when flow is stopped for brief periods (<30 s).

6. Observation of cell interactions with substrate molecule(s) are made in real time and recorded via the camera/VCR for later analysis. Typically, two types of analyses are performed. First, a "scanning" analysis is performed by moving the microscope stage to view the entire length of the lane where proteins have been resolved while maintaining a constant shear rate. Tethering and rolling interactions are usually most prominent at the leading edge of a "band" on the blot, with rolling cells evident across the band and discontinuation of rolling interactions observed at the downstream edge. Once bands or areas of the blot of interest are identified, we typically perform additional studies comparing control cell preparations (mock-transfected or cells incubated with function blocking antibodies) and varying shear conditions while viewing a fixed area or band on the blot. In this analysis, tethering is observed at low physiologic wall shear stresses (0.5–1.5 dyne/cm^2) for short periods (one to several minutes). Bound cells are then subjected to timed stepwise increases in wall shear stress to assess shear resistance as an estimate of relative strength of binding. The dependence of interactions on the presence of calcium (typical of selectin-mediated binding) can be confirmed by perfusion of the chamber with H/H with 10% glycerol and 5 mM EDTA and observing release of bound cells, or by repeating the analysis using H/H with 10% glycerol and 5 mM EDTA for both dilution of the cells and perfusion through the chamber.
7. Antibody inhibition experiments utilizing antibodies to the membrane bound ligand can be performed in situ by first observing interactions on an area of defined interest, then perfusing the chamber with a solution of antibody (e.g., 1–10 μg/mL for 30 min) and repeating the observation on the same site. If the antibody blocks binding interactions, then the blot is unusable for further studies of that ligand. However, if the applied antibody does not block adhesive interactions, the blot may be reused, i.e., successive rounds of antibodies or reagents may be screened in this fashion.

8. For selectin-mediated binding, cells must tether and roll in shear flow to be considered specific. Tethering in our studies is defined as reduction of forward motion below the hydrodynamic velocity lasting a minimum of two video frames (0.07 s), and rolling is defined as >5 cell diameters of lateral translation below the hydrodynamic velocity. The majority of tethered cells in such studies are observed to roll smoothly across the entire band or field of view. Nonspecific interactions (i.e., cellular collisions with the substrate that did not lead to tethering and/or rolling) are defined as interactions lasting less than 0.07 s and are not included in the analysis. Nonspecifically bound cells (not rolling and/or not released by perfusion with EDTA) are also discounted from analysis. In general, there is a time-dependent increase in nonspecific (nonrolling) attachments in the absence of flow, though it is our experience that very few cells form nonspecific attachments to the blot in continuous shear flow. Tethering rate is calculated as the number of cells that tether per field per time at a defined shear stress and adjusted to per minute values. As stated above, wall shear stress (t) values are calculated according to the formula τ (dynes/cm^2) = $6\mu Q/wh^2$ where μ is the viscosity of the solution in the chamber in poise, Q is the volumetric flow rate in mL/s, w is the channel width (e.g., 0.5 cm for chamber we employ), and h is the channel height (e.g., 0.0127 cm for chamber we employ) *(8)*. A value of 0.009 poise is used for the viscosity (μ) of water at 25°C *(9)* and a value of 0.0123 poise is used for the viscosity of 10% glycerol (v/v) in water at 25°C *(10)*. By these values, wall shear stress in 10% glycerol at 25°C is approximately 1.37-fold greater than in water at the same temperature.
9. Regions (bands) of interest can be aligned with molecular weight markers run in adjacent lanes to estimate molecular weight. Additionally, one may subsequently probe the blot with antibody to specific proteins of interest to determine whether band(s) supporting binding represent known proteins. This approach has been used to confirm PSGL-1 and identify CD43 as selectin ligands *(3, 6)*. One can also excise regions of blot supporting binding and submit for mass spectrometry and/or protein microsequencing. This approach was used to identify HCELL, a glycoform of CD44 that functions as a high affinity L-selectin and E-selectin ligand *(4, 5)*. **Figure 3** is a representative histogram of a blot rolling assay of T cell lysates probed with stably transfected Chinese hamster ovary (CHO) cells expressing P-selectin (CHO-P) and E-selectin (CHO-E). A videotape of these studies is available for viewing on the Harvard Skin Disease Research Center web page, under the Leukocyte Migration Core subheading (http://dermatology.bwh.harvard.edu/leukocyte.html).

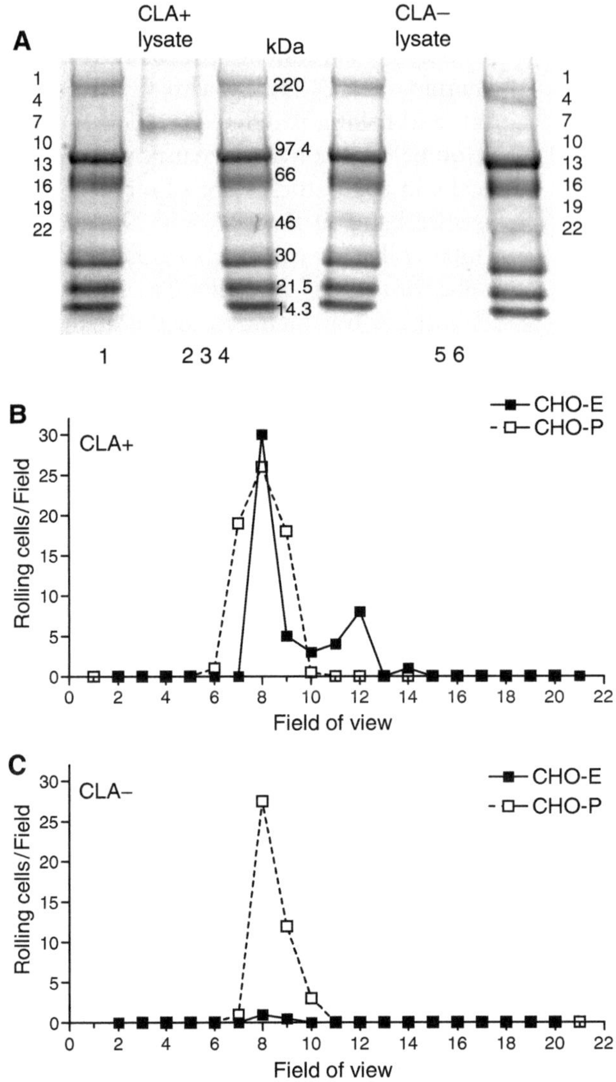

Fig. 3. Cutaneous lymphocyte antigen (CLA) immobilized on western blots supports both E-selectin and P-selectin mediated rolling. (**A**) Western blots of T cell lysates stained for CLA expression. 50 mg of cell lysate protein from CLA-positive (*left panel*) or CLA-negative (*right panel*) T cells were subjected to SDS-PAGE, electroblotted onto PVDF membranes and immunostained with HECA-452 mAb. The single major reactive glycoprotein identified at approximately 140 kDa in CLA-positive cells (lane 2) corresponds to the monomer form of CLA/PSGL-1. Similar blots stained with anti-PSGL-1 mAb show approximately equal quantities of 140 kDa PSGL-1 protein in each sample (data not shown). (**B** and **C**) Rolling cells per visual field (10× objective) at 1.75 dyne/cm^2 wall shear stress. The number of cells observed to bind and roll was observed in overlapping fields of view extending from approximately 250 to 40 kDa (identified alongside blot images in panel (**A**)). Both CHO-E and CHO-P cells are observed to tether and roll across the area of the western blot corresponding to CLA/PSGl-1 from CLA-positive T cells (panel (**B**)) while only CHO-P cells are noted to form significant rolling adhesions on PSGL-1 from CLA-negative T cells (panel (**C**)). Both CHO-E and CHO-P cells were noted to tether primarily over the area corresponding to CLA/PSGL-1. Mock-transfected CHO cells did not form rolling tethers on any areas of the blots observed (data not shown). Results presented are the means of two independent experiments on a single substrate and are representative of observations on numerous blots of CLA-positive and CLA-negative T cell lysates (reproduced with permission from *(3)*, Copyright 2002, The American Association of Immunologists, Inc.).

4. Notes

1. We typically avoid immunoblotting with function blocking antibodies, since these antibodies may interfere with cell binding to the membrane (*see* **Subheading 3.4, Step 7**).
2. Any uniform percent gel or gradient gel can be used so long as sufficient resolution of protein bands is achieved in the molecular weight range of interest.
3. If one wishes to measure the relative strength of interactions between component bands of samples loaded in different lanes of the same gel, it is important to load samples in lanes an appropriate distance apart such that the membrane-attached flow chamber does not cover another sample lane in order to avoid damaging the latter. For our flow chamber and the Criterion precast gel system, we load samples in every third lane, using molecular weight markers in between.
4. Surface irregularities can result in disruption of cell interactions on the membrane. Some membrane surface irregularities due to manufacture are readily visible when the membrane is immersed in methanol prior to the transfer step, while others are only observable when the membrane is immersed in 10% glycerol. Avoid the use of visibly scored or scratched membranes, as these defects may prevent a proper vacuum seal of the chamber to the membrane or provide a "trough" in which perfused cells may collect and subsequently adhere nonspecifically.
5. In all incubations with mAb or serum, it is critical to avoid using bacterially contaminated solutions in order to prevent digestion of binding epitopes by bacterial proteases and/or glycosidases.
6. It is important that cells are not stored in glycerol, as a 10% glycerol solution will significantly decrease cell (e.g., CHO) viability within 15 min.
7. Viscosity is temperature dependent. If the assay temperature is adjusted from 25 to 37°C, make sure to recalculate flow rates using the appropriate viscosity values for cell perfusion at specific shear stresses (*see* **Subheadings 2** and **3**).
8. We use Dow Corning High-Vacuum Grease to seal the silicon rubber gasket to the flow chamber to create the flow channel. In order to prevent the gasket from slipping after placement, we allow the grease to set for 24 h before use in experiments. It is critical to remove all excess vacuum grease from the channel, gasket surface, and other flow chamber surfaces completely (using a cotton swab and 70% ethanol) before use in blot rolling assays to prevent the transfer of grease onto the membrane, which will damage the lanes and render them unusable.

9. All bubbles and debris from the channel must be flushed to clear all potential for flow disturbances.
10. The cell concentration can be adjusted to increase or decrease assay sensitivity as desired (i.e., higher cell concentration increases sensitivity).

Acknowledgements

This work was supported by National Institutes of Health grants HL60528 (RS), CA84156 (RS), HL073714 (RS), and AR02115 (RCF). The authors thank Drs. Sandra L. King and Monica M. Burdick for critical review of the manuscript.

References

1. Stamper, H.B., and Woodruff, J.J. (1976) Lymphocyte homing into lymph nodes: *in vitro* demonstration of the selective affinity of recirculating lymphocytes for high endothelial venules. *J Exp Med* 144, 828–833.
2. Lawrence, M.B., and Springer, T.A. (1991) Leukocytes roll on P-selectin at physiologic flow rates: distinction from and prerequisite for adhesion through integrins. *Cell* 65, 859–873.
3. Fuhlbrigge, R.C., King, S.L., Dimitroff, C.J., and Sackstein, R. (2002) Direct, real-time observation of E- and P- selectin mediated rolling on cutaneous lymphocyte-associated antigen immobilized on Western blots. *J Immunol* 168, 5645–5651.
4. Dimitroff, C.J., Lee, J.Y., Fuhlbrigge, R.C., and Sackstein, R. (2000) A distinct glycoform of CD44 is an L-selectin ligand on human hematopoietic cells. *Proc Natl Acad Sci USA* 97, 13841–13846.
5. Dimitroff, C.J., Lee, J.Y., Rafii, S., Fuhlbrigge, R.C., and Sackstein, R. (2001) CD44 is a major E-selectin ligand on human hematopoietic progenitor cells. *J Cell Biol* 153, 1277–1286.
6. Fuhlbrigge, R.C., King, S.L., Sackstein, R., and Kupper, T.S. (2006) CD43 is a ligand for E-selectin on human skin homing T cells. *Blood* 107, 1421–1426.
7. Hanley, W.D., Burdick, M.M., Konstantopoulos, K., and Sackstein, R. (2005) CD44 on LS174T colon carcinoma cells possesses E-selectin ligand activity. *Cancer Res* 65, 5812–5817.
8. Glasby, M.A. (1995) Principles of haemodynamics. In *Applied physiology for surgery and critical care* (M.A. Glasby, and C.L.-H. , eds. Huang), Oxford & Associates, Boston, MA, p. 131.
9. Sengers, J.V. and Watson, J.T.R. (1986) Improved international formulations for the viscosity and thermal conductivity of water substance. *J Phys Chem Ref Data* 15, 1291.
10. Borchers, H. (ed.) Material values and mechanical behaviour of non-metals (1995) In *Landolt-Bornstein Numerical Data and Functional Relationships in Phsyics, Chemistry, Astronomy, Geophysics, and Technology*, Vol. 4 (Part 1), Springer, Berlin.

Chapter 37

ELISPOT Assay on Membrane Microplates

Alexander E. Kalyuzhny

Summary

Membranes used for western blotting can be also used for ELISPOT, an enzyme-linked immunospot assay, which allows determining frequencies of cytokine-secreting immune system cells. In addition to their high antibody-retaining capacity PVDF and NC membranes provide good support to immune system cells cultured in vitro and do not affect their physiology. ELISPOT assays utilizing membrane-backed microplates are used in many areas of research including vaccine development, HIV research, cancer and infection disease research, autoimmune disease, and allergy research.

ELISPOT utilizes the same antibody "sandwich" technique as enzyme-linked immunosorbent assay, but unlike the latter ELISPOT belongs to state-of-the-art techniques when outcome of the assay depends on skills and accuracy of the operator, a thorough selection of matched pairs of capture and detection antibodies, and using appropriate staining reagents. This review covers basics of ELISPOT assay including its immunochemical design, selection of reagents and membrane microplates, and some troubleshooting recommendations.

Key words: ELISPOT, Membrane microplates, Detection antibodies, Capture antibodies, Chromogenic substrates, Spot forming cells, Quantification of spots, ELISPOT readers, Background staining, Membrane removal device

1. Invention of ELISPOT Assay and Areas of its Application

It appears that the first mentioning of ELISPOT assay occurred in 1983 when Sedgwick and Holt *(1, 2)* published a paper describing a very simple novel technique allowing quick and accurate quantification of antibody producing cells in vitro. This new technique resembled ELISA in that immobilized antigen was capturing antibodies secreted by cultured immune system cells. Later in 1981 another paper describing a similar antibody detection technique was published by Czerkinsky and colleagues *(3)*,

Biji T. Kurien and R. Hal Scofield (eds.), *Methods in Molecular Biology, Protein Blotting and Detection, vol. 536*

DOI: 10.1007/978-1-59745-542-8_37

who called this assay Enzyme-Linked Immunospot or ELISPOT. A fascinating historic overview of ELISPOT assay invention was recently published *(2)*. ELISPOT assays can be used for enumeration of either (1) antibody-producing cells or (2) cells secreting antigens (e.g. cytokines, growth factors, enzymes). The latter is sometimes referred to as "reversed ELISPOT." According to Tanguay and Killion, ELISPOT is much more sensitive than ELISA (enzyme-linked immunosorbent assay) for the detection of cell-secreted cytokines: ELISA cannot detect cytokines released by less than 10^4 cells, whereas ELISPOT allows detecting cytokines secreted by 10–100 cells in a well *(4)*. Such a high sensitivity makes ELISPOT a technique of choice for the detection of spontaneous and antigen-induced secretion of cytokines from peripheral blood lymphocytes *(5, 6)*. To mention few applications, ELSIPOT *(7–9)* is used for AIDS research *(10, 11)*, cancer research (*see review* **ref.** *12)*, infectious diseases monitoring *(13)*, autoimmune disease studies *(14)*, and allergy and transplantation research *(15, 16)*. ELISPOT can be also used for diagnostic applications as well. For example, it was reported that ESAT-6/CFP-10-based ELISPOT assay can be used to detect active tuberculosis in HIV-positive individuals *(17)*. ELISPOT may be also used for allergy diagnostics: peripheral blood mononuclear cells from nickel-allergic individuals responded to Ni^{2+} with significantly greater production of interleukin (IL)-4, IL-5, IL-13 and interferon (IFN)-gamma, compared with the healthy controls *(18)*.

A comprehensive collection of articles on various topics of ELISPOT assay can be found in a recently published Handbook of ELISPOT *(19)*.

2. Composition and Protocol of ELISPOT Assay

Like ELISA, ELISPOT is a sandwich immunoassay (**Figs. 1** and **2**) but, unlike ELISA, ELISPOT does not allow determining the real concentration of secreted cytokine *(20)*: ELISPOT only allows to quantify the number of cytokine-secreting cells *(2, 3)*. ELISA is just an immunoassay mostly used to measure cytokines in cell-free media *(20)* and its performance entirely depends on the quality of antibodies, enzyme conjugates, and color reagents. ELISPOT is the combination of both immunoassay and bioassay and its performance depends not only on the quality of antibodies, enzyme conjugates, and color reagents, but also on the viability of cells cultured in the ELISPOT plate. Four major components contribute to the success of ELISPOT assay include (1) capture and detection of antibodies, (2) enzyme conjugates,

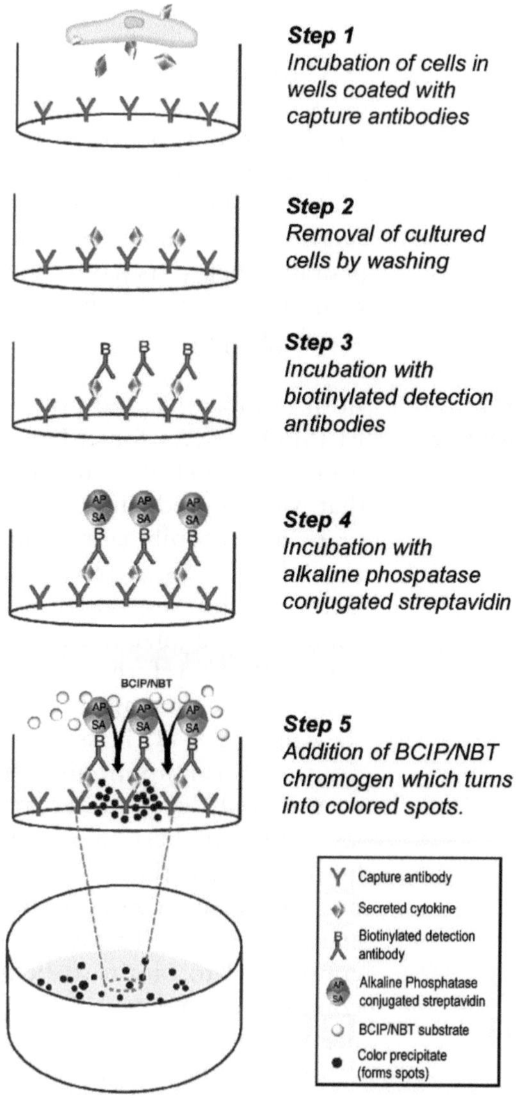

Fig. 1. Typical ELISPOT assay procedure.

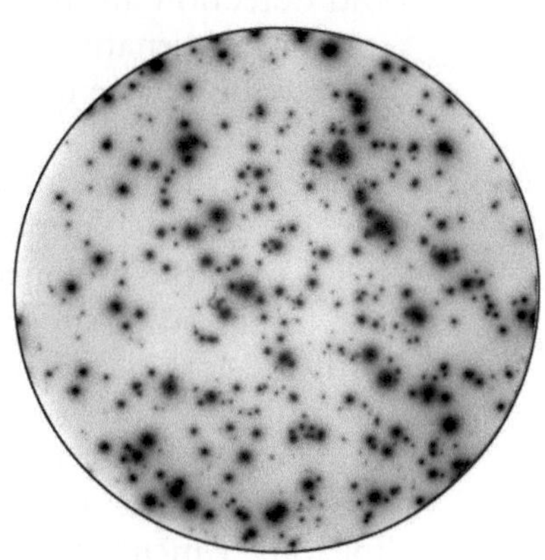

Fig. 2. Typical single well ELISPOT image captured with QHub elispot reader (courtesy MVS Pacific).

(3) chromogenic substrates, and (4) microplate membranes. Since the secretion activity of cells in ELSIPOT is determined by the number of spots on the bottom of the plate *(2, 3)*, it appears that all four components should be optimized to facilitate formation of detectable spots. Spots should have strong staining intensity (high signal-to-noise ratio) and have well-defined edges. It is also desirable that spots have a small diameter to avoid their merging with each other: a few merged spots may be erroneously counted as a single spot.

2.1. Capture and Detection Antibodies

Both monoclonal and polyclonal antibodies can be utilized in ELISPOT assays for either antigen capture or antigen detection. ELISPOT can utilize capture and detection antibodies that were raised either against the entire antigen molecule (e.g., antirecombinant protein antibodies) or against a portion of the antigen (e.g., antipeptide antibodies). The critical factor in choosing capture and detection antibodies is their ability to recognize nonoverlapping epitopes of the target antigen *(20)*. Therefore, it may not be recommended to use the same monoclonal antibody for both capture and detection in the same ELISPOT assay. Suitability of antibodies for applications such as immunohistochemistry and western blotting and even ELISA does not necessarily guarantee that these antibodies will be the best choice for ELISPOT. Usually capture antibodies work within a wide range of concentration (1–30 μg/mL), whereas detection antibodies work at much lower concentrations (0.1–0.5 μg/mL). Detection antibodies need to be conjugated to biotin to be able to interact with streptavidin-conjugated enzymes *(21)*. The reason detection antibodies need to be biotinylated is to avoid their cross-reactivity with capture antibodies. If both capture and detection antibodies are raised in the same species (e.g., mouse), antibodies (e.g., antimouse) conjugated to enzyme will bind to both capture and detection antibodies rather binding to detection antibodies only. Alternatively, detection antibodies directly conjugated to enzyme can be used as well.

2.2. Enzyme Conjugates

Horseradish peroxidase (HRP) or alkaline phosphatase (AP) can be used as streptavidin conjugates *(21)*. HRP (optimum pH 7.6) in the presence of hydrogen peroxide (H_2O_2) catalyzes the oxidation of substrates causing them to change their color with the loss of electrons. The advantage of using HRP is its high turnover rate (spots develop faster), whereas the drawback is increased background. Unlike HRP, AP (optimum pH 9.0–9.6) has a linear reaction rate (spots develop slower) allowing for longer incubations with chromogenic substrates *(21)* without a risk of developing background staining. Longer incubation may be done if it is necessary to increase the sensitivity of

AP-based assay. By combining HRP and AP it is possible to develop an ELISPOT assay for simultaneous detection of two different cell-secreted molecules *(22–24)*. The major drawback of multianalyte systems is the loss of sensitivity for each of the antigens. We have observed that a number of spots formed by IL-2 and IFN-gamma secreted from peripheral blood mononuclear cells in the plate coated with anti-IL-2 and anti-IFN-gamma antibodies were noticeably lower in comparison with corresponding single-cytokine assays. We have found that the drop in sensitivity becomes even more profound if ELISPOT plate is coated with more than two capture antibodies *(25)*. The mechanism underlying this phenomenon is not known and additional research is needed to find the ways of building high sensitivity multianalyte ELISPOT assays.

2.3. Enzyme Substrates

Regardless of which enzyme conjugate is used, their corresponding substrates should produce intense and stable colors. Such HRP substrate as AEC (3-amino-9-ethylcarbazole, $C_{14}H_{14}N_2$) forms intense red color spots *(21)*. However, AEC is unstable *(21)* and spots may lose their color resulting in a loss of primary data. Another HRP substrate, DAB (3,3-diaminobenzidine, $C_{12}H_{14}N_4$), produces brown color spots, which are less intense than spots formed by AEC precipitate *(21)*. Unlike AEC, DAB is stable, but is carcinogenic, which may be a major drawback in using this chromogen. It appears that the most frequently used chromogen is a mixture of BCIP (5-bromo-4-chloro-3-indolylphosphate *p*-toluidine salt, $C_{15}H_{15}N_2O_4PClBr$) with NBT (nitroblue tetrazolium chloride, $C_{40}H_{30}N_{10}O_6Cl_2$). BCIP/NBT is converted into insoluble black–blue spots by alkaline phosphatase *(21)*. Spots formed by BCIP/NBT, do not fade over long periods of time allowing reanalyze primary ELISPOT data when needed.

2.4. Fluorescence ELISPOT Assay

In addition to chromogenic substrates, which develop colors that can be seen using bright-field microscopy, detection can be also done using fluorescent probes. This modification of ELISPOT was given a name Fluorospot and employs anticytokine antibodies conjugated to fluorescent probes. Fluorospot allows simultaneous detection of two cytokines secreted by the same cultured immune system cells *(26, 27)*.

2.5. Assay Developing Procedures

The secretion capacity of cells may be tested in two ways (1) cells are cultured in a designated plate and then transferred into ELISPOT plates *(28–30)* or (2) cells are stimulated and cultured directly in ELISPOT plates *(31, 32)*. When cells are cultured and stimulated outside ELISPOT plate need to be washed with fresh culture media before plating them into ELISPOT plate to avoid background staining (*see* **Notes 1** and **2**).

2.6. Membrane Microplates

ELISPOT assays can be performed using either all plastic plates *(4, 33)* or plastic plates backed with either PVDF *(34, 35)* or nitrocellulose membranes *(33, 36)*. Membranes support the growth of the cultured cells, have a much higher retaining capacity for capture antibodies than plastic plates, and white color of membranes provides an excellent backdrop for enumeration of colored spots [characteristics and attributes of membranes are reviewed in details by Weiss *(37)*]. In ELISPOT assay, the flow of reagents through or across the membrane is not required, but rather a diffusion of cell-secreted molecules toward capture antibodies immobilized on the membrane. Membrane plates are manufactured by different vendors including Millipore Corporation, Pall Corporation, and Whatman. It appears that Millipore plates are more popular for ELISPOT assay, due to the fact that after finishing the assay, membranes can be easily removed from the plate for archiving purposes.

2.7. Archiving ELISPOT Data

After finishing the experiment, ELISPOT plates considered primary experimental data and it may be required to store them in a safe place. Unfortunately 96-well plates are bulky and their storage requires a lot of space. To solve this problem, membranes with spots can be from ELIPSOT plates, laminated, bar-coded, and compactly stored in a binder. **Figure 3** shows a membrane removal device (http://www.mvspacific.com), which allows quick simultaneous removal of all membranes from the ELISPOT plate. Removed membranes can be laminated using heat laminators for additional protection.

Fig. 3. Membrane removal device (courtesy MVS Pacific) for simultaneous removal of all 96 membranes from ELISPOT plates. Removed membranes can be laminated for additional protection and stored in a binder.

2.8. ELISPOT Assay Formats

There are two major commercial formats of ELISPOT assay (1) fully developed and optimized ready-to-use kits and (2) a set of reagents which the researcher needs to optimize by himself. ELISPOT assay standardization and validation techniques (*see* **Note 3**) are covered in detail by Janetzki et al. *(32)* It appears that ready-to-use ELISPOT kits are the best choice for large-scale clinical trial projects which require convenience and a high degree of accuracy *(38)*. R&D Systems, Inc. (http://www.rndsystems.com) was the first company which designed and introduced ready-to-use ELISPOT kits which include precoated PVDF membrane microplates, wash buffers, detection antibody, enzyme conhugate, and a color reagent.

3. ELISPOT Data Analysis

In ELISA assay, the concentration of the molecules in the sample is determined by measuring the optical density of the color substrate solution filling the wells *(20)*, whereas in ELISPOT, cell-secretion capacity is measured by counting colored spots at the bottom of the well *(2, 3, 32)*. The term "spot forming cells," or SFC, is used as a quantitative measure of the cell secretion activity in ELISPOT assay *(39, 40)*. After finishing the assay, spots can be counted either manually or by using computer-aided image analysis *(41)*. Manual counting can be done using, for example, a dissecting microscope but is very inefficient and time consuming. Computer-aided quantification of spots is thought to be more reliable than the manual counting *(41, 42)* and currently several automated ELISOPT readers are available from different vendors (e.g., Zeiss, C.T.L., and MVS Pacific) and they allow fast and accurate quantification of spots. **Figure 2** describes a typical single-well image from ELISPOT plate recorded with automated QHub ELISPOT reader manufactured by MVS Pacific (http://www.mvspacific.com).

4. Notes

1. *Spot staining*. The quality of staining in ELISPOT assay is of key importance in obtaining quantifiable spots. The ultimate goal of ELISPOT operator is to obtain strongly stained spots and a low background hindering the detection and counting of spots *(43)*. Background may be caused by using a low affinity capture antibodies: cytokine molecules secreted from cells dissociate from capture antibodies surrounding the releasing

cell diffuse in the well and bind to capture antibodies in the cell-free zone. Background can be also caused by disturbing ELISPOT plate with cells during the incubation: frequent opening and closing of the incubator's door will shake the ELISPOT plate and facilitate diffusion of secreted cytokines away from the secreting cell. Another source of background is the adsorption of detection antibodies, enzyme conjugate, and precipitating substrate onto the membrane: preliminary experiments need to be done to determine whether reagents can bind to membranes in ELISPOT plates even without cells. We have found that wrapping ELISPOT plates into aluminum foil reduces background staining, improves contrast of stained spots across the entire membrane, and improve well-to-well reproducibility *(44)*. It appears that aluminum foil facilitates even distribution of heat across the bottom of ELISPOT plate in CO_2 incubator.

2. *Washing procedures.* The ultimate purpose of washing ELISPOT plates is to remove cultured cells and unbound reagents (detection antibodies, enzyme conjugate, enzyme substrate) form ELISPOT plates to minimize background staining. Plates can be washed, for example, with phosphate buffered saline (PBS) of various pH and molarity. It is necessary to remove as many cells as possible by washing since stained cells may be confused with specific spots and thus affect the accuracy of quantification. However, if washing of cells with buffer does not help, treatment of plates with enzymatic cell-detachment reagents may be an option *(53)*.

3. *Viability of cultured cells and their removal from the plate.* The quality of staining also depends on the quality of cultured cells. It is of critical importance to determine the percent of dead cells (for example by trypan blue exclusion). We have found that having a high number of dead cells in the ELISPOT plate (30–50% and more) may result in a high background staining obscuring specific spots and making their quantification challenging. It is likely that cytokine-secreting capacity of cells is not known in advance, and therefore it is recommended to do serial cell dilutions: 10^3, 10^4, 10^5, 10^6 cells per well. This will help to determine the optimal cell dilution, which produces quantifiable spots. ELISPOT can be also performed with cells stored in liquid nitrogen: freezing of peripheral blood lymphocytes did not impair their physiology *(45, 46)* and cytokine-secreting activity *(35, 47–49)*. We have reported previously that cryopreserved peripheral blood lymphocytes are suitable for studying secretion of multiple cytokines *(50)*. Other reports indicate that cryopreserved cells can be more active in secreting some cytokines *(51, 52)*. This may be due to elimination of platelets (cytokine secretion inhibitors) which do not survive freezing *(35)*.

References

1. Sedgwick, J.D., and Holt, P.G. (1983). A solid-phase immunoenzymatic technique for the enumeration of specific antibody-secreting cells. *J Immunol Methods* 57, 301–309.
2. Sedgwick, J.D. (2005) ELISPOT assay: a personal retrospective. *Methods Mol Biol* 302, 3–14.
3. Czerkinsky, C.C., Nilsson, L.A., Nygren, H., Ouchterlony, O., and Tarkowski, A. (1983) A solid-phase enzyme-linked immunospot (ELISPOT) assay for enumeration of specific antibody-secreting cells. *J Immunol Methods* 65, 109–121.
4. Tanguay, S., and Killion, J.J. (1994) Direct comparison of ELISPOT and ELISA-based assays for detection of individual cytokine-secreting cells. *Lymphokine Cytokine Res* 13, 259–263.
5. Bienvenu, J., Monneret, G., Fabien, N., and Revillard, J.P. (2000) The clinical usefulness of the measurement of cytokines. *Clin Chem Lab Med* 38, 267–285.
6. Mashishi, T., and Gray, C.M. (2002) The ELISPOT assay: an easily transferable method for measuring cellular responses and identifying T cell epitopes. *Clin Chem Lab Med* 40, 903–910.
7. Pass, H.A., Schwarz, S.L., Wunderlich, J.R., and Rosenberg, S.A. (1998) Immunization of patients with melanoma peptide vaccines: immunologic assessment using the ELISPOT assay. *Cancer J Sci Am* 4, 316–323.
8. Asai, T., Storkus, W.J., and Whiteside, T.L. (2000) Evaluation of the modified ELISPOT assay for gamma interferon production in cancer patients receiving antitumor vaccines. *Clin Diagn Lab Immunol* 7, 145–154.
9. Kamath, A.T., Groat, N.L., Bean, A.G., and Britton, W.J. (2000) Protective effect of DNA immunization against mycobacterial infection is associated with the early emergence of interferon-gamma (IFN-gamma)-secreting lymphocytes. *Clin Exp Immunol* 120, 476–482.
10. Eriksson, K., Nordstrom, I., Horal, P., Jeansson, S., Svennerholm, B., Vahlne, A., et al. (1992) Amplified ELISPOT assay for the detection of HIV-specific antibody-secreting cells in subhuman primates. *J Immunol Methods* 153, 107–113.
11. Howell, D.M., Feldman, S.B., Kloser, P., and Fitzgerald-Bocarsly, P. (1994) Decreased frequency of functional natural interferon-producing cells in peripheral blood of patients with the acquired immune deficiency syndrome. *Clin Immunol Immunopathol* 71, 223–230.
12. Schmittel, A., Keilholz, U., Thiel, E., and Scheibenbogen, C. (2000) Quantification of tumor-specific T lymphocytes with the ELISPOT assay. *J Immunother* 23, 289–295.
13. Smith, S.M., Brookes, R., Klein, M.R., Malin, A.S., Lukey, P.T., King, A.S., et al. (2000) Human CD8 + CTL specific for the mycobacterial major secreted antigen 85A. *J Immunol* 165, 7088–7095.
14. Pelfrey, C.M., Cotleur, A.C., Lee, J.C., and Rudick, R.A. (2002) Sex differences in cytokine responses to myelin peptides in multiple sclerosis. *J Neuroimmunol* 130, 211–223.
15. Schmid-Grendelmeier, P., Altznauer, F., Fischer, B., Bizer, C., Straumann, A., Menz, G. et al. (2002) Eosinophils express functional IL-13 in eosinophilic inflammatory diseases. *J Immunol* 169, 1021–1027.
16. Groth, R., and Aanonsen, L. (2002) Spinal brain-derived neurotrophic factor (BDNF) produces hyperalgesia in normal mice while antisense directed against either BDNF or trkB, prevent inflammation-induced hyperalgesia. *Pain* 100, 171–181.
17. Chapman, A.L., Munkanta, M., Wilkinson, K.A., Pathan, A.A., Ewer, K., Ayles, H., et al. (2002) Rapid detection of active and latent tuberculosis infection in HIV-positive individuals by enumeration of Mycobacterium tuberculosis-specific T cells. *Aids* 16, 2285–2293.
18. Jakobson, E., Masjedi, K., Ahlborg, N., Lundeberg, L., Karlberg, A.T., and Scheynius, A., et al. (2002) Cytokine production in nickel-sensitized individuals analysed with enzyme-linked immunospot assay: possible implication for diagnosis. *Br J Dermatol* 147, 442–449.
19. Kalyuzhny, A.E. (2005) Handbook of elispot: methods and protocols, in: Methods in *Molecular Biology*, ed. J.M. Walker, 1 edition, vol. 302. Totowa, NJ: Humana Press, p. 321.
20. Kemeny, D.M. (1997) Enzyme-linked immunoassays, in: *Immunochemistry* 1, eds. A.P. Johnstone and M.W. Turner. Oxford: Oxford University Press, pp. 147–175.
21. Savage, M.D., Mattson, G., Desai, S., Nielander, G.W., Morgensen, S., and Conklin, E.J. (1992) Avidin-Biotin Chemistry: A Handbook. Rockford, IL: Pierce Chemical Company, p. 467.

22. Okamoto, Y., Abe, T., Niwa, T., Mizuhashi, S., and Nishida, M. (1998) Development of a dual color enzyme-linked immunospot assay for simultaneous detection of murine T helper type 1- and T helper type 2-cells. *Immunopharmacology* 39, 107–116.
23. Okamoto, Y., Gotoh, Y., Tokui, H., Mizuno, A., Kobayashi, Y., and Nishida, M. (2000) Characterization of the cytokine network at a single cell level in mice with collagen-induced arthritis using a dual color ELISPOT assay. *J Interferon Cytokine Res* 20, 55–61.
24. Okamoto, Y., and Nishida, M. (2005) Dual-color ELISPOT assay for analyzing cytokine balance. *Methods Mol Biol* 302, 263–272.
25. Palzer, S., Bailey, T., Hartnett, C., Grant, A., Tsang, M., and Kalyuzhny, A.E. (2005) Simultaneous detection of multiple cytokines in ELISPOT assays. *Methods Mol Biol* 302, 273–288.
26. Gazagne, A., Malkusch, W., Vingert, B., Fridman, W.H., and Tartour, E. (2005) Fluorospot assay: methodological analysis. *Methods Mol Biol* 302, 289–296.
27. Malkusch, W. (2005) High resolution as a key feature to perform accurate ELISPOT measurements using Zeiss KS ELISPOT readers. *Methods Mol Biol* 302, 133–150.
28. Favre, N., Bordmann, G., and Rudin, W. (1997) Comparison of cytokine measurements using ELISA, ELISPOT and semi-quantitative RT-PCR. *J Immunol Methods* 204, 57–66.
29. Herr, W., Schneider, J., Lohse, A.W., Meyer zum Buschenfelde, K.H., and Wolfel, T. (1996) Detection and quantification of blood-derived CD8 + T lymphocytes secreting tumor necrosis factor alpha in response to HLA-A2.1-binding melanoma and viral peptide antigens. *J Immunol Methods* 191, 131–142.
30. Arlen, P., Tsang, K.Y., Marshall, J.L., Chen, A., Steinberg, S.M., Poole, D., et al. (2000) The use of a rapid ELISPOT assay to analyze peptide-specific immune responses in carcinoma patients to peptide vs. recombinant poxvirus vaccines. *Cancer Immunol Immunother* 49, 517–529.
31. Janetzki, S., Song, P., Gupta, V., Lewis, J.J., and Houghton, A.N. (2000) Insect cells as HLA-restricted antigen-presenting cells for the IFN-gamma elispot assay. *J Immunol Methods* 234, 1–12.
32. Janetzki, S., Cox, J.H., Oden, N., and Ferrari, G. (2005) Standardization and validation issues of the ELISPOT assay. *Methods Mol Biol* 302, 51–86.
33. Ronnelid, J., and Klareskog, L. (1997) A comparison between ELISPOT methods for the detection of cytokine producing cells: greater sensitivity and specificity using ELISA plates as compared to nitrocellulose membranes. *J Immunol Methods* 200, 17–26.
34. Schielen, P., van Rodijnen, W., Tekstra, J., Albers, R., and Seinen, W. (1995) Quantification of natural antibody producing B cells in rats by an improved ELISPOT technique using the polyvinylidene difluoride membrane as the solid support. *J Immunol Methods* 188, 33–41.
35. McCutcheon, M., Wehner, N., Wensky, A., Kushner, M., Doan, S., Hsiao, L., et al. (1997) A sensitive ELISPOT assay to detect low-frequency human T lymphocytes. *J Immunol Methods* 210, 149–166.
36. Taguchi, T., McGhee, J.R., Coffman, R.L., Beagley, K.W., Eldridge, J.H., Takatsu, K., et al. (1990) Detection of individual mouse splenic T cells producing IFN-gamma and IL-5 using the enzyme-linked immunospot (ELISPOT) assay. *J Immunol Methods* 128, 65–73.
37. Weiss, A.J. (2005) Membranes and membrane plates used in ELISPOT. *Methods Mol Biol* 302, 33–50.
38. Klencke, B., Matijevic, M., Urban, R.G., Lathey, J.L., Hedley, M.L., Berry, M., et al. (2002) Encapsulated plasmid DNA treatment for human papillomavirus 16-associated anal dysplasia: a Phase I study of ZYC101. *Clin Cancer Res* 8, 1028–1037.
39. McGhee, M.L., Ogawa, T., Pitts, A.M., Moldoveanu, Z., Mestecky, J., McGhee, J. R., et al. (1989). Cellular analysis of functional mononuclear cells from chronically inflamed gingival tissue. *Reg Immunol* 2, 103–110.
40. Merville, P., Pouteil-Noble, C., Wijdenes, J., Potaux, L., Touraine, J.L., and Banchereau, J. (1993) Detection of single cells secreting IFN-gamma, IL-6, and IL-10 in irreversibly rejected human kidney allografts, and their modulation by IL-2 and IL-4. *Transplantation* 55, 639–646.
41. Herr, W., Linn, B., Leister, N., Wandel, E., Meyer zum Buschenfelde, K.H., and Wolfel, T. (1997) The use of computer-assisted video image analysis for the quantification of CD8 + T lymphocytes producing tumor necrosis factor alpha spots in response to peptide antigens. *J Immunol Methods* 203, 141–152.
42. Lehmann, P.V. (2005) Image analysis and data management of ELISPOT assay results. *Methods Mol Biol* 302, 117–132.

43. Kalyuzhny, A.E. (2005) Chemistry and biology of the ELISPOT assay. *Methods Mol Biol* 302, 15–31.
44. Kalyuzhny, A., and Stark, S. (2001) A simple method to reduce the background and improve well-to-well reproducibility of staining in ELISPOT assays. *J Immunol Methods* 257, 93–97.
45. Merker, R., Check, I., and Hunter, R.L. (1979) Use of cryopreserved cells in quality control of human lymphocyte assays: analysis of variation and limits of reproducibility in long-term replicate studies. *Clin Exp Immunol* 38, 116–126.
46. Lewalle, P., Rouas, R., Lehmann, F., and Martiat, P. (2000) Freezing of dendritic cells, generated from cryopreserved leukaphereses, does not influence their ability to induce antigen-specific immune responses or functionally react to maturation stimuli. *J Immunol Methods* 240, 69–78.
47. Keane, N.M., Price, P., Stone, S.F., John, M., Murray, R.J., and French, M.A. (2000) Assessment of immune function by lymphoproliferation underestimates lymphocyte functional capacity in HIV patients treated with highly active antiretroviral therapy. *AIDS Res Hum Retroviruses* 16, 1991–1996.
48. Smith, J.G., Liu, X., Kaufhold, R.M., Clair, J., and Caulfield, M.J. (2001) Development and validation of a gamma interferon elispot assay for quantitation of cellular immune responses to varicella-zoster virus. *Clin Diagn Lab Immunol* 8, 871–879.
49. Sobota, V., Bubenik, J., Indrova, M., Vlk, V., and Jakoubkova, J. (1997) Use of cryopreserved lymphocytes for assessment of the immunological effects of interferon therapy in renal cell carcinoma patients. *J Immunol Methods* 203, 1–10.
50. Bailey, T., Stark, S., Grant, A., Hartnett, C., Tsang, M., and Kalyuzhny, A. (2002) A multidonor ELISPOT study of IL-1beta, IL-2, IL-4, IL-6, IL-13, IFN-gamma and TNF-alpha release by cryopreserved human peripheral blood mononuclear cells. *J Immunol Methods* 270, 171–182.
51. Venkataraman, M. (1994) Effects of cryopreservation on immune responses: VII. Freezing induced enhancement of IL-6 production in human peripheral blood mononuclear cells. *Cryobiology* 31, 468–477.
52. Venkataraman, M. (1995) Effects of cryopreservation on immune responses. VIII. Enhanced secretion of interferon-gamma by frozen human peripheral blood mononuclear cells. *Cryobiology* 32, 528–534.
53. Grant, A., Palzer, S., Hartnett, C., Bailey, T., Tsang, M., and Kalyuzhny, A.E. (2005) A cell-detachment solution can reduce background staining in the ELISPOT assay. *Methods Mol Biol* 302, 87–94.

Chapter 38

A Brief Review of Other Notable Protein Blotting Methods

Biji T. Kurien and R. Hal Scofield

Summary

A plethora of methods have been used for transferring proteins from the gel to the membrane. These include centrifuge blotting, electroblotting of proteins to Teflon tape and membranes for N- and C-terminal sequence analysis, multiple tissue blotting, a two-step transfer of low and high molecular weight proteins, blotting of Coomassie Brilliant Blue (CBB)-stained proteins from polyacrylamide gels to transparencies, acid electroblotting onto activated glass, membrane-array method for the detection of human intestinal bacteria in fecal samples, protein microarray using a new black cellulose nitrate support, electrotransfer using square wave alternating voltage for enhanced protein recovery, polyethylene glycol-mediated significant enhancement of the immunoblotting transfer, parallel protein chemical processing before and during western blot and the molecular scanner concept, electronic western blot of matrix-assisted laser desorption/ionization (MALDI) mass spectrometry-identified polypeptides from parallel processed gel-separated proteins, semidry electroblotting of peptides and proteins from acid-urea polyacrylamide gels, transfer of silver-stained proteins from polyacrylamide gels to polyvinylidene difluoride (PVDF) membranes, and the display of K^+ channel proteins on a solid nitrocellulose support for assaying toxin binding. The quantification of proteins bound to PVDF membranes by elution of CBB, clarification of immunoblots on PVDF for transmission densitometry, gold coating of nonconductive membranes before MALDI tandem mass spectrometric analysis to prevent charging effect for analysis of peptides from PVDF membranes, and a simple method for coating native polysaccharides onto nitrocellulose are some of the methods involving either the manipulation of membranes with transferred proteins or just a passive transfer of antigens to membranes. All these methods are briefly reviewed in this chapter.

Key words: Centrifuge blotting, Black cellulose nitrate, Protein microarray, Square wave alternating voltage, Acid electroblotting, Electronic western blotting, Gold coating of membranes

1. Centrifuge Blotting

Hermansen et al. *(1)* describe a method to elute and transfer proteins, detected in the gel by visualization with 1 M potassium chloride (KCl), to a polyvinylidene diflouride (PVDF)

Biji T. Kurien and R. Hal Scofield (eds.), *Methods in Molecular Biology, Protein Blotting and Detection, vol. 536*

DOI: 10.1007/978-1-59745-542-8_38

(Immobilon-P, BioRad Laboratories, Hercules, CA, USA) membrane by centrifugation instead of electroblotting. The SDS polyacrylamide gel was immersed in 1 M KCl for 2 min. Protein bands that appear as a clear zone against an opaque background were cut out and immersed in distilled water to remove excess KCl (3–5 min). The gel pieces were then used for centrifuge blotting or stored moist in a microcentrifuge tube at −20°C.

PVDF and dialysis membranes (10 kD cutoff) were excised into 12-mm disks. The PVDF disk was moistened in methanol for 5 min for activation. The gel bit was soaked in an eluant for 15 min at 37°C. The dialysis and PVDF membranes were soaked in the eluant (3–5 min) prior to centrifugation.

The centrifuge receptacle assembly consists of an outer chamber that supported the entire structure, an inner cylinder (serving as the reservoir for the eluant), a 12-mm sinter base support made of polyethane, which holds the dialysis and PVDF membranes in place, a polystyrene tube for eluate collection, and a flat O-ring gasket made of silicon to prevent leakage in the receptacle during centrifugation.

The dialysis membrane was positioned on the base support of the receptacle followed by the O-ring. An eluant was added on the dialysis membrane prior to placing the PVDF membrane. This prevents formation of vacuum between the PVDF and dialysis membranes. The inner cylinder is screwed in the receptacle after the PVDF membrane is placed on top of the O-ring. The eluant was added into the receptacle after placing the gel bit on top of the PVDF membrane. Centrifugation was carried out at 3,000 × *g* at 20°C for 1 h in a table top centrifuge. Following centrifugation the PVDF membrane with the transferred protein adhering to it was vacuum dried and stored at −20°C for N-terminal sequence analysis or other analyses. The dialysis membrane placed beneath the PVDF membranes helps retain the nonimmobilized proteins *(1)*.

2. Electroblotting of Proteins to Teflon Tape and Membranes for N- and C-Terminal Sequence Analysis

Several membranes have been used in electroblotting since the inception of the electrotransfer of proteins from polyacrylamide gels to membranes and the subsequent characterization of the blotted proteins *(2)*. The chemical inertness of a membrane has been one of the critical properties that determined the usefulness of a particular membrane in electroblotting. Proteins eletrotransferred to nitrocellulose cannot be subjected to Edman sequencing but are amenable to in situ proteolytic digestion. PVDF (polyvinylidene diflouride) membranes have been widely used for N-terminal sequencing applications on account of the fact that they are inert to Edman degradation chemistry.

This study, to evaluate the electrotransfer of proteins to Teflon tape and membranes, was initiated on account of the fact that PVDF membranes were not inert to the chemistry used on a Hewlett-Packard G1009A C-terminal sequencer. Burkhart et al. *(3)* report the optimized conditions for electrotransfer of proteins from polyacrylamide gels to low-density Teflon tape and GORE-TEX expanded polytetrafluorethylene membranes that were discovered to provide performance similar to that obtained with PVDF. Teflon was found to be a suitable Edman sequencing support in an earlier study *(4)*. Teflon blots were found to be amenable to amino acid analysis, in situ proteolytic digestion, and a combination of N-terminal sequencing followed by C-terminal sequencing in addition to being inert to the chemistry used on Hewlett-Packard G1009A C-terminal sequencer.

In this procedure, electrophoretically separated proteins were electrotransferred onto Teflon membrane. The Teflon tape or GORE-TEX was moistened thoroughly with absolute ethanol prior to assembling the blotting sandwich. The proteins were stained with 0.005% sulforhodamine B in 30% methanol for 10 min following transfer. The blots were washed several times with distilled water, prior to drying, to remove excess stain *(3)*.

Blotted protein bands were cut out from dry blots and placed directly into the reaction cartridge without additional washing for automated sequencing purposes. They were found to be amenable to both N- and C-terminal sequencing. As long as the described wetting procedures (it was important to moisten the membranes with absolute ethanol or isopropanol and not with methanol and to keep the membrane wet with ethanol and not equilibrated with transfer buffer prior to being placed next to the gel on the blotting sandwich) were followed, the ease of preparing samples on Teflon blots was similar to that on PVDF membranes. However, Teflon tapes were not useful for western blotting owing to their low porosity.

3. Multiple Tissue Western Blot

An important part of protein blotting is the study of comparative protein levels in various tissues. Analysis of this kind permits characterization of tissue-specific protein isoforms, detection of immunologically related proteins, and examination of posttranslational modifications leading to changes in molecular weight. Immunoblot analysis of this kind requires considerable expenditure of money, time, and energy in terms of obtaining hard-to-get human tissues, protein content normalization, preparation of protein samples, and electrophoretic separation and transfer. The availability of prefabricated immunoblots would allow the investigator to proceed directly to the antigen detection phase of the experiment.

Human multiple tissue western (MTW) blots provide a new immunological tool for the investigation of tissue-specific protein expression. MTW blots are premade immunoblots *(5)* and are prepared utilizing proteins isolated from adult human tissue. The proteins are obtained from whole tissue under conditions of minimal proteolysis and ensuring maximal representation of tissue-specific proteins. Proteins are solubilized in sodium dodecyl sulfate (SDS) and fractionated by SDS polyacrylamide gels. They are then electroblotted onto polyvinylidene fluoride membranes to generate blots ready for incubation with specific antibodies. Following detection of antigen, the blots can be reused several times using a stripping protocol (*see* Chapter "Sequential Use of Immunoblots for Characterization of Autoantibody Specificities") that permits the selective removal of both primary and secondary antibodies in a single incubation. Multiple reprobing makes this protocol very useful to study human-tissue-specific proteins.

4. A Two-Step Transfer of Low and High Molecular Weight Proteins

Transfer of high molecular weight proteins has been the bane of investigators everywhere *(6)*. When higher percentage gels are used this problem is accentuated. Interestingly, investigators have sought to increase the transfer of large proteins by enhancing the degree of protein migration out of the gel during the transfer. Researchers have used prolonged electrotransfer (16–21 h) at high current density along with inclusion of SDS in the transfer buffer to enhance protein elution and efficiently transfer high molecular weight proteins *(7, 8)*.

Adding SDS to a final concentration of 0.1% in the transfer buffer and transferring for 21 h helps to quantitatively transfer proteins *(7)*. Some investigators have used novel gels and blotting buffers *(6)* to efficiently transfer high molecular weight proteins. These investigators made gels using a 50:1 ratio of acrylamide:bisacrylamide in all experiments, since they were mainly interested in transferring high molecular weight proteins. In addition, they also used a different blotting buffer. Others *(9)* have cleaved high molecular weight proteins with periodate or alkali, before transfer, to facilitate transfer of proteins. Partial proteolytic digestion of high molecular weight proteins prior to transfer *(10)* has also been attempted. Bigger pore sizes provided by agarose gels have been used to advantage in *(11)* a composite agarose–polyacrylamide gel containing SDS and urea for this same purpose.

Otter et al. *(8)* used a two-step procedure to electrotransfer both high molecular weight (greater than 400,000) and low molecular weight (less than 20,000) proteins from polyacrylamide gels to nitrocellulose sheets. The salient features of this method included a two-step electrotransfer. The low molecular weight polypeptides were eluted at a low current density (approximately 1 mA/cm^2) for 1 h. This was followed by prolonged electrotransfer (16–20 h) at high current density (approximately 3.5–7.5 mA/cm^2) under conditions that favored the transfer of high molecular weight proteins. SDS (0.01%) was added to the transfer buffer to enhance protein transfer. The nitrocellulose was air-dried following the transfer to eliminate protein loss during subsequent processing. This transfer procedure was found to work well with all polyacrylamide gel systems tested and with proteins prepared from many different cell types.

5. Blotting of Coomassie Brilliant Blue-Stained Proteins from PAGE Gels to Transparencies

The transfer of proteins from fixed SDS gels, stained with Coomassie, to nitrocellulose for purposes of immunostaining has been documented *(12)*. The procedure required the complete removal of the stain followed by equilibration of the gel in SDS running buffer prior to transfer. The efficiency of transfer was comparable with that obtained with unfixed gels. The protein pattern of the transfer was found to be sharper, probably due to reduced diffusion during the transfer process, and protein antigenicity was found to be unaffected.

The transfer of proteins stained with Coomassie Blue from polyacrylamide gels to transparencies of the type used in plain-paper copiers have also been demonstrated *(12, 13)*. The details of the original electropherogram were retained on transfer, and this did not fade over a period of 3 years. The protein transferred along with the dye, but failed to transfer in absence of the dye. Protein patterns present in SDS-polyacrylamide gel electrophoresis (PAGE), nondenaturing PAGE, SDS-agarose, and isoelectric focusing PAGE were found to be transferred successfully following staining with the dye. Proteins visualized with other organic dyes such as Uniblue A 3, Fast Green FCF 2, and Procion Blue MX-R 4 were also found to transfer. However, proteins stained with Stains-all 5 or silver did not transfer. This transfer method, thus, provides a simple, economical way to preserve data from slab gel electrophoresis and a convenient method to display data using an overhead projector.

6. Toxin Binding to Chimeric K^+ Channels Immobilized on a Solid Nitrocellulose Support

Chip technology is playing an important role in pharmaceutical and biological research. Microarrays that are used currently involve DNA. Peptides are also used to some extent. Large proteins, however, have not been used extensively mainly owing to difficulties associated with purification and structural integrity *(14, 15)*. It is even more problematic in the case of membrane proteins that are often stabilized by the lipid moiety. There have been reports of the functional immobilization on gold/glass *(16)* or sensor surface of BIAcore chip *(17)* of a ligand-activated G-protein-coupled receptor. K^+ channels are membrane proteins making up the largest and most ubiquitous family of ion channels that control excitability in a number of cell types. Several neurological diseases have been thought to be involved owing to the dysfunction of these channels and as a result are potential targets for therapeutic drugs *(15)*. Therefore, displaying K^+ channel proteins on a solid surface is valuable, potentially leading to clinical applications. Such an approach also has the potential for drug screening methodologies.

7. Development of a Membrane-Array Method for the Detection of Human Intestinal Bacteria in Fecal Samples

All mammals, including human beings, are adapted to life in a world of microbes. For every gram weight of human intestinal contents there are 10^{10}–10^{12} colony forming units of bacteria (this is 10–20 times the total number of tissue cells in the whole body) *(18)*. At least 400 different species of bacteria (of which 30–40 species account for 99% of the total microflora) colonize the human gastrointestinal tract (GIT) *(19, 20)*. The indigenous intestinal bacteria play important roles in food digestion, production of vitamins and other essential nutrients, metabolism of endogenous and exogenous compounds, and prevention of pathogenic microflora from colonizing the GIT. Therefore, the knowledge relating to the numbers and bacterial species found in the human GIT is important *(21)*. The safety implications of antibiotic resistant bacteria in foods, contamination of foods by fecal material, the effect of diet, food additives, and veterinary drug residues on the ecosystem of the intestine, and the use of probiotics in the prevention and treatment of GI disorders have stimulated the interest in intestinal microflora *(20)*.

The population of anaerobic bacteria in the human GIT has been characterized traditionally by biochemical, microscopic, physiological, and selective culture plating methods of fecal samples from human beings. A number of molecular techniques have been used in recent years to analyze the bacterial flora in human fecal samples. Such analysis can detect changes in the human GIT flora rapidly and precisely *(19, 20)*.

Wang et al. *(20)* have used a nitrocellulose membrane-array method to detect human GIT bacteria in fecal samples without the use of expensive microarray arrayer and laser scanner. Three 40-mer oligonucleotides specific for each of 20 important human GIT bacterial species (total 60 probes) were designed and synthesized, based on comparison of the 16S rDNA sequences in the GenBank data library.

The oligos were diluted and xylene cyanol was added as an indicator. The diluted oligos were heated in a boiling water bath for 2 min and then cooled immediately in ice water for 1 min. The cooled oligos were applied to the nitrocellulose membrane with a micropipetman in a 6 × 10 array. The membranes were heated at 80°C for 2 h after air-drying and UV cross-linking. Digoxigenin (DIG)-labeled 16S rDNAs were amplified by polymerase chain reaction from human fecal samples or pure cultured bacteria with the help of two universal primers and hybridized to the membrane array. Hybridization signals were read by NBT (nitro blue tetrazolium)/BCIP (5-bromo-4-chloro-3-indolyl phosphate) color development. The two universal primers were found to amplify full size 16S rDNA from all the 20 bacterial species that were tested. The membrane-array method was thus found to be a reliable method for the detection of important human intestinal bacteria in the fecal samples *(20)*.

8. A New Black Cellulose Nitrate Support for Protein Microarray

Protein microarray is still in an early stage of development compared with the well-established DNA microarray technology. This is due to the different physical and chemical properties of nucleic acids and proteins. There is no comparable amplification method for proteins like PCR used for nucleic acids, and also proteins are much more complicated to purify. The irreversible denaturation of proteins during the process of immobilization and the insufficient stability of purified proteins are still problematic *(22, 23)*. An optimal surface for all proteins has not been discovered yet owing to the varying properties of different proteins.

Reck et al. *(24)* have used a modified nitrocellulose membrane for the optimization of a microarray sandwich-enzyme-linked immunosorbant assay (ELISA) against hINF (human interferon)-γ. This membrane was found to provide an excellent signal-to-noise ratio (SNR) and very little autofluorescence. The novel microarray slide used by these investigators was a self-made prototype coated with a black microporous cellulose nitrate substrate. The porous nitrocellulose substrate was produced by Sartorius Stedim Biotech (GmbH, Gottingen, Germany) using an evaporation technology. Porous nitrocellulose substrates that are available commercially tend to produce a high background fluorescence through their overall thickness even though they provide high protein binding capacity. Such background fluorescence may be caused by the substrate itself or by the adhesives used to attach the cellulose nitrate to glass. Sartorius developed a novel recipe and adhesive-free manufacturing procedure to overcome this problem. The white substrate made by Sartorius was already optimized for high binding capacity and low autofluorescence (made possible by the choice of various cellulose nitrate grades). The black substrate was made to provide an additional benefit by adsorbing background fluorescence originating in the depth of the structure. The added coloring material is essential for the low autofluorescence of the black substrate.

Reck et al. *(24)* used this self-made black nitrocellulose membrane with a high SNR and low autofluorescence as a microarray substrate. For spotting automation, an affymetrix 417 contact printer was used. Probes were spotted using a spotting buffer containing phosphate buffered saline and 0.5% trehalose to which 40% glycerol was added to prevent denaturation and improve the stability of protein probes. Using this black nitrocellulose membrane the authors have optimized a microarray sandwich-ELISA against hINF-γ *(24)*.

9. Quantification of Proteins Bound to PVDF Membranes by Elution of Coomassie Brilliant Blue

Proteins transferred to PVDF membranes can be stained with a variety of stains including Coomassie Brilliant Blue (CBB), amido black, and colloidal gold *(25)*. It becomes essential to determine the amount of protein present on the membrane in order to accurately determine the sensitivity of immunostaining, initial yields from protein microsequencing, and the specific activity of enzymes bound to PVDF. Since the extent and rate of electroblotting is likely to change between samples, such values cannot be obtained from the amount of loaded proteins. Therefore, Kain and Henry *(26)* developed a method to quantify proteins bound to PVDF membrane by elution of CBB, in order to quantify proteins for microsequencing. Proteins were resolved on a 13% gel and transferred to PVDF.

The membrane was stained with CBB and rapidly destained. The stained membrane was air-dried and individual protein bands were excised with a scalpel. The protein from each PVDF piece was eluted with 0.1% SDS/50% isopropanol. The colored liquid was removed and read at 595 nm using a Beckman Model DU-40 spectrophotometer after calibrating with CBB eluted from a PVDF fragment derived from a region of the blot that did not have a protein band. The proteins are not eluted along with the dye. The authors eluted CBB from ^{14}C labeled proteins electrotransferred to PVDF membranes and subjected the eluted solution and the PVDF membrane containing the radiolabeled protein to liquid scintillation counting. Virtually all of the protein was found to be retained on the PVDF membrane fragment during CBB elution. The authors also found that each of the protein that they tested gave a linear response with respect to the change in absorbance in the protein range of 500 ng to 10 µg. The slope of each curve was found to differ considerably, confirming the well-known fact regarding the variance of CBB bound by different proteins. Thus, in any assay of protein content a standard curve with the protein of interest needs to be obtained to get the most accurate quantitative results. This protocol is useful for the analysis of multiple samples since the absorbance readings obtained from the eluted dye stay stable for up to 1 h. This procedure is not dependent on bandshape, since it could be useful in the quantitation of proteins separated by two-dimensional gel electrophoresis. Densitometric scanning of 2-D gels is complicated by the lack of uniformity in the shape of protein spots. On account of the fact that the proteins are not extracted from the membrane, this procedure is useful for protein quantitation before protein microsequencing, for analysis of amino acid composition and for immunological procedures.

10. Enhanced Protein Recovery After Electrotransfer Using Square Wave Alternating Voltage

Bienvenut et al. *(27)* studied the efficiency of protein transfer using the conventional continuous current (as in Towbin transfer system) and the use of an unusual square wave alternative voltage (SWAV).

In this procedure, immediately following SDS-PAGE, the gels were soaked in deionized water for 5 min and then equilibrated two times (5 min each time) in the cathodic blotting buffer. Transblot PVDF membranes were equilibrated with the anodic buffer for 5 min. The standard blotting technique used a continuous current (1 mA/cm^2) using 3-(cyclohexylamino)-1-propane-sulfonique acid (CAPS) buffer. The voltage used in the SWAV method of transfer was an asymmetrical alternating voltage also using CAPS buffer. This method delivered +12 V for 125 ms followed by −5 V for 125 ms repetitively. This corresponded to a 4-Hz frequency signal and an average tension of 3.5 V. Following transfer the gels were stained with Coomassie Blue and the membranes were stained with amido black and destained by repeatedly washing with deionized water. An average 65% increase of protein recovery was observed using the SWAV technology in combination with CAPS buffer compared with that obtained with standard immunoblotting conditions *(27)*.

11. Polyethylene Glycol-Mediated Significant Enhancement of the Immunoblotting Transfer

Zeng et al. *(28)* have studied the effects of polyethylene glycol (PEG) on protein fixation, electrotransfer from SDS-polyacrylamide gels onto PVDF membranes, and immunoblotting. Serum proteins were resolved by sodium dodecyl sulfate polyacrylamide gel electrophoresis (SDS-PAGE) and the gel was then immersed in a 30% PEG 2000 buffer for 2 h for reversibly fixing the proteins (unlike trichloroacetic acid–sulfosalicylic acid or acetic acid–methanol systems that irreversibly fix proteins). PEG 1000, 1500, or 2000 (at 30% level) were found to have almost equal ability to preserve the protein bands in the gels very well. PEG less than 1000 was not found to have significant effect. The proteins in the gel were electrotransferred to PVDF for 24–48 h (in a −20°C freezer at 200 mA/120 V) using the same buffer and visualized finally by the indirect HRP-antibody method. The total protein detected (lowest level) per lane was 25 pg. It was found that when the transfer was carried out in the absence of PEG, a similar immunostain revealed a significantly lower sensitivity and the bands obtained were found to be slightly blurred. PEG 1000–2000 was capable of increasing the sensitivity of immunoblotting by 10–100-fold. This increase is thought to be brought about by its nondenaturing hydrophobicity with (a) self-association of protein molecules that could diminish protein blow-through (through the PVDF membrane), (b) miniaturization caused by PEG, possibly increasing the intraband protein density, and (c) possible enhancement of the interaction between free antigen and antibody by PEG treatment.

12. Acid Electroblotting onto Activated Glass

Data obtained from partial amino acid sequencing have been used frequently in assisting the isolation of a gene coding for a specific protein and also for confirming that the right gene has been isolated. To establish the structure of the mature gene translation product and to cross-check sequences determined at the nucleic acid level more extensive amino acid sequence data can be employed. Aebersold et al. *(29)* have used a procedure to electroblot proteins to activated glass to isolate subpicomolar levels of proteins for microsequencing. They considered nitrocellulose and nylon membranes for this purpose. However, nitrocellulose was found to dissolve during the sequencing chemistry while the charge-modified nylon membranes were found to collapse into a solid pellet during the procedure. The glass fiber paper support was found to be completely stable to the sequencing conditions. But, untreated glass fiber sheets possess a very limited ability to adsorb proteins. However, the authors found that the glass fiber sheets developed considerable capacity to adsorb proteins (7–10 μg/cm^2) following treatment with triflouroacetic acid (TFA). The mechanism of protein adsorption was a result of ionic interaction of the net positively charged proteins (owing to the acidic transfer buffer) to the negative charges on the glass fiber sheet. The TFA treatment of the glass sheet apparently unveiled these negative charges on the sheet.

12.1. Activation of Glass Fiber Sheets (Acid Etching)

Whatman GF/C or GF/F circles or sheets were placed inside neat TFA in a covered Petri dish and incubated for 1 h at room temperature. Extreme care was taken to avoid air bubbles if more than one sheet was used. The glass fiber sheets were dried completely until there was no trace of TFA.

12.2. Acid Blotting

The SDS from the proteins, following SDS-PAGE, was displaced by immersing the gels in 0.5% (v/v) acetic acid containing 0.5% Nonidet P-40 for 10 min at room temperature. Before assembling the sandwich for blotting, each of the blotting components was incubated with the blotting buffer. However, in order to minimize acid-catalyzed protein cleavage, acid concentration was kept as low as possible. Blotting was carried out. At this low pH of protein transfer, the proteins have a net positive charge and migrate toward the cathode onto the glass fiber paper.

Protein bands, following detection with Coomassie or fluourescent staining, were excised out of the glass fiber sheets and placed in the cartridge of a gas-phase sequenator without further treatment *(29)*.

13. Clarification of Immunoblots on PVDF for Transmission Densitometry

For protein quantitation, methods such as radioimmunoassay or enzyme-linked immunosorbent assay have been used. However, these assays cannot provide much information regarding the characteristics of the protein, nor can they distinguish between cross-reactive species including different protein components that react with the probe. Consequently, SDS PAGE followed by immunoblotting is required to quantify events such as cleavage of polypeptide fragments utilizing polyclonal sera or the determination of enzyme activation status consequent to a loss of proenzyme domains *(30)*.

While results are compromised by the limited linearity of photographic methods, immunoblot quantification has been done indirectly by densitometry carried out on the autoradiograph of the blot *(31)* as well as by making a transparent copy of the blot *(30)*. Tagami et al. *(32)* have performed densitometry of a dry, color-stained blot with a laser on account of the intensity and collimation of the light source. However, it was found that the absorbance of even the unstained parts of the membrane was approximately 2 and also leading to variable baselines since they were found to be very sensitive to warping *(33)*. However, it was found that such densitometry was possible if the membranes could be rendered transparent like a polyacrylamide gel. Nitrocellulose could be made transparent by treating with three-in-one lubricating oil *(33)*, thus allowing the blot to be quantitated using a conventional densitometer, analogous to the manner in which gels are scanned. However, PVDF (on account of its more hydrophilic character) was not made transparent by treatment with oil.

Tarlton and Knight tried several solvents differing in refractive index to make PVDF transparent *(30)*. Their idea was that if PVDF could be made transparent like nitrocellulose it could be used for several experiments that were not possible with nitrocellulose, such as adsorbing lipids for the detection of anticholesterol antibodies, to be used in conjunction with transmission densitometry. They found that ethylene glycol was the most effective. It was found that ethylene glycol/glycerol mixture (9:1; v/v; refractive index of 1.433) was found to bring about the most optimal clarification of PVDF membranes. For this procedure dry PVDF membrane was first moistened in methanol and then immersed in a solvent mixture with a minimum of one change of the solvent, and optical absorbance of PVDF was measured with a laser densitometer. Immunoblots were taken from storage in water and immersed in the clarification mixture with two changes.

The authors found that the PVDF immunoblots could be examined over long periods of time without loss of transparency, owing to the low volatility of both ethylene glycol and glycerol. The authors found no fading of bands after repeated clarification of dried PVDF blots by ethylene glycol/glycerol over a period of 2 months. However, clarification of nitrocellulose for the purpose of quantitative densitometry was found to produce some fading when oil was used.

14. Parallel Protein Chemical Processing Before and During Western Blot and the Molecular Scanner Concept

Following gel electrophoresis, the proteins are stained with either CBB or silver, and spots containing the protein of interest are excised either manually or using robotic methods. Following several rounds of washing, the gel slices are subjected to in-gel tryptic digestion. The tryptic peptides are extracted from each gel slice and concentrated, purified, and subjected to mass spectrometry. The robots that carry out the initial sample preparation are expensive. Furthermore, the investigator has to keep track of how each sample was related to the original gel during all of these steps, thus introducing opportunities for confusion. Matrix-assisted laser desorption/ionization (MALDI) mass spectrometry directly on a polymeric surface has been performed as early as 1990 *(34)*. Specifically, proteins transferred to a capture membrane can be analyzed by scanning the surface with the laser beam of a MALDI machine. The concept of molecular scanning *(35)* involves the use of enzymatic digestion during the blotting from an SDS-PAGE gel or two-dimensional gel to a PVDF capture membrane (a trypsin-coated membrane being placed between the gel and the PVDF membrane). This procedure does not require the use of gel staining, spot excision, or extraction. These authors found that a combination of in-gel digestion (prior to western blotting) together with the positioning of a trypsin-coated membrane between the gel and the PVDF capture membrane (during western blotting) resulted in obtaining the best digestion efficiency (digestion of high molecular weight and basic proteins without loss of low molecular weight polypeptides due to diffusion) compared with that obtained either by the use of in-gel digestion or by the use of trypsin-coated membrane alone during transfer. The capture membrane absorbs the peptide fragments from the digested proteins, following which the membrane is treated with a MALDI ionizing matrix and mass spectrometry analysis is carried out directly on the membrane. All the proteins that were originally present in the gel are processed at the same time using methods compatible with mass spectrometry. To obtain maximum digestion and transfer to the capture membrane the authors used a pulsed or alternating electric current (an unusual SWAV – *see* **Ref.***36)*. The time of contact between the proteins migrating out of the gel, with the trypsin immobilized on the membrane was maximized by this oscillating current. This present study highlighted a positive influence of the "shaking" effect of the asymmetric alternative voltage on gel protein extraction.

15. Electronic Western Blot of MALDI Mass Spectrometric-Identified Polypeptides from Parallel Processed Gel-Separated Proteins

As described earlier, proteins separated by one-dimensional or two-dimensional gel electrophoresis are transferred directly through a trypsin-coated membrane onto a membrane ready for MALDI *(37)*. The authors found that the protein transfer and efficiency of digestion were more than 95%. They identified 110 unique proteins obtained from an *Escherichia coli* extract and 149 proteins from a mouse liver homogenate resolved by one-dimensional SDS-PAGE. Furthermore, they used a Visual Basic Program to plot the identified proteins according to where they were found on the gel. Thus, the presence and distribution of any of the identified proteins could be visualized as in a western blot without the use of an antibody.

Nadler et al. *(37)* used a special membrane and procedure to covalently couple trypsin to obtain a high activity, compared with that used by Bienvenut et al. *(35)*. Thus, they could avoid starting the digestion process in the separating gel before electroblotting and also the use of the special oscillating electroblotting apparatus as done earlier *(35*). The aldehyde-activated polyethersulfone membrane, Gelman US450, reacts with nucleophiles (such as primary amines) to form nucleophiles that are then reduced with sodium cyanaborohydride. The authors determined that the amount of trypsin covalently attached to this membrane ranged from 14–22 $\mu g/cm^2$ of membrane. This amount of trypsin had sufficiently high activity to bring about a quick digestion of proteins in a single pass *(37)*.

16. Gold Coating of Nonconductive Membranes Before MALDI Tandem Mass Spectrometric Analysis Prevents Charging Effect

As seen earlier, it is possible to perform MALDI-mass spectrometry directly on polyvinylidene difluoride (PVDF) membrane. However, it is not possible to carry out tandem mass spectrometry (MS/MS) directly on this polymeric surface. The reason is that the isolating material is not able to dissipate the charge made by the MALDI process *(34, 38)*. This charging effect has been postulated to create local perturbations in the electric field between the acceleration plate and the sample. To overcome this problem and allow MS/MS analysis on tandem TOF instruments, the sample has to be conductive to dissipate the charges. By depositing a thin gold layer on the surface of a nonconductive membrane such as PVDF, Scherl et al. *(34)* describe a positive effect. The thin gold layer is applied by anodic vaporization following matrix deposition. The conductive material permits the dissipation of the charges, resulting in the first successful MS/MS analysis of peptides from PVDF membranes using a MALDI-TOF/TOF instrument under normal operating conditions *(34)*.

17. Semidry Electroblotting of Peptides and Proteins from Acid-Urea Polyacrylamide Gels

Electroblotting protocols have been mostly designed for protein transfer from SDS-containing gels, using tank-type apparatus typically requiring 10–18 h for transfer. Semidry transfer methods have been developed for SDS-PAGE not long ago *(39)*. Owing to its speed and convenience this procedure has been adopted for DNA and RNA electroblotting.

Polyacrylamide gels without SDS provide an important avenue for separating proteins under partially denaturing or native conditions. Separation of proteins and peptides as a function of their combined size and charge has been made possible with the use of low pH PAGE systems. The acidic, urea-containing (AU)-PAGE system *(40)* enables excellent resolution of several proteins and peptides that cannot be resolved by SDS-PAGE.

Model polypeptides such as calf thymus histone (21.5 kD), ribonuclease A (13.7 kD), human lysozyme (14 kD), pancreatic trypsin inhibitor (6.5 kD), and others with molecular weights ranging from 6.5 to 3.3 kD were used to test transfer parameters. Using a power setting of 115 mA and 5 V, a transfer solution of 5% acetic acid (the same solution was used for electrophoresis), and a transfer time of 15 min it was possible to transfer the polypeptides almost completely. PVDF (0.2 μm) was found to be a superior membrane, compared with nitrocellulose (0.2 μm) for efficient transfer.

The authors studied the effect of adding methanol to the transfer buffer. Methanol addition has been recommended for increasing protein binding capacity of nitrocellulose membrane *(41)*. Wang et al. (*42*) found that the addition of methanol (10%) brought about precipitation of protein on the gel, impeding the electrophoretic transfer. Also, they found no increase in protein binding to either PVDF or nitrocellulose membranes upon addition of methanol to the buffer.

18. Transfer of Silver-Stained Proteins from Polyacrylamide Gels to PVDF Membranes

Wise and Lin *(43)* showed that they could transfer almost all the silver-stained proteins from a polyacrylamide gel to PVDF by rinsing the gel in 2× SDS Laemmli sample before transfer. Some silver-stained proteins were also found to be directly transferred without a rinse with the Laemmli buffer. The antigenicity of the transferred proteins was found to be retained when transferred either way.

19. A Simple Method for Coating Native Polysaccharides onto Nitrocellulose

Lipid-free or protein-free polysaccharides (PS) have been reported to have very low affinity for plastic and polystyrene. Carbohydrate molecules have been modified in an effort to increase their binding affinity. These modifications have involved the covalent attachment of poly-L-lysine biotin or tyramine to the PS. The modification has the possibilities of being limited by high backgrounds, loss of specificity, loss of antigenicity, and lack of reproducibility.

Feng et al. describe a method to immobilize PS to nitrocellulose without using any modification and permitting antibody analysis by enzyme-linked immunoassays. Bacterial levan (a $\beta(2 \rightarrow 6)$-polyfructosan and dextran (a polymer of $\alpha(1 \rightarrow 6)$ glucose), both neutral PS and a highly charged PS (*Neisseria meningitides* group C polysaccharide; a polymer of $\alpha(2 \rightarrow 9)$ sialic acid) have been coated onto nitrocellulose through filtration devices. Various blotting assays can be used with the PS-coated membrane to investigate specific antibodies *(44)*.

References

1. Hermansen, L.F., Pedersen, O., and Sletten, K. (1993) Centrifuge-blotting of proteins after separation by sodium dodecyl sulfate-polyacrylamide gel electrophoresis. *Electrophoresis.* 14, 1302–1306.
2. Patterson, S.D. (1994) From electrophoretically separated protein to identification: strategies for sequence and mass analysis. *Anal Biochem.* 221, 1–15. Review.
3. Burkhart, W.A., Moyer, M.B., Bailey, J.M., and Miller, C.G. (1996) Electroblotting of proteins to Teflon tape and membranes for N- and C-terminal sequence analysis. *Anal Biochem.* 236, 364–367.
4. Burkhart, W., Moyer, M., Bodnar, W., Everson, A., Valladares, V., et al. (1995) Direct collection into Zitex and PVDF for Edman sequencing: elimination of polybyrene. In: Techniques in Protein Chemistry VI (Crabb, J., ed.), Academic Press, San Diego, pp. 169–176.
5. Kain, S.R., Mai, K., and Sinai, P. (1994) Human multiple tissue western blots: a new immunological tool for the analysis of tissue-specific protein expression. *Biotechniques.* 17, 982–987.
6. Bolt, M.W. and Mahoney, P.A. (1997) High efficiency blotting of proteins of diverse sizes following sodium dodecyl sulfate-polyacrylamide gel electrophoresis. *Anal Biochem.* 247, 185–192.

7. Erickson, P.F., Minier, L.N., and Lasher, R.S. (1982) Quantitative electrophoretic transfer of polypeptides from SDS polyacrylamide gels to nitrocellulose sheets: a method for their re-use in immunoautoradiographic detection of antigens. *J Immunol Methods.* 51, 241–249.
8. Otter, T., King, S.M., and Witman, G.B. (1987) A two-step procedure for efficient electro transfer of both high-molecular weight (greater than 400,000) and low-molecular weight (less than 20,000) proteins. *Anal Biochem.* 162, 370–377.
9. Renart, J., Reiser, J., and Stark, G.R. (1979) Transfer of proteins from gels to diazobenzyloxymethyl paper and detection with and detection with anti-sera: a method for studying antibody specificity and antigen structure. *Proc Natl Acad Sci USA.* 76, 3116–3120.
10. Gibson, W. (1981) Protease-facilitated transfer of high-molecular-weight proteins during electrotransfer to nitrocellulose. *Anal Biochem.* 118, 1–3.
11. Elkon, K.B., Jankowski, P.W., and Chu, J.L. (1984) Blotting intact immunoglobulins and other high-molecular-weight proteins after composite agarose-polyacrylamide gel electrophoresis. *Anal Biochem.* 140, 208–213.
12. Phelps, D.S. (1984) Electrophoretic transfer of proteins from fixed and stained gels. *Anal Biochem.* 141, 409–412.
13. Chu, Y.H., and Whitesides, G.M. (1993). A convenient procedure for transfer blotting of Coomassie Blue stained proteins from PAGE gels to transparencies. *Biotechniques.* 14, 925–930.
14. Tomizaki, K.Y., Usui, K., and Mihara, H. (2005) Protein-detecting microarrays: current accomplishments and requirements. *Chembiochem.* 6, 782–799. Review.
15. Legros, C., Martin-Eauclaire, M.F., Pongs, O., and Bougis, P.E. (2007) Toxin binding to chimeric K+ channels immobilised on a solid nitrocellulose support. *Biochem Biophys Res Commun.* 353, 1086–1090.
16. Neumann, L., Wohland, T., Whelan, R.J., Zare, R.N., and Kobilka, B.K. (2002) Functional immobilization of a ligand-activated G-protein-coupled receptor. *Chembiochem.* 3, 993–998.
17. Bieri, C., Ernst, O.P., Heyse, S., Hofmann, K.P., and Vogel, H. (1999) Micropatterned immobilization of a G protein-coupled receptor and direct detection of G protein activation. *Nat Biotechnol.* 17, 1105–1108.
18. Suau, A., Bonnet, R., Sutren, M., Godon, J.J., Gibson, G.R., Collins, M.D., et al. (1999) Direct analysis of genes encoding 16S rRNA from complex communities reveals many novel molecular species within the human gut. *Appl Environ Microbiol.* 65, 4799–4807.
19. Moore, W.E. and Moore, L.H. (1995) Intestinal floras of populations that have a high risk of colon cancer. *Appl Environ Microbiol.* 61, 3202–3207.
20. Wang, R.F., Kim, S.J., Robertson, L.H., and Cerniglia, C.E. (2002) Development of a membrane-array method for the detection of human intestinal bacteria in fecal samples. *Mol Cell Probes.* 16, 341–350.
21. Falk, P.G., Hooper, L.V., Midtvedt, T., and Gordon, J.I. (1998) Creating and maintaining the gastrointestinal ecosystem: what we know and need to know from gnotobiology. *Microbiol Mol Biol Rev.* 62, 1157–1170. Review.
22. Glökler, J., and Angenendt, P. (2003) Protein and antibody microarray technology. *J Chromatogr B Analyt Technol Biomed Life Sci.* 797, 229–240. Review.
23. Kusnezow, W., and Hoheisel, J.D. (2003) Solid supports for microarray immunoassays. *J Mol Recognit.* 16, 165–176. Review.
24. Reck, M., Stahl, F., Walter, J.G., Hollas, M., Melzner, D., and Scheper, T. (2007) Optimization of a microarray sandwich-ELISA against hINF-gamma on a modified nitrocellulose membrane. *Biotechnol Prog.* 23, 1498–1505.
25. LeGendre, N. and Matsudaira, P. (1988) Direct protein microsequencing from Immobilon-P transfer membrane. *Biotechniques.* 6, 154–159.
26. Kain, S.R., and Henry, H.L. (1990) Quantitation of proteins bound to polyvinylidene difluoride membranes by elution of Coomassie brilliant blue R-250. *Anal Biochem.* 189, 169–172.
27. Bienvenut, W.V., Déon, C., Sanchez, J.C., and Hochstrasser, D.F. (2002) Enhanced protein recovery after electrotransfer using square wave alternating voltage. *Anal Biochem.* 307, 297–303.
28. Zeng, C., Suzuki, Y., and Alpert, E. (1990) Polyethylene glycol significantly enhances the transfer of membrane immunoblotting. *Anal Biochem.* 189, 197–201.
29. Aebersold, R.H., Teplow, D.B., Hood, L.E., and Kent, S.B. (1986) Electroblotting onto activated glass. High efficiency preparation of proteins from analytical sodium dodecyl sulfate-polyacrylamide gels for direct sequence analysis. *J Biol Chem.* 261, 4229–4238.

30. Tarlton, J.F. and Knight, P.J. (1996) Clarification of immunoblots on polyvinylidene difluoride (PVDF) membranes for transmission densitometry. *J Immunol Methods.* 191, 65–69.
31. Likavcanova, E. and Lagacé, J. (1992) Quantitative analysis of immunoglobulin G subclass responses to *Pseudomonas aeruginosa antigens in cystic fibrosis. J Med Microbiol.* 36, 437–444.
32. Tagami, T., Nakamura, H., Sasaki, S., Miyoshi, Y., and Imura, H. (1993) Estimation of the protein content of thyroid hormone receptor alpha 1 and beta 1 in rat tissues by western blotting. *Endocrinology.* 132, 275–279.
33. Maciewicz, R.A. and Knight, P.J. (1988) Transmission densitometry of stained nitrocellulose paper. *Anal Biochem.* 175, 85–90.
34. Scherl, A., Zimmermann-Ivol, C.G., Di Dio, J., Vaezzadeh, A.R., Binz, P.A., Amez-Droz, M., et al. (2005) Gold coating of non-conductive membranes before matrix-assisted laser desorption/ionization tandem mass spectrometric analysis prevents charging effect. *Rapid Commun Mass Spectrom.* 19, 605–610.
35. Bienvenut, W.V., Sanchez, J.C., Karmime, A., Rouge, V., Rose, K., Binz, P.A., et al. (1999) Toward a clinical molecular scanner for proteome research: parallel protein chemical processing before and during western blot. *Anal Chem.* 71, 4800–4807.
36. Bienvenut, W.V., Déon, C., Sanchez, J.C., and Hochstrasser, D.F. (2002) Enhanced protein recovery after electrotransfer using square wave alternating voltage. *Anal Biochem.* 307, 297–303.
37. Nadler, T.K., Wagenfeld, B.G., Huang, Y., Lotti, R.J., Parker, K.C., and Vella, G.J. (2004). Electronic western blot of matrix-assisted laser desorption/ionization mass spectrometric-identified polypeptides from parallel processed gel-separated proteins. *Anal Biochem.* 332, 337–348.
38. Knochenmuss, R. (2004) Photoionization pathways and free electrons in UV-MALDI. *Anal Chem.* 76, 3179–3184.
39. Jacobson, G. and Kårsnäs, P. (1990) Important parameters in semi-dry electrophoretic transfer. *Electrophoresis* 11, 46–52.
40. Panyim, S., and Chalkley, R. (1969) High resolution acrylamide gel electrophoresis of histones. *Arch Biochem Biophys.* 130, 337–346.
41. Johnson, T.K., Yuen, K.C., Denell, R.E., and Consigli, R.A. (1983) Efficient transfer of proteins from acetic acid-urea and isoelectric-focusing gels to nitrocellulose membrane filters with retention of protein antigenicity. *Anal Biochem.* 133, 126–131.
42. Wang, M.S., Pang, J.S., and Selsted, M.E. (1997) Semidry electroblotting of peptides and proteins from acid-urea polyacrylamide gels. *Anal Biochem.* 253, 225–320.
43. Wise, G.E. and Lin, F. (1991) Transfer of silver-stained proteins from polyacrylamide gels to polyvinylidene difluoride membranes. *J Biochem Biophys Methods.* 22, 223–231.
44. Feng, S.H., Rubinstein, L.J., and Stein, K.E. (1991) A simple method for coating native polysaccharides onto nitrocellulose. *J Immunol Methods.* 137, 261–266.

Chapter 39

Phosphoprotein Detection on Protein Electroblot Using a Phosphate-Specific Fluorophore

Lee Broderick Bockus and R. Hal Scofield

Summary

The reversible phosphorylation of phosphoproteins is a vital regulatory process for many cellular pathways. A reliable and simple fluorescent detection technique for phosphoproteins has been developed using a small-molecule organic fluorophore, Pro-Q Diamond dye. This was originally developed for use in gel staining, but a new formulation has allowed for its use in protein blotting. The dye binds noncovalently and selectively to the phosphate moiety, so proteins lacking phosphate groups and other macromolecules such as DNA or RNA are not detected. It uses a standard electrophoresis and electroblotting technique, which can blot the sample onto nitrocellulose membranes or polyvinylidene fluoride (PVDF). The electroblotting is followed by staining with the dye and destaining. The blot can then be read by multiple types of imaging devices such as a laser-based gel scanner. This process is compatible with matrix-assisted laser desorption time-of-flight mass spectrometry (MALDI-TOF-MS) and Edman sequencing. It can also be followed by standard chemiluminescent, colorimetric, and fluorogenic detection techniques used in immunoblotting.

Key words: Phosphoprotein detection, Phosphoproteome, Phosphopeptides, Protein blotting, Pro-Q diamond, Fluorescent detection

1. Introduction

Protein phosphorylation is one of the most commonly occurring posttranslational modifications (PTM). This reversible modification changes the structure and consequently the function of the target protein. Kinases transfer the phosphate group to the target phosphoprotein and phosphatases remove it, creating a molecular on–off switch *(1)*. This modification is a necessary

Biji T. Kurien and R. Hal Scofield (eds.), *Methods in Molecular Biology, Protein Blotting and Detection, vol. 536*

DOI: 10.1007/978-1-59745-542-8_39

action in various biological processes including metabolism, signal transduction, cell division, gene regulation, and many others. It is one of the few types of reversible PTMs, and about one out of three proteins are phosphorylated at some point during their life cycle *(2)*. Out of the four types of phosphorylation, O-phosphorylation is the most common, and the amino acids usually bound by the phosphate moiety are serine, threonine, or tyrosine *(3)*.

The widespread dependence on phosphoproteins among cellular processes necessitates reliable detection techniques. The oldest method for studying phosphorylation is isotopic labeling of proteins using an inorganic ^{32}P as a radiolabel for integration into phosphoproteins. After incorporation of the isotopes, this technique generally involves protein separation by gel electrophoresis and autoradiography *(4)*. Although a reliable technique for many in vitro studies, radiolabeling has many disadvantages. Perhaps most importantly, it can only be used in the study of metabolically active cells, which allows for the incorporation of the radioactive isotopes and therefore is not applicable in studies of isolated protein systems or postmortem samples. Also, since this technique actively phosphorylates proteins over a given time frame, the results can only apply to that time period and not necessarily to the native state of the cells. Lastly, the standard amount of radioactivity introduced into cells has been shown to induce apoptosis *(5)*, which would greatly affect the phosphorylation patterns of the studied system.

Western blotting is another technique that may be used for the detection of phosphoproteins *(6)*. Phosphoamino acid-specific antibodies and 2-D gel electrophoresis have allowed the advancement in this technique, and it has become very sensitive with the development of anti-phospho antibodies and ECL detection *(7)*. Although quality phosphotyrosine antibodies are available, excellent antibodies for phosphoserine and phosphothreonine are lacking. The variability in enzyme quality and specificity of western blotting can make any quantitative attempts inaccurate.

The most convenient way to identify the phosphoproteins in a sample is to use phosphoprotein-specific stains, but this has been previously ineffective on account of low sensitivity as well as poor specificity. The advancement from colorimetric dyes to fluorescent dyes has allowed for more widespread and accurate detection of phosphoproteins. The most noteworthy of these is the Pro-Q Diamond dye, which is a fluorophore specifically formulated for phosphoprotein detection directly in the gel *(8)*. This has shown the sensitivity and specificity required to make it a leading technique and as a result has opened the field of multiplexed proteomics. One drawback is that the fluorescent dye is not compatible with phosphoprotein detection on an electroblot membrane because of high background staining.

Consequently, a more recent formulation of Pro-Q Diamond dye is made specifically for detection on PVDF or nitrocellulose membranes *(9)*. Similar to the previous dye, it also binds directly to the phosphate moiety but the background staining is very low, giving clear detection on protein blots. This method avoids the need to extensively fix the gel, previously required for removal of SDS before staining, and also has much shorter staining and destaining times because of the more easily accessible protein on the blot. A disadvantage of this method is that quantitative methods of detection through multiplex proteomics become somewhat inaccurate because of the variability in levels of protein transferred to the membrane although a heated transfer or longer transfer times can alleviate this problem for the most part. Although this second fluorescent method is also slightly less sensitive, it allows one to use the various techniques of immunoblotting directly after the fluorescent staining is completed. This allows the stain and the immunoblot to use the same membrane, and thus eliminates the need for cross-referencing with polyacrylamide gels or film.

The procedure itself is very simple and straightforward compared with the previous radiolabeling and western blotting techniques. First the sample is separated by one- or two-dimensional gel electrophoresis and transferred to either a nitrocellulose or PVDF membrane. The electroblot is quickly fixed before staining with Pro-Q Diamond phosphoprotein blot stain and then subsequently destained to maintain minimal background interference. The fluorescent blot can then be viewed with a variety of imaging instruments, preferably a laser-based gel scanner with appropriate lasers and filters. After the phosphoprotein imaging is complete, SYPRO Ruby protein stain may also be used for total protein detection. The combination of the two fluorescent dyes allows the differentiation between a nonphosphorylated protein in high abundance that has some nonspecific staining and a less abundant highly phosphorylated protein. This technique is able to detect phosphoproteins at levels as low as 8–16 ng, and also has a linear dynamic range of about 15-fold. Also, since Pro-Q Diamond dye bonds noncovalently, it is compatible with matrix-assisted laser desorption time-of-flight mass spectrometry (MALDI-TOF-MS) and Edman sequencing. As long as an adequate imaging device is available, this protein blot technique provides a quick, cheap, and reliable technique for the detection of phosphoproteins, while also allowing for the subsequent employment of other colorimetric, fluorogenic, and chemiluminescent techniques used in immunoblotting.

2. Materials

2.1. SDS-Polyacrylamide Gel Electrophoresis (SDS-PAGE)

1. PeppermintStick phosphoprotein molecular weight standard (Molecular Probes, Eugene, OR, USA): For long-term storage keep at –20°C, for short-term storage keep at 2–6°C (*see* **Note 1**).
2. BenchMark prestained protein ladder (Gibco BRL, Gaithersburg, MD, USA): Store at –20°C and avoid repetitive freezing (*see* **Note 2**).

2.2. Electroblotting

1. Transfer buffer (10×): 250 mM Tris, 1.92 M glycine.
2. BioTrace NT Nitrocellulose transfer membrane (Pall, Pensacola, FL, USA): Avoid pure organic solvents (for PVDF, *see* **Note 3**).

2.3. Staining of Phosphoproteins

1. Fixing Solution: 7% (v/v) concentrated acetic acid, 10% (v/v) methanol.
2. Pro-Q Diamond phosphoprotein blot stain kit (Molecular Probes, Eugene, OR): Blot stain reagent (Component A) (store at < –20°C, protect from light); blot stain buffer (Component B) (store at 2–25°C), and keep above freezing.
3. Pro-Q Diamond phosphoprotein gel destaining solution (Molecular Probes, Eugene, OR, USA).

2.4. Phosphoprotein Detection and Analysis

1. Molecular Imager PharosFX (Bio-Rad Laboratories, Hercules, CA, USA) (for other examples of imagers, *see* **Note 4**).
2. Quantity One 1-D Analysis Software (Bio-Rad): The Analysis software was provided with the imaging instrument.

2.5. Total Protein Stain, Detection, and Analysis

1. SYPRO Ruby protein blot stain (Molecular Probes): Store at room temperature and protect from light.

3. Methods

This detection technique has a relatively uncomplicated procedure that provides an easy way to detect phosphoproteins on a blot membrane and also allows immunoblotting of the same blot. All equipment must be very clean and all solutions pure because the sensitivity of the method can pick up any contaminants. The procedure needs to be followed carefully especially with respect to staining and destaining times so that consistent blots are produced, allowing better comparison between separate blots. Once the staining procedure is finished, the blots only need to be washed before either pursuing any colorimetric, chemiluminescent, or fluorogenic immunoblotting techniques, MALDI-TOF-MS, or pursuing Edman sequencing.

3.1. SDS-PAGE

1. Sample and marker preparation: For each sample, mix 4 μL of SDS lysis buffer (5×) with 16 μL of sample solution, which should previously have been made to the desired concentration using PBS. Adjust accordingly if repeating one sample in multiple wells. Prepare the PeppermintStick phosphoprotein marker by mixing 1 μL of the protein standard with 6 μL of SDS gel loading buffer, which must be diluted from its stock concentration to its standard concentration before addition. The BenchMark standard, once thawed, is gel ready.
2. Heat the samples and the PeppermintStick phosphoprotein marker by placing them in a hot water bath at 95°C for 4 min (*see* **Note 5**). Do not heat the BenchMark standard.
3. SDS-PAGE was carried according to Laemmli *(10)*. The proteins were separated on 10% SDS-PAGE precast gels (10-well) (ISC Bioexpress, Kaysville, UT, USA).

3.2. Electroblotting

1. Protein transfer was carried out according to a standard procedure *(6)* (*see* Chapter "Detection of La/SS-B by western blot using nanogold-tagged antibodies and silver enhancement").

3.3. Staining of Phosphoproteins

1. Wash the container for the membrane rigorously and rinse with 70% ethanol (*see* **Note 6**).
2. Fixing: Immerse the membrane, with the protein-bound side facing down, in 25 mL (or more if required to comfortably submerge the membrane) of the fixing solution for 10 min (*see* **Note 7**).
3. While fixing is taking place, make the phosphoprotein blot stain by adding 25 μL of the blot stain reagent to 25 mL of the blot stain buffer (*see* **Note 8**).
4. Pour out the rest of the fixing solution and wash the membrane with at least 25 mL dH_2O four times for 5 min.
5. Staining: Incubate the membrane in 25 mL of the prepared phosphoprotein blot stain for 15 min (*see* **Note 9**).
6. Destaining: Pour out the stain and gently tap out most of what is remaining, and then wash the membrane with three changes of 30 mL destaining solution for 5 min each. Make sure to keep the time of destaining consistent between blots since it will directly affect the signal strength.
7. Wash the membrane three times with 30 mL dH_2O for 5 min each, and let it air-dry completely before imaging (*see* **Note 10**).

3.4. Phosphoprotein Detection and Analysis

1. Open the tray cover and pull out the loading tray of the imager. Clean the surface with 70–100% ethanol, followed by a water wash as well.
2. Once the surface is clean and dry, place the blot on the tray making sure that the fluorescent signal is directed upward (*see* **Note 11**).
3. Take note of the coordinates on the tray that the blot covers (*see* **Note 12**), carefully push in the tray, and close the tray cover.
4. Open the Quantity One window and either select the Pro-Q Diamond stain from the computer options or manually select the 532 nm excitation and the 555 nm longpass filter (*see* **Note 13**).
5. After entering the coordinates for the scanning area, choose a desired resolution from 50 to 800 mm and click Acquire (*see* **Note 14**).
6. Use the PeppermintStick marker to adjust the settings so that the optimal gel image can be acquired. An example of results is shown in **Fig. 1**.
7. Quantity One will help with image export and any further analysis needed, including automatic band detection and

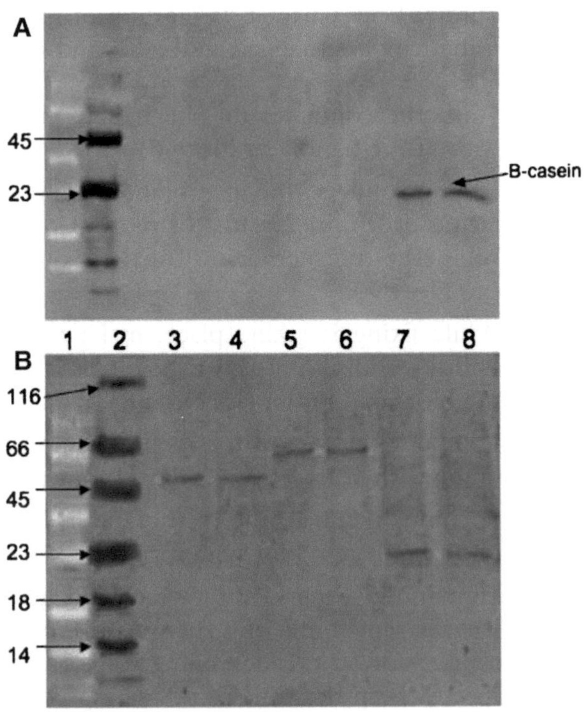

Fig. 1. Image of Pro-Q Diamond (**A**) and SYPRO Ruby (**B**) stained electroblots showing phosphoprotein and total protein detection, respectively. *Lane 1*, BenchMark protein standard. *Lane 2*, PeppermintStick phosphoprotein standard. *Lanes 3–4*, an unphosphorylated phosphoprotein (La). *Lanes 5–6*, bovine serum albumin. *Lanes 7–8*, phosphorylated β-casein. Two hundred nanograms of each protein was used.

image stacking so that multiple blots can be aligned and compared.

3.5. Total Protein Stain, Detection, and Analysis

1. The blot must first be refixed for 10 min in a solution of 45% (v/v) methanol, 5% (v/v) concentrated acetic acid, using the same method as before.
2. Wash the membrane three times with 30 mL dH_2O for 5 min each.
3. After pouring out the water, immerse the membrane in 20 mL SYPRO Ruby protein blot stain.
4. Pour the stain out or into a storage container for reuse, and then gently tap out as much else as possible. Again wash the membrane three times with 30 mL dH_2O for 5 min each (*see* **Note 15**).
5. Dry and handle the blot as before.
6. Image using the same methods as before, except selecting the SYPRO Ruby from the computer options or manually selecting the 488 nm excitation and the 640 nm bandpass filter (*see* **Note 16**). An example of results is shown in **Fig. 1**.
7. If the membrane is going to be used for immunoblotting, immerse it three times in 25 mL of PBS for 15 min. After the washing is finished begin the immunoblotting protocol as is standard.

4. Notes

1. These markers also serve as a positive and negative control. The standards include two phosphorylated and four nonphosphorylated proteins that range from 14–116 kDa.
2. Inclusion of this marker is unnecessary but convenient since it can be seen, aiding in positioning the gel and the subsequent blot as well as confirming the quality of the electrophoretic separation before proceeding with staining. Also, when measuring the proper amount of marker from the stock product, it is best to measure numerous aliquots once and freeze them for future use so that the thawing and freezing of the stock protein marker can be kept to a minimum.
3. PVDF membranes can be used as well with only a few changes to the procedure: The PVDF membrane must be wetted using methanol before fixing. Instead of the destaining solution used for nitrocellulose membranes use 50 mM sodium acetate, pH 4, 20% acetonitrile. Lastly, after the destaining step, it is not necessary to wash the PVDF membrane before air-drying.

4. The following imaging instruments are also commonly used: FluorImager and the Typhoon series of imagers by Amersham Biosciences, the FLA-3000G and FLA-5100 by Fuji Photo Film Co., and the ProXPRESS by PerkinElmer LifeSciences. Although laser-based or xenon arc lamp-based gel scanners are recommended for optimum sensitivity, UV epi- or transilluminators in conjunction with a charge-coupled device (CCD) camera can be used as well but with significantly diminished results.
5. Be careful not to let any water get into the samples. Keep the tops of the tubes above the water level. It may also be necessary to cover the top with parafilm.
6. For a more rapid transfer, run at 100 V for 1 h. Since this produces more heat, surround the tank with ice to keep it from overheating.
7. The best container to use is one slightly larger than the membrane with the walls at right angles to a smooth bottom, so that the membrane will not scrape on an angled wall during agitation.
8. All fixing, washing, staining, and destaining incubation peri ods should use gentle agitation, such as 50–60 rpm on an orbital shaker, with the membrane immersed facedown in the solution.
9. After this point, take care to minimize the exposure of the fluorophore to light since both time and exposure to light will decrease its optimal signal. Once the stain is added to the membrane, cover the container with aluminum foil for all subsequent steps, when not adding or decanting solutions.
10. For a more economical approach, the Pro-Q Diamond phosphoprotein blot stain can be reused for up to four blots with only a slight drop in performance, as long as such blots are completed in a relatively close time to each other. The SYPRO Ruby protein blot stain can also be reused for the same number of times, but it will store as long as the original stock stain if it is protected from light and kept at room temperature.
11. When still wet, do not touch the face of the blot and try not to touch it with gloves; use tweezers instead. The air-drying is best done by hanging the membrane vertically by one corner so that it will not develop an uneven drying pattern. A binder clip is convenient for clasping the corner of the membrane for suspended air-drying. The drying process must be done away from light and for further storage it is best to keep between 2 and 6°C.
12. Dry blots will often lie unevenly on the surface of the tray, so it is best to pin the sides of the blot down to make it as flat as possible.

13. It is best to place the blot on an edge of the tray to ensure proper alignment and minimize any movement of the blot during scanning.
14. If the 555 nm longpass is not available, the 605 nm bandpass filter is also effective, but a significant decrease in sensitivity is shown with the 640 nm bandpass filter. If an additional filter is required, installation is quite straightforward.
15. The scanner must be placed on a firm surface so that it will not shake so vigorously. A freestanding table will probably shake too much causing the blot to move and distorting the image.
16. During the final wash, the membrane may be monitored by a UV-B epi-illuminating light source to make sure that any residual background stain has been washed off. This stain is very stable, so this will not inhibit its fluorescence. For other imaging instruments, try to match the excitation maxima of 450 nm and the emission maxima of 618 nm as closely as possible. For UV light, the excitation maxima of 280 nm may also be used.

Acknowledgement

This work was supported by NIH grant ARO1844 and Oklahoma Center for the Advancement of Science and Technology to RHS.

References

1. Hunter, T. (1995) Protein kinases and phosphatases: the yin and yang of protein phosphorylation and signaling. *Cell* 80, 225–236.
2. Zolnierowicz, S. and Bollem, M. (2000) Protein phosphorylation and protein phosphatases. *EMBO J* 19, 483–488.
3. Sickmann, A. and Meyer, H.E. (2001) Phosphoamino acid analysis. *Proteomics* 1, 200–206.
4. Wettenhall, R.E., Aebersold, R.H., and Hood, L.E. (1991) Solid-phase sequencing of 32P-labeled phosphopeptides at picomole and subpicomole levels. *Methods Enzymol* 201, 186–199.
5. Yeargin, J. and Haas, M. (1995) Elevated levels of wild-type p53 induced by radiolabeling of cells leads to apoptosis or sustained growth arrest. *Curr Biol* 5, 423–431.
6. Towbin, H., Staehelin, T., and Gordon, J. (1979) Electrophoretic transfer of proteins from polyacrylamide gels to nitrocellulose sheets: procedure and some applications. *Proc Natl Acad Sci USA* 76, 4350–4354.
7. Kaufmann, H., Bailey, J.E., and Fussenegger, M. (2001) Use of antibodies for detection of phosphorylated proteins separated by two-dimensional gel electrophoresis. *Proteomics* 1, 194–199.
8. Schulenberg, B., Aggeler, R., Beechem, J.M., Capaldi, R.A., and Patton, W.F. (2003) Analysis of steady-state protein phosphorylation in mitochondria using a novel fluorescent phosphosensor dye. *J Biol Chem* 278, 27251–27255.
9. Goodman, T., Schulenberg, B., Steinberg, T.H., and Patton, W.F. (2004) Detection of phosphoproteins on electroblot membranes using a small-molecule organic fluorophore. *Electrophoresis* 25, 2533–2538.
10. Laemmli, U.K. (1970) Cleavage of structural proteins during assembly of the head of bacteriophage T4. *Nature* 227, 680–685.

Chapter 40

Rapid, Antibody-Free Detection of Recombinant Proteins on Blots Using Enzyme Fragment Complementation

Neil W. Charter, Joe Horecka, Chin-Yee Loh, Albert Doan, Tom Wehrman, and Keith R. Olson

Summary

Alternative, antibody-free techniques to western analysis of protein blots can offer reduced assay times for routine analysis of expression of recombinant proteins. We have adapted the commercially available enzyme fragment complementation technology to provide a rapid protein detection method for protein blots based on significantly reducing the number of incubation and washing steps used in traditional approaches, and eliminating the requirement for antibodies. In this article, we highlight the use of this assay for measuring recombinant protein expressed in mammalian cells for a range of applications, including dot blot screening of large numbers of different cell samples, assessment of protein integrity through detection of degradation bands, and characterization of post-translational protein modifications such as glycosylation.

Key words: Protein blot, Recombinant protein, Enzyme fragment complementation, Chemiluminescent detection

1. Introduction

Western blot analysis and enzyme-linked immunosorbant assay have provided valuable tools for characterization of protein expression in recombinant systems *(1)*. Both techniques rely on the use of antibodies specific for the protein of interest or to epitope tags (e.g., c-*myc*, HA) when the protein is expressed recombinantly and tagged *(2, 3)*. Antibody-based systems offer good performance and sensitivity; however, they require multiple incubation and wash steps that significantly extend the time required to perform the analyses. This is because the detection enzyme is directly

Biji T. Kurien and R. Hal Scofield (eds.), *Methods in Molecular Biology, Protein Blotting and Detection, vol. 536*

DOI: 10.1007/978-1-59745-542-8_40

conjugated to a secondary antibody and all antibodies present will generate signal, not just the antibody bound to the antigen.

Enzyme fragment complementation (EFC, also known as α-complementation) is a proprietary, high-sensitivity, enzyme-amplified detection approach that has been described previously and can be used routinely to detect expressed proteins in bacterial and mammalian cells, as well as for in vitro clinical and diagnostic applications *(4–8)*. We have adapted this technology with increased overall productivity and efficiency for the detection of ProLabel™ and ProLink™-tagged recombinant proteins on protein blots, which is called EAstern™ detection *(9)*. We will describe here a number of specific adaptations of this approach for detection of proteins labeled with our unique complementation peptide tags after they are bound to solid supports such as membrane blots.

One of the key enabling features of EFC is the lack of requirement for specific antibodies or multiple incubation and wash steps in the detection process, resulting in reduced assay time and minimal handling. This is possible because in our EFC-based approach we replace the antibody steps required in traditional approaches with detection based on β-galactosidase complementation. This method uses a unique protein (EA, enzyme acceptor) that shows no cross-reactivity with other cellular proteins and binds exclusively to the ProLabel or ProLink peptides added to the expressed protein of interest. When these components come together on the support membrane it results in formation of β-galactosidase enzyme, and generation of chemiluminescence signal. We have developed a portfolio of cell lines expressing ProLabel- and ProLink-tagged proteins, and the EFC blot assay is routinely used for characterization of these cells. This provides valuable information of the level of gene expression and posttranslational modifications, as well as information on protein integrity based on the extent of short degradation products. This has been particularly enabling for proteins such as orphan G-protein-coupled receptors (GPCRs) where it is not always possible to obtain a commercially available antibody.

2. Materials

2.1. Cell Culture and Lysis

1. Culture media: All culture media, additives, and reagents were obtained from Invitrogen, Carlsbad, CA, USA. The culture media were supplemented with 10% heat-inactivated fetal bovine serum (FBS), 1% penicillin/streptomycin/glutamate (from a combined 100× reagent).
2. Cell dissociation: Either a solution of 0.05% trypsin/1 mM ethylene diamine tetraacetic acid (EDTA) (Invitrogen) or

enzyme-free cell dissociation buffer (Invitrogen), depending on the cell line and the expressed protein of interest.

3. Cell lines and cell pools: All cell lines (either as stable or transient pools or clonal cell lines) were generated by DiscoveRx Corporation (Fremont, CA, USA) by expressing fusion proteins containing the gene of interest fused to a ProLabel (PL) or ProLink (PK) reporter peptide sequence and placing those cells under appropriate antibiotic selection. Nuclear factor of activated cells (NFAT)-PL, c-Jun-PL, and Cyclin-D-PL expression plasmids were generated in the ProLabel C1 Expression Vector (DiscoveRx) using standard molecular biology techniques. These plasmids were transfected into CHO-K1 cells using Lipofectamine transfection reagent (Invitrogen). An NFAT-PL clonal cell line was isolated by single-cell dilution cloning in the presence of 800 μg/mL G418 (Invitrogen). All other cell lines and constructs were generated by cloning into the pCMV-ProLink™ mammalian cloning vector (DiscoveRx). For further information on the vector systems please contact DiscoveRx directly.
4. Cell Lysis: 1× SDS dye loading buffer containing 1× denaturing reagent (Invitrogen) or 1× cell lysis buffer [phosphate buffered saline (PBS), 0.5% CHAPS, 1× complete mini EDTA-free protease inhibitor cocktail (Roche, Nutley, NJ, USA)] or Glyko PNGase F Kit denaturing buffer (ProZyme, San Leandro, CA, USA). Protein concentration was determined using the bicinchoninic acid (BCA) protein assay (Pierce, Rockford, IL, USA).

2.2. SDS-Polyacrylamide Gel Electrophoresis (SDS-PAGE)

1. All buffers, markers, and precast gels were obtained from Invitrogen and used according to the manufacturer's instructions. They consisted of the following:
   (a) 4–12% NuPAGE Bis-Tris gel.
   (b) 20× MES running buffer.
   (c) 4× NuPAGE PDS sample buffer.
   (d) 10× NuPAGE reducing agent.
   (e) NuPAGE antioxidant.

2.3. Protein Blotting

1. All buffers and materials for protein blotting were obtained from Invitrogen and used according to the manufacturer's directions. They consisted of the following:
   (a) Precut nitrocellulose membrane with filter pads.
   (b) NuPAGE transfer buffer.
   (c) NuPAGE antioxidant.
   (d) Methanol.

2.4. Western Analysis

1. Blocking buffer: SuperBlock (Pierce) containing 0.05% Tween-20.

2. Antibodies: Mouse anti-NFAT monoclonal antibody (Cell Signaling Technologies, Danvers, MA, USA) and goat anti-mouse IgG–horse radish peroxidase (HRP) conjugate (Pierce).
3. Washing buffer: PBS, 0.05% Tween-20 (PBST).
4. Chemiluminescence detection reagent: Super Signal West Dura substrate (Pierce).
5. Digital imaging system: UVP Epi Chem II Station.

2.5. EFC Blot Detection

1. Detection reagents consist of EA reagent, CL chemiluminescent substrate (lyophilate and reconstitution buffer), and positive control peptide. These are commercially available in kit form (EAstern™ blot assay). X-ray film (Kodak BioMax AR) can be used to capture image.

2.6. PNGase Treatment

1. The Glyko PNGase F kit (ProZyme) was used for enzymatic removal of N-linked glycans from glycoproteins. The kit contains the following:
   (a) Glyko PNGase F Kit denaturing buffer.
   (b) PNGase F (2 units/mL).

3. Methods

Preparation of protein blots for detecting ProLabel- or ProLink-tagged recombinant proteins is identical to that for western blots. Thus, it is easy to perform both assays in parallel on duplicate blots or even consecutively on the same blot after a brief washing step. Unless noted otherwise, all manipulations are performed at room temperature (RT).

3.1. Preparation of Samples for Protein Blots

1. Cells are cultured in media and passaged when approaching confluence. Clonal cell lines are maintained in media containing 800 µg/mL G418. Antibiotic selection is not necessary for transiently transfected cell pools.
2. Cells for analysis can be harvested directly after dissociation during routine passaging or seeded into 6-well plates or 35-mm-diameter dishes. If cells are used directly from passaging, we recommend that they be released from the plate with trypsin-free cell dissociation buffer to eliminate the potential for proteolytic damage to the protein of interest. Cells are pelleted, washed in PBS, and then lysed in either 1× sample loading buffer, cell lysis buffer, or Glyko PNGase F denaturing buffer. Cells plated into separate plates or dishes can be washed with PBS and lysed with 1× sample loading buffer or cell lysis buffer. As a guideline, 1

mL of lysis reagent can be used to lyse 4–10 million cells. The amount of sample loaded in each SDS-PAGE gel lane will vary with the level of fusion protein expression and should be determined empirically. Transiently transfected cells typically express higher levels of tagged-fusion proteins than stably transfected cells (*see* **Note 1**). Lysates can be aliquoted and stored at –20°C for long-term storage. Alternatively, cells can be resuspended with Glyko PNGase F denaturing buffer at a concentration of 10 million cells/mL.

3. When determining total protein concentration using the BCA protein assay, cells should be lysed in cell lysis buffer to minimize interference with the BCA assay. Follow the manufacturer's instructions and use BSA standards as a reference for quantification.
4. For samples prepared in cell lysis buffer, 5 μL 4× LDS loading buffer and 2 μL 10× denaturing reagent are added to 14-μL sample. The mixture is heated to 70°C for 10 min prior to loading onto SDS-PAGE gels. Samples in sample loading buffer or Glyko PNGase F denaturing buffer only require heating at 70–95°C for 10 min before loading onto gels.

3.2. Sds-page

1. Precast 4–20% SDS-PAGE gels are prepared and installed into running chambers according to the manufacturer's instructions. A 2-(*N*-morpholino)ethanesulfonic acid (MES) running buffer system is typically used because it provides good separation of proteins within the 5–200-kDa range.
2. 5–20-μL sample is loaded into each well. Prestained molecular weight markers are loaded in adjacent wells.
3. Gels are run according to the manufacturer's instructions, typically at 200 V for 45 min or until the leading loading dye front has reached the bottom of the gel.

3.3. Protein Blotting

1. Samples that have been separated by SDS-PAGE are transferred to nitrocellulose electrophoretically using the NOVEX transfer system. The gel is prepared according to the manufacturer's instructions and the separated proteins are transferred at 35 V for 1 h.
2. After transfer, the protein blot can be stored in PBS in a sealed dish for at least 24 h at 4°C if needed.

3.4. Western Analysis

The following protocol is used for detection of NFAT-PL fusion protein from stably transfected Chinese hamster ovary (CHO)-Kl cells *(10, 11)*.

1. Rinse the blot with water for 2 min.
2. Incubate for 1 h with 10-mL SuperBlock containing 0.05% Tween-20 to block nonspecific sites.

3. Incubate for 1 h with mouse anti-NFAT monoclonal antibody diluted 1:1,000 in 5-mL blocking buffer.
4. Wash extensively (six washes, 10-min each) in PBST.
5. Incubate the blot for 1 h with a goat anti-mouse IgG–HRP conjugate diluted 1:1,000 in 5-mL blocking buffer.
6. Wash extensively (six washes, 10-min each) in PBST.
7. Add 5 mL of Super Signal West Dura substrate to the blot and incubate for 5 min.
8. Drain excess liquid from the blot and then place it between two sheets of clear plastic.
9. Detect the chemiluminescent signal using a digital imaging system or by exposure to X-ray film. Exposure times will vary. Typically 1–30-min exposures are performed.
10. A comparison of western and EFC blots is shown in **Fig. 2**.

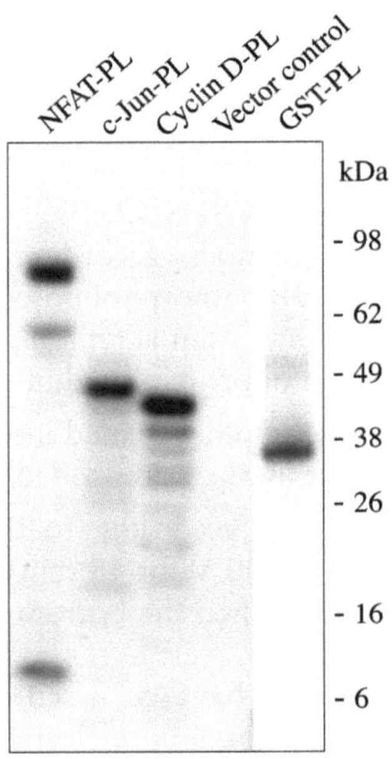

Fig. 1. EFC blot detection of PL-tagged fusion proteins in cell lysates. Samples representing 4-μg NFAT-PL, 0.5-μg c-Jun-PL, 0.35-μg cyclin D-PL, or 0.5-μg CHO-KI lysates were separated by SDS-PAGE and then transferred to a nitrocellulose membrane. Purified GST-PL control protein was run in an adjacent lane. The presence of ProLabel-tagged protein was detected by incubation with EA reagent followed by CL chemiluminescent substrate. The signal resulting from the complementation of EA with membrane-bound PL was detected using a digital imaging system. Shown is an inverted image acquired with a 5-min exposure, which produced sufficient signal to detect NFAT-PL in a cell lysate sample equivalent to 4-μg total protein. Major bands with apparent molecular weights of 85, 45, 42, and 38 kDa were detected for NFAT-PL, c-Jun-PL, cyclin D-PL, and GST-PL, respectively, consistent with the expected masses of the mature proteins fused to ProLabel. Minor bands were also detected, which represent proteolytic cleavage products of full-length proteins, and were pronounced in lysates from transiently transfected cells, where high expression was expected. Therefore, this technique is extremely useful in rapidly determining the integrity of recombinant proteins. There is likely no particular limitation in the size of fusion proteins detected by EFC blot, since we have detected a wide range of fusion proteins using the technique (reproduced from (*9*) with permission from Informa Healthcare: http://www.biotechniques.com/default.asp?page=contact).

3.5. EFC Blot Analysis

1. Prepare detection reagents prior to running the assay (*see* **Note 2**).
2. A 10 cm × 10 cm square dish is ideal for the incubation steps.
3. Rinse the blot with water for 2 min.
4. Decant water and add 5-mL EA reagent. Cover the dish with a foil and incubate the blot for 1 h with gentle rocking to ensure even coverage.
5. Decant EA reagent and add 2.5 mL of CL substrate. Cover the dish with the foil to protect it from light and incubate for 15 min with gentle rocking to ensure even coverage.
6. Drain excess liquid from the blot and place it between two sheets of clear plastic.
7. Detect the chemiluminescent signal as described earlier for western analysis. Exposure times will vary. We typically perform 1, 5, 10, and 30-min exposures to obtain optimal results.
8. Examples of tagged proteins detected by EFC blot are shown in **Fig. 1**. A comparison of western and EFC blots is shown in **Fig. 2**.

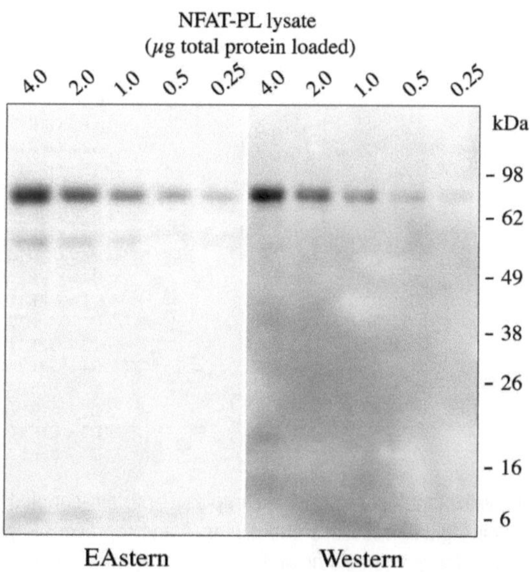

Fig. 2. Comparison of EFC and western blot detection sensitivities. An NFAT-PL cell lysate was serially diluted in lysis buffer, separated by SDS-PAGE, and transferred to a single nitrocellulose membrane. The blot, containing duplicate samples sets, was divided into two and probed separately for PL and NFAT using EFC or western methods, respectively. The resulting signals were detected using a digital imaging system. Using identical exposure times to capture chemiluminescent signal, both EFC and western blot techniques were able to detect NFAT-PL in lysate containing 0.25-μg total protein. For the anti-NFAT antibody, pilot titration experiments were required to determine the concentration required to provide sensitivity equivalent to that obtained by EFC blot. While sensitivity was comparable, a higher background was associated with the western blot compared with the EFC blot. This was likely due to nonspecific interactions of the primary antibody at the high concentrations used. Similar results were seen in side-by-side EFC and western blot comparisons with a number of other PL-tagged fusion proteins and a variety of monoclonal and polyclonal antibodies (data not shown). This experiment also demonstrates that EFC blotting is entirely exchangeable with the current protocols for western blotting, since both methods use the same gel electrophoresis and blotting steps (reproduced from (*9*) with permission from Informa Healthcare).

3.6. Repeat Use of Blots for Additional Analysis

1. Protein blots used for EFC detection can be reused for western analysis without apparent loss in sensitivity.
2. Rinse the blots in PBST for 15 min. The blots can be stored overnight at 4°C for subsequent analysis the following day.

3.7. Use of EFC in Dot Blot Format

The following protocol describes the steps involved in conducting an EFC dot blot. In this example, cells were lysed with the Glyko ProZyme Kit denaturing buffer at 10 million cells/mL. For convenience, SDS-PAGE sample buffer can be substituted.

1. Spot 10-μL cell lysate samples onto a clean, dry nitrocellulose membrane. Place an equivalent volume of positive and negative control lysates next to the samples. Space the spots 2-cm apart so that signals from adjoining spots do not overlap. Allow the spots to dry before proceeding to next step.
2. Rinse the nitrocellulose membrane once with 5-mL PBS buffer.
3. Add 5-mL EA reagent to the membrane, cover with foil, and rock gently to ensure even coverage. Incubate for 1 h at RT or overnight at 4°C. The data shown in **Fig. 3** were prepared

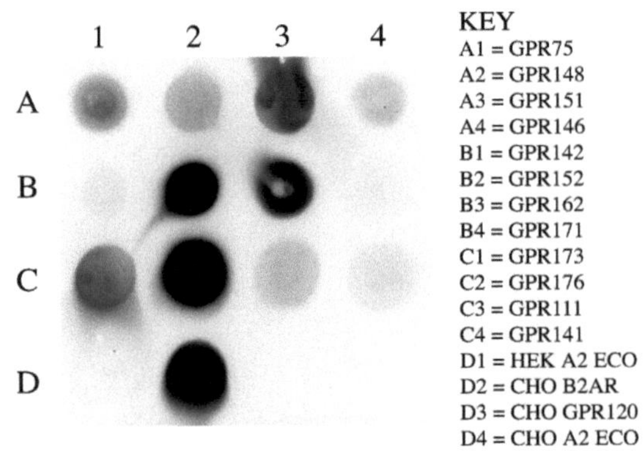

Fig. 3. EFC dot blot analysis of multiple cell lysate samples. G-protein-coupled receptors (GPCRs) are involved in a wide range of physiological processes and diseases and represent important pharmaceutical targets. Because "orphan" GPCRs, those without known ligands, represent a significant percentage of all GPCRs, we cloned orphan GPCRs and expressed them in mammalian cells fused to a C-terminal PK tag. EFC dot blot analysis allowed us to rapidly determine the relative levels of the protein expression of these orphan receptors. Lysates from a number of cell lines expressing variety of PK-tagged orphan GPCRs were spotted onto nitrocellulose and probed using the EFC blot assay. A number of cell lines were identified that express strong EFC signal (e.g., GPCRs 151, 152, 162, 173, 176). Two parental, negative control cell lines, HEK A2 ECO and CHO A2 ECO, do not express PK and produced no discernable signal. In contrast, the positive control CHO B2AR cell line expressing PK-tagged ADR2B receptor produced a strong signal. Since the EFC dot blot procedure does not require SDS-PAGE and blotting steps, results can be obtained in less than 2 h, enabling rapid confirmation of fusion protein expression for such activities as clonal selection and receptor comparison.

using EA reagent at high concentration. However, similar results are obtained using the EA reagent provided in the EAstern Blot Assay kit.

4. Remove the membrane from the EA reagent and rinse once with PBS.
5. Add 3 mL of CL substrate to the membrane and then incubate, protected from light, for 30 min with gentle rocking.
6. Drain excess liquid from the blot and place it between two sheets of clear plastic.
7. Detect the chemiluminescent signal as described earlier for western analysis. Exposure times will vary. Typically 1, 2, 4, 6, and 12-min exposures are performed.
8. An example of the use of EFC dot blots for examining orphan GPCR expression is shown in **Fig. 3**.

3.8. Deglycosylation Treatment to Examine Posttranslational Modification

The following protocol describes the steps involved in using PNGase F to treat cell lysates. PNGase F is used for the enzymatic removal of N-linked glycans from glycoprotein to leave the core protein intact. This is a useful approach to elucidate the type and level of glycosylation. Similarly, a variety of enzymes are available with a range of specificities that can be applied to different deglycosylation strategies.

1. Lyse cells at a concentration of 10 million cells/mL with Glyko PNGase F Kit denaturing buffer, the optimal buffer for PNGase F. Heat samples for 95°C for 10 min.
2. Prepare a reaction master mix:
   (a) 4-μL reaction buffer.
   (b) 9-μL deionized water.
   (c) 2-μL NP-40.
   (d) 2-μL PNGase F (2 units/μL)
3. Dilute the lysate from earlier **step 1** in water to 2,000 cells/μL.
4. Add 15 μL of the master mix to 5 μL of the diluted lysate.
5. Incubate the reaction at 37°C for 1 h.
6. Following incubation, a sample of the reaction can be used for EFC blot or EFC dot blot analysis (*see* earlier).
7. An example of the use of deglycosylation enzymes in conjunction with the EFC blot is shown in **Fig. 4**.

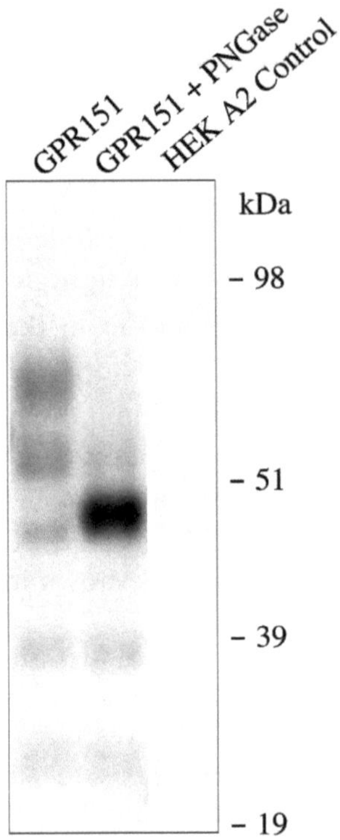

Fig. 4. Examination of glycosylation status using EFC blot. Many GPCRs are N-glycosylated, which complicates their analysis using protein blots because glycosylation results in a population of glycoprotein bands with a broad range of apparent molecular weights. GPCR 151, a class A orphan receptor, was tagged with ProLink and expressed in HEK cells. In EFC blots the PK-tagged glycoprotein migrates as a broad range of bands with weak overall signal. The amidase PNGase F, also known as *N*-glycosidase F, was used to cleave the oligosaccharides from the glycosylated GPCR. Cell lysates were preincubated with PNGase F or control buffer before EFC blot analysis. As a result of PNGase F treatment, the ProLink-tagged protein migrated as a single band of 47 kDa, which is consistent with the expected molecular weight of the full-length GPCR tagged with ProLink. We have used this EFC blot technique to validate the correct expression of a number of orphan GPCRs for which there is currently little functional data. This type of EFC blot analysis can easily be extended to saccharide sequencing strategies to further understanding the complexity of glycoprotein composition.

4. Notes

1. The substrate used for EFC blots is very sensitive. Therefore, it is important to optimize the amount of sample loaded per lane to prevent high levels of signal that might obscure adjacent lanes. As a starting point, load 1–10 μg of total protein from cells stably expressing ProLabel- or ProLink-tagged proteins. For cells transiently expressing tagged proteins, use approximately one-tenth this amount.
2. The EFC detection reagents and protocols used for this assay are commercialized as the EAstern™ Blot Assay Kit (DiscoveRx). EA is provided as a ready-to-use reagent and should be

thawed in a RT water bath prior to use. The CL chemiluminescent substrate is provided as two components. Thaw the reconstitution buffer in a RT water bath and pour contents into the substrate bottle. Invert several times to mix and store at RT for 5 min. Any unused reagents should be aliquoted and stored at −20°C for long-term use. The reconstituted reagents retain activity for at least 1 month if kept frozen.

References

1. Kurien, B.T. and Scofield, R.H. (2003) Protein blotting: a review. *J Immunol Methods.* 274, 1–15.
2. McIlhinney, R.A. (2004) Generation and use of epitope-tagged receptors. *Methods Mol Biol.* 259, 81–98.
3. Patton, W.F. (2002) Detection technologies in proteome analysis. *J Chromatogr B Analyt Technol Biomed Life Sci.* 771, 3–31.
4. Langley, K.E., Villarejo, M.R., Fowler, A.V., Zamenhof, P.J., and Zabin, I. (1975) Molecular basis of beta-galactosidase alpha-complementation. *Proc Natl Acad Sci USA* 72, 1254–1257.
5. Eglen, R.M. (2002) Enzyme fragment complementation: a flexible high throughput screening assay technology. *Assay Drug Dev Technol* 1, 97–104.
6. Zhao, X., Vainshtein, I., Gellibolian, R., Shu, Y., Dotimas, H., Wang, X.M., et al. (2003) Homogeneous assays for cellular protein degradation using beta-galactosidase complementation: NF-kappaB/IkappaB pathway signaling. *Assay Drug Dev Technol* 1, 823–833.
7. Vainshtein, I., Silveria, S., Kaul, P., Rouhani, R., Eglen, R.M., and Wang, J. (2002) A high-throughput, nonisotopic, competitive binding assay for kinases using nonselective inhibitor probes (ED-NSIP). *J Biomol Screen.* 7, 507–514.
8. Naqvi, T., Lim, A., Rouhani, R., Singh, R., and Eglen, R.M. (2004) Beta galactosidase enzyme fragment complementation as a high-throughput screening protease technology. *J Biomol Screen.* 9, 398–408.
9. Horecka, J., Charter, N.W., Bosano, B.L., Fung, P., Kobel, P., Peng, P., et al. (2006) An antibody-free method for protein detection on blots using enzyme fragment complementation. *BioTech.* 40, 381–383.
10. Macian, F. (2005) NFAT proteins: key regulators of T-cell development and function. *Nat Rev Immunol.* 5, 472–484.
11. Horsley, V. and Pavlath, G.K. (2002) NFAT: ubiquitous regulator of cell differentiation and adaptation. *J Cell Biol.* 156, 771–774.

Chapter 41

Fluorescent Labeling of Proteins and Its Application to SDS-PAGE and Western Blotting

F. Javier Alba, Salvador Bartolomé, Antonio Bermúdez, and Joan-Ramon Daban

Summary

This chapter describes very simple fluorescent methods developed in our laboratory allowing the rapid monitoring of total protein patterns on both sodium dodecyl sulfate (SDS) polyacrylamide gels and western blots. The noncovalent dye Nile red (9-diethylamino-5*H*-benzo[α]phenoxazine-5-one) is used for the sensitive staining of proteins in SDS gels. This method is compatible with the electroblotting of protein bands and with the staining of the resulting blot with the covalent dye MDPF (2-methoxy-2,4-diphenyl-3(2*H*)-furanone). These staining procedures are applied sequentially; there is no need to run a duplicate unstained gel for protein blotting. Furthermore, since only the adduct formed by the reaction of MDPF with proteins is fluorescent, there is no need to destain the membrane after protein labeling. In addition, MDPF staining is compatible with further immunodetection of specific bands with polyclonal antibodies. Finally, using the adequate conditions described later, MDPF staining does not preclude the N-terminal sequence analysis of proteins in selected bands.

Key words: Fluorescent protein detection, Immunodetection, MDPF, Nile red, N-terminal sequencing, SDS-PAGE, Western blotting

1. Introduction

The typical procedures for the detection of protein bands after SDS-PAGE using the visible dye Coomassie Blue and silver staining have several time-consuming steps and require the fixation of proteins in the gel *(1)*. Here, we describe a very simple and sensitive method developed in our laboratory *(2–4)* for the rapid staining of unfixed proteins in SDS gels. The method is based on the fluorescent properties of the hydrophobic dye Nile red.

Biji T. Kurien and R. Hal Scofield (eds.), *Methods in Molecular Biology, Protein Blotting and Detection, vol. 536*

DOI: 10.1007/978-1-59745-542-8_41

This dye is nearly insoluble in water, but is soluble and shows a high increase in the fluorescence intensity in nonpolar solvents and in the presence of lipid droplets *(5)*, SDS micelles, and SDS–protein complexes *(6, 7)*.The enhancement of Nile red fluorescence observed with different SDS–protein complexes occurs at SDS concentration lower than the critical micelle concentration of this detergent in the typical Tris-glycine buffer used in SDS-PAGE. Thus, for Nile red staining of SDS-polyacrylamide gels, electrophoresis is performed in the presence of 0.05% SDS instead of the typical SDS concentration (0.1%) used in current SDS-PAGE protocols. This concentration of SDS is high enough to maintain the stability of the SDS–protein complexes in the bands but is lower than SDS critical micelle concentration and consequently precludes the formation of pure detergent micelles in the gel. The staining of these modified gels with Nile red produces very high fluorescence intensity in the SDS–protein bands and low background fluorescence. Nile red staining is completed in about 5 min.

On the other hand, we have shown previously that, in contrast to the current staining methods, Nile red staining does not preclude the direct electroblotting of protein bands *(8)* and does not interfere with the staining of total protein patterns on polyvinylidene difluoride (PVDF) membranes using the fluorogenic dye MDPF *(9, 10)*. This staining method is based on the observation that this dye and its hydrolysis product are nonfluorescent; only the adduct formed after the reaction with proteins is fluorescent *(11)*. This makes unnecessary the destaining of the PVDF membrane after protein labeling *(9)*. The whole process of staining with MDPF is completed in about 20 min. Wet membranes are translucent, allowing the detection with a high sensitivity of MDPF-labeled protein bands by transillumination with UV light. In addition, MDPF staining allows further immunodetection and N-terminal sequence analysis of specific bands. A review about the physicochemical basis of Nile red and MDPF staining procedures has been published elsewhere *(12)*. A flow diagram showing the sequential staining of SDS gels and western blots is presented in **Fig. 1**.

2. Materials

All solutions should be prepared using electrophoresis-grade reagents and deionized water and stored at room temperature (exceptions are indicated). Wear gloves to handle all reagents and solutions. Collect and dispose all waste according to good laboratory practice and waste disposal regulations.

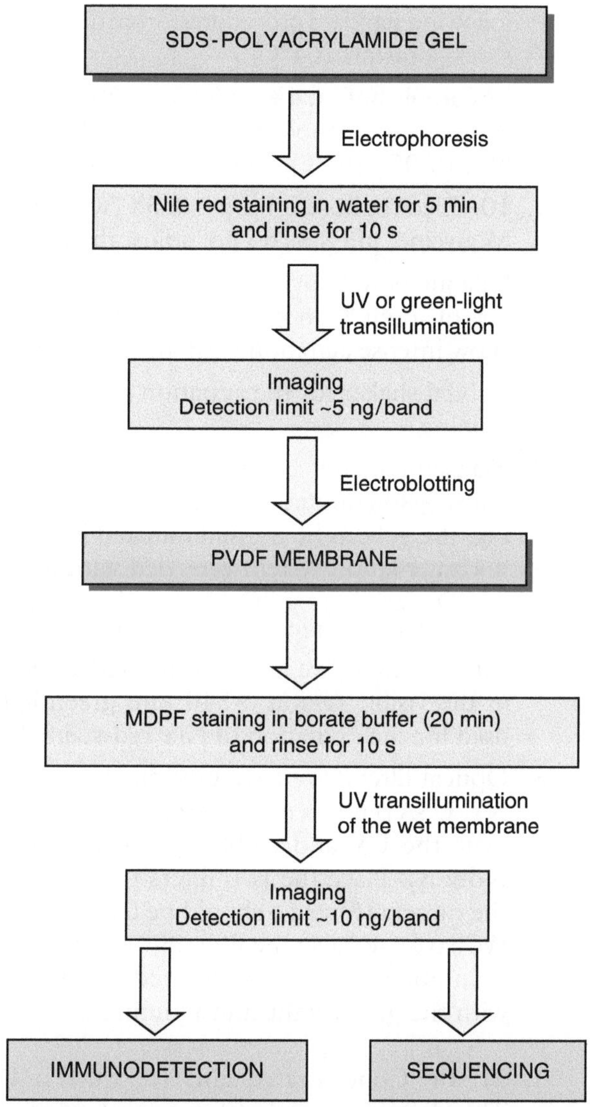

Fig. 1. Scheme summarizing the sequential fluorescent staining of the SDS gel with Nile red and the corresponding western blot with MDPF (reproduced from (*12*) with permission). The time required in each step and the observed sensitivities are indicated. Technical details about green-light transillumination can be found in (*13*).

2.1. Electrophoresis

1. Nile red: Concentrated stock (0.4 mg/mL) in dimethyl sulfoxide (DMSO; *see* **Note 1**). This solution is stable for at least 3 months when stored at room temperature in a glass bottle wrapped in aluminum foil to prevent damage by light. Nile red can be obtained from Sigma Chemical Co. (St. Louis, MO, USA).
2. The 30% (w/v) acrylamide–0.8% (w/v) *bis*-acrylamide stock solution and the resolving (1.5 M Tris–HCl, pH 8.8) and stacking (0.5 M Tris–HCl, pH 6.8) gel stock buffers are prepared

following standard procedures. Store the acrylamide-*bis*-acrylamide stock solution at 4°C.

3. 2× Sample buffer: 4% (w/v) SDS, 20% (v/v) glycerol, 10% (v/v) 2-mercaptoethanol, 0.125 M Tris–HCl, pH 6.8; bromophenol blue (0.05% [w/v]) can be added as a tracking dye.
4. 10× Electrophoresis buffer: 0. 5% (w/v) SDS, 0.25 M Tris, 1.92 M glycine, pH 8.3 (do not adjust the pH of this solution).
5. Opaque polypropylene container (~12 cm × 7.5 cm × 7 cm for gels of about 8 cm × 6 cm × 0.075 cm) with a close-fitting lid to allow intense agitation without spilling the staining solution.
6. Orbital shaker for the agitation of the plastic boxes during gel staining.
7. Transilluminator equipped with midrange ultraviolet (UV) bulbs (~300 nm; *see* **Note 2**) to excite Nile red *(2, 3)*. For imaging, the gel can be transilluminated inside the dark cabinet of a documentation system provided with a charge-coupled device (CCD) camera (e.g., Gel Doc 1000 from Bio-Rad Laboratories [Hercules, CA, USA]). To avoid the safety problems associated with UV light, we have constructed a transilluminator working in the visible region (~540 nm; green-light) that can also be used for the excitation of Nile red-stained bands *(13)*.
8. Optical filters: We have used the Wratten (Kodak [Rochester, NY, USA]) filters number 9 (yellow) and 16 (orange) to eliminate the UV and visible light from the transilluminator (*see* **Note 3**). Place the two filters together in the filter holder of the camera (filter 16 should be facing toward the camera lens). We have used also the filter 590DF100 (Bio-Rad). Wratten filter number 26 (red) is required when the bands are visualized with the green-light transilluminator.

2.2. Western Blot

1. MDPF: Concentrated stock (35 mM) in DMSO (*see* **Note 1**). This solution is stable for 1 week at room temperature in a glass bottle wrapped in aluminum foil. MDPF can be obtained from Fluka (Bunch, Switzerland).
2. PVDF membranes (Bio-Rad).
3. Transfer buffer: 25 mM Tris, 192 mM glycine, pH 8.3.
4. Borate buffer: 10 mM sodium borate, pH 9.5.
5. Transfer apparatus (e.g., Mini-Trans-Blot Cell [Bio-Rad]).
6. Opaque plastic box for membrane equilibration and staining.
7. Orbital shaker.
8. UV transilluminator (~300 nm; *see* **Note 2**) to excite MDPF-labeled protein on blots *(9)*.
9. Photography: We have used the CCD camera of the Gel Doc 1000 system (Bio-Rad) with the Wratten (Kodak) filters numbers

3 (yellow) and 47 (blue) (*see* **Note 3**). Place filter 47, facing toward the camera lens, on top of filter 3.

10. Immunodetection: ECL detection kit, including horseradish peroxidase-labeled secondary antibodies (Amersham-GE Healthcare [Buckinghamshire, UK]).
11. Phosphate buffered saline containing Tween-20: 80 mM Na_2HPO_4, 20 mM NaH_2PO_4, 100 mM NaCl, pH 7.4, and 0.1% (v/v) Tween-20.

3. Methods

Unless otherwise indicated, all operations are performed at room temperature.

3.1. Nile Red Staining of SDS-Polyacrylamide Gels

1. Typically we prepare 15% acrylamide–0.4% *bis*-acrylamide separating gel containing 0.05% SDS (*see* **Note 4**), 0.375 M Tris–HCl, pH 8.8. The stacking gel contains 6% acrylamide–0.16% *bis*-acrylamide, 0.05% SDS (*see* **Note 4**), 0.125 M Tris–HCl, pH 6.8.
2. Dissolve proteins in the appropriate volume of deionized water, add 1 volume of the 2× sample buffer (*see* **Note 5**), and incubate the resulting samples in a boiling-water bath for 3 min. The samples are kept at room temperature before loading them into the gel (*see* **Note 6**).
3. Place the gel sandwiched between the glass plates in the electrophoresis apparatus, fill the electrode reservoirs with 1× electrophoresis buffer (*see* **Note 4**), and rinse the wells of the gel with this buffer.
4. Load the protein samples, carry out electrophoresis, and at the end of the run remove the gel sandwich from the electrophoresis apparatus and place it on a flat surface.
5. Wearing gloves, remove the upper glass plate (use a spatula) and excise the stacking gel and the bottom part of the separating gel (*see* **Note** 7).
6. Add quickly 0.25 mL of the concentrated Nile red (0.4 mg/mL) staining solution in DMSO to 50 mL of deionized water previously placed in a plastic box (*see* **Note 8**).
7. Immediately after the addition of concentrated solution of Nile red to water, agitate the resulting solution vigorously for 3 s (*see* **Note 8**).
8. Immerse the gel very quickly in the staining solution, put the lid on, and agitate vigorously using an orbital shaker (at about 150 rpm) for 5 min (*see* **Note 8**).

9. Discard the staining solution (*see* **Note 1**) and rinse the gel with deionized water (four times; about 10 s) to remove completely the excess Nile red precipitated during the staining of the gel (*see* **Note 9**).
10. Wearing gloves, remove the gel from the plastic box and place it on the UV or green-light transilluminator. Turn off the room lights, turn on the transilluminator, and examine the protein bands, which fluoresce light red (*see* **Note 10**). Turn off the transilluminator immediately after the visualization of the bands (*see* **Note 11**).
11. Focus the CCD camera, place the optical filters (*see* **Subheading 2.1, item 8**) in front of the camera lens and, in the dark, turn on the transilluminator and photograph the gel (*see* **Note 11**). Finally, turn off the transilluminator.

3.2. MDPF Staining of Western Blots

Handle the PVDF membranes by their edges using stainless steel forceps.

1. After electrophoresis and Nile red staining, equilibrate the gel in 100 mL of transfer buffer for 15 min.
2. Immerse sequentially the PVDF membrane in 20 mL of methanol for 5 s, in 100 mL of water for 2–3 min, and finally in 100 mL of transfer buffer for 10 min.
3. Assemble the gel and the membrane in the blotting apparatus and fill the tank with transfer buffer. Perform electroblotting at 100 V for 1 h.
4. Following transfer, equilibrate the blot twice (for 5 min each time) in 100 mL of borate buffer. Use an orbital shaker at about 75 rpm.
5. Incubate the blots for 10 min in a staining solution (prepared just before use) containing 40 mL of borate buffer and 0.2 mL of the concentrated stock (35 mM) of MDPF in DMSO (*see* **Note 12**); use an orbital shaker (~75 rpm). The plastic box containing the blot should be covered with aluminum foil during staining. The staining conditions compatible with N-terminal sequencing of selected bands are indicated in **Note 13**.
6. After staining (*see* **Note 1**), rinse the blot briefly (for about 10 s) with borate buffer.
7. Place the wet membrane (*see* **Note 14**) on the UV transilluminator. Focus the CCD camera, place the optical filters indicated in **Subheading 2.2, item 9** (*see* **Note 15**) in front of the camera lens and, in the dark, turn on the transilluminator and photograph the blot (*see* **Note 16**). Finally, turn off the transilluminator.
8. After photography, if specific bands have to be immunodetected (*see* **Note 17**), equilibrate the stained blot for 15 min

in phosphate buffered saline containing 0.1% Tween-20, and then perform the ECL immunodetection according to the manufacturer's instructions.

9. Use a UV transilluminator to visualize the fluorescent bands and mark with a soft pencil the stained protein bands (*see* **Note 13**) to be sequenced (*see* **Note 18**). Cut out of the membrane the selected bands and apply them to the sequencer.

4. Notes

1. Handle this solution with care; DMSO is flammable and, in addition, this solvent may facilitate the passage of potentially hazardous chemicals such Nile red and MDPF through the skin. Used staining solutions must be collected and disposed of in accordance with waste disposal regulations.
2. UV light is dangerous to skin and particularly to eyes. UV-blocking goggles and a full face shield, and protective gloves and clothing, should be worn when the stained gel or blot is examined by naked eye using an UV transilluminator.
3. Store the filters in the dark and protect them from heat, intense light sources, and humidity.
4. To reduce the background fluorescence after the staining with Nile red it is necessary to preclude the formation of pure SDS micelles in the gel (*see* **Subheading 1**). Therefore, use 0.05% SDS to prepare both the separating and stacking gels and the electrophoresis buffer. This concentration is lower than the critical micelle concentration of this detergent (~0.1%, **ref.** *6*), but is high enough to allow the formation of the normal SDS–protein complexes that are specifically stained by Nile red *(2, 12)*.
5. Use 2% SDS in the sample buffer to be sure that all protein samples are completely saturated with SDS. Lower concentrations of SDS in the sample buffer can produce only a partial saturation of proteins (in particular in highly concentrated samples), and consequently the electrophoretic bands could have anomalous electrophoretic mobilities. The excess SDS (uncomplexed by proteins) present in the sample buffer migrates faster than the proteins and forms a broad band at the bottom of the gel (*see* **Note** 7 and **ref.** *2*).
6. Storage of protein samples prepared as indicated in **Subheading 3.1, step 2** at 4°C (or at lower temperatures) causes the precipitation of the SDS present in the solution. These samples should be incubated in a boiling-water bath to redissolve SDS before using them for electrophoresis.

7. The bottom part of the gel (i.e., from few millimeters above the bromophenol blue band to the end) should be excised before the staining of the gel. Otherwise, after the addition of Nile red, the lower part of the gel produces a broad band with intense fluorescence. This band is presumably caused by the association of Nile red with the excess SDS used in the sample buffer (*see* **Note 5**). In the case of long runs, bromophenol blue and the excess SDS band diffuse into the buffer of the lower reservoir and it is not necessary to excise the gel bottom.
8. Nile red is very stable when dissolved in DMSO (*see* **Subheading 2.1, item 1**), but this dye precipitates in aqueous solutions. Since the precipitation of Nile red in water is a rapid process and this dye is only active for the staining of SDS–protein bands before it is completely precipitated *(2, 3)*, to obtain satisfactory results, it is necessary (a) to perform all the agitations indicated in these steps (in order to favor as much as possible the dispersion of the dye) and (b) to work very rapidly in the **steps 6–9** of **Subheading 3.1**. Furthermore, to obtain a homogeneous staining the gel has to be completely covered with the staining solution.
9. After the water rinsing indicated in **step 9** of **Subheading 3.1**, the staining process is completely finished and the gel can be imaged immediately. It is not necessary, however, to examine and photograph the gel just after staining. Nile red-stained bands are stable and the gel can be kept in the plastic box immersed in water for 1–2 h before imaging.
10. About 100 ng of protein per band can be seen by naked eye (when using the green-light transilluminator it is necessary to place a Wratten filter number 26 in front of the eyes to visualize the bands). Faint bands that are not visible by direct observation of the transilluminated gel can be clearly seen in the CCD image. Long integration times with a CCD camera allow the detection of 5 ng of protein per band (using both UV and green-light transilluminators). To obtain this high sensitivity it is necessary to have sharp bands; broad bands considerably reduce the sensitivity.
11. Nile red is sensitive to intense UV irradiation *(6)*, but has a photochemical stability high enough to allow gel staining without being necessary to introduce complex precautions in the protocol (*see* **Subheading 3.1, steps 6–9**). For long-term storage, solutions containing this dye are kept in the dark (*see* **Subheading 2.1, item 1**). Transillumination of the gel for more than a few minutes produces a significant loss of fluorescence intensity. Thus, transillumination time must be reduced as much as possible both during visualization and photography. However, the relative short exposure times (4–12 s) required for the CCD documentation system allow taking several photographs with different exposure times if necessary.

12. The reaction of MDPF with different proteins in solutions containing 10 mM sodium borate, pH 9.5, is completed in less than 10 min *(14)*.
13. When the protein bands have to be used for sequencing (*see* **Note 18**), use a lower concentration of MDPF (0.1 mL of the concentrated solution of MDPF in 100 mL of borate buffer) and reduce the incubation time to 2 min. The sensitivity obtained under these conditions is lower than that obtained using the normal staining conditions (*see* **Note 16**), but it is high enough to detect bands containing protein amounts suitable for automatic sequencing by Edman degradation.
14. Membranes must be wet during visualization and imaging. Wet membranes are highly translucent, but dry membranes are opaque and reduce dramatically the sensitivity *(9)*.
15. Wratten filters numbers 3 and 47 allow the visualization of the blue fluorescence emission from MDPF-labeled bands. In addition, these filters eliminate the light coming from the transilluminator and the background fluorescence produced by Nile red adsorbed on the surface of the PVDF membrane during the electrotransfer.
16. Electrophoretic bands containing 5–10 ng of protein can be seen in the images obtained with the CCD system. However, considering that the yield of protein transfer from the gel to the membrane is relatively low, the actual sensitivity is presumably higher. In fact, we have observed that when proteins are transferred directly to the membrane using a slot-blotting device, 0.5 ng of protein per slot can be detected *(9)*.
17. Our results *(9)* have shown that the covalent modification of blotted proteins produced by the reaction with MDPF does not alter the antigenic properties that allow the binding of polyclonal antibodies.
18. Our results *(9)* have shown that the staining conditions indicated in **Note 13** do not preclude further sequencing reactions with a high yield. Since MDPF reacts with primary amino groups of proteins *(11)*, probably including N-terminal groups, these results indicate that under these staining conditions the reaction with MDPF is not complete.

Acknowledgement

This work was supported in part by grants BMC2002-3948 (MEC-FEDER) and BFU2005-3883 (MEC).

References

1. Patton, W.F. (2001) Detecting proteins in polyacrylamide gels and on electroblot membranes, In: *Proteomics: From Protein Sequence to Function* (Pennington, S.R. and Dunn, M.J., eds), BIOS Scientific, Oxford, UK, pp. 65–86.
2. Daban, J.-R., Bartolomé, S., and Samsó, M. (1991) Use of the hydrophobic probe Nile red for the fluorescent staining of protein bands in sodium dodecyl sulfate-polyacrylamide gels. *Anal Biochem.* 199, 169–174.
3. Alba, F.J., Bermúdez, A., Bartolomé, S., and Daban, J.-R. (1996) Detection of five nanograms of protein by two-minute Nile red staining of unfixed SDS gels. *Biotechniques.* 21, 625–626.
4. Daban, J.-R., Bartolomé, S., Bermúdez, A., and Alba, F.J. (2002) Rapid and sensitive staining of unfixed proteins in polyacrylamide gels with Nile red, In: *The Protein Protocols Handbook* (Walker, J.M., ed.), Humana, Totowa, NJ, pp. 243–249.
5. Greenspan, P., Mayer, E.P., and Fowler, S.D. (1985) Nile red: a selective stain for intracellular lipid droplets. *J Cell Biol.* 100, 965–973.
6. Daban, J.-R., Samsó, M., and Bartolomé, S. (1991) Use of Nile red as a fluorescent probe for the study of the hydrophobic properties of protein–sodium dodecyl sulfate complexes in solution. *Anal Biochem.* 199, 162–168.
7. Sansó, M., Daban, J.-R., Hansen, S., and Jones, G.R. (1995) Evidence for sodium dodecyl sulfate/protein complexes adopting a necklace structure. *Eur J Biochem.* 232, 818–824.
8. Bermúdez, A., Daban, J.-R., Garcia, J.R., and Mendez, E. (1994) Direct blotting, sequencing and immunodetection of proteins after five-minute staining of SDS and SDS-treated IEF gels with Nile red. *Biotechniques.* 16, 621–624.
9. Alba, F.J. and Daban, J.-R. (1998) Rapid fluorescent monitoring of total protein patterns on sodium dodecyl sulfate-polyacrylamide gels and western blots before immunodetection and sequencing. *Electrophoresis.* 19, 2407–2411.
10. Alba, F.J. and Daban, J.-R. (2002) MDPF staining of proteins in Western blots, In: *The Protein Protocols Handbook* (Walker, J.M., ed.), Humana, Totowa, NJ, pp. 375–379.
11. Weigele, M., DeBernardo, S., Leimgruber, W., Cleeland, R., and Grunberg, E. (1973) Fluorescent labeling of proteins: a new methodology. *Biochem Biophys Res Commun.* 54, 899–906.
12. Daban, J.-R. (2001) Fluorescent labeling of proteins with Nile red and 2-methoxy-2,4-diphenyl-3(2*H*)-furanone: physicochemical basis and application to the rapid staining of sodium dodecyl sulfate polyacrylamide gels and western blots. *Electrophoresis.* 22, 874–880.
13. Alba, F.J., Bermúdez, A., and Daban, J.-R. (2001) Green-light transilluminator for the detection without photodamage of proteins and DNA labeled with different fluorescent dyes. *Electrophoresis.* 22, 399–403.
14. Alba, F.J. and Daban, J.-R. (1997) Nonenzymatic chemiluminescent detection and quantitation of total protein on western and slot blots allowing subsequent immunodetection and sequencing. *Electrophoresis.* 18, 1960–1966.

Chapter 42

Whole and Strip Nitrocellulose Membrane as well as New Line Immunoblotting of Antigen Using the Chemiluminescence Technique

Yaser Dorri, Ali W. Khalili, and R. Hal Scofield

Summary

There are a number of techniques in the scientific world that researchers use to detect specific antigens. One such technique that has provided many advantages over typical immunochemical staining is chemiluminescence. The emission of visible radiation by compounds once exposed to sunlight has been known for centuries and currently is the main principle for chemiluminescence. Here, we introduce three different chemiluminescence techniques that are widely used in immunodetection of antigens: (a) whole membrane chemiluminescence detection, (b) strip membrane chemiluminescence detection, and (c) new line blotting chemiluminescence.

Key words: Strip blotting chemiluminescence, ECL chemiluminescence, Dot-blotting, Line blotting, Chemiluminescence, Immunodetection of proteins, Whole membrane blotting chemiluminescence

1. Introduction

The term "luminescence" was first introduced by Wiedemann in 1888 *(1)* as he classified luminescence phenomena based on their excitations to six different stimuli: (a) photoluminescence, caused by absorption of light; (b) electroluminescence, produced in gases by electric discharges; (c) thermoluminescence, produced by slight heating; (d) triboluminescence, caused by friction; (e) cristalloluminescence, caused by crystallization; and (f) chemiluminescence (CL), caused by a chemical reaction *(1)*. The basic mechanism of CL can be described in two reactions (*see* **Fig. 1**). One such reaction requires a direct interaction of two reagents,

Biji T. Kurien and R. Hal Scofield (eds.), *Methods in Molecular Biology, Protein Blotting and Detection, vol. 536*

DOI: 10.1007/978-1-59745-542-8_42

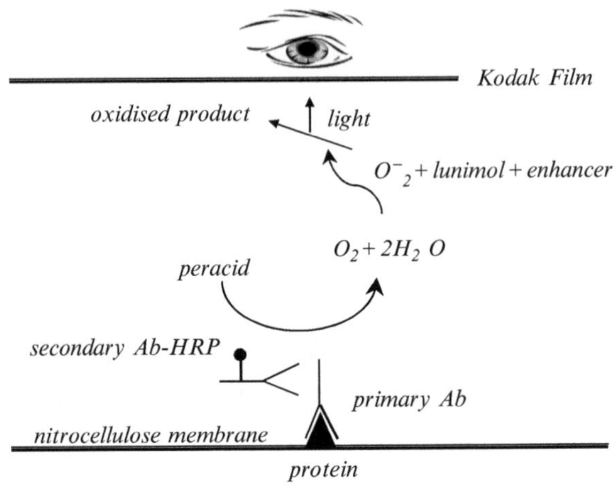

Fig. 1. Demonstrating the immunodetection mechanism of antigens using chemiluminescence in the ECL system [adapted from *(2)*].

a substrate and an oxidant, with the addition of some cofactors, and in some instances, a catalyst, which reduces the activation energy (E_A) and provides an adequate environment for efficient CL. The addition of cofactors is sometimes needed to convert one or more of the substrates into a form capable of reacting and interacting with the catalyst or to provide an efficient leaving group if bond cleavage is required to produce the excited emitter *(1)*. Typically, a portion of the product formed will be in an electronically excited state, and when relaxed to its ground state will emit energy in the form of photons. The substrate is the CL precursor, which when converted into the electronically excited molecule, is responsible for light emission or acting as the energy transfer donor in indirect CL *(1)*.

CL has been used over a wide range of scientific fields, is one of the most highly adaptable and efficient methods of antigen detection, and has provided many advantages over typical immunochemical staining. One such CL technique is based on the conversion of a substrate into a luminescent product by peroxidase–antibody conjugates. This specific technique is highly useful owing to the fact that the resulting emitted light of infected cell foci can be easily recorded by video imaging or autoradiography giving a hard copy for documentation, something that has not been possible in the past *(3)*. CL also provides higher detection sensitivity due to the inherent magnification effect of luminescence resulting in an increase in intensity and a reduction in incubation time along with the required amount of antibody needed for antigen detection *(3)*. This method has been specifically used for a focus reduction assay for the serotyping of hantavirus-specific neutralizing antibodies in infected persons and also the determination of activity of antiviral agents against hantavirus *(3)*. Because of their inherent stability, CL substances have also been used as labels, eliminating the need of inconvenient radiolabels. By using CL substances, researchers are able to use simple oxidation reactions, which produce light with high quantum yield in a matter of seconds *(4)*. Several homogeneous assays for biotin and progesterone as well as heterogeneous assays for thyroxine and testosterone take advantage of this technique *(4)*. In addition, this method has aided in the development of a model double antibody, solid phase immunoassay *(4)*. The use of CL has also helped in the development of an immune microanalysis system. The use of peroxyoxalate CL, which takes advantage of the strong CL emitted through the energy transfer from peroxide intermediate that is generated by the reaction between oxalate reagent and hydrogen peroxide (H_2O_2), was paramount in developing the microanalysis system *(5)*. The system is capable of determining, with high selectivity and sensitivity, human serum albumin (HSA) or immunosuppressive acidic protein (IAP) as a cancer marker in human serum *(5)*.

There are many methods used to detect antigens using CL. Here, three forms of CL immunodetection of antigens are introduced: (a) whole membrane, (b) strip membrane, and (c) new line blotting. In whole membrane blotting, only one primary and secondary antibody is used to detect antigens, and therefore there is no need for dividing the membrane. In strip membrane blotting, each membrane is divided into different strips for the use of different antibodies. In the new line blotting, protein samples are directly applied to a membrane and immunodetection is followed using CL.

2. Materials

After running the first dimensional SDS polyacrylamide gel and/or two-dimensional IEF gel, the proteins in the gel need to be transferred to a nitrocellulose membrane. The following materials are needed for a nonelectrophoretic transfer prior to CL detection.

2.1. Immunoblotting and Nonelectrophoretic Transfer Component

1. Nitrocellulose membranes (Gelman Sciences, Fisher Scientific, Dallas, TX, USA).
2. Mini PROTEAN® 3 System Glass plates (Catalog number 1653311) (Bio-Rad Laboratories, Hercules, CA, USA).
3. Western blot transfer buffer: 0.025 M Tris, 0.192 M glycine, 20% methanol (*see* **Note 1**).
4. Tris buffered saline (TBS, 10×): 1.5 M NaCl, 0.1 M Tris–HCl, pH 7.4.
5. TBS containing 0.05% Tween-20 (TBST).
6. Blocking solution: 5% milk in TBS (*see* **Note 2**). Store at 4°C.

7. Diluent solution: 5% milk in TBST (*see* **Note 2**). Store at 4°C.
8. Wypall X-60 reinforced paper (Kimberly-Clark, Neenah, WI, USA).
9. Medium binder clips (1¼ in.).
10. Plastic container.
11. 1.7 mL PGC tube (Catalog number 505-111) (PGC Scientific, Frederick, MD, USA).

2.2. Antigens, Conjugate, and Chemiluminescence Components

1. Ro autoantigen (Immunovision, Springdale, AK, USA).
2. Systemic lupus erythematosus (SLE) serum samples were obtained from Clinical Immunology Laboratory at Oklahoma Medical Research Foundation (Oklahoma City, OK, USA).
3. ECL Plus western blotting conjugate (GE Healthcare Bioscience Corp, Piscataway, NJ, USA).
4. Kodak Biomax Film (8 in. × 11 in.) (Sigma-Aldrich, Inc., St. Louis, MO, USA).
5. *Scienceware®* autoradiography pens (Cardinal health, Dublin, OH, USA).
6. Clear plastic wrap, purchased from local store.

3. Methods

3.1. Nonelectrophoretic Transfer

Nonelectrophoretic transfer is a very useful and easy technique that allows proteins to be transferred from the gel to more than one membrane through diffusion. Using this technique, there will be proteins left on the gel, and therefore, after staining the gel with Coomassie Brilliant Blue (CBB), protein spots detected through CL antigen immunodetection can be tracked on the gel and excised for protein identification. After running the first dimensional SDS-PAGE gel, once the last band of standard protein marker and/or indicator dye reaches the bottom of the gel, it is time to turn off the power supply and start the next step, transferring the proteins to a membrane. This step can also be done after running a second dimensional IEF gel as well. Here, a nonelectrophoretic transfer is introduced.

1. After turning off the power supply, gel plates need to be separated carefully using forceps or a similar tool (*see* **Note 3**). Remove the top plate first.
2. Rinse the gel using deionized water to remove any remaining SDS-PAGE running buffer on the gel.
3. Cut several nitrocellulose membranes to the size of the SDS-PAGE gel and gently set a membrane on top of the gel, while the gel is still supported by the bottom (glass/plastic) plate (*see* **Note 4**).

4. Place a precut Whatman No. 3 filter paper on top of the nitrocellulose membrane. Remember, do this step while the gel is still supported by the bottom (glass/plastic) plate (*see* **Note 5**).
5. Gently place the third piece, PROTEAN® 3 System alumina or glass plates on the filter paper. Again, this step needs to be done while the other side of the gel is still supported by the bottom (glass/plastic) plate.
6. Gently place one hand on the bottom glass plate (in **step 5**) and flip the gel upside down into the free hand. Gently separate the other glass/plastic plate from the gel (*see* **Note 6**).
7. Repeat **steps 3** and **4** for this side of the gel.
8. Using a 5 or 10 mm pipette, gently roll over the filter paper to remove any possible air bubbles in between the gel and membrane. This needs to be done very gently as the gel can be broken.
9. Repeat **step 5**.
10. Secure the two PROTEAN^{R} 3 System glass plates with clamps (*see* **Note 7**).
11. Place the assembly in a humidified plastic container and incubate overnight at 37°C (*see* **Note 8**).
12. Remove the membranes for immunoblotting and label each membrane as "top" or "bottom" membranes (*see* **Note 9**).
13. Stain the gel with CBB for 1 h or overnight and then destain.

3.2. Whole Membrane Blotting Chemiluminescence

For whole membrane blotting CL detection, only one primary antibody is used and therefore this technique allows the use of whole membrane without dividing it into strips. To do so, a container similar in the size of the membranes being used needs to be prepared to house them.

1. Block the membranes using the western blotting solution for 2 h at room temperature (RT) on a shaker (*see* **Note 10**).
2. Drain the blocking solution out.
3. Add the appropriate amount of primary antibody to the membrane any where from 1:2,000 dilution to 1:20,000, diluted in western diluent solution. Incubate for 2 h.
4. Wash 3× with TBST, 5 min each time. After washing, block the membranes with western blotting solution for 5 min at RT. Then drain the blocking solution out.
5. Add the appropriate amount of HRP conjugate solution to the membrane (usually 1:10,000 dilution, diluted in diluent) and incubate for 1 h (*see* **Note 11**).
6. Wash 3× with TBST, 5 min each time. While waiting for the last washing, lie a 16 in. × 22 in. clear plastic wrap on a working bench (membranes will be laid on this plastic wrap after washing). Also, prepare a 5–10 mL glass test tube and

wrap it with aluminum foil. Take 2 mL of solution A and add it into the glass tube. Do not add solution B until the membrane is ready (*see* **Note 12**).

7. Drain the TBST solution out and lie the membranes on the clear plastic wrap, facing up (*see* **Note 13**). Turn off the lights in the working area, as the ECL conjugate solution is light sensitive. Quickly add 50 μL of solution B to the glass tube, already containing 2 mL of solution A (*see* **Note 14**). Mix the solution. Using a P1000 pipette, draw the conjugate solution from the glass tube and eject it slowly, all over the membrane. Make sure that the conjugate solution covers the surface of the membranes for 5 min (*see* **Note 15**).
8. Soak up the solution by placing several paper towels on the corner of the clear plastic wrap (*see* **Note 16**).
9. Cover the membrane with a single layer of clean clear plastic wrap so that the signals can be detected. Make sure that there are no bubbles between the membrane and plastic wrap. Formation of the bubbles can be avoided by wrapping the membrane slowly from one corner to the other.
10. Place the wrapped membranes in a Kodak X-ray film cassette or film holder. Attach the wrapped membrane to the film holder using a tape or other tools (*see* **Fig. 2**).
11. Right above and below where the membranes are placed on the film cassette, stick a piece of white laboratory labeling tape (about 1 in. long). Using Scienceware® autoradiography pen, write a letter on the top and bottom tape ("L" or "M," for example) (*see* **Fig. 2**). This letter may help later if the X-ray film is upside down and/or is rotated; therefore, choose a unique letter and do not use letters such as "X" that are indistinguishable when rotated. Use a pen and draw a line around the letter so that it can be seen better later on for data analysis (*see* **Fig. 3**).
12. This step needs to be done in a dark room. Place an X-ray film over the membranes (*see* **Fig. 3**), top and bottom tapes. Lock the film and make sure it does not move. Close the film holder. Using a timer, take a 10 s reading first. After the 10 s period is over, open the film holder and develop the X-ray film. Take more readings as needed. Depending on the signals and results, different readings may be required, anywhere from 4 s to 10 or 15 min (*see* **Note 17**).
13. Once protein signals are detected, it is time to analyze and identify the signals to see what bands they represent. To do so, carefully wrap the destained gel with a clear plastic wrap and slowly place it over the membrane (*see* **Note 18**). The gel should fully overlap the membrane. This can be done by moving the gel back and forth until the stained standard

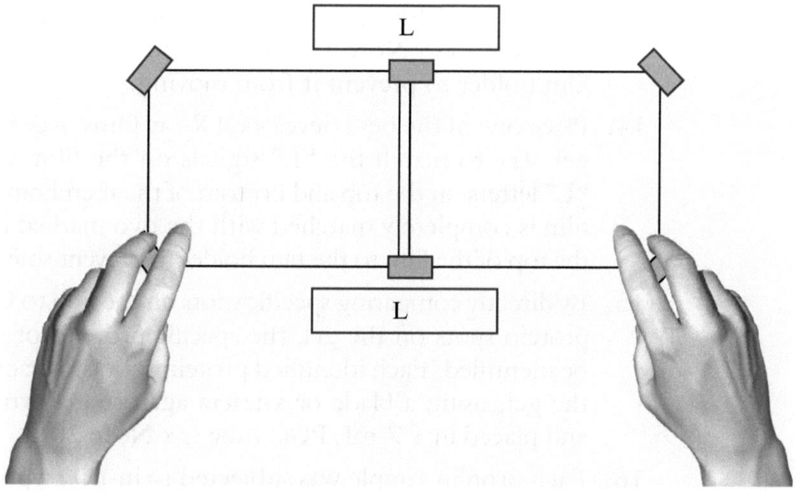

Fig. 2. In this figure, up to four membranes 7 cm × 10 cm can be placed side by side for a one-time reading, using a 20.3 cm × 25.4 cm X-ray Kodak film. After wrapping the membranes with clear plastic wrap, stick all sides of the membrane to the X-ray film holder as shown earlier and write two letters above and below the membrane on a piece of transparent tape, using a *Scienceware*® autoradiography pen.

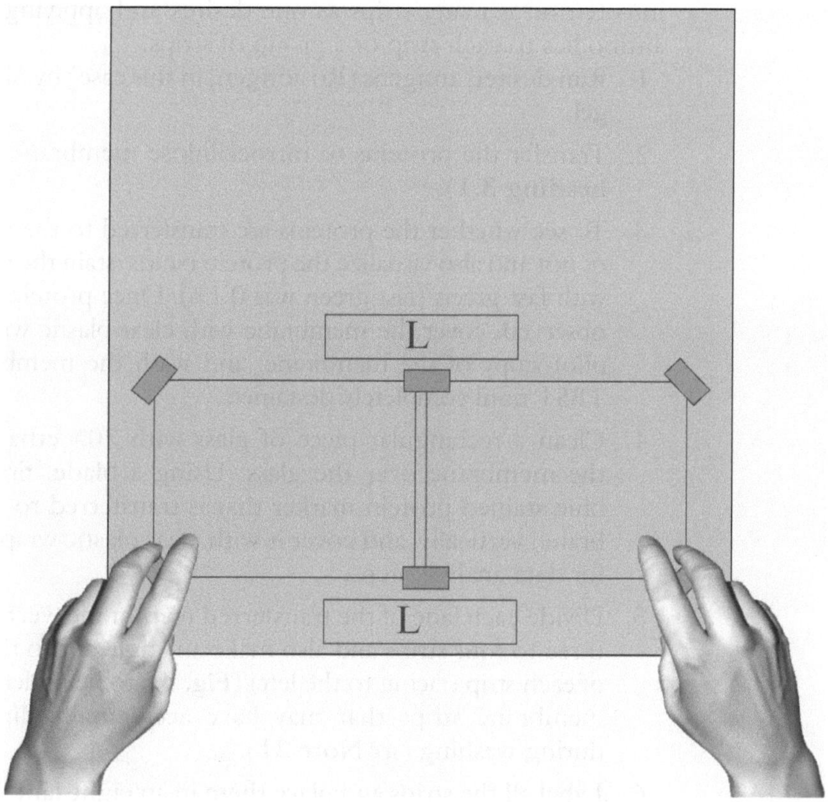

Fig. 3. By holding only half of the X-ray film over the membrane, after exposing the membrane with ECL conjugate, simply rotate the X-ray film to use the other half of the X-ray film. This can be done when one or two small membranes are being studied or a large X-ray film is being used. In some cases, four corners of the X-ray film can be used one at a time.

marker on CBB gel is fully overlapped with that on the membrane (*see* **Note 19**). Then, tape the wrapped gel to the film holder to prevent it from moving.

14. Place one of the best-developed X-ray films over the wrapped gel. Try to match the "L" signals on the film with marked "L" letters on the top and bottom of the membrane. Once the film is completely matched with the two marked letters, tape the top of the film to the film holder to prevent shifting.
15. By directly comparing specific spots on the film to CBB stained protein spots on the gel, the specific proteins of interest can be identified. Each identified protein spot can be excised from the gel, using a blade or x-tracta agarose gel extraction tool and placed in 1.7-mL PGC tube (*see* **Note 20**).
16. Each protein sample was subjected to in-gel tryptic digestion followed by MALDI-TOFMS for protein identification.

3.3. Strip Immunoblotting Chemiluminescence

One of the great advantages of strip blotting CL is that unlike whole membrane blotting, which allows the use of only one primary antibody, one can use a number of antibodies and check for different antigen responses. This can be done by dividing a membrane into ten or as many strips as one desires and applying different antibodies to each strip or a group of strips.

1. Run desired antigens (Ro antigen, in this case) by SDS-PAGE gel.
2. Transfer the proteins to nitrocellulose membrane (*see* **Subheading 3.1**).
3. To see whether the proteins are transferred to the membrane or not and also visualize the protein bands, stain the membrane with fast green (fast green was 0.1%). Once protein bands are observed, cover the membrane with clear plastic wrap, take a photocopy of the membrane, and wash the membrane with TBST until completely destained.
4. Clean a rectangular piece of glass with 70% ethanol. Place the membrane over the glass. Using a blade, first cut the blue stained protein marker that is transferred to the membrane, vertically, and cover it with clear plastic wrap and keep for data analysis steps.
5. Divide each lane of the transferred membrane, vertically, into three to four strips and also make an angle cut to the bottom of each strip (facing to the left) (**Fig. 4**), to help identify those membrane strips that may have accidentally flipped over during washing (*see* **Note 21**).
6. Label all the strips and place them in an eight-lane strip incubation reservoir. From now on, make sure that when the membrane strips are placed into the incubation reservoir or when they are being washed, the strips are not facing down.

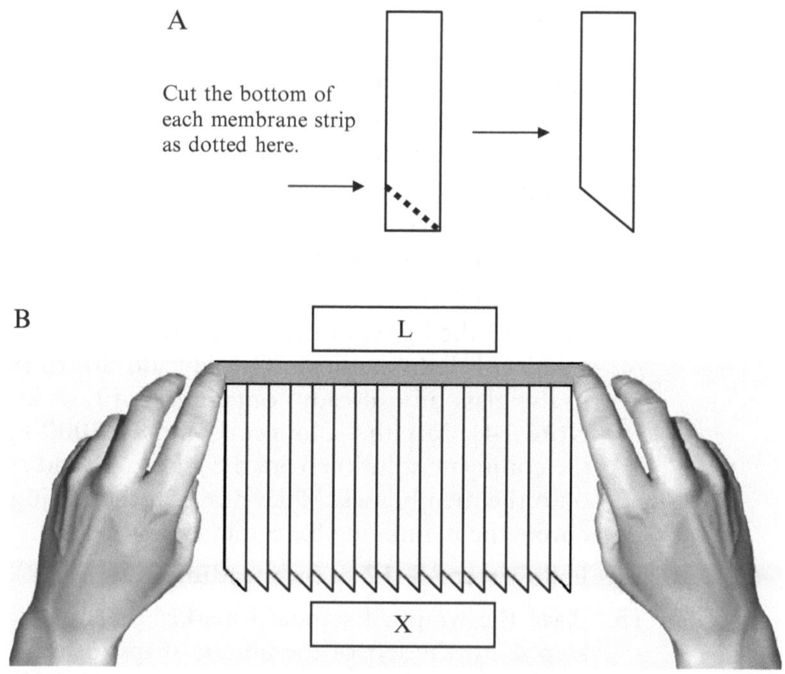

Fig. 4. (**a**) Using a blade, cut the dotted area for all the strips, to make sure which membrane has flipped over and which has not. Make sure that the cut is facing toward left and the transferred membranes are facing up. (**b**) After placing each strip side by side, carefully tape the top portion of the membranes. The bottom of the membranes can also be taped if needed. Label the top and bottom of the membranes with a letter using a *Scienceware*® autoradiography pen.

This can be checked by looking at the cuts that were made at the bottom of the strips.

7. Add 1 mL of blocking solution to each lane and incubate the plate for 2 h on a shaker.
8. Drain the blocking solution out.
9. Make appropriate dilutions of all primary antibodies anywhere from 1:1,000 to 1:20,000 (here, anti-Ro human serum dilution was made to 1:1,000 and anti-HNE anti-rabbit antibody was made to 1:20,000) using western diluent. Add 1 mL to each well (*see* **Note 11**).
10. Wash 3× with TBST, 5 min each time. When washing, fill all the wells equally with TBST. Make sure that the strips do not rotate.
11. Add 1 mL of blocking solution to each well and incubate for 5 min. Then drain the solution out and follow **steps 5** and **6** of **Subheading 3.2** (for **step 5**, add 1 mL of HRP conjugate solution to each well).
12. Drain the TBST solution out of the wells at the end of the third wash. Using forceps, carefully move each strip one by one, in order, from each lane and lie them on the clear plastic wrap side by side, vertically (*see* **Note 22**). Only move the

membrane strips with forceps either from the tip of the strip or from the bottom, where the angled cut was made, and not from the middle. Once all the strips are transferred to the clear plastic wrap, and laid side-by-side, tape the top portion of them (horizontally) to the clear plastic wrap, using transparent tape. The bottom of the strips can also be taped to the clear plastic wrap if needed. Make sure that the tape does not cover more than 2 mm of the membrane strip's top or bottom portion (*see* **Note 23**).

13. Turn off the lights in the work area, as the ECL conjugate solution is light sensitive. Quickly add 50 μL of solution B to the glass tube already containing 2 mL of solution A (*see* **Note 14**). Mix the solution. Using a P1000 pipette, draw the conjugate solution from the glass tube and eject it slowly onto the membranes. Make sure that the conjugate solution covers the membrane for 5 min (*see* **Note 15**).
14. Follow **steps 8–12** of **Subheading 3.2**.
15. Place the wrapped standard marker that was kept aside, in **step 4**, to the left of membrane strips, using a clear paper tape. This will help to show the molecular weight of each of the detected protein bands.
16. Follow **step 14** of **Subheading 3.2**.
17. By directly corresponding specific spots on the X-ray film to the strip membranes, the specific proteins of interest can be tracked for molecular weight identification using the standard marker strip. Each identified protein spot can be cut from the membrane strip, using a blade and can be placed in a 1.7 mL PGC tube.
18. Each protein sample was subjected to on-membrane tryptic digestion followed by MALDI-TOFMS for protein identification.

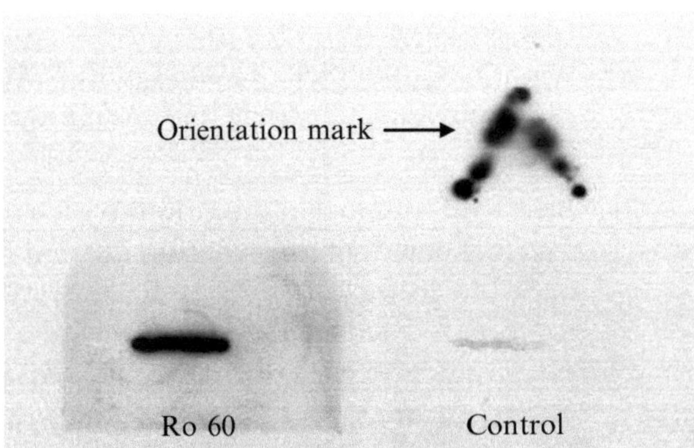

Fig. 5. Antigen coated on the membrane on the left is Ro antigen. Anti-human Ro antibody was used as the primary antibody and anti-rabbit HRP conjugate was used as a secondary antibody. The sample on the right is a control.

3.4. New Line Blotting Chemiluminescence

Line blotting CL is one of the easiest techniques available for immunodetection of antigens. This method is applicable to only purified proteins. This technique is very easy because proteins do not have to be run through a gel or be transferred to a membrane as protein samples can be applied directly onto a nitrocellulose membrane. Primary and secondary antibodies can then be added serially and proteins of interest are detected through the CL technique (*see* **Fig. 5**).

1. Cut the sufficient number of nitrocellulose membranes needed to the size of 1 cm × 2 cm (*see* **Note 24**).
2. Using a pencil, label the right bottom corner of each membrane by a number or letter according to the sample number that is going to be used for that membrane.
3. Using a P2 pipette, load 1 μL of protein sample into the pipette tip and while holding the small membrane on a clean surface, slowly expel the protein sample onto the membrane from the left to the right of the membrane, forming a centimeter long line.
4. Let the membrane dry for 10 min or as needed. Once the protein line is dry, the line formed on the membrane is easily observed.
5. Place each membrane in a small Petri dish. The smaller the dish, less solution it needs, however, the Petri dish should be large enough to cover the membrane with blocking and diluent solutions in the following steps.
6. Add enough western blocking solution onto the membrane to cover it. Make sure that the membrane is covered with the blocking solution at all times and does not float. If the membrane floats on top, flip the membrane so that it is facing down and therefore is coated with solution at all times. Incubate for 2 h.
7. Follow **steps 2–12** of **Subheading 3.2**.

4. Notes

1. To avoid precipitation, first dilute 100 mL of 10× native buffer to 700 mL with deionized water and then add 200 mL of methanol. Avoid adding methanol directly to the 10× native buffer, since methanol precipitates buffer ingredients. Even in such a scenario the precipitate can be redissolved by the addition of 700 mL water, but takes more time.
2. Add 100 mL of 10× TBS to a 1 L graduated cylinder and make it to 800 mL with deionized water. Weigh out 50 g of skim milk powder and add into the cylinder. Mix and stir until dissolved. Once the solution is dissolved, make to 1 L with deionized water. Separate 500 mL as the western blot blocking solution. To the remaining 500 mL, add 250 μL of Tween-20

(use a 1 mL syringe, or cut end of a 1 mL pipette tip to aspirate Tween-20 easily), dissolve, and use it as western blot diluent solution.

3. In the old SE 250 vertical electrophoresis units there is a glass plate and an alumina back plate but today many gels are premade and both gel plates are plastic. These two plates need to be separated carefully using a spatula, a small forceps, or a similar tool. It is very easy to break the gel at this point; therefore, unlocking all sides of the plates is advised.
4. Although it has been reported to place the membrane on the gel when the gel is completely dry this can create bubbles. Bubbles can be avoided by placing the nitrocellulose membrane on the gel right after rinsing the gel with water. This not only prevents bubbles but also creates very good contact between the gel and the membrane. As a result this has shown to be helpful for better transfer. Also, do not move the gel from the bottom glass/plastic support instead place on the nitrocellulose membrane while the gel is on the bottom plate support. This prevents the gel from moving or breaking. Another important suggestion is that the membrane can be slowly laid down onto the gel starting on either side and while holding the opposite side in your free hand, slowly lowering it onto the gel, as opposed to simply laying the membrane onto the gel all at once. By doing this, the junction of the gel and membrane can be observed for bubbles.
5. Cut the filter paper a little bigger than the membrane (1–2 mm bigger on all sides), so that it will protect the membrane.
6. To separate the gel from the last glass plate, a spatula or forceps is needed to unseal the bottom of the gel from the glass plate. This makes it easier for this separation to happen.
7. Use the same number of clamps for all the gels as the condition for all the gels must be the same. Do not use too many clamps either as it may damage the gel. For PROTEAN^{R} 3 System, we use two large clamps.
8. Different incubation times are available but we use overnight incubation. To humidify a container, place several moist Terri Wipes paper towels (Kimberly-Clark, Neenah, WI, USA) at the bottom of a large plastic container. There should not be any running water in the container. After placing the assembly in the container, place the lid and seal it with clear plastic wrap or Parafilm.
9. After removing the container from the incubator and before opening the lid, it is important to remember that the bottom and top membranes need to be labeled, as it is believed that more proteins transferred to the bottom membrane due to gravity. Therefore, bottom membranes must be compared together

in studies. Rotating the assembly at this point can confuse this labeling. Even though the container is humid, gels get dry and it is normal to have membranes stuck to the gels. Since nonelectrophoretic transfer is mostly used to compare the blots with the stained gels (unlike the complete electrophoretic transfer in which the gel is not used anymore), the gel and membrane need to be separated carefully to preserve the gel. This can be done by removing the assembly from the container and removing the clamps. Remove the two PROTEAN^{R} 3 System glass plates that are on both sides of the assembly first by handing the assembly to the left hand and then to the right. Then keep the assembly on a clean bench and gently remove the membrane with gloves starting from one corner of the gel. With the opposite hand keep the assembly from moving. In case the membrane is stuck to the membrane, have a forceps handy. Sometimes the top wells of the gel (where the samples are loaded into the wells) stick to the membrane. In such cases, one can rinse off the top portion of the membrane with deionized water but avoid washing the whole membrane. Now that the gel is on the top and the top membrane is taken off, the gel can be placed in a separate container for staining. This helps for a better separation of gel and second membrane. To do this, simply keep the gel facing up (with the filter paper facing down) in hand and gently lay down the gel into a large container (large enough for staining the gel-filter paper facing up and the gel facing down). Remove the filter paper from the top. Remove the membrane slowly as described earlier starting from one corner. Label this membrane as "bottom" or "top" membrane.

10. Every set of two membranes can be placed in the same container together. This can be done by placing the two membranes back-to-back (one membrane faces down and one up). Put them in the container and add the solution. This can be done for washing as well as the conjugate step. Another way of saving solution is to place the membranes into a plastic bag and add the solutions into the bag. After carefully removing the bubbles, the bag can be placed on a shaker, which can also save considerable antibody compared with the use of a container.
11. If the antibody was made from rabbit, then the conjugate must be an anti-rabbit conjugate or if a patient serum with antibodies was used as primary antibody (for example anti-Ro) then the conjugate must be an anti-human conjugate. Also, do not drain blocking solution out of wells until primary antibody is ready to be added to the membranes. Remember, membrane strips should be kept wet at all times.
12. At this point, it is good to make sure that the X-ray film developing machine is in working readiness as sometimes it

takes about 20 min for some machines to warm up and in such cases, the experiment will give very weak signals due to the waste of time. Therefore, this point is the best time to check this equipment and make sure it is working. It is also important to be close to the machine as any delay can lead to very weak signals.

13. To know the orientation of the membrane, look for the blue-stained protein markers. Blue stained protein markers are noticeable even after the transfer, and the side that has the blue stain should face up and it has the transferred proteins.
14. For every 2 mL of solution A (ECL Plus western blotting conjugate), 50 μL of solution B is needed.
15. For the solutions to cover the membranes for 5 min, either there should be a large amount of solution or the following should be done. For two membranes, set the membranes side by side. Eject equal amounts of conjugate to each membrane. Then by slowly picking up the top clear plastic wrap from all sides of the membrane and creating slope, try to guide the solution toward all sides of the membrane during that 5 min. If there are four membranes, set the membranes in the shape of a square on the clear plastic wrap and follow the aforementioned procedure. Another way is to recover the conjugate solution once ejected on the membranes using a P1000 pipette and to redistribute it over the membranes but be careful not to make a hole in the clear plastic wrap and not to touch the membrane.
16. Once the conjugate is soaked up with paper towels, the rest of the steps need to be done quickly but carefully. The membranes should not become dry at any time; therefore, **steps 8–11** need to be done in less than 5 min, which is more than enough time. Also, if too much time is wasted, weaker signals may be given.
17. For one or two membranes, it is possible to use only one X-ray film for two readings by placing the top part of the X-ray film on the two membranes and not allowing the bottom part to be exposed to the membranes, saving them for another reading. Once done with one reading, rotate the film and using the unexposed end, make another reading (need to hold with hands). If there is only one membrane, only one corner of the X-ray film can be exposed at a time and four readings can be done with only one X-ray film (*see* **Fig. 3**).
18. It is usually hard to hold the gel and very easy to break it. Therefore, to wrap the gel, add about 100–150 mL of destaining solution or more to the gel, if the gel was originally in destaining solution. Add the same amount of deionized water if the gel was originally in deionized water. There must

be enough solution in the container to make the gel float in the solution. Wearing gloves, gently move fingers under the gel, going from the corners of the gel (using both hands) and once the four fingers are holding the gel from the bottom, hold the gel with the thumb from the top, very gently. Slowly remove the gel and lay it over a clear plastic wrap that has been laid over the bench and cover the gel quickly preventing it from becoming dry.

19. It is always hard to overlap the CBB stained gel to the membrane, but there are always little things that can assure that this gets done. One is a protein standard marker. Protein standard marker (molecular weight standards) is blue stained and can be seen on the membrane as well as a CBB stained gel. Therefore, by carefully looking at specific bands in the gel and comparing them on the membrane, the gel and the membrane can be matched and placed over one another. Another sign to look for is the shape of the wells on the membranes. When membranes are placed over the gel, usually the shape of the well is imprinted over the membrane and can be easily observed. It is easy to track each well and by comparing each well from the CBB stained gel and placing it exactly on the membrane, it is simple to match the membrane and the gel together. One thing that should be remembered is that after staining and destaining the gel, the gel has more water in it and the size of the gel is a little bigger. Therefore, it may not be fully comparable to the membrane and do not expect the gel to fully overlap the membrane. However, it is still very easy to identify and excise the protein from the X-ray film.
20. It is always good to label each protein spot separately in a lab book along with the estimated molecular weight of that spot, before excising it. It is also possible to dry the gel before cutting the protein spots as it may give distinct protein spots, however, it is easier to cut the gel when it is not dry.
21. When cutting the membrane with a blade, always keep the membrane wet with deionized water. This will give nice and sharp membrane strip cuts. As soon as each membrane strip is cut, place it in an eight-lane strip incubation reservoir and label it accordingly to avoid possible mixing of the strips.
22. When transferring the membrane strips to the clear plastic wrap, always keep the membrane strips wet at all time to avoid them rotating, moving, or mixing.
23. Once all of the membrane strips are transferred to the clear plastic wrap, place a Kimwipe at the top of the membrane strips and let it soak up the water. Then tape the top portion

of membranes with a transparent tape to avoid their movement.

24. Instead of cutting small membranes, it is more convenient to apply all the antigen samples to one rectangular membrane with 2–3cm distance to one another and add blocking and diluent solution to just over one membrane in one container. This can be done only when there is one primary and secondary antibody. However, multiple membranes must be cut if more than one primary and/or secondary antibodies are going to be used for an experiment.

References

1. Garcia-Campaña, A.M., and Baeyens, W.R.G. (eds), (2001) *Chemiluminescence in Analytical Chemistry*. Marcer Dekker. New York.
2. Crisp, J.S. and Dunn, J.M. (1994) Detection of protein blots using chemiluminescent systems, in *Methods in Biology* (Walker, J.M., ed.), Humana, Totowa, NJ, pp. 233–237.
3. Heider, H., Ziaja, B., Priemer, C., Lundkvist, A., Neyts, J., Krüger, D.H., et al. (2001) A chemiluminescence detection method of hantaviral antigens in neutralisation assays and inhibitor studies. *J Virol Methods*. 96, 17–23.
4. Deluca, M.A. and Mcelroy, W.D. (1981) Bioluminescence and Chemiluminescence. Basic Chemistry and Analytical Applications. Academic, New York.
5. Tsukagoshi, K., Tsuge, K., and Nakajima, R. (2007) Development of an immune microanalysis system by use of peroxyoxalate chemiluminescence detection. *Anal Sci*. 23, 739–741.

Chapter 43

Protein Stains to Detect Antigen on Membranes

Anil D'souza and R. Hal Scofield

Summary

Western blotting (protein blotting/electroblotting) is the gold standard in the analysis of complex protein mixtures. Electroblotting drives protein molecules from a polyacrylamide (or less commonly, of an agarose) gel to the surface of a binding membrane, thereby facilitating an increased availability of the sites with affinity for both general and specific protein reagents. The analysis of these complex protein mixtures is achieved by the detection of specific protein bands on a membrane, which in turn is made possible by the visualization of protein bands either by chemical staining or by reaction with an antibody of a conjugated ligand. Chemical methods employ staining with organic dyes, metal chelates, autoradiography, fluorescent dyes, complexing with silver, or prelabeling with fluorophores. All of these methods have differing sensitivities and quantitative determinations vary significantly. This review will describe the various protein staining methods applied to membranes after electrophoresis. "Detection" precedes and is a prerequisite to obtaining qualitative and quantitative data on the proteins in a sample, as much as to comparing the protein composition of different samples. Detection is often synonymous to staining, i.e., the reversible or irreversible binding by the proteins of a colored organic or inorganic chemical.

Key words: Blotting, Protein staining, Electrophoresis, Western

1. Introduction

The selection of the appropriate protein detection method is most important for detection of proteins and for quantitative assessment of expression levels of these proteins in biological samples. A number of very different techniques exist. The following criteria need to be evaluated in the decision making process: (a) The detection limit should be as low as possible with a high signal to noise ratio, (b) the detection method should also have a wide linear relationship between the quantity of protein and the staining intensity, (c) the procedure should be easy and fast

Biji T. Kurien and R. Hal Scofield (eds.), *Methods in Molecular Biology, Protein Blotting and Detection, vol. 536*

DOI: 10.1007/978-1-59745-542-8_43

to perform, (d) the procedure should be nontoxic, environment-friendly, and most importantly (e) the procedure should not be too expensive.

1.1. Staining Blots with Ponceau S

PonceauS(3-hydroxy-4-[2-sulfo-4-(sulfo-phenylazo)phenylazo]-2,7-naphtalene disulfonic acid) is a relatively insensitive stain wherein greater than 1 μg or more of protein can be detected *(1)*. Accordingly, only the most abundant proteins will be visible. However, it is also a reversible stain that can be removed completely with water prior to processing the blots. After staining, a regular pencil can be used to record the presence of visible proteins and molecular-weight markers, which will help when aligning the proteins detected on the membrane by western analysis with those in a total protein-stained gel or membrane. Ponceau S is compatible with both nitrocellulose and PVDF membranes. This is a quick and easy way to visualize proteins transferred to membranes following SDS PAGE but it is about half as sensitive as Coomassie Blue staining (*see* **Notes 1–3**).

1.2. Staining Blots with India Ink

India ink is an alternative staining protocol to the Ponceau S staining method and is a recommended method for staining blots *(2)*. Staining with India ink depends on the preferential adherence of the colloidal carbon particles composed within the ink to the immobilized protein on the filter. It is cheap, reliable, and sensitive; it yields a permanent record; and it does not interfere with subsequent binding of antibody to the antigen. The India ink method is generally more useful when low amounts of protein are transferred, such as when purified protein preparations are being studied. It is also very useful when staining a duplicate blot or when using radioactive antibodies for antigen detection. However, India ink is not compatible with enzyme-based chromogenic detection methods, in which case staining can be carried out after immunoenzymatic detetction is complete. Care must be taken though to see that all the remaining sites on the membrane have been blocked with a blocking solution such as Tween-20 (*see* **Notes 4–5**).

1.3. Staining Blots with Amido Black

Amido black is yet another alternative staining protocol to the Ponceau S staining method and is also a recommended method for staining blots *(3)*. Amido black staining solution is designed for rapid staining of protein bands on nitrocellulose membranes. Amido black staining solution facilitates visualization of low concentration proteins with a low background. Proteins can be easily destained with 25% (v/v) isopropanol and 10% (v/v) acetic acid (*see* **Notes 6** and 7).

1.4. Staining Membrane-Bound Proteins with Colloidal Gold

Colloidal gold is the most sensitive staining technique for proteins bound on membranes, detecting as little as 1–3 ng of protein *(4)*. Protein spots are permanently stained dark red after incubation with the colloidal gold solution. Colloidal gold staining can detect proteins on both nitrocellulose and PVDF membranes, but it is not recommended for nylon membranes.

1.5. Coomassie Brilliant Blue Staining

Since its introduction in 1963 by Fazekas de St. Groth et al. *(5)* staining with the organic dye Coomassie Brilliant Blue (CBB) is still the most frequently employed method for protein detection in polyacrylamide electrophoresis gels and membranes. CBB is a disulfonated triphenylmethane textile dye of which two modifications exist: Coomassie R-250 (reddish tint) and G-250 (greenish tint, dimethylated form, which has two additional methyl groups). In acidic solutions the dye binds to the amino groups of the proteins by electrostatic and hydrophobic interactions. CBB staining and its various modifications *(6)* are generally applied to acrylamide gels but they are applicable to PVDF membranes. The blots are stained at 0.1% Coomassie dye dissolved in 45% (v/v) methanol, 45% (v/v) water, and 10% acetic acid. The background is destained with 25% (v/v) methanol, 65% (v/v) water, and 10% acetic acid (*see* **Notes 8** and **9**).

1.5.1. Alternate Method to Stain with CBB

This staining procedure is employed when low background staining and a high degree of sensitivity are required. This staining procedure has a sensitivity threshold close to that of silver staining; however, one needs to keep in mind that the time required for staining is substantially greater than that of either standard silver or Coomassie staining. Additionally, in order to see optimal results from this staining protocol one should allow 3–4 days for completion (*see* **Note 9**).

1.6. Fast Green Staining

Considered an "old generation" stain, the organic dye Fast Green FCF does not stain proteins multichromatically (*see* **Note 10**). It can be used for protein detection after native PAGE, SDS-PAGE, and IEF, and for alkaline proteins *(7, 8)*. It also belongs to the triphenylmethane dye family, but, unlike CBB R-250, it does not bind to ampholytes and can, therefore, be used as a stain for IEF gels without a separate protein fixation and ampholyte removal step. In quantitation, Fast Green proved slightly less sensitive than CBB but showed a broader linear range *(9)* (*see* **Note 10**).

2. Materials

2.1. Staining Blots with Ponceau S

1. 1× Ponceau stain: 0.1% Ponceau-S in 1% acetic acid. Mix 0.5-g Ponceau-S (Fisher Scientific, Dallas, TX, USA; BP103-10) in 5-mL acetic acid and make it up to 500 mL with distilled water (dH_2O).
2. Ponceau destain: 1% acetic acid, 5-mL acetic acid made up to 500 mL with dH_2O.

3. India ink suspension (Becton Dickinson, USA, 261194).
4. Phosphate buffered saline (PBS), pH 7.4.
5. Tween-20 (0.3%).

2.2. Staining Blots with Amido Black

1. Water (250 mL).
2. Methanol (200 mL).
3. Acetic acid (50 mL).
4. Amido black (also called Napthol Blue Black): 0.1% (w/v; Sigma Chemical Company, St. Louis, MO, USA; A8181).
5. Methanol (20%), acetic acid (7.5%)

2.3. Staining Membrane-Bound Proteins with Colloidal Gold

1. Colloidal gold stain solution (Amersham Biosciences, GE Health-care, Piscataway, NJ, USA or Bio-Rad, Hercules, CA, USA).
2. PVDF membrane containing bound proteins.
3. Tween-20.

2.4. Staining with CBB

1. Brilliant Blue R-250 (BBR) (Fisher BP101-25).
2. HPLC water or Milli-Q water.
3. Fixing solution: 50:10:40 (methanol:acetic acid:H_2O).
4. Destaining solution: 45:10:45 (methanol:acetic acid:H_2O).
5. Coomassie concentrated stain solution: 12-g CBB + 300-mL methanol dissolved, then add 60-mL acetic acid. Stir well.
6. Coomassie working solution: 500-mL methanol + 30-mL Coomassie concentrated stain solution + 400-mL milli Q water + 100-mL acetic acid. Mix and filter using 0.22-μm presterilized filter.

2.4.1. Alternate Method to Stain with CBB

1. CBB G250 (Biorad).
2. HPLC water or Mill-Q water.
3. Fixing solution: 50:3:47 (ethanol:phosphoric acid:H_2O).
4. Neuhoffs solution: 16:25:5:54 (ammonium sulfate:methanol: phosphoric acid:H_2O).

2.5. Fast Green Staining

1. Fast Green (F7252 Sigma).
2. Methanol.
3. Acetic acid.

3. Methods

3.1. Staining Blots with Ponceau S

1. Submerge the transfer membrane in Ponceau S stain solution with gentle agitation for 5 min.

2. Decant the stain, and rinse the membrane several times with distilled water (D/W) until the protein bands are visible.
3. Use a soft lead pencil to mark the major protein spots and molecular weight markers even when the membrane is wet.
4. Continue rinsing the membrane with D/W with gentle agitation until the Ponceau S is removed.

3.2. Staining Blots with India Ink

1. Wash the blot in 100 mL of PBS/Tween 20 (0.3%), with two changes at 5 min each.
2. Place the blot in ~100 mL of India ink suspension.
3. Incubate at room temperature (RT) for 15 min to 18 h. Longer incubations will increase sensitivity.
4. Destain by washing the blot in multiple changes of PBS.
5. Photograph the blot, or record lane positions and molecular-weight markers as required.

3.3. Staining Blots with Amido Black

1. Wash the blot in 100 mL of PBS/Tween 20 (0.3%), with two changes at 5 min each.
2. Place the blot in ~100 mL of amido black suspension.
3. Incubate at room temperature (RT) for 5 min.
4. Destain by washing the blot in multiple changes of destaining solution.
5. Photograph the blot, or record lane positions and molecular-weight markers as required.

3.4. Staining Blots with Colloidal Gold

1. Submerge the membrane in Tween-20 protein staining solution for 45 min at 37°C with gentle agitation.
2. Wash the membrane with Tween-20 protein staining solution for 5 min at RT with gentle agitation.
3. Discard the wash solution, and repeat **step 2** two more times.
4. Stain the membrane in the colloidal gold stain solution for 2 h at RT with gentle agitation.
5. Rinse the membrane with H_2O for 5 min with gentle agitation.
6. Proceed with protein characterization or analysis.

3.5. Method for Staining Blots with CBB

1. After electroblotting, transfer the membrane to a clean glass container.
2. Add 0.25% CBB R-250, 50% methanol, and 10% acetic acid, 25–45 min.
3. Decant the stain.
4. Destain the membrane with repeated washings of the membrane in 5% methanol, 7.5% acetic acid at RT and shaking.

3.5.1. Alternate Staining with CBB

1. Allow gels/membranes to fix in this solution overnight (as little as 4 h will do and fixation can be prolonged to 4 days if need be).
2. Remove excess fixative by washing gels three times for 30 min per wash.
3. Equilibrate gels in Neuhoff's solution for 1 h.
4. Add CBB G-250 powder (1 g/L) to each staining tray and stain for 3 days. Normally spots can be seen by 24–48 h but for optimal staining stain for 3–4 days.
5. Store the gel in 5% acetic acid solution at 4°C until in-gel digestion is performed (The gel can be stored for several weeks).

3.6. Staining with Fast Green

1. Electrotransfer onto nitrocellulose or polyvinylidene fluoride (PVDF).
2. Wet the dry membrane.
3. Stain for 10 min with 0.001% Fast Green (F7252, Sigma) in 30% methanol and 7% acetic acid.
4. Destain the membrane for 10 min with 30% methanol and 7% acetic acid.
5. Wash in water for 10 min.

4. Notes

1. Advantages of Ponceau S staining: (a) Staining is rapid. (b) Binding to proteins is reversible. (c) Can quickly verify transfer of proteins. (d) Storage at room temperature. (e) Stock solution stable for over a year.
2. Disadvantages of Ponceau S staining: (a) Staining is not permanent. (b) Staining is not very sensitive. (c) Difficult to record photographically.
3. Ponceau S is a kind of an odd stain in that in order to visualize the staining, one needs to briefly stain (5–10 min) and then very briefly wash with water (just until a little of the red comes off) 5–10 s at a time, discard the water and look, add more water to clear the background a bit, and then swish around and discard the water. If you wash too long, the stain will all be removed, since it is a completely reversible stain, which is why it is the *only* stain that can be used if one wants to subsequently use the blot for immunodetection. Do not reuse the stain; it will result in nonreproducible results because of depletion of the dye after the first use.
4. Advantages of staining with India ink: (a) Staining is rapid if required, 15-min turnaround time. (b) The materials are cheap

yet reliable. (c) Binding to proteins is irreversible, therefore a permanent record could be obtained. (d) Can quickly verify transfer of proteins. (e) Storage at room temperature. (f) Stock solution stable for over a year. (g) Does not interfere with poststaining steps such as binding of antibody to antigen.

5. Disadvantages of staining with India Ink: (a) Staining is quite sensitive (upto 100 ng). (b) Difficult to record photographically due to lack of contrast as the protein bands appear as black bands on a gray background. (c) Correct India Ink needs to be used (Pelikan Fount India Drawing Ink).
6. Advantages of staining with amido black: (a) Staining is rapid if required, 5-min turnaround time. (b) Binding to proteins is irreversible, therefore a permanent record could be obtained. (c) Can quickly verify transfer of proteins. (d) Storage at room temperature. (e) Stock solution stable for over a year.
7. Disadvantages of staining with amido black: (a) Staining is of low sensitivity. (b) Difficult to record photographically due to lack of contrast. (c) Not compatible with immunological detection. (d) May cause distortion of the membrane.
8. Advantages of staining with CBB: (a) The gel can remain in staining or stabilizing solution for several days without any effect on the results. (b) The optimal sensitivity and detection limit are low of several nanograms per spot. (c) Other advantages of Coomassie staining methods are the quantitative binding of the dye to proteins, the low price, and the good reproducibility. (d) The stain can be reused.
9. Disadvantages of staining with CBB: (a) One of the drawbacks of this staining method is that because of the presence of alcohol some of the proteins release the dye during the background destaining process. (b) Some proteins, such as collagen, for example, destain faster than the polyacrylamide gel/membrane and likewise some proteins, such as bovine serum albumin, for example, stain a lot faster than others do. (c) The other downside to this procedure is that the total procedure takes long. (d) The sensitivity is relatively low and stains in a narrow dynamic range. (e) Staining times can be reduced by using hot stain and destain solutions. (f) Kimwipes can be placed in the destaining container, which will absorb the dye, to increase the destaining procedure and decrease the time of destaining. (g) Sensitivity can be increased by staining the membrane with 0.25% CBB in 50% trichloroacetic acid.
10. Advantages of staining with Fast Green: (a) Fast Green is available in high purity. (b) Protein can be quantitated on blots by staining with Fast Green before blocking and immunodetection. (c) The fluorescence response on membranes is proportional to the amount of protein.

References

1. Salinovich, O. and Montelaro, R.C. (1986) Reversible staining and peptide mapping of proteins transferred to nitrocellulose after separation by sodium dodecyl sulfate-polyacrylamide gel electrophoresis. *Anal Biochem.* 156, 341–347.
2. Klarskov, K. and Naylor, S. (2002) India ink staining after sodium dodecyl sulfate polyacrylamide gel electrophoresis and in conjunction with western blots for peptide mapping by matrix-assisted laser desorption/ionization time-of-flight mass spectrometry. *Rapid Commun Mass Spectrom.* 16, 35–42.
3. Yonan, C.R., Duong, P.T., and Chang, F.N. (2005) High-efficiency staining of proteins on different blot membranes. *Anal Biochem.* 338, 159–161.
4. Moeremans, M., Daneels, G., and De Mey, J. (1985) Sensitive colloidal metal (gold or silver) staining of proteins blots on nitrocellulose membranes. *Anal Biochem.* 145, 315–321.
5. Fazekas de St Groth, S., Webster, R.G., and Datyner, A. (1963) Two new staining procedures for quantitative estimation of proteins in electrophoresis strips. *Biochim Biophys Acta.* 71, 377–391.
6. Anderson, N.L., Esquer-Blasco, R., Hofmann, J.-P., and Anderson, N.G. (1991) A two-dimensional gel database of rat liver proteins useful in gene regulation and drug effect studies. *Electrophoresis* 12, 907–930.
7. Gorovsky, M.A., Carlson, K., and Rosenbaum, J.L. (1970) Simple method for quantitive densitometry of polyacrylamide gels using fast green. *Anal Biochem.* 35, 359–370.
8. Allen, R.E., Masak, K.C., and McAllister, P.K. (1980) Staining protein in isoelectric focusing gels with fast green. *Anal Biochem.* 104, 494–498.
9. Syrovy, I. and Hodny, Z. (1991) Staining and quantification of proteins separated by polyacrylamide gel electrophoresis. *J Chromatogr.* 569, 175–196.

Chapter 44

Use of Nonradioactive Detection Method for North- and Southwestern Blot

Claudia Franke, Daniel Gräfe, Holger Bartsch, and Michael Bachmann

Summary

Many proteins bind to nucleic acids. For the first characterization of novel proteins, a fast and simple technique for testing their nucleic acid binding capabilities is desirable. Here, we describe the use of a northwestern and southwestern blot protocol for the evaluation of the DNA and RNA binding abilities of a novel putative methyl transferase HSPC133 (METTL5).

Key words: DNA binding, RNA binding, Northwestern blot, Southwestern blot

1. Introduction

Proteins interact with different compounds within the living cells. Often they can bind to nucleic acids, e.g., nuclear DNA for initiation or inhibition of RNA transcription. To have an easy and simple tool to evaluate such binding capabilities, one can use standard western blotting and combine this technique with an additional incubation step involving labeled nucleic acids. These procedures are commonly known as northwestern or southwestern blot when RNA or DNA is used as probe, respectively. Here, we describe such a nonradioactive procedure.

La/SS-B is an autoantigen known to bind to RNA *(1)* as well as to DNA *(2)*. We have used this protein as a control for the evaluation of the binding capacities of a novel, so far not characterized protein HSPC133 (METLL5) that is supposed to have methyl transferase activities *(3)*. Therefore, it was of interest to figure out, if DNA or RNA or both could be possible substrates.

Biji T. Kurien and R. Hal Scofield (eds.), *Methods in Molecular Biology, Protein Blotting and Detection, vol. 536*

DOI: 10.1007/978-1-59745-542-8_44

2. Materials

2.1. Preparing DIG-Labeled Nonradioactive RNA Probe

1. DEPC (diethylpyrocarbonate)-treated water: Dissolve 0.1% (v/v) DEPC solution (Carl Roth, Karlsruhe, Germany) in water and incubate over night. Destroy remaining DEPC by autoclaving (*see* **Note 1**).
2. Phenol/chloroform/isoamyl alcohol (25/24/1) and chloroform/isoamyl alcohol (24:1) (both supplied from Carl Roth, Karlsruhe, Germany).
3. DIG-NTP labeling mixture (10×, 10 mM ATP, 10 mM CTP, 10 mM GTP, 6.5 mM UTP, 3.5 mM DIG-11-UTP, pH 7.5).
4. Transcription buffer (10× conc.).
5. DNase I (RNase free, 10 U/μL).
6. Protector RNase inhibitor (20 U/μL).
7. SP6 RNA polymerase (20 U/μL) and all solutions (**Subheading 2.1, items 3–7**) are provided by Roche, Mannheim, Germany (*see* **Note 2**).
8. 200 mM EDTA (pH 8), prepare in DEPC-treated water.
9. DIG-labeled actin control RNA (Roche, Mannheim, Germany).
10. Nylon membrane, uncharged (Micron Separations Inc. (MSI), Westboro, MA, USA) (*see* **Note 3**).
11. TBST (10×): 1.5 M NaCl, 0.5 M Tris–HCl, pH 8, 1% Tween-20. Dilute 100 mL with 900 mL water for use.
12. Blocking buffer: 5% (w/v) Blocking Reagent (Roche, Mannheim, Germany) in TBST.
13. Detection buffer: 100 mM Tris–HCl, pH 9.5, 50 mM $MgCl_2$, 100 mM NaCl.

2.2. Preparation of Nonradioactive DNA Probe

1. Polymerase (Expand High Fidelity) (3.5 U/μL), PCR buffer with 15 mM $MgCl_2$ (10× conc.), dNTP stock solution (10× conc., dATP, dCTP, dGTP, dTTP (2 mM each), pH 7) (Roche, Mannheim, Germany) (*see* **Note 4**).
2. PCR DIG Probe Synthesis Mix (10× conc., dATP, dCTP, dGTP (2 mM each), 1.3 mM dTTP, 0.7 mM DIG-11-dUTP, pH 7) (Roche, Mannheim, Germany).

2.3. Western Blotting

1. Transfer Buffer: Roti-Blot A and Roti-Blot K (Carl-Roth, Karlsruhe, Germany).
2. Nitrocellulose membrane: Hybond-C extra (Amersham/GE-HealthCare Life Science, Munich, Germany) (*see* **Note 5**).
3. 3MM chromatography paper (Whatman, Maidstone, UK).
4. Ponceau S staining solution: 0.1% (w/v) Ponceau S in 5% (v/v) acetic acid.

2.4. Development of Northwestern/ Southwestern Blot

1. Blocking buffer: 5% (v/v) Blocking Reagent (Roche, Mannheim, Germany) in TBST.
2. Tris-buffered saline with Tween-20 (TBST) (10×): 1.5 M NaCl, 0.5 M Tris–HCl, pH 8, 1% Tween-20. Dilute 100 mL with 900 mL water for use.
3. Antibodies: Anti-penta-HIS mouse IgG (Qiagen, Hilden, Germany), anti-DIG IgG coupled to alkaline phosphatase (Roche, Mannheim, Germany), anti-mouse IgG coupled to peroxidase (POD) (Sigma-Aldrich, Taufkirchen, Germany).
4. Detection buffer: 100 mM Tris–HCl, pH 9.5, 50 mM $MgCl_2$, 100 mM NaCl.
5. NBT (*p*-nitrobluetetrazoliumchloride) stock: 77 mg/mL in 70% dimethylformamide (DMFA).
6. BCIP (5-bromo-4-chloro-3-indolyl-phosphate) stock: 50 mg/mL in 100% DMFA.

3. Methods

For characterization of novel proteins, often the binding to nucleic acids has to be addressed. A fast and reliable method, therefore, is the use of a combination of western- and northern- and accordingly western- and Southern-blot techniques to a northwestern or southwestern blot. Here, we have used the method in order to characterize the binding capabilities of a novel protein, a putative methyl transferase, called HSPC133/METTL5.

Figs. 1 and **2** give examples of a northwestern (**Fig. 1**) and southwestern blot (**Fig. 2**) using purified HSPC133 and La/SS-B protein, which is able to bind to both DNA and RNA.

3.1. Preparation of DIG-Labeled Nonradioactive RNA Probe

1. Clone the appropriate cDNA sequence into a vector containing the T7, T3, or Sp6 promoter sequence depending on the RNA polymerase you are going to use (e.g., pGemTeasy). We used the reporter gene *CAT* in our experiments for the detection of general RNA binding (*see* **Note 6**).
2. Linearize 10 μg of this vector using a restriction enzyme that cuts nearby the 3′ end of your sequence to prevent excessive 3′ transcription of unwanted sequences.
3. Extract the DNA once with phenol/chloroform/isoamyl alcohol and once with chloroform/isoamyl alcohol. Precipitate the DNA with pure ethanol. Redissolve the DNA in water at a concentration of approximately 1 μg/μL.
4. Use 1 μg of linearized DNA to set up your in vitro transcription experiment. Add 2 μL DIG-NTP labeling mixture, 2 μL

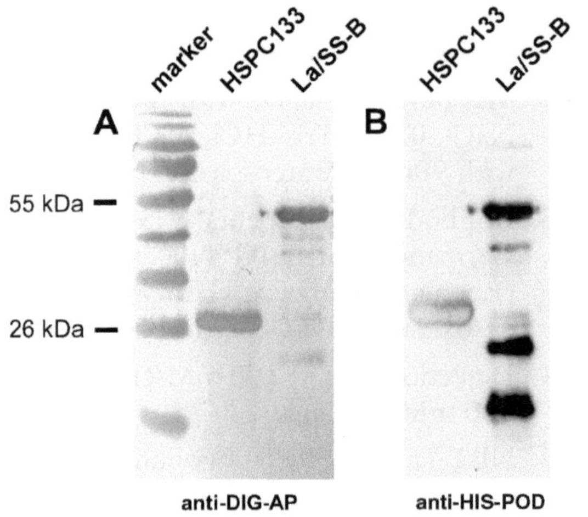

Fig. 1. Northwestern blotting analysis of recombinant HSPC133 and La/SS-B protein to investigate the binding capability to RNA. Blot (**A**) was incubated with DIG-labeled RNA (1 μg/mL) and developed with an anti-DIG antibody coupled to AP. RNA is binding to La/SS-B and HSPC133. After detection of the bound RNA, the same blot was processed with an anti-His antibody and a secondary antibody coupled to peroxidase in order to visualize the recombinant proteins themselves [blot (**B**)].

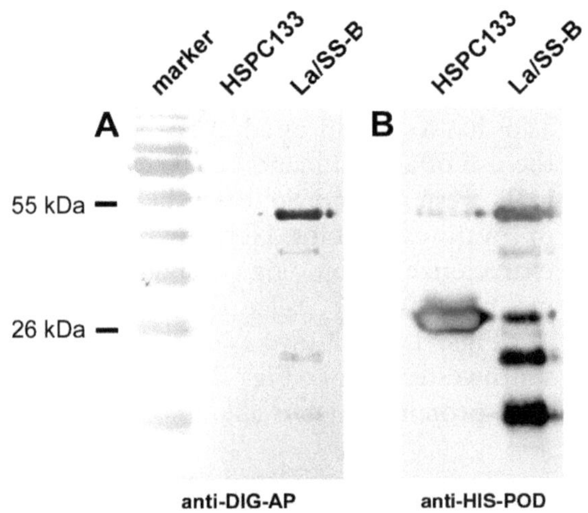

Fig. 2. Southwestern blotting analysis of recombinant HSPC133 and La/SS-B protein to investigate the binding capability to DNA. Blot (**A**) was incubated with DIG-labeled DNA (diluted 1/1,000) and developed with a anti-DIG antibody coupled to AP. The DNA is binding only to the La/SS-B protein. After detection of the bound DNA, the same blot was processed with an anti-His antibody and a secondary antibody coupled to peroxidase in order to visualize the recombinant proteins themselves [blot (**B**)].

transcription buffer, 1 μL RNase inhibitor, and 2 μL SP6 RNA-polymerase (40 U) and adjust to 20 μL with water.

5. Incubate at 37°C for 2 h. You could expect a yield of approximately 10–20 μg DIG-labeled RNA (0.5–1 μg/μL final concentration).
6. Add 2 μL RNase-free DNaseI (Roche, Mannheim, Germany) and incubate for 15 min at 37°C in order to destroy the template

DNA. Stop the reaction by adding 2 μL EDTA (200 mM, pH 8, made of DEPC-treated water).

7. For quantification prepare a serial dilution of your labeled RNA (as a starting point estimate 1 μg/μL for your concentration) and corresponding dilutions of the DIG-labeled actin RNA.
8. Spot 1, 10, and 100 pg, and 1 ng onto a nylon membrane. UV cross-link the RNA in a Strata-Linker (90 mJ/cm^2) and wash briefly in TBST.
9. Block the membrane for 20 min in a blocking buffer at room temperature (RT). Add the anti-DIG alkaline phosphatase coupled antibody (dilute 1 to 10,000) directly to the blocking buffer and incubate for additional 30 min at RT on a rocking platform.
10. Wash twice with TBST for 5 min each at RT. Equilibrate the membrane with detection buffer for 5 min at RT.
11. Prepare substrate solution by mixing 50 μL NBT stock solution and 37.6 μL BCIP stock solution in 10 mL detection buffer; prepare freshly and protect from light.
12. Incubate the membrane with substrate solution in the dark without shaking until clear bands are visible. Stop the reaction by emerging the blot in water.
13. Calculate the amount of your DIG-labeled RNA by comparison with the actin standard. Store the RNA in aliquots at –20°C.

3.2. Preparation of DIG-Labeled Nonradioactive DNA Probe

1. Choose the DNA sequence you want to use. We used a CAT containing plasmid in order to amplify a CAT construct (approximately 1,000 bp) for our experiments.
2. Set up the PCR reaction in a 0.2 mL PCR tube as follows: 5 μL 10× PCR buffer (including $MgCl_2$), 2.5 μL each primer (0.5 μM each), 5 μL 10× PCR-DIG mix (Roche, Mannheim, Germany), 2.5 U polymerase, 1 ng template DNA. Fill up the reaction mix to 50 μL final volume with PCR grade water. In parallel set up a control reaction using unlabeled dNTPs instead of the DIG labeled ones.
3. Mix the reagents and centrifuge briefly to collect the liquid at the bottom of the PCR tube.
4. Put your sample in a thermal cycler and perform a normal PCR according to your primer and template DNA with 30 cycles (*see* **Note 7**).
5. Check the outcome of the PCR in a normal agarose gel electrophoresis. In comparison to the unlabeled PCR, the labeled PCR product shows a slightly higher molecular weight (*see* **Fig. 3**).
6. Store in aliquots at –20°C.

3.3. SDS-PAGE

1. Carry out SDS-PAGE (1-mm thick mini gels) according to Laemmli *(4)* (*see* Chapter "Immunoblotting using Radiolabeled Reagents for Detection).

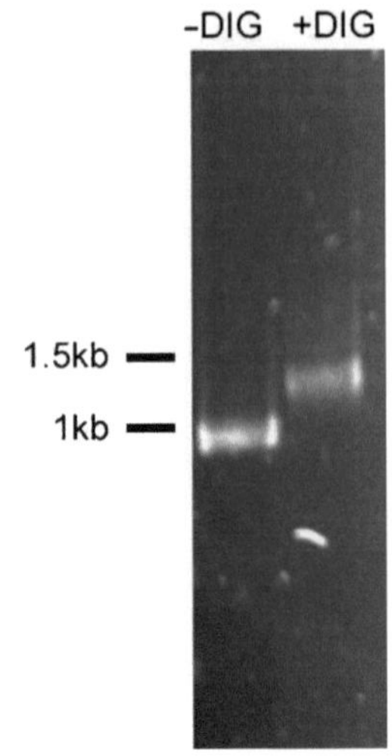

Fig. 3. PCR result after DIG labeling. This ethidium bromide stained agarose gel gives the result of the DIG labeling PCR (+DIG) compared with the normal control PCR with the same template, but without the labeling mix (–DIG). Notice the size difference between the different PCR bands.

2. Load 30 μL (approximately 3 μg protein) of each sample in a well. Include one well for prestained molecular weight markers, Page Ruler (Fermentas, St. Leon-Roth, Germany). Either load the samples for the DNA-binding assay (southwestern blot) or for the RNA-binding assay (northwestern blot) (*see* **Note 8**).

3.4. Blotting of the Protein

1. After separation by SDS-PAGE, the samples are transferred to nitrocellulose membranes electrophoretically. These directions assume the use of a semidry system provided by Bio-Rad (Trans-Blot Semi-Dry Transfer Cell, Bio-Rad, Munich, Germany). Two trays with transfer buffer Roti-Blot A and transfer buffer Roti-Blot K, respectively, are prepared with a size slightly bigger than the dimension of the gel.
2. Cut the nitrocellulose membrane and eight pieces of Whatman 3MM paper in the size of the separating gel. Cut one edge of the membrane for later orientation.
3. Four sheets of 3MM are moistened in transfer buffer Roti-Blot A and transferred to the anode plate of the blotting device. On top of this stack, the nitrocellulose membrane is added.
4. The gel unit is disconnected from the power supply and disassembled. The stacking gel is removed and discarded and one corner (corresponding to the membrane) is cut from the separating gel to allow its orientation to be tracked. The separating gel is then laid on top of the nitrocellulose membrane.
5. Another four sheets of 3MM paper are wetted in transfer buffer Roti-Blot K and carefully laid on top of the gel, ensuring that no bubbles are trapped in the resulting sandwich. You can remove air bubbles by carefully rolling a glass pipette on top of the pile.

6. The lid is put on top of the stack and the power supply is activated. Transfers can be accomplished at 0.8 mA/cm^2 membrane size for 1 h.
7. Once the transfer is complete the stack is carefully disassembled. The 3MM paper and the gel can then be discarded. If you want, you can stain the gel in Coomassie Blue staining reagent for 1 h and destain thereafter in order to check for complete transfer of the separated proteins. On the nitrocellulose membrane the colored molecular weight markers should be clearly visible.
8. Stain the membrane with Ponceau S staining solution for 5 min at RT. Destain the blot with water until the protein bands are clearly visible.

3.5. Development of Northwestern/ Southwestern Blot

1. Block the membrane in 10 mL blocking buffer overnight at 4°C. Alternatively, blocking is done for 1.5 h at RT on a rocking platform.
2. Continue by discarding the blocking buffer. Add the respective labeled nucleic acid diluted in blocking buffer (DIG-labeled DNA 1:1,000 diluted or DIG-labeled RNA at 1 μg/mL final concentration) and incubate for 2 h at RT on a rocking platform.
3. Remove the labeling solution and wash the membrane four times for 5 min each with 10 mL TBST.
4. The detection antibody (anti-DIG conjugated with alkaline phosphatase, Roche, Mannheim, Germany) is freshly prepared for each experiment as 1:1,000-fold dilution in antibody incubation buffer and added to the membrane for 1 h at RT on a rocking platform.
5. The antibody is discarded and the membrane washed five times for 5 min each with 10 mL TBST.
6. Equilibrate the membrane with detection buffer for 5 min at RT.
7. Prepare substrate solution by mixing 50 μL NBT stock solution and 37.6 μL BCIP stock solution in 10 mL detection buffer; prepare freshly and protect from light.
8. Incubate the membrane with substrate solution in the dark without shaking until clear bands are visible. Stop the reaction by immersing the blot in water.
9. Document the result by scanning the blots with a normal flatbed scanner. Pouch the blot between two sheets of transparent foil in order to prevent drying of the blot.
10. Continue with the ECL detection of the blotted proteins. Incubate the membrane in 10 mL blocking buffer for 30 min at RT on a rocking platform. Alternatively, the membrane could be blocked overnight at 4°C.

11. Discard the blocking buffer and incubate with an anti-Tag antibody (in our case anti-penta His-antibody diluted 1:2,500 in blocking buffer) and incubate for 2 h at RT on a rocking platform.
12. Remove the labeling solution and wash the membrane four times for 5 min each with 10 mL TBST.
13. The secondary antibody (anti-mouse conjugated with POD) is freshly prepared for each experiment as 1:80,000-fold dilution in antibody incubation buffer and added to the membrane for 1 h at RT on a rocking platform.
14. The secondary antibody is discarded and the membrane is washed five times for 5 min each with 10 mL TBST.
15. During the final wash, for each blot warm 1 mL solution A and 25 μL solution B of the ECL Plus solutions to RT. Once the final wash is removed from the blot, the ECL Plus reagents are mixed together and then immediately added to the blot, which is then incubated for 5 min at RT. Take care that the membrane is always covered with liquid.
16. The blot is removed from the ECL Plus reagents, dried briefly, and analyzed in a ChemiDoc XRS system (BioRad, Munich, Germany) or equivalent system.

4. Notes

1. Unless stated otherwise, all solutions should be prepared in water that has a conductivity of 0.056 μS/cm and total organic content of less than 5 ppb. This standard is referred to as "water" in this text.
2. All solutions had been purchased as a complete kit (DIG RNA labeling kit, Roche, Mannheim, Germany). But as there are many in vitro transcription systems available, they might end up in identical results. Only the NTP labeling mix should be obtained from Roche.
3. We have very good experience with the nylon membrane from MSI, but other nylon membranes (either charged or uncharged) used for northern- or southern-blot techniques will work equally.
4. Any PCR system that is established in your lab might work. Only the 10× PCR-DIG mix must be used instead of the normally used dNTP mix in the labeling PCR reaction. For your convenience you might also use the PCR-DIG Probe Synthesis Kit available from Roche (Mannheim, Germany) that contains all needed materials.

5. Any nitrocellulose membrane might work with this system.
6. If you want to evaluate a specific substrate, you have to clone your cDNA sequence in antisense direction in order to obtain labeled sense RNA as a probe.
7. If you want to amplify a sequence longer than 3 kb increase the elongation time after the first 10 cycles by adding 20 s to each missing cycle, or use a long-run PCR mix instead.
8. Alternatively, the membrane could be cleaved after the blotting in order to provide two pieces for incubation with DNA or RNA, respectively.

Acknowledgements

Special thanks for excellent technical support to Martina Franke (Inst. Immunology, Carl-Gustav Carus TU-Dresden, Germany).

References

1. Horke, S., Reumann, K., Schulze, C., Grosse, F., and Heise, T. (2004) The La motif and the RNA recognition motifs of human La autoantigen contribute individually to RNA recognition and subcellular localization. *J Biol Chem* 279, 50302–50309.
2. Castro, A., Faura, M., Agell, N., Renau-Piqueras, J., and Bachs, O. (1996) The autoantigen La/SSB is a calmodulin-binding protein. *Cell Calcium* 20, 493–500.
3. Zhang, Q.H., Ye, M., Wu, X.Y., Ren, S.X., Zhao, M., Zhao, C.J., et al. (2000) Cloning and functional analysis of cDNAs with open reading frames for 300 previously undefined genes expressed in CD34+ hematopoietic stem/progenitor cells. *Genome Res* 10, 1546–1560.
4. Laemmli, U.K. (1970) Cleavage of structural proteins during assembly of the head of bacteriophage T4. *Nature* 227, 680–685.

Chapter 45

Immunoblotting Using Radiolabeled Reagents for Detection

Holger Bartsch, Claudia Franke, and Michael Bachmann

Summary

Development of immunoblots is commonly performed using enzyme-labeled antibodies, which convert soluble substrates into insoluble colored products. A simple, rapid, and sensitive alternative method that produces low background and allows a rapid quantitative evaluation is the use of radiolabeled antibodies or protein A conjugates. Here, we describe the use of iodinated secondary antibodies for immunodetection of an autoantigen during HPLC purification.

Key words: Systemic autoimmunity, Autoantibodies, Immunoblotting, Radiolabeled reagents

1. Introduction

Biochemical purification of proteins from total extracts is a time-consuming multistep process, which commonly includes fractionated ammonium sulfate precipitation, ion-exchange chromatography using diethylaminoethyl (DEAE)- and/or phosphocellulose, gel filtration steps such as those using Sephadex or Sepharose, and affinity chromatography. Thereby, a series of samples are collected and analyzed to identify those fractions containing the desired protein. Many proteins are posttranslationally modified. Consequently, a rapid, sensitive, and quantitative procedure is required for the analysis of such a series of samples. The use of radiolabeled reagents can become a rapid and simple useful alternative, especially in cases where the antibodies are limited or expensive.

Here, we describe such a procedure. In this study we analyzed the copurification of the autoantigen La/SS-B and RNA polymerase III *(1)*. The purification included a series of high-performance

Biji T. Kurien and R. Hal Scofield (eds.), *Methods in Molecular Biology, Protein Blotting and Detection, vol. 536*

DOI: 10.1007/978-1-59745-542-8_45

liquid chromatography (HPLC) steps. To identify the La antigen we used the monoclonal anti-La antibody La4B6 *(2)* in combination with an iodinated secondary anti-Ig antibody.

2. Materials

2.1. Cell Culture and Lysis

1. Dulbecco's Modified Eagle's Medium (DMEM) (Invitrogen, Karlsruhe, Germany) supplemented with 10% fetal bovine serum (FBS, Biochrom, Berlin, Germany).
2. HeLa cells from a 2 L culture are harvested and total extract is prepared. The extract was separated by HPLC on a diethylaminoethyl (DEAE) resin using a linear salt gradient as described *(1)*.
3. Modified Laemmli *(3)* buffer for gel loading: 200 mM dithiothreitol (DTT), 4% (w/v) sodium dodecyl sulfate (SDS), 20% (v/v) glycerol, 0.2% (w/v) bromphenol blue, 100 mM Tris–HCl (pH 6.8). Store in aliquots at –20°C (*see* **Note 1**).

2.2. SDS-Polyacrylamide Gel Electrophoresis (SDS-PAGE)

In principle, SDS-PAGE was performed according to Laemmli *(3)* using the following modifications:

1. Separating buffer (4×): 1.5 M Tris–HCl, pH 8.8, 0.5% (w/v) SDS. Store at room temperature (RT).
2. Stacking buffer (4×): 0.5 M Tris–HCl, pH 6.8, 0.5% (w/v) SDS. Store at RT.
3. Thirty percent acrylamide solution (Rotiphorese® Gel A, Carl-Roth, Karlsruhe, Germany) (this is a neurotoxin when unpolymerized and so care should be taken not to receive exposure).
4. 2% Bis-acrylamide solution (Rotiphorese® Gel B, Carl Roth, Karlsruhe, Germany).
5. *N,N,N,N'*-tetramethyl-ethylenediamine (*see* **Note 2**).
6. Ammonium persulfate: 10% (w/v) solution in water and freeze in single use (200 μL) aliquots at –20°C.
7. Water-saturated isobutanol: Shake equal volumes of water and isobutanol in a glass bottle and allow to separate. Use the top layer. Store at RT.
8. Running buffer (10×): 250 mM Tris (do not adjust pH), 1.92 M glycine, 1% (w/v) SDS. Store at RT.
9. Prestained molecular weight marker: Page Ruler™ (Fermentas, St. Leon-Roth, Germany).

2.3. Western Blotting for Detection of La/SS-B

After SDS-PAGE, transfer proteins to nitrocellulose:

1. Transfer buffer: Roti-Blot A and Roti-Blot K (Carl-Roth, Karlsruhe, Germany).

2. Nitrocellulose membrane: Hybond-C extra (Amersham/GE-HealthCare Life Science, Munich, Germany) (*see* **Note 3**).
3. 3MM chromatography paper (Whatman, Maidstone, UK).
4. Tris-buffered saline with Tween (TBST) (10×): 1.5 M NaCl, 0.5 M Tris–HCl, pH 8, 1% Tween-20. Dilute 100 mL with 900 mL water for use.
5. Blocking buffer: 5% (w/v) Blocking Reagent (Roche, Mannheim, Germany) in TBST.
6. Antibody incubation buffer: TBST supplemented with 1% (w/v) fraction V bovine serum albumin (BSA).
7. Secondary antibody: Iodinated anti-mouse IgG 500,000 counts/min (CPM) per 5 mL of blocking buffer (Perkin-Elmer, Rodgau, Germany) (*see* **Notes 4** and **5**).

3. Methods

Separation of La protein containing fractions from HeLa cell extract was performed as described *(1)*. The collected samples were separated by SDS-PAGE, transferred to nitrocellulose, incubated with a previously described anti-La monoclonal antibody La4B6, and detected by radiolabeled iodinated anti-IgG antibodies using a phosphoimager system (**Fig. 1**).

3.1. Preparation of Samples for Detection of La/SS-B by Western Blotting

1. A labeled microcentrifuge tube for each sample with a hole poked in the cap using a 26-gauge syringe needle, a heat-block at 100°C, and sample buffer preheated to 100°C are held ready.
2. Mix collected samples (20 μL) with 5 μL of sample buffer.
3. The tubes are closed and then boiled for further 10 min. After cooling to room temperature, they are ready for separation by SDS-PAGE.

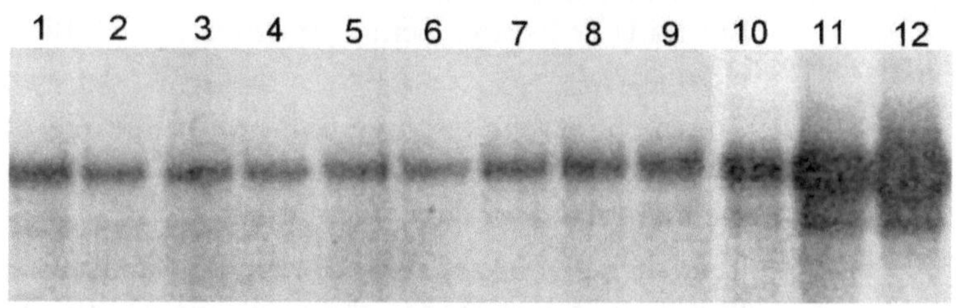

Fig. 1. Samples (1–12) obtained by HPLC chromatography were collected and analzyed for the presence of the La protein. The samples were separated by SDS-PAGE, transferred to nitrocellulose, and after blocking the membrane was incubated with the previously described anti-La mab La4B6 followed by an iodinated secondary antibody. The blot(s) were developed using a phosphoimager system.

3.2. SDS-PAGE

1. These instructions assume the use of a 1-mm thick mini gel system (Bio-Rad, Munich, Germany). It is critical that the glass plates for the gels are scrubbed clean with a detergent before and after use and rinsed extensively with distilled water.
2. Prepare the 1-mm thick, 10% separating gel by mixing 3.2 mL Roti Gel A, 1.3 mL Roti Gel B, 2.6 mL of separating buffer, 2.5 mL water, 50 μL ammonium persulfate solution, and 7.5 μL TEMED. Pour the gel, leaving space for a stacking gel, and overlay with water-saturated isobutanol. The gel should polymerize in about 30 min.
3. Pour off the isobutanol and rinse the top of the gel twice with separating buffer.
4. Prepare the 3.5% stacking gel by mixing 375 μL Roti Gel A, 150 μL Roti Gel B, 500 μL of stacking buffer, 1.3 mL water, 25 μL ammonium persulfate solution, and 2.5 μL TEMED. Pour the stack and insert the comb. The stacking gel should polymerize within 30 min.
5. Prepare the running buffer by diluting 100 mL of the 10× running buffer with 900 mL of water in a measuring cylinder. Cover with Parafilm and invert to mix.
6. Once the stacking gel has set, carefully remove the comb and use a 3 mL syringe fitted with a 22-gauge needle to wash the wells with running buffer.
7. Add the running buffer to the upper and lower chambers of the gel unit and load 30 μL of each sample in each well. Include one well for prestained molecular weight markers.
8. Complete the assembly of the gel unit and connect to a power supply. The gel is run at 80 V until the dye front reaches the separating gel. Then the voltage is raised to 100 V and the gel runs for approximately 1.5 h until the dye front reach the bottom of the gel.

3.3. Western Blotting for Detection of La/SS-B

1. After the separation by SDS-PAGE, the protein samples are transferred to nitrocellulose membrane electrophoretically. These directions assume the use of a semidry system provided by Bio-Rad (TransBlot™ Semi-Dry Transfer Cell, Bio-Rad, Munich, Germany).
2. Cut the nitrocellulose membrane and eight pieces of Whatman 3MM paper to the size of the separating gel. Cut one edge of the membrane for later orientation.
3. Four sheets of 3MM are moistened in the transfer buffer Roti-Blot A and transferred to the anode plate of the blotting device. Place the nitrocellulose membrane on top of this stack.
4. The gel unit is disconnected from the power supply and disassembled. Discard the stacking gel, and one corner (corresponding to the membrane) is cut from the separating gel to allow its

orientation to be tracked. The separating gel is then laid on top of the membrane.

5. Moisten another four sheets of 3MM paper in the transfer buffer Roti-Blot K and carefully place on top of the gel, ensuring that no bubbles are trapped in the resulting sandwich. One can remove air bubbles by carefully rolling a glass pipette on top of the pile.
6. The lid is placed on top of the stack and the power supply is activated. Transfers can be accomplished at 0.8 mA/cm^2 membrane size for 1 h.
7. Once the transfer is complete the stack is carefully disassembled. The 3MM paper and the gel can then be discarded. One can stain the gel in Coomassie Blue for 1 h and destain thereafter in order to check for completeness of transfer. The colored molecular weight markers should be clearly visible on the membrane.
8. Block the membrane in 10 mL blocking buffer overnight at 4°C. Alternatively, blocking is done for 1.5 h at RT on a rocking platform.
9. After transfer and blocking the nitrocellulose sheets were incubated with a dilution of the primary antibody of 1 to 100 for 1 h.
10. Wash the membrane twice with blocking solution.
11. Seal the membrane in plastic wrap and add iodinated secondary antibody equivalent to 500,000 CPM per 5 mL of blocking solution. Incubate further for 1 h (*see* **Notes 4** and **5**).
12. After washing with PBS (4 × 5 min) the blots are ready for detection using either autoradiography or a phosphoimager system.

4. Notes

1. Unless stated otherwise, all solutions should be prepared in water that has a conductivity of 0.056 μS/cm and total organic content of less than 5 ppb. This standard is referred to as "water" in this text.
2. TEMED is best stored at RT in a desiccator. Buy small bottles as it may decline in quality (gels will take longer to polymerize) after opening.
3. Any nitrocellulose membrane might work with this system.
4. Instead of iodinated anti-mouse antibody, iodinated protein A or G preparations can be used equally.
5. The best suitable amount of radioactively labeled secondary antibody must be determined experimentally. Usually 500,000 CPM per 5 mL gave low background and resulted in good sensitivity.

Acknowledgements

Special thanks for excellent technical support to K. Joanne and Tim Gross (Oklahoma Medical Research Foundation, Oklahoma City, USA).

References

1. Weser, S., Bachmann, M., Seifart, K.H., and Meissner, W. (2000) Transcription efficiency of human polymerase III genes *in vitro* does not depend on the RNP-forming autoantigen La. *Nucleic Acids Res.* 15, 3935–3942.
2. Tröster, H., Bartsch, H., Klein, R., Metzger, T.E., Pollak, G., et al. (1995) Activation of a murautoreactive B cell by immunization with human recombinant autoantigen La/SS-B: characterization of the autoepitope. *J Autoimmun.* 8, 825–842.
3. Laemmli, U.K. (1970) Cleavage of structural proteins during the assembly of the head of bacteriophage T4. *Nature* 227, 680–685.

Chapter 46

Detection of Protein Carbonyls by Means of Biotin Hydrazide–Streptavidin Affinity Methods

Kenneth Hensley

Summary

Oxidative posttranslational protein modifications occur as a normal process of cell biology and to a greater extent during pathogenic conditions. The detection and quantitation of protein oxidation has posed a continuing challenge to bioanalytical chemists because the products of oxidative protein damage are chemically diverse, protein oxidation generally occurs at low background levels, and the complexity of biological samples introduces high background noise when standard techniques such as immunolabeling are applied to "dirty" tissue extracts. A refinement of classic reductive amination methods has been developed, which circumvents these difficulties by incorporating a biotin label at sites of protein carbonylation. Biotin hydrazide-labeled proteins are detectable using standard streptavidin-coupled detection techniques such as peroxidase-catalyzed chemiluminescence of immunoblots. Advantages of the biotin hydrazide-labeling technique are its sensitivity and its lack of reliance upon antibodies that inevitably suffer from nonspecific background noise and contaminating endogenous immunoglobulins.

Key words: Protein carbonyl, Free radical, Biotin, Hydrazide, Oxidation

1. Introduction

The routine study of protein oxidative damage has depended heavily upon techniques to assay relatively stable protein (i.e., mostly irreversible) carbonyl groups formed by contact with harsh oxidants, particularly oxygen-centered free radicals *(1)*. Early methods for measuring protein carbonyls relied upon reductive treatment of biological samples with tritiated borohydride *(2–4)*. Refinements of this strategy produced the dinitrophenylhydrazine technique of reductively coupling a 2,4-dinitrophenol (DNP) derivative to protein carbonyls, which allowed spectrophotometric or immunochemical detection of the DNP label *(5–7)*.

Biji T. Kurien and R. Hal Scofield (eds.), *Methods in Molecular Biology, Protein Blotting and Detection, vol. 536*

DOI: 10.1007/978-1-59745-542-8_46

The DNP technique has been notably successful in many cases. For instance the DNP technique has demonstrated age-related increase in protein carbonyl accumulation within aging and Alzheimer's disease-afflicted human brain *(6)* and in mouse models of Alzheimer's disease-like histopathology *(7)*. Unfortunately, however, this technique suffers from severe limitations when applied to tissue extracts from mammalian sources. One of the most severe limitations is that of contaminating immunoglobulins in the tissue. Inevitably the secondary antibodies that typically are used to detect DNP:anti-DNP complexes will detect nonspecific IgG present in the mammalian tissue. If serum is present this problem can be rather severe, even when "species-specific" secondary IgG is employed. Furthermore, one typically notices substantial anti-DNP immunoreactivity in control samples wherein DNP labeling steps are omitted, suggesting a degree of nonspecificity in anti-DNP affinity.

These problems can be circumvented somewhat by modifying the classic reductive hydrazide-labeling chemistry to employ biotin hydrazide or biotin-LC-hydrazide (containing an alkyl spacer) in place of DNP. Reaction of biotin hydrazide with a protein carbonyl yields a stable hydrazone derivative (**Fig. 1**). The biotin-tagged protein carbonyl can be visualized by standard streptavidin-dependent detection methods such as chemiluminescent detection of derivatized proteins on western blots using horseradish peroxidase-conjugated streptavidin (HRP-streptavidin). This method displays the known specificity of streptavidin detection chemistries while being completely insensitive to endogenous immunoglobulin. Our laboratory has employed the biotin hydrazide technique successfully in the study of oxidative damage in the central nervous system, particularly

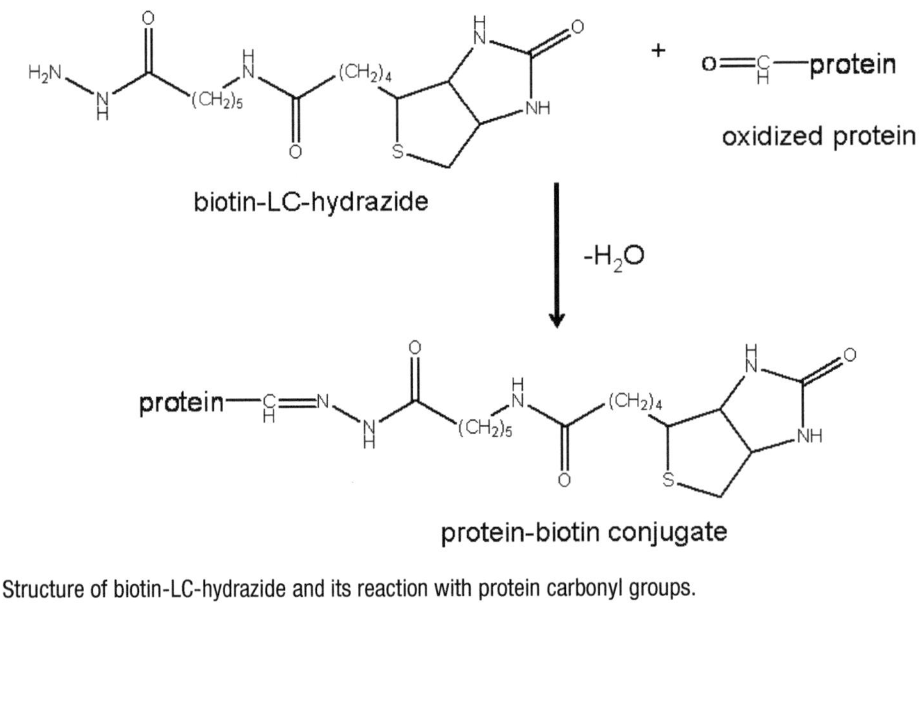

Fig. 1. Structure of biotin-LC-hydrazide and its reaction with protein carbonyl groups.

utilizing the SOD1^{G93A} mouse model of amyotrophic lateral sclerosis *(8–10)*.

Methods and examples of biotin hydrazide labeling are described wherein protein carbonyl groups are visualized in immunoblots of mammalian tissue or cell lysates. These methods are intended only as a guideline that can be adapted for specific needs.

2. Materials

1. Biotin hydrazide or biotin-LC-hydrazide (Pierce Biotechnology, Rockford, IL, USA): 50 mM in dimethyl sulfoxide (DMSO) and freeze at –20°C (stable for at least 6 months). The term "biotin hydrazide" is used generically throughout this discussion to refer to either biotin hydrazide or biotin-LC-hydrazide. The choice to use the reagent with or without the alkyl spacer is determined by the researcher's needs and should be determined empirically by the researcher (*see* **Note 1**).
2. Streptavidin-conjugated HRP (stock solution): 0.5 mg/mL in sterile saline and frozen in aliquots at –20°C; stable for at least 3 months.
3. *N*-morpholino-ethanesulfonate (MES buffer): 20 mM in water, adjusted to pH 5.5.
4. Butylated hydroxytoluene (BHT): 100 mM in absolute ethanol.
5. Desferoxamine mesylate (DFO), an iron chelator, 100 mM in saline.
6. SDS-PAGE loading dye: 10% Tris–HCl (pH 8.8), 50% glycerol, 1% sodium dodecyl sulfate, and 0.01% bromphenol blue plus or minus 2% mercaptoethanol or 100 mM dithiothreitol.
7. Wash buffer: 24-g Tris base, 80-g NaCl, 15-mL Tween-20 in 1-L H_2O.

3. Methods

1. Lyse sample in MES buffer. For extraction of tissue lysates and lipid-rich samples the buffer can be supplemented with 0.1% Triton X-100 or other detergents or chaotropic agents at similar concentrations. Protease inhibitors and phosphatase inhibitors can be included as desired. Samples should be assayed for total protein concentration and adjusted to the same uniform protein concentration.

2. To small volumes of each sample (100 μL is sufficient) add a 10% additional volume of biotin hydrazide. To slow or prevent metal-catalyzed oxidation during the incubation step, desferoxamine and BHT can be added in 1:100 dilution from the concentrated stock solutions. If desired, samples can be prepared in duplicate in which case one sample (a reagent blank) omits treatment with biotin hydrazide. This sample would receive DMSO only, or alternatively the sample would receive biotin in lieu of biotin hydrazide.
3. Incubate samples at 37°C with gentle agitation. The incubation time and temperature can be varied to accommodate specific experiments and must be determined empirically for each project. Typically 2-h incubation at 37°C is sufficient to ensure near complete labeling of protein carbonyls.
4. After labeling samples can be stored at −80°C indefinitely. Samples are processed for immunoblotting using standard protocols either with, or without reducing agents. Typically 50 μL of sample is mixed with 50 μL of SDS-PAGE loading dye. Samples are boiled thoroughly and 20-μL aliquots are then loaded onto 4–20% gradient polyacrylamide gels. Samples are transferred by electroblot onto polyvinylidene difluoride (PVDF) membranes. Samples containing biotin hydrazide are alternated with "blanks" omitting the biotin hydrazide as described earlier.
5. Protein blots are developed as follows. The PVDF membrane is blocked overnight at 4°C in 4% fatty acid-free bovine serum albumin (BSA). Blots are incubated for 1 h at room temperature with HRP-streptavidin diluted to 0.1 μg/mL in 4% BSA. Diluted HRP-streptavidin must be prepared fresh each day. Blots are rinsed thoroughly in wash buffer to remove excess unbound HRP-streptavidin. Blots are developed using any standard commercially available chemiluminescence reagents and visualized instrumentally or by exposure to photographic film. **Figure 2** shows typical results from a biotin hydrazide blot of astrocyte cell lysate. In this experiment cells were cultured from $SOD1^{G93A}$ modeling amyotrophic lateral sclerosis, or nontransgenic mice *(9, 10)*. Cells were stimulated with recombinant tumor necrosis factor (TNFα) plus interferon gamma (IFNγ) for 48 h and blotted for protein carbonyls using biotin-LC hydrazide as described earlier.

4. Notes

1. Hydrazides react rapidly with carbonyl groups to yield stable hydrazone derivatives. This has allowed the development of various assays for monitoring carbohydrates after in

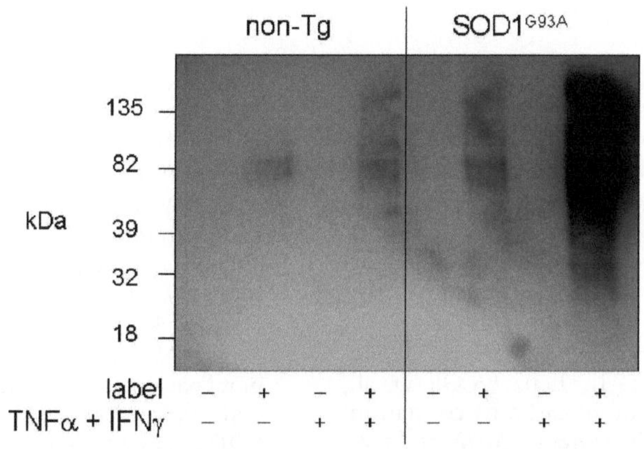

Fig. 2. Typical results obtained in an immunoblotting experiment using biotin hydrazide to visualize protein carbonyl groups in cell culture lysates. Primary glia from nontransgenic neonatal mice or mice carrying mutant superoxide dismutase (*SOD1*) genes were cultured as described elsewhere *(9, 10)*. Cells were stimulated 48 h with 50 U/mL IFNγ plus 40 ng/mL TNFα, lysed in the presence or absence of biotin-LC-hydrazide (+ or – label as indicated), blotted onto a PVDF membrane and probed with streptavidin-conjugated horseradish peroxidase as described in the text.

vitro oxidation, and more recently has allowed techniques for measuring low levels of protein carbonylation that occur in the course of oxidative stress-related events. Biotin hydrazide reagents with and without alkyl spacers have been utilized by our laboratory with some success. Biotin hydrazide labels protein carbonyl groups selectively. Using biotin–streptavidin affinity methods one can discern the biotinylated proteins with very low background signals that generally, in our experience, cannot be obtained using immunologically based assays such as the dinitrophenylhydrazine technique.

2. The basic biotin labeling approach described in this article can be amended and adapted for specific uses. Currently we are investigating the utility of biotin labeling for qualitative determination of oxidized protein sequences by means of mass spectrometry. In general the biotin-labeling approach should be amenable to inclusion in any concentration or purification strategy that relies upon classical avidin:biotin interactions. Therefore, it is our hope that improvements to the biotin hydrazide method will accelerate research into oxidative stress-related pathologies across a broad range of disease states and experimental models.

Acknowledgements

This work was supported in part by grants from the National Institutes of Health (NS044154), the ALS Association, and the Oklahoma Center for Advancement of Science and Technology (OCAST).

References

1. Stadtman, E.R. and Levine, R.L. (2003) Free radical-mediated oxidation of free amino acids and amino acid residues in proteins. *Amino Acids* 25, 207–218.
2. Lenz, A.G., Costabel, U., Shaltiel, S., and Levine, R.L. (1989) Determination of carbonyl groups in oxidatively modified proteins by reduction with tritiated sodium borohydride. *Anal Biochem* 177, 419–425.
3. Levine, R.L., Garland, D., Oliver, C.N., Amici, A., Climent, I., Lenz, A.G., et al. (1990) Determination of carbonyl content in oxidatively modified proteins. *Methods Enzymol* 186, 464–478.
4. Amici, A., Levine, R.L., Tsai, L., and Stadtman, E.R. (1989) Conversion of amino acid residues in proteins and amino acid homopolymers to carbonyl derivatives by metal-catalyzed oxidation reactions. *J Biol Chem* 264, 3341–3346.
5. Levine, R.L., Wehr, N., Williams, J.A., Stadtman, E.R., and Shacter, E. (2000) Determination of carbonyl groups in oxidized proteins. *Methods Mol Biol* 99, 15–24.
6. Smith, C.D., Carney, J.M., Starke-Reed, P.E., Oliver, C.N., Stadtman, E.R., Floyd, R.A., et al. (1991) Excess brain protein oxidation and enzyme dysfunction in normal aging and Alzheimer disease. *Proc Natl Acad Sci USA* 8, 10540–10543.
7. Lim, G.P., Chu, T., Yang, F., Beech, W., Frautschy, S.A., and Cole, G.M. (2001) The curry spice curcumin reduces oxidative damage and amyloid pathology in an Alzheimer transgenic mouse. *J Neurosci* 21, 8370–8377.
8. Hensley, K., Floyd, R.A., Gordon, B., Mou, S., Pye, Q.N., Stewart, C.A., et al. (2002) Temporal patterns of cytokine and apoptosis-related gene expression in spinal cords of the G93A-SOD1 mouse model of amyotrophic lateral sclerosis. *J Neurochem* 82, 365–374.
9. Hensley, K., Abdel-Moaty, H., Hunter, J., Mhatre, M., Mou, S., Nguyen, K., et al. (2006) Primary glia expressing the G93A-SOD1 mutation present a neuroinflammatory phenotype and provide a cellular system for studies of glial inflammation. *J Neuroinflammation* 3, 2.
10. Hensley, K., Mhatre, M., Mou, S., Pye, Q.N., Stewart, C., West, M., et al. (2006) On the relationship of oxidative stress to neuroinflammation: Lessons learned from the G93A-SOD1 mouse model of amyotrophic lateral sclerosis. *Antioxid Redox Signal* 8, 2075–2087.

Chapter 47

"Rainbow" Western Blotting

Stan (Stanislaw) Krajewski

Summary

The "Rainbow western" method permits detection of multiple antigens on a single protein blot. The procedure utilizes horseradish peroxidase (HRPase)-based detection with both a chemiluminescent and colorimetric substrate. In the "Rainbow western" procedure four different HRPase-colorimetric substrates that produce black, brown, red, and green colors are employed sequentially for detection and simultaneous display of four different antigens on the same blot. The Rainbow western methods have the potential to consolidate the work to analyze the expression levels of several proteins in studies of signaling pathways within biological samples. This technique could be particularly valuable for analysis of comigrating proteins, isoforms, and/or facilitating studies on phosphorylation, acetylation, and oligomerization of proteins tagged by the same epitope.

Key words: Western blotting, Immunoblotting, Nonstripping quenching, ECL and colorimetric

1. Introduction

Western blotting, aka immunoblotting, is widely used to detect and compare the relative levels of particular proteins of interest in cells, tissue homogenates and extracts, serum, as well as other biological materials. Western blotting consists of four parts: (a) Extraction of proteins from biological samples, (b) separation of proteins by sodium dodecyl sulfate-polyacrylamide gel electrophoresis (SDS-PAGE), (c) transfer of the electrophoretically separated proteins from the gel to a membrane by electroblotting, and (d) immunodetection of specific antigens immobilized on the membrane. Since several articles and recent reviews describe the first three steps *(1–3)*, we will focus on the step that we propose to modify – antigen immunodetection.

Biji T. Kurien and R. Hal Scofield (eds.), *Methods in Molecular Biology, Protein Blotting and Detection, vol. 536*

DOI: 10.1007/978-1-59745-542-8_47

Protein blots are prepared using standard methods and nitrocellulose (0.45 μm; Bio-Rad Labs, Hercules, CA) or polyvinyldifluoridine (PVDF) membranes *(4, 5)*. The proposed modification permits multiple antigens to be sequentially detected on a single blot as originally described by Krajewski et al. *(6)*. The "Rainbow western" emphasizes colorimetric display of the results and allows four different antigens to be discriminated (**Fig. 1A–C**). To our knowledge, previous attempts to achieve multicolor detection in immunoblotting methods as well as current immunofluorescent methodology, including quantum dot tags, have not exceeded three colors to differentiate biomarkers on a single membrane *(7–16)* (Invitrogen Dychrome Western: http://probes.invitrogen.com/lit/catalog/1/sections/3940.html).

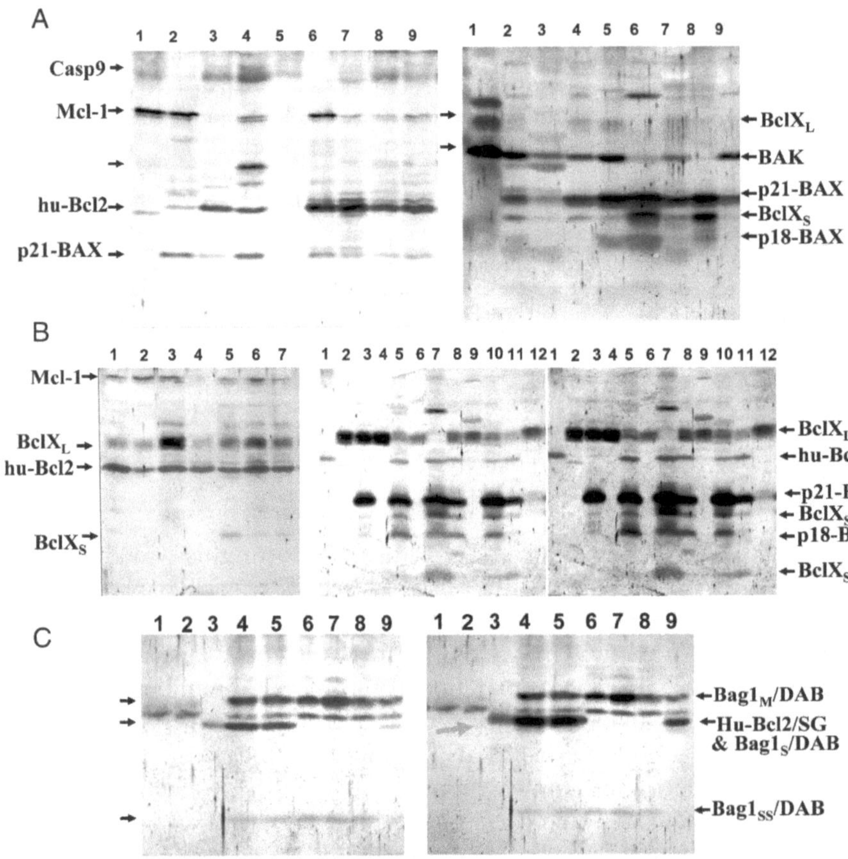

Fig. 1. Examples of *Rainbow western blots* to assess the physiological content of Bcl2 family proteins in normal human tissues. (**A**) The sequential tagging of antibodies with different HRP substrates resulted in 3–4 colors specifically linked to each primary antibody. Experimentally established order of chromogen developments prevented color "overwriting" [(**B**) *left, middle*]. Changing the order led to color overwriting [(**B**) *right*; *green* TMB color was turned into *red* by AEC reagent]. Panel **C** demonstrates strong immunoreactions for two proteins with similar molecular weight, Bcl2 and Bag-1 using Vector® SG and DAB chromogens. All human samples were loaded at 100 μg total protein per lane. The cell culture lysates were normalized for protein content to apply 30–50 μg per lane (*see Color Plates*).

The basis for the Rainbow western procedure lies in the observation that colorimetric detection of HRPase-containing antigen–antibody complexes using either DAB, Vector® SG, or AEC renders these complexes incapable of participating in HRPase-based reactions during subsequent exposures to HRPase substrates.

In our research, we developed a protocol for probing protein blots with multiple independent primary antibodies, involving both ECL-based detection and the colorimetric readout of the data *(6)*. Specifically, we have tested and applied this method to investigate apoptosis proteins, such as Bcl-2 family members, thus providing examples of how this method can be applied for immunoblot analysis of multiple proteins on western blots.

2. Materials

2.1. Buffers, Immunoreagents, and Detection Systems

1. Samples: Any type of cell and tissue lysates transferred into nitrocellulose or PVDF membranes according to standard western blotting protocol.
2. W-I (western blot buffer-I; pH 7.8–8.2): 10 mM Tris Base, 150 mM NaCl, 0.05–0.1% (v/v) Triton X-100 or Tween-20, and 5% skim milk, with 0.01% thimerosal. Mix 30 mL of 1 M Tris, 90 mL of 5 M NaCl, 3 mL of Tween-20 (add about 20 μL extra), 150-g skim milk, and 900 μL of 10% thimerosal. Adjust the final pH to 7.8–8.2 by adding 1N HCL or NaOH accordingly and make to a final volume of 3 L with double-distilled water (ddH$_2$0). Distribute 40-mL aliquots in 50-mL tubes → store at −20°C.
3. W-II (western blot buffer-II; pH 7.8–8.2): Mix 5 mL of 1 M Tris Base, 15 mL of 5 M NaCl, 10 g of 2% (w/v) bovine serum albumin (BSA) fraction V, 250 μL of 10% thimerosal stock, and 800 μL of Tween-20 (add about 20 μL extra). Adjust final pH to 7.9–8.2 by adding 1N HCL or NaOH accordingly. Add ddH$_2$0 to a total volume of 500 mL. Distribute 40-mL aliquots in 50-mL tubes store at −20°C.
4. Sodium acetate buffer (SAB, 0.05 M, pH 5.5): Combine 150 mL of 0.2 M glacial acetic acid (solution A) and 350 mL of 0.2 M sodium acetate (solution B). Adjust the final pH to 5.5 → store at +4°C. Make up volume to 500 mL with ddH$_2$0.
   (a) Solution A (1 L): 11.55 mL of 0.2 M acetic acid in 1,000-mL ddH$_2$0.
   (b) Solution B (1 L): 0.2 M sodium acetate, 27.2 g in 1,000-mL ddH$_2$0.

5. 0.1 M Tris Base (pH 8.2): Weigh 12.1-g Tris Base and dissolve in less than 1,000-mL ddH_20. Adjust the final pH with HCl to 8.2 → store at +4°C. Make up to 1 L with ddH_20.
6. Modified PBS (pH 7.6): Phosphate-buffer saline containing 120 mM NaCl, 11.5 mM NaH_2PO_4, 30 mM K_2HPO_4, prepared fresh each week. Weigh 14.026-g NaCl, 2.76-g $NaH2PO_4$, and 10.88-g K_2HPO_4 and dissolve in about 1 L of ddH_20. Adjust final pH to 7.6 by adding 1N HCL or NaOH accordingly. Make up to 2 L with ddH_20.
7. Primary Antibodies: Any type could be applied. We standardized the detection methods primarily for rabbit and mouse antibodies. In these studies we employed rabbit polyclonal antisera generated against synthetic peptides corresponding to sequences in the human (h) Bcl-2, hBax, hBcl-X, hMcl-1, and hBak proteins, mouse (m) Bax, Bcl2, and BAG-1 proteins *(17–21)*, caspases 3, 6, 9 *(22)* as well as mouse monoclonal antibodies specific for hBAG-1 (IgG1) *(23)* and actin [JLA20; IgM (Oncogene Science, Inc., Cambridge, MA, USA)].
8. Secondary Antibodies: Biotinylated goat anti-rabbit Ig (No. BA-1000), horse anti-mouse IgG/H&L chains (No. BA-2000), and goat anti-mouse IgM (No. BA-2021), avidin–biotin complex (ABC) with horseradish peroxidase (HRPase; No. PK-6100) were purchased from Vector Labs, Inc. (Burlingame, CA, USA). HRPase-conjugated goat anti-rabbit IgG (No. 172-6515) and goat anti-mouse IgG (No. 170-6516) were purchased from Bio-Rad, Inc. (Richmond, CA, USA). HRPase-conjugated sheep anti-mouse IgM was obtained from Amersham, Inc. (Buckinghamshire, England, No. NA-931).
9. ECL (enhanced chemiluminescence) and chromogenic substrates: The ECL western blotting detection reagent (Amersham, Inc., No. RPN2106) was prepared according to the manufacturer's instructions. The chromogenic HRPase substrate Vector® SG (No. SK-4700) was purchased from Vector Labs, Inc. The chromogen, 3,3′-diaminobenzidine (DAB: No. D-5637), 3-amino-9-ethylcarbasole (AEC: No. A-5754) was purchased from Sigma, Inc., St. Louis, MO, USA. Alternatively, DAB stock solution from DakoCytomation, Inc. (Carpenteria, CA, USA; No. K3468) was used. The chromogen 3,3′,5,5′-tetramethylbenzidene stabilized substrate for HRP (TMB) was purchased as a solution from Promega, Inc. (San Luis Obispo, CA, USA; No. W4121).
10. DAB working solution:
    (a) DAB (3,3′-diaminobenzidine-tetrahydrochloride-dihidrate; Sigma: No. D-5637): dissolve 15–30 min earlier at 0.5 mg/mL in 0.1 M Tris Base (pH 8.2). Immediately before use, add 0.1 mL of 1% H_2O_2 to 10 mL of DAB

solution followed by filtration through a 0.45-µm filter directly on the membrane (Acrodisk; Millipore, Inc., USA). 1% H_2O_2; 333 µL of 30% H_2O_2 mixed with 9.7 mL of ddH_2O.

(b) Alternatively, use DAB solution prepared from stock concentrate (DakoCytomation, No. K3468). Immediately before use dilute 1:10 (one drop in 10 mL of manufacturer activating buffer). In place of manufacturer's buffer, 10 mL of 0.1 M Tris Base, pH 8.2 buffer and activate by adding 35 µL of 30% H_2O_2 may be used. After use, DAB should be collected in the appropriately labeled biohazard waste containers.

11. Vector® SG working solution (chromogenic substrate; Vector Lab; No. SK-4700): Dissolve five drops of each component in 20-mL PBS as described by the manufacturer. Add Vector® SG mix to membranes, and shake by hand at RT for 10–15 min.
12. AEC (3-amino-9-ethylcarbasole) is first dissolved at 25 mg/mL in *N,N*-dimethyl formamide and then added to 50 mL of SAB (0.05 M; pH 5.5) and filtered through Whatmann 3MM paper. Immediately before use, 150 µL of 30% H_2O_2 is added to 50 mL of the AEC-containing solution.
13. TMB (3,3′,5,5′-tetramethyl-benzidine; Promega, Inc., No. W4121) is provided as a ready to use solution.
14. ECL-II-Rb-HRP: HRPase-conjugated goat anti-rabbit IgG (Biorad, No. 172-6515). Dilute *ECL-II-Rb-HRP* (1:5,000) and NGS (1:2,000) v/v in W-II buffer. Freeze and store at −20°C after use. This solution, along with those described in **steps 11–13** can be reused up to five times.
15. ECL-II-M-HRP: HRP-goat anti-mouse IgG (H&L; Biorad, No. 170-6516). Dilute 1:5,000 with 1:2,000 NGS v/v in W-II buffer. Freeze and store at −20°C after use.
16. Col-II-Rb-Biot [Vector No. BA-1000; biotinylated goat anti-rabbit IgG (H + L)]: Dilute 1:5,000–8,000 with 1:2,000 NGS v/v in W-II buffer. Freeze and store at −20°C after use.
17. Col-II-M-Biot (Vector No. BA-2000; biotinylated horse anti-mouse IgG, and Vector No. BA-2021; goat anti-mouse IgM). Dilute each reagent 1:5,000–8,000 and equine normal serum (EquN) or NGS 1:2,000 v/v, respectively, in W-II buffer. Freeze and store at −20°C after use.
18. A–B complex working solution (ABC Vector-Elite HRPase Kit, No. PK-6100): Prepare in advance at least 15–30 min before use to allow A–B complexes to form and stabilize. Using ABC Vector-Elite Kit, take 1 drop from vial A and 1 drop from vial B for each 5 mL of 0.1 M Tris Base; shake well by hand, apply to membrane, and incubate for 10 but no longer than 15 min.

3. Methods

3.1. Day 1

3.1.1. Transfer and Membrane Blocking

1. After routine blotting/transfer, submerge the membranes at RT for 30 min in W-I standard western blocking buffer containing skim milk and NGS 1:500 V/V (*see* **Note 1**).
2. Without an intervening washing step, proceed with rabbit primary antibody incubation in W-1, ON at RT. Add normal goat serum (NGS) 1:5,000 v/v to the working antibody buffer if the antibody is a full serum. The preblocking and primary antibody solutions can be saved and reused (*see* **Note 2**).

3.2. Day 2

3.2.1. Washing and Developing Procedure

1. Decant the primary antibody solution and save for later use.
2. Rinse the blot briefly in PBS and then wash three times for 10 min each with ≥0.5 mL/cm^2 of PBS.
3. Incubate 15 min with secondary biotinylated antibody anti-rabbit IgG (H + L) [Vector Lab, Catolog # BA-1000) diluted 1:5,000 with addition of 1:2,000 v/v NGS in W-II buffer.
4. Rinse the blot briefly in PBS and then wash two times for 10 min each with ≥0.5 mL/cm^2 of PBS.
5. Prepare the ABC–HRP complex 0.5–1 h in advance. For each 5 mL of 0.1 M Tris add 2 drops from vials A and B of the Vector-Elite Kit (No. PK-6100) and shake well.
6. Incubate the washed membrane in 0.25–0.5 mL/cm^2 HRPase–ABC solution for 20 min. After regular washing in PBS, proceed with the routine ECL development *(6)*. After successful luminogram generation, develop colorimetric products as follows.
7. Four-color blot development is accomplished by developing the first primary antibody with Vector® SG substrate using an acetate buffer (pH 5.5) instead of PBS, which generates a black-colored reaction product (*see* **Note 3**). Discard the substrate solution and rinse the blot in tap H_2O to stop the reaction.
8. Photograph the wet blot under good lighting using a vertically mounted camera.
9. Wash the membrane two times for 5 min each in PBS and reblock for 15–20 min using saved "preblocking" solution described earlier, followed by repetition of the **steps 1–5** for the next primary rabbit antibody.
10. Perform routine ECL development and then develop the blot using DAB, which generates a brown-colored product (*see* **Note 4**). Place the blot in a clean container and cover with H_2O_2/DAB solution for 10 min, when DAB solution is self-prepared or 3–5 min when commercial DAB stock from DakoCytomation is used. Stop the reactions by rinsing

in tap H_2O, photograph the blot while wet, and transfer the blot to PBS.

11. For the third antibody, reblock the membrane, incubate with the next primary antibody ON, and perform ECL as described earlier. Then, develop the blot with AEC, which generates a brick-red color. Transfer the blot to a fresh tray, cover the blot with SAB (pH 5.5), and incubate for 10 min. Decant the acetate buffer and submerge the blot in H_2O_2/AEC solution for ~10–20 min (*see* **Note 5**). Rinse in tap H_2O to stop reaction and photograph. Before transferring to PBS, wash the blot once for 5 min in acetate buffer to retain red color.
12. Reblock the blot, incubate with the fourth primary antibody ON, and perform ECL development as described earlier, followed by three washes in dH_2O. Transfer the blot facedown to a fresh container, cover with TMB solution, and incubate for 1–3 min. Photographs can be taken at various times during development with TMB (*see* **Note 6**).
13. Stop the reaction with dH_2O and photograph wet. Air-dry the blot and store in plastic wrap. Do not transfer to PBS because the TMB reaction product (green) will fade. Rewet the blot in water, if additional photographs are desired. Note that the TMB product is unstable and will fade with time (*see* **Note 7**).
14. You may end here or proceed to the multiple antibody detection (MAD) method for further sequential multi-antibody developments. To do the latter, first execute points in **Subheading 3.1.2** of the MAD protocol with DAB quenching (*see* **Note 7** and Chapter "Multiple Antigen Detection Western Blotting").

4. Notes

1. All incubation steps described here are performed with gentle agitation on a rotating or rocking platform shaker at RT. The preblocking and antibody working solution can be stored at −20°C and reused typically ~5 times.
2. Nonimmune serum from the host species in which the secondary antibody was raised, e.g., NGS saturates the membrane and assures that additional nonspecific binding of goat IgG will not take place when incubated with any *ECL-II-Rb-HRP* and *ECL-II-M-HRP* (made in goat). To obtain better results for your primary antibody, perform 2–3 h or overnight incubation at RT rather than at 4°C, adding the antibacterial agent thimerosal to the working antibody buffer.

3. We have successfully used Vector® SG in SAB, modified PBS, or Tris-based buffers. It works well in every environment regardless of the pH values.
4. *DAB* is a recognized *carcinogen (24, 25)* and precautions should be taken when handling it. Use gloves and collect the used DAB as biohazard waste for proper disposal. Do not pour the DAB solution down the drain to avoid water contamination.
5. For AEC development, it is important to monitor the progress of the colorimetric reaction since when overdeveloped the color will be closer to brown than red and therefore difficult to distinguish from DAB. The blot can be briefly removed from AEC solution and photographed a few times during the course of developing if desired.
6. *The order of chromogen application (Vector® SG→ DAB→ AEC → TMB)* is critical for optimal results (*see* **Fig. 1A, B**). Dispense three drops of Vector® SG substrate into 15-mL acetate buffer and mix well. Add three drops H_2O_2 (0.5%) from the kit and mix. Place the blot into a clean container, cover with the substrate solution, and incubate for 15–20 min. Note that we have modified the procedure for this application, using an acetate solution (pH 5.5) instead of the citrate buffer (pH 6.0) recommended by the manufacturer. The acetate buffer reduces background and consistently generates a more stable, intense, and black reaction product compared with the citrate buffer, which yields a blue-gray stain.
7. *AEC and TMB* are weakly saturating chromogens, whose precipitates could be washed out during lengthy procedure. Unlike Vector® SG and DAB, the AEC and TMB chromogens do not completely block the activity of HRPase in antigen–antibody complexes on blots, and thus these HRPase substrates are not suitable for simultaneous ECL-based multidetections. This will cause the reappearance of the ECL signal for primary antibodies for which these chromogens were used. Additional treatment with DAB or Vector® SG will secure the permanent "sealing" and saturation of epitopes after AEC or TMB use, particularly if proceeding further with multiple antigen detection (MAD) method (*see* "Multiple Antigen Detection" Western Boltting.

Acknowledgements

The author would like to thank Prof. J.C. Reed for the many years of inspiration in the study of apoptosis and would like also to acknowledge NIH grant NS36821, supporting his research.

References

1. Kurien, B.T. and Scofield, R.H. (2003) Protein blotting: a review. *J. Immunol. Methods* 274, 1–15.
2. Kurien, B.T. and Scofield, R.H. (2006) Western blotting. *Methods* 38, 283–293.
3. Lee, C. (2007) Western blotting. *Methods Mol. Biol.* 362, 391–399.
4. Reed, J., Meister, L., Cuddy, M., Geyer, C., and Pleasure, D. (1991) Differential expression of the Bcl-2 proto-oncogene in neuroblastoma and other human tumor cell lines of neural origin. *Cancer Res.* 51, 6529–6538.
5. Towbin, H., Staehelin, T., and Gordon, J. (1979) Electrophoretic transfer of proteins from polyacrylamide gels to nitrocellulose sheets: procedure and some applications. *Proc. Natl. Acad. Sci. USA* 76, 4350–4354.
6. Krajewski, S., Zapata, J.M., and Reed, J.C. (1996) Detection of multiple antigens on Western blots. *Anal. Biochem.* 236, 221–228.
7. Geysen, J., De Loof, A., Vandesande, F. (1984) How to perform subsequent or "double" immunostaining of two different antigens on a single nitrocellulose blot within one day with and immunoperoxidase technique. *Electrophoresis* 5, 129–131.
8. Lin, J.C. and Pagano, J.S. (1986) Sequential detection of different antigens induced by Epstein–Barr virus and herpes simplex virus in the same western blot by using dual antibody probes. *J. Virol.* 59, 522–524.
9. Theisen, N., Lohoff, E.S., von Figura, K., and Hasillik, A. (1986) Sequential detection of antigens in western blots with differently colored products. *Anal. Biochem.* 152, 211–214.
10. Lee, N., Yang, J., and Testa, D. (1988) A simplified high speed multicolor immunoblotting method. *Anal. Biochem.* 175, 30–34.
11. Poor, M.L., Santa, P.F., and Sittampalam, G.S. (1988) Visualization of multiple protein bands on the same nitrocellulose membrane by double immunoblotting. *Anal. Biochem.* 175, 191–195.
12. Schilling, K. and Aletsee-Ufrecht, M.C. (1989) An immunoblot assay for the simultaneous quantification of several antigens. *Anal. Biochem.* 177, 203–206.
13. Sundaram, P. (1996) A simplified 'coupled' probing approach for the detection of proteins on dot and western blots. *J. Immunol. Methods* 195, 149–152.
14. Martin, K., Hart, C., Liu, J., Leung, W.Y., and Patton, W.F. (2003) Simultaneous trichromatic fluorescence detection of proteins on western blots using an amine-reactive dye in combination with alkaline phosphatase- and horseradish peroxidase-antibody conjugates. *Proteomics* 3, 1215–1227.
15. Bakalova, R., Zhelev, Z., Ohba, H., and Baba, Y. (2005) Quantum dot-based western blot technology for ultrasensitive detection of tracer proteins. *J. Am. Chem. Soc.* 127, 9328–9329.
16. Makrides, S.C., Gasbarro, C., and Bello, J.M. (2005) Bioconjugation of quantum dot luminescent probes for western blot analysis. *Biotechniques* 39, 501–506.
17. Krajewski, S., Bodrug, S., Gascoyne, R., Berean, K., Krajewska, M., and Reed, J.C. (1994) Immunohistochemical analysis of Mcl-1 and Bcl-2 proteins in normal and neoplastic lymph nodes. *Am. J. Pathol.* 145, 515–525.
18. Krajewski, S., Krajewska, M., Shabaik, A., Wang, H.-G., Irie, S., Fong, L., and Reed, J.C. (1994) Immunohistochemical analysis of in vivo patterns of Bcl-X expression. *Cancer Res.* 54, 5501–5507.
19. Krajewski, S., Bodrug, S., Krajewska, M., Shabaik, A., Gascoyne, R., Berean, K., and Reed, J.C. (1995) Immunohistochemical analysis of Mcl-1 protein in human tissues: differential regulation of Mcl-1 and Bcl-2 protein production suggests a unique role for Mcl-1 in control of programmed cell death in vivo. *Am. J. Pathol.* 146, 1309–1319.
20. Krajewski, S., Blomvqvist, C., Franssila, K., Krajewska, M., Wasenius, V.-M., Niskanen, E., and Reed, J.C. (1995) Reduced expression of pro-apoptotic gene *Bax* is associated with poor response rates to combination chemotherapy and shorter survival in women with metastatic breast adenocarcinoma. *Cancer Res.* 55, 4471–4478.
21. Krajewska, M., Moss, S., Krajewski, S., Song, K., Holt, P., and Reed, J.C. (1996) Elevated expression of Bcl-X and reduced Bak in primary colorectal adenocarcinomas. *Cancer Res.* 56, 2422–2427.
22. Krajewska, M., Rosenthal, R.E., Henning, J.M., Stennicke, R., Wiesenthal, T., Mai, J., Naito, M., Salvesen, G.S., Reed, J.C., Fiskum, G., et al. (2004) Early processing of Bid and caspase 6, 8, 10, 14 in the canine brain during cardiac arrest and resuscitation. *Exp. Neurol.* 189, 261–279.
23. Takayama, S., Sato, T., Krajewski, S., Kochel, K., Irie, S., Millan, J.A., and Reed, J.C. (1995) Cloning and functional analysis of BAG-1: a novel Bcl-2 binding protein with anti-cell death activity. *Cell* 80, 279–284.

24. Ashby, J., Elliott, B.M., Lefevre, P.A., Styles, J., and Longstaff, E. (1983) Initiation/promotion versus complete carcinogenicity in the rodent liver. *Environ. Health Perspect.* 50, 339–346.

25. Kaufmann, S.H. (2001) Reutilization of immunoblots after chemiluminescent detection. *Anal. Biochem.* 296, 283–286.

Chapter 48

Multiple Antigen Detection Western Blotting

Stan Krajewski, Xianshu Huang, and Maryla Krajewska

Summary

A variation of ECL immunodetection method permits sequential detection of multiple antigens (MAD) on a single protein blot without stripping previously bound antibodies. Because antibody stripping is not involved, immobilized proteins are not lost from the membrane, which permits multiple sequential reprobings of the same membrane with different primary antibodies (≥12) and retention of strong signal intensities for all antibody probings. This procedure utilizes horseradish peroxidase (HRPase)-based detection with both chemiluminescent and colorimetric substrates. Initial incubation of the blot with secondary antibody followed by colorimetric development prior to probing the blot with primary antibodies markedly reduces background intensities in ECL-based detection procedures and permits sequential use of antibodies derived from a single species. By allowing large amounts of data to be obtained from a single blot, MAD immunoblotting has the potential to markedly streamline the work required to compare the expression levels of several proteins within biological samples. This technique could be particularly valuable for analyzing cellular populations that are difficult to isolate in large numbers or clinical specimens where the amount of protein samples is limited or available on a one-time basis.

Key words: Western blotting, Immunoblotting, Nonstripping quenching, Multiple antigen detection, ECL

1. Introduction

In our research, we developed a protocol for probing protein blots with multiple independent primary antibodies that involves ECL-based detection and colorimetric quenching. The proposed modification permits measurement of 12 or more antigens to be sequentially detected on a single blot (**Fig. 1A–J**). This method does not require removal of antibodies from filters between cycles of reprobing *(1, 2)*, thus avoiding loss of immobilized proteins from blots upon successive rounds of antigen detection.

Biji T. Kurien and R. Hal Scofield (eds.), *Methods in Molecular Biology, Protein Blotting and Detection, vol. 536*

DOI: 10.1007/978-1-59745-542-8_48

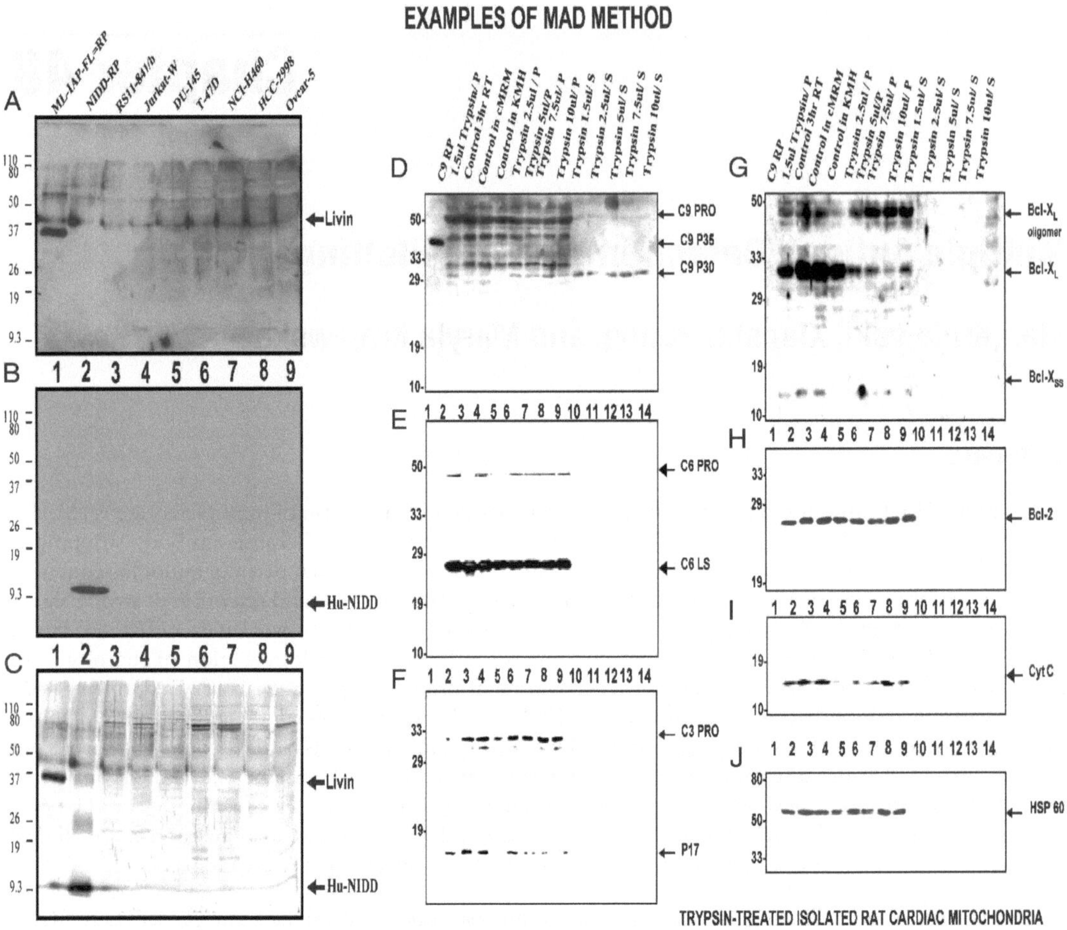

Fig. 1. Panels (**A–C**) show the colorimetric (DAB/ Vector® SG) background reducing procedure for antibodies raised to apoptotic proteins Livin [panel (**A**); first ECL probing) and NIDD [panel (**B**); second probing]. The blot contains a variety of human cancer cell lines and recombinant protein positive controls. Panel (**C**) demonstrates the abundance of the cross-reacting lanes after chromatographic development. Panels (**D–J**) depict results obtained from a single membrane sequentially probed with 11 antibodies in an experiment with isolated trypsin-treated rat heart mitochondria. Selected blots are shown for antibodies specific to the following proteins: caspase-9 (1st probing), Bcl-X_L (3rd probing), active caspase-6 (4th probing), proform and active caspase-3 (5th), mouse Bcl-2 (8th), cytochrome *c* (10th) and HSP60 (11th). The results indicate early digestion of Bcl-X_L, release and activation of caspases with late changes in Bcl2 levels in mitochondrial membranes (*see Color Plates*).

A helpful step in the MAD immunoblotting procedure to minimize unspecific background is to first incubate the blot with secondary antibody alone (or in conjunction with preimmune serum or/and a negative control monoclonal antibody) and develop the membranes with DAB or SG, before probing the blot with the primary antibodies of interest. This secondary antibody pretreatment can markedly reduce background in ECL-based detection procedures (**Fig. 1A–C**). The color development step not only provides independent confirmation of the ECL results, but also prevents the antibodies on the blots from reacting in subsequent reprobings with secondary antibodies, which eliminates the need for stripping. The striking effect that this secondary antibody treatment step has on reducing unspecific background suggests that this approach could also be used in experiments where immunoprecipitations are performed and the resulting antigen–antibody complexes are subjected to SDS–PAGE/immunoblotting analysis.

Unlike other approaches to detect two and no more than three antigens on a single blot *(3–5)*, the MAD immunoblotting method does not depend on the use of primary antibodies derived from different species, nor does it require that different enzymatic methods of detection such as HRPase in combination with alkaline phosphatase be employed. Instead, our method permits sequential use of antibodies derived from a single species and relies on a single enzymatic reporter (HRPase) for ECL-based detection.

2. Materials

2.1. Buffers, Immunoreagents, and Detection Systems

1. Samples: Any type of cell and tissue lysates transferred into nitrocellulose or PVDF membranes according to standard western blotting protocol.
2. W-I (western blot buffer-I; pH 7.8–8.2): 10 mM Tris Base, 150 mM NaCl, 0.05–0.1% (v/v) Triton X-100 or Tween-20, and 5% skim milk, with 0.01% thimerosal. Mix 30 mL of 1 M Tris, 90 mL of 5 M NaCl, 3 mL of Tween-20 (add about 20 μL extra), 150-g skim milk, and 900 μL of 10% thimerosal. Adjust the final pH to 7.8–8.2 by adding 1N HCL or NaOH accordingly and make to a final volume of 3 L with double-distilled water (ddH_20). Distribute 40-mL aliquots in 50-mL tubes → store at –20°C.
3. W-II (western blot buffer-II; pH 7.8–8.2): Mix 5 mL of 1 M Tris Base, 15 mL of 5 M NaCl, 10 g of 2% (w/v) bovine serum albumin (BSA) fraction V, 250 μL of 10% thimerosal stock, and 800 μL of Tween-20 (add about 20 μL extra). Adjust the final pH to 7.9–8.2 by adding 1N HCL or NaOH accordingly. Add ddH_20 to a total volume of 500 mL. Distribute 40-mL aliquots in 50-mL tubes → store at –20°C.
4. Tris (1 M): 60.6 g in 500-mL ddH_20.
5. NaCl (5 M): 146.1 g in 500-mL ddH_20.
6. Modified PBS (pH 7.6): Phosphate-buffer saline containing 120 mM NaCl, 11.5 mM NaH_2PO_4, 30 mM K_2HPO_4, prepared fresh each week. Weigh 14.026-g NaCl, 2.76-g NaH2PO_4, and 10.88-g K_2HPO_4 and dissolve in about 1 L of ddH_20. Adjust the final pH to 7.6 by adding 1N HCL or NaOH accordingly. Make up to 2 L with ddH_20.

7. 0.1 M Tris Base (pH 8.2): Weigh 12.1-g Tris Base and dissolve in less than 1,000-mL ddH_20. Adjust the final pH with HCl to 8.2 → store at +4°C. Make up to 1 L with ddH_20.
8. Tris–HCl–glycine buffer: 25 mM Tris–HCl (pH 8.3), 192 mM glycine, 20% methanol, and 0.01% sodium dodecyl sulfate.
9. Primary antibodies: Any type of antibody could be applied. We standardized the detection methods primarily for rabbit and mouse antibodies. In these studies we employed rabbit polyclonal antisera generated against synthetic peptides corresponding to sequences in the apoptotic protein Livin and NIDD, mouse (m) Bcl2, mBax, Bcl-X, and BAG-1 proteins *(6–10)*, caspases 3, 6, 9 *(11)* as well as mouse monoclonal antibodies specific for actin [JLA20; IgM (Oncogene Science, Inc., Cambridge, MA)] and cytochrome *c*, HSP60 MAB (Stressgen, Inc., Ann Arbor, MI, USA).
10. Secondary antibodies: HRPase-conjugated goat anti-rabbit IgG (No. 172-6515) and goat anti-mouse IgG (No. 170-6516) were from Bio-Rad, Inc. (Richmond, CA). HRPase-conjugated sheep anti-mouse IgM was obtained from Amersham, Inc. (Buckinghamshire, England, No. NA-931).
11. ECL (enhanced chemiluminescence) and chromogenic substrates: The ECL western blotting detection reagent (Amersham, Inc., No. RPN2106) was prepared according to the manufacturer's instructions. The chromogenic HRPase substrate Vector® SG (No. SK-4700) was purchased from Vector Labs, Inc., and, 3,3′-diaminobenzidine (DAB) chromogen stock solution was purchased from DakoCytomation, Inc. (Carpenteria, CA, USA, No. K3468).
12. DAB working solution: 3,3′-diaminobenzidine-tetrahydrochloride-dihydrate (DAB) working solution is prepared from stock concentrate (DakoCytomation, No. K3468). Immediately before use dilute 1:10 (one drop in 10 mL of manufacturer's activating buffer). *DAB* is a recognized *carcinogen (12, 13)* and precautions should be taken when handling this compound. Use gloves and collect the used DAB as biohazard waste for proper disposal. Do not pour the DAB solution down the drain to avoid water contamination.
13. Vector® SG working solution (chromogenic substrate; Vector Lab, No. SK-4700): Dissolve five drops of each component in 20 mL of modified PBS as described by the manufacturer. Add Vector® SG mix to membranes; shake by hand at RT for 10–15 min.
14. ECL-II-Rb-HRP: HRPase-conjugated goat anti-rabbit IgG (Biorad, No. No. 172-6515): Dilute *ECL-II-Rb-HRP* (1:5,000) and NGS (1:2,000) v/v in W-II buffer. Freeze and store at –20°C after use. It can be reused up to five times.

15. ECL-II-M-HRP: HRP-goat anti-mouse IgG (H&L; Bio-rad, No. 170-6516): Dilute 1:5,000 with 1:2,000 NGS v/v in W-II buffer. Freeze and store at –20°C after use. It can be reused up to five times.
16. X-ray film (Eastman–Kodak, Inc., XOMAT-AR).

3. Methods

3.1. Day 1

3.1.1. Transfer and Membrane Blocking

1. After completing protein transfer to the membranes in Tris–HCl–glycine buffer, place the membranes in a Tupperware container with PBS.
2. Wash the membranes for 5 min in PBS (≥0.5 mL/cm^2 blot) (*see* **Note 1**).
3. Decant PBS and cover the nitrocellulose/PVDF membranes with western blocking solution W-1 (~0.25–0.5 mL/cm^2 membrane blocking solution). Incubate for 15 min.

3.1.2. Preimmune Blocking and Universal Quenching of Secondary Cross-Reactions for Mouse (MABs) and Rabbit (PABs) Detection Systems (see Note 2)

1. Prepare universal preimmunoblocking solution containing normal/or preimmune serum from rabbit (NRbS) and mouse (NMS) at 1:8,000 dilution in W-1. Incubate the membranes for 20 min at room temperature (RT).
2. Decant off blocking solution and save (*see* **Note 3**). Wash for 5 min in PBS.
3. Apply a mixture of secondary antibodies, e.g., HRPase-conjugated goat anti-rabbit IgG (ECL-II-Rb) and goat anti-mouse IgG (ECL-II-M) diluted 1:5,000 v/v in W-II for 30 min at RT. Decant and save the reagents. Wash for 5 min in PBS (*see* **Note 4**).

3.1.3. First Colorimetric Quenching (DAB)

1. Place the membranes facedown in DAB working solution and incubate for 3–5 min. Transfer the used DAB solution into the biohazard waste container (*see* **Note 5**). Stop the reaction in two changes of tap water, followed by PBS rinse for 5 min at RT. The membrane is now ready for the first primary antibody incubation.
2. Incubate the membranes at RT overnight (ON) with the first primary antibody at the working dilution in W-1 buffer. Add normal goat serum (NGS) 1:5,000 v/v to the buffer if the antibody is a full serum (*see* **Note 6**).

3.2. Day 2

3.2.1. Washing and Developing Procedure

1. Decant the primary antibody solution and save for subsequent use (*see* **Note 7**).
2. Rinse the blot briefly in PBS and wash three times for 10 min each in ≥0.5 mL/cm^2 PBS.

3. Incubate with appropriate secondary antibody conjugated to horseradish peroxidase (ECL-II-Rb-HRP or ECL-II-M-HRP) for 30 min at RT.
4. Decant the secondary antibody solution and save at −20°C (*see* **Note 4**). Rinse the blots once (PBS) and wash three times for 10 min each with ≥0.5 mL/cm² of PBS.
5. Detect secondary antibody using a chemiluminescent substrate reagent (Amersham, Inc.) with exposure to X-ray film according to the manufacturer's instructions. Exposure times vary widely, from a few seconds to 1 min. We typically develop after 15–30 s initially and then adjust exposure time based on this result. After completion the original membrane is returned to a tray with PBS.
6. Quantitate luminograms using any type of laser scanning densitometer or analyze scanned images of blots with the Image-Pro plus 4.1 program (Media Cybernetics LP, Silver Spring, MD, USA).

3.2.2. Second Colorimetric Quenching (Vector® SG)

1. Add 0.25 mL/cm² activated Vector® SG (Vector Laboratories, Cat. # SK-4700) chromogenic substrate (*see* **Note 9**). To prepare the Vector® SG substrate, dispense five drops of Vector® SG substrate into 20-mL PBS, mix well, and then add five drops of Vector-H_2O_2 (0.5%) substrate reagent and mix again. Incubate for 10 min (*see* **Notes5** and **8**).
2. Discard substrate solution and rinse the blot in tap H_2O to stop reaction.
3. Transfer the membrane to a clean container and submerge in PBS. Rinse and wash the blot twice for 5 min each in PBS, and then (optional) preblock and rejuvenate the membrane by soaking for 15 min in W-1 to prepare for the next cycle of antibody probing.

3.3. Next Day(s)

1. Perform steps under **Subheading 3.1** once. To continue with development of multiple primary antibodies repeat steps under **Subheading 3.2** for each antibody. Proceed with the second primary antibody ON incubation (*see* **Note 7**).
2. If time does not permit additional probings, membranes may be stored in PBS + thimerosal (1–2 µg /mL of 10% stock) at 4°C up to 1 month.
3. When the experiment can be resumed, wash the membrane in fresh PBS and start again following steps described earlier in **Subheading 3.2**.
4. For troubleshooting advice, e.g., no results, blank luminogram, or glowing membrane, *see* **Note 9**.

4. Notes

1. All incubation steps described here are performed with gentle agitation on a rotating or rocking platform shaker at RT.
2. All HRP-conjugated secondary antibodies used for detection in a variety of systems (ECL, ABC, polymers, etc.) display 0.5–5% cross-reactivity with serum proteins (mainly albumins and IgGs H&L chains) of multiple species, which can result in secondary unspecific bands. The proposed procedure eradicates this problem.
3. Universal preimmunoblocking solution can be saved, stored at −20°C, and reused approximately five times.
4. This protocol is optimized for rabbit/mouse detection systems. Best results were obtained using the following secondary antibodies: (a) goat anti-rabbit IgG (H&L) HRP conjugated; Biorad, cat. # 172-6515 and (b) goat anti-mouse IgG (H&L; HRP conjugated); Biorad, cat. # 170-6516 (*ECL-II-Rb-HRP and ECL-II-M-HRP*). All secondary antisera were used at a standard 1:5,000 v/v (~0.2 μg/mL) in the W-II western buffer. Adjust this step accordingly if your experiment utilizes primary antibodies raised in the same species. The working mixture of *ECL-II-Rb-HRP and ECL-II-M-HRP* can be saved, stored at −20°C, and reused approximately five times. For the MAD immunoblotting procedure, either secondary antibodies directly conjugated with HRPase or biotinylated secondary antibodies in combination with HRPase-containing avidin–biotin complex (ABC) reagents can be employed. In many cases, however, the ECL signals generated with the ABC approach are too strong, presumably because of the extreme sensitivity of ECL and the signal amplification that occurs with the ABC procedure. For this reason, the MAD immunblotting methodology was optimized for directly HRP-conjugated secondary antibodies.
5. DAB seems to provide more complete blocking than Vector® SG. For this reason, we typically use DAB when performing the universal preblocking procedure where the blots are incubated initially with secondary antibody alone (after preimmune serum or other suitable negative control antibody reagents), and then use Vector® SG for all subsequent incubations and quenching steps with all other primary antibodies of interest. The Vector® SG substrate is preferable to DAB for blocking subsequent immunoreactivity of antigen–antibody complexes on blots, in terms of the number of sequential reprobings of the blot that are attainable. With DAB, fewer reprobings are generally possible, implying that repeated DAB-based stainings are somehow detrimental to either the antigenicity or retention of proteins immobilized on nitrocellulose. With Vector® SG, however,

there is sometimes (but not routinely) some residual reactivity that appears during the next round of antibody probing.

6. Nonimmune serum from the host species in which the secondary antibody was raised; e.g., NGS saturates membrane and assures that no more unspecific binding of goat IgG will take place when incubated with any *ECL-II-Rb-HRP and ECL-II-M-HRP* (made in goat). To obtain better results for your primary antibody perform 2–3 h or overnight incubation at RT rather than at 4°C, adding the antibacterial agent thimerosal to the working antibody buffer.
7. The antibody working solution can be stored at –20°C and reused typically approximately five times.
8. We have successfully used Vector® SG in SAB, modified PBS, or Tris-based buffers. It works well in every environment regardless of the pH values.
9. *MAD troubleshooting*: (a) Blank luminogram – *Cause*: Inactive secondary antibody; *Solution*: replace and repeat incubation with a fresh reagent. *Cause*: Container(s) contaminated with Vector® SG; *Solution*: clean container with 5% bleach and repeat incubation with secondary antibody in a clean container. Devote a designated container for DAB/Vector® SG developments. (b) The entire membrane is glowing during ECL recording – *Cause*: Excessively long incubation with a secondary antibody; primary antibody concentration is too high; bacterial growth in primary antibody solution (acidification of the membrane). *Solution*: half-a-day wash of the membranes in fresh PBS buffer, followed by 45-min soaking in W-1 buffer and ECL development on the same day. Alternatively, dilute the primary antibody and repeat ON incubation.

Acknowledgement

We would like to acknowledge NIH grant NS36821 for supporting our research.

References

1. Legocki, R.P. and Verma, D.P. (1981) Multiple immunoreplica technique: screening for specific proteins with a series of different antibodies using one polyacrylamide gel. *Anal. Biochem.* 111, 385–392.
2. Kain, S.R., Mai, K., and Sinai, P. (1994) Human multiple tissue western blots: a new immunological tool for the analysis of tissue-specific protein expression. *Biotechniques* 17, 982–987.
3. Lin, J.C. and Pagano, J.S. (1986) Sequential detection of different antigens induced by Epstein-Barr virus and herpes simplex virus in the same Western blot by using dual antibody probes. *J. Virol.* 59, 522–524.
4. Theisen, N., Lohoff, E.S., von Figura, K., and Hasillik, A. (1986) Sequential detection of antigens in western blots with differently colored products. *Anal. Biochem.* 152, 211–214.

5. Lee, N., Yang, J., and Testa, D. (1988) A simplified high speed multicolor immunoblotting method. *Anal. Biochem.* 175, 30–34.
6. Krajewski, S., Bodrug, S., Gascoyne, R., Berean, K., Krajewska, M., and Reed, J.C. (1994) Immunohistochemical analysis of Mcl-1 and Bcl-2 proteins in normal and neoplastic lymph nodes. *Am. J. Pathol.* 145, 515–525.
7. Krajewski, S., Krajewska, M., Shabaik, A., Wang, H.-G., Irie, S., Fong, L., and Reed, J.C. (1994) Immunohistochemical analysis of in vivo patterns of Bcl-X expression. *Cancer Res.* 54, 5501–5507.
8. Krajewski, S., Bodrug, S., Krajewska, M., Shabaik, A., Gascoyne, R., Berean, K., and Reed, J.C. (1995) Immunohistochemical analysis of Mcl-1 protein in human tissues: differential regulation of Mcl-1 and Bcl-2 protein production suggests a unique role for Mcl-1 in control of programmed cell death in vivo. *Am. J. Pathol.* 146, 1309–1319.
9. Krajewski, S., Blomvqvist, C., Franssila, K., Krajewska, M., Wasenius, V.-M., Niskanen, E., and Reed, J.C. (1995) Reduced expression of pro-apoptotic gene *Bax* is associated with poor response rates to combination chemotherapy and shorter survival in women with metastatic breast adenocarcinoma. *Cancer Res.* 55, 4471–4478.
10. Krajewska, M., Moss, S., Krajewski, S., Song, K., Holt, P., and Reed, J.C. (1996) Elevated expression of Bcl-X and reduced Bak in primary colorectal adenocarcinomas. *Cancer Res.* 56, 2422–2427.
11. Krajewska, M., Rosenthal, R.E., Henning, J.M., Stennicke, R., Wiesenthal, T., Mai, J., Naito, M., Salvesen, G.S., Reed, J.C., Fiskum, G., et al. (2004) Early processing of Bid and caspase 6, 8, 10, 14 in the canine brain during cardiac arrest and resuscitation. *Exp. Neurol.* 18, 261–279.
12. Ashby, J., Elliott, B.M., Lefevre, P.A., Styles, J., and Longstaff, E. (1983) Initiation/promotion versus complete carcinogenicity in the rodent liver. *Environ. Health Perspect.* 50, 339–346.
13. Kaufmann, S.H. (2001) Reutilization of immunoblots after chemiluminescent detection. *Anal. Biochem.* 296, 283–286.

Chapter 49

Detection of Calcium Binding by Ro 60 Multiple Antigenic Peptides on PVDF Membrane Using Quin-2

Biji T. Kurien and Michael Bachmann

Summary

Systemic lupus erythematosus is associated with the production of antibodies to self-constituents, particularly those that target certain specific ribonucleoprotein (RNP) particles. Among these is the Ro RNP particle, composed of a 60,000 molecular weight protein (Ro 60 or SS-A) that is noncovalently associated with at least one of four short uridine-rich RNAs (the hY RNAs). Our earlier work demonstrated that multiple antigenic peptides (MAPs) constructed from the sequence of the Ro 60 autoantigen could be used, using double immunodiffusion studies, enzyme-linked immunosorbant assay, affinity chromatography, and surface plasmon resonance (SPR), to show intramolecular and intermolecular protein–protein interaction within the Ro 60 RNP particle. We also found that calcium is important in mediating this interaction. We hypothesized, therefore, that the Ro 60 antigen is a calcium binding protein. To investigate this we used Ro 60 MAPs and assayed calcium binding using the Quin-2 system. Several Ro 60 MAPs were found to bind calcium using this assay, as well as bovine serum albumin, another calcium binding protein. However, a MAP constructed from the Sm autoantigen did not bind to calcium. These data, along with our observation regarding the involvement of calcium in the protein–protein interaction occurring between Ro 60 antigen and Ro 60 MAPs, make us propose that the Ro 60 antigen is a calcium binding protein.

Key words: Quin-2, Calcium binding, Multiple antigenic peptides, Ro ribonucleoprotein, Systemic lupus erythematosus

1. Introduction

Protein–protein interactions between various structural proteins are very important in maintaining the structural and functional integrity of the cell. In addition, molecular recognition is a vital component in almost all biochemical processes, many of which involve protein–protein interaction. The association between

Biji T. Kurien and R. Hal Scofield (eds.), *Methods in Molecular Biology, Protein Blotting and Detection, vol. 536*

DOI: 10.1007/978-1-59745-542-8_49

antigens and antibodies *(1)* is a classic example of protein–protein interaction. Interactions among proteins producing the phenomena of muscle contraction, blood clotting, signal transduction, or complement fixation reveal the importance of such interactions.

Systemic lupus erythematosus (SLE) is a chronic autoimmune disease with autoantibodies as a near universal feature of the disease. The Ro ribonucleoprotein (RNP) particle composed of a 60,000 molecular weight protein is noncovalently associated with one of four short uridine-rich human cytoplasmic RNA (hYRNA). This antigen is the target of antibodies in up to 40% of lupus patients as well as most patients with Sjögren's syndrome and subacute cutaneous lupus, and mothers of infants with congenital lupus. Anti-Ro 60 is associated with several clinical features of SLE, including the presence of anti-La, which is rarely if ever found without anti-Ro *(2, 3)*.

Recently we showed, using multiple antigenic peptides (MAPs) constructed from the Ro 60 sequence, localized protein–protein interaction occurring within the Ro 60 molecule *(4)*. We also showed that Ro and La interact with each other through protein–protein interaction *(4)*. It was observed that the interaction was mediated through calcium ions *(5)*, and therefore we hypothesized that the Ro 60 antigen may be a metalloprotein.

The interaction of calcium with proteins on a membrane support (following transfer to membrane from SDS polyacrylamide gel) has been most commonly detected using ^{45}Ca autoradiography *(6)*. However, this assay takes over a week for enough exposure to X-ray film and requires handling of radioisotopes with its concomitant hazards of radiation exposure. We have used the fluorescent dye Quin-2 for detecting calcium binding to Ro 60 MAPs and as seen in **Fig. 1** several Ro 60 MAPs were found to bind. On the basis of these results and our observation that calcium mediates protein–protein interaction within the Ro RNP particle, we propose that Ro 60 is a calcium binding protein.

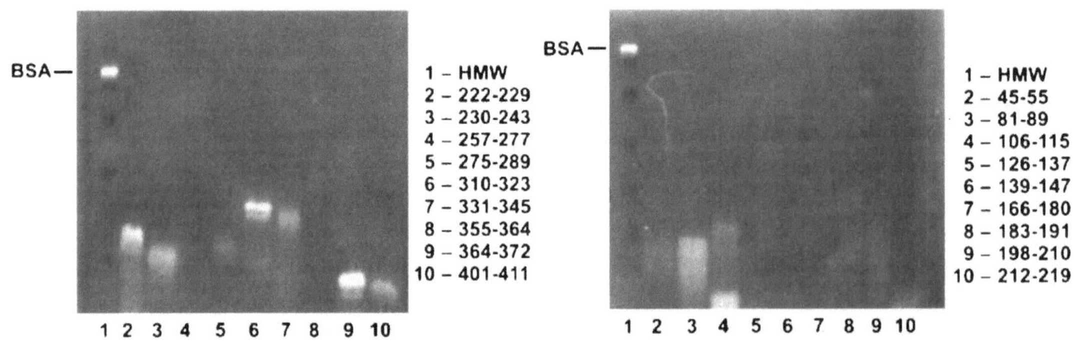

Fig. 1. Binding of calcium ions by Ro 60 MAPs. Each MAP (5 μg) was analyzed on SDS PAGE and transferred to PVDF membrane. Calcium binding was determined using the fluorophore Quin-2. High-range molecular weight (HMW) markers are shown in *lane 1* of each figure. Bovine serum albumin (calcium binding protein) (*lane 1*) fluoresces following binding Quin-2 on the blot (reproduced from *(5)* with permission from Elsevier).

2. Materials

Prepare all solutions using ultrapure water (prepared by purifying deionized water, to attain a sensitivity of 18 MΩ cm at 25°C) and analytical grade reagents. Prepare and store all reagents at room temperature (unless indicated otherwise). Diligently follow all waste disposal regulations when disposing waste materials. We do not add sodium azide to our reagents.

1. Ro 60 MAPs: Twenty-one Ro 60 MAPs were synthesized from the sequence of the Ro 60 autoantigen *(7, 8)* (Molecular Biology Resource Facility, University of Oklahoma Health Sciences Center, Oklahoma City, OK, USA) by a manual stepwise solid phase procedure (*see* **Note 1**). An unrelated MAP with the sequence PPPGRRPP from the Sm autoantigen *(9)* was also synthesized (*see* **Table 1**).
2. Quin-2(Tetrapotassiumsalt(2-{[2-bis(Carboxymethyl)amino-5-methylphenoxy] methyl}-6-methoxy-8-bis(carboxymethyl) aminoquinoline, potassium) (Calbiochem, San Diego, CA, USA).
3. High and low molecular weight protein standards (BRL Life Technologies, Bethesda, MD, USA).
4. Polyvinylidene difluoride membrane (Millipore, Bedford, MA, USA).
5. 10 mM imidazole buffer (pH 6.8) containing 60 mM potassium chloride and 5 mM magnesium chloride.
6. 1 mM calcium chloride (*see* **Note 2**).
7. 0.05% Coomassie Brilliant Blue (*see* **Note 3**).
8. Whatman 3MM paper.
9. Transfer buffer: 0.025 M Tris, 0.192 M glycine, 20% methanol.
10. Western blot transfer apparatus: Mini Trans-Blot apparatus (Bio-Rad, Hercules, CA, USA).
11. 20% ethanol.
12. Photodyne UV Transilluminator.
13. Polaroid camera.
14. Polaroid film, Type 57 or 55.
15. Red filter.
16. Shaker.
17. Coomassie Brilliant Blue: 0.05% in 25% methanol, 10% acetic acid.
18. Destaining solution: 25% methanol, 10% acetic acid.

Table 1
The sequences, amino acid position, and the molecular weight (as measured by mass spectrometry) of the different Ro 60 multiple antigenic peptides constructed from the sequence of the Ro 60 protein

Amino acid sequence of 60 kD Ro MAPs used	Location on the Ro protein	Mol. Wt. by mass spectrometry (in kD)
1. TYYIKEQKLGL	45–55	11.69
2. SQEGRTTKQ	81–89	9.12
3. STKQAAFKAV	106–115	9.25
4. TFIQFKKDLKES	126–137	12.7
5. MKCGMWGRA	139–147	9.2
6. MWGRALRKAIA	143–153	11.02
7. LAVTKYKQRNGWSHK	166–180	15.37
8. LRLSHLKPS	183–191	9.25
9. VTKYITKGWKEVH	198–210	13.55
10. LYKEKALS	212–219	8.45
11. TEKLLKYL	222–229	8.9
12. EAVEKVKRTKDELE	230–243	14.23
13. HLLTNHLKSKEVWKAL	257–272	16.18
14. ALLRNLGKMTA	280–290	10.34
15. NEKLLKKARIHPFH	310–323	14.69
16. YKTGHGLRGKLKWRP	331–345	15.22
17. AAFYKTFKTV	355–364	10.25
18. VEPTGKRFL	364–372	9.21
19. MVVTRTEKDSY	401–411	11.4
20. LPMIWAQKTNTP	449–460	12.04
21. ALREYRKKMDIPAK	482–495	14.59
22. PPPGRRPP	Sm	7.63

3. Methods

3.1. Sodium Dodecyl Sulfate Polyacrylamide Gel Electrophoresis (SDS-PAGE)

1. Carry out SDS-PAGE (15%) according to Laemmli *(10)* (*see* Chapter "Non-Electrophoretic Bidirectional Transfer of a Single SDS-PAGE gel with Multiple Antigens to Obtain Twelve Immunoblots"). Electrophorese the various MAPs along with the high and low molecular weight protein standards (*see* **Note 4**).

3.2. Western Blotting

1. Carefully extricate the SDS polyacrylamide gel from between the plates and immerse in western blot transfer buffer.
2. Cut a PVDF membrane to the size of the gel and immerse in methanol. Rinse once with water and once with transfer buffer.
3. Cut four sheets of absorbant filter paper (Whatman 3MM) to the size of the gel and transfer to the transfer buffer. Two absorbant pads are also placed in the buffer (*see* **Note 5**).
4. Assemble the transfer sandwich on a Saran plastic wrap (about 12 in. long; placed on the workbench) in this order: absorbant pad, two filter papers, the PVDF membrane, SDS polyacrylamide gel, two filter papers, absorbant pad. Fold the Saran wrap over the sandwich, roll out air bubbles with a 10 mL pipette and place in a transfer cassette. Position the cassette in the transfer apparatus in such a way that the PVDF membrane is in between the gel and the anode (*see* **Note 6**). Fill with chilled transfer buffer (4°C).
5. Carry out electrotransfer overnight (~18 h) at 30–40 V (constant voltage).
6. Disconnect power supply and disassemble the sandwich. Rinse membrane with water. Stain the polyacrylamide gel with Coomassie Brilliant Blue stain (*see* **Note** 7) to discern efficiency of transfer.

3.3. Calcium Binding Assay

1. Add 200 mL of imidazole wash buffer to the membrane and incubate with gentle shaking for 2 h.
2. Add 200 mL of water and rinse the membrane once.
3. Incubate with 200 mL of 1 mM $CaCl_2$ for 1 h.
4. Wash the membrane thrice with 100 mL of 20% ethanol each time (2 min/wash).
5. Add 200 mL of water and rinse the membrane once to remove ethanol.
6. Incubate the membrane with 1 mM Quin-2 for 1 h. Use enough volume to cover the membrane (*see* **Note 8**).
7. Add 200 mL of water and rinse the membrane once to remove Quin-2.

8. Visualize with UV light using a Photodyne UV Transilluminator (*see* **Note 9**).
9. Take picture (*see* **Note 10**).

3.4. Staining Blot with Coomassie Brilliant Blue

The blot can be stained with Coomassie after photography of calcium binding (*see* Chapter "Protein stains to Detect Antigen on Membranes"). This can reveal the amount of protein in each lane of the blot.

1. Add enough Coomassie (*see* **Note 11**) to cover the membrane.
2. Stain for 30 min.
3. Remove the stain (*see* **Note 12**).
4. Add enough destain to cover the membrane. Add couple of Kimwipes into the destain (*see* **Note 13**).
5. After destaining, the membrane is ready to be imaged.

4. Notes

1. A MAP is made up of a seven-lysine backbone upon which are built eight copies of the same peptide sequence. It is possible to contruct four copies as well. We synthesize eight-branched MAPs at 0.1 mM level with a purity of 95%. As a control it is possible to make just the lysine backbone. A 16-mer MAP runs with an apparent molecular weight of 25,000 (*11*).
2. We dilute from 1 M stock solution of calcium chloride.
3. Prepare a 0.5% stock of Coomassie solution in 25% methanol and 10% acetic acid. Filter with a filter paper in a sintered glass funnel fitted to a side-arm flask attached to an in-house vacuum. Dilute it 10 times with 25% methanol and 10% acetic acid solution.
4. Do not boil the MAP in SDS lysis buffer. If boiled, it runs as an aggregate. Just add SDS lysis buffer and load on gel. Since they are short peptides, there is no need to denature them with heat.
5. Exclude air bubbles from the support pads by pressing down on the pads several times, while the pads are immersed in the buffer.
6. In the Bio-Rad transfer apparatus, we place the gel side of the transfer cassette facing the black side of the transfer cassette holder and the membrane side facing the red side.
7. Microwave or water-bath heat the closed container containing the gel and 0.05% Coomassie until the solution becomes hot (about 70°C).

Incubate for 10 min with shaking *(12)*. Decant Coomassie stain into a bottle (this can be filtered and reused except for gels or membrane meant for tryptic digestion or microsequencing). Place the gel between two or three folds of Kimwipes, add destain, and microwave as before and incubate for 15 min. The Kimwipe will adsorb the Coomassie that exudes out of the gel. The dye remains tightly bound to the Kimwipe, which can now be discarded as solid trash. The destain can be reused.

8. Dissolve Quin-2 in water. Incubate the membrane in sealed plastic bags to reduce the volume of Quin-2 used. Cut a corner of a polyethylene bag, slightly bigger than the size of the membrane. Add the Quin-2 solution and seal the edges with Quick heat seal. Leave some air in the bag to allow the Quin-2 solution to move around the membrane when shaken on a laboratory shaker. For good seal using the Quick heat sealer, it is important not to have any liquid along the edge to be sealed.
9. Wear ultraviolet-proof goggles. Place the membrane on transilluminator box. Cover naked area (area not covered by the membrane) with multifold paper towels. Cover the entire area with ultraviolet-proof plexiglass.
10. Use #57 or #55 polaroid film to take pictures. Use red filter. If negatives are needed, then use #55 film. The pictures can also be taken with an UVP digital system.
11. We make a 0.5% Coomassie stock and then dilute it 1:10 with 25% methanol and 10% acetic acid. To make the stock dissolve Coomassie in methanol first, and then add acetic acid and water to make up the volume. Filter using two Whatman 1 (circular filter) filters, Buchner funnel, side-arm conical flask, and in-house laboratory vacuum set up.
12. Do not dispose used Coomassie down the drain. Save and refilter for reuse.
13. Use of Kimwipes reduces the amount of destain needed. The Kimwipes adsorbs the Coomassie and can be disposed as solid waste.

Acknowledgements

This work was supported by NIH grant ARO1844 and Oklahoma Center for the Advancement of Science and Technology to Dr. Hal Scofield, OMRF, OKC, OK, USA.

References

1. Davies, D.R. and Padlan, E.A. (1990) Antibody–antigen complexes. *Annu Rev Biochem* 59, 439–473.
2. Harley, J.B., Scofield, R.H., and Reichlin, M. (1992) Anti-Ro in Sjögren's syndrome and systemic lupus erythematosus. *Rheum Dis Clin North Am* 18, 337–358. Review.
3. Scofield, R.H. and Harley, J.B. (1991) Autoantigenicity of Ro/SSA antigen is related to a nucleocapsid protein of vesicular stomatitis virus. *Proc Natl Acad Sci USA* 88, 3343–3347.
4. Scofield, R.H., Kurien, B.T., Zhang, F., Mehta, P., Kaufman, K., Gross, T., et al. (1999). Protein–protein interaction of the Ro- ribonucleoprotein particle using multiple antigenic peptides. *Mol Immunol* 36, 1093–1106.
5. Scofield, R.H., Kurien, B.T., Gross, J.K., Reed, N.B., Taylor, A.K., Dominguez, N., et al. (2004) Interaction of calcium and Ro60: increase of antigenicity. *Mol Immunol* 41, 809–816.
6. Maruyama, K., Mikawa, T., and Ebashi, S. (1984) Detection of calcium binding roteins by 45Ca autoradiography on nitrocellulose membrane after sodium dodecyl sulfate gel electrophoresis. *J Biochem* 95, 511–519.
7. Deutscher, S.L., Harley, J.B., and Keene, J.D. (1988) Molecular analysis of the 60 kD human Ro ribonucleoprotein. *Proc Natl Acad Sci USA* 85, 9479–9483.
8. Ben-Chetrit, E., Gandy, B.J., Tan, E.M., and Sullivan, K.F. (1989) Isolation and characterization of a cDNA clone encoding the 60 kDa component of the human SS-A/Ro ribonucleoprotein autoantigen. *J Clin Invest* 83, 1284–1292.
9. James, J.A., Gross, T., Scofield, R.H., and Harley, J.B. (1995) Immunoglobulin epitope spreading and autoimmune disease after peptide immunization: Sm B/B′-derived PPPG-MRPP and PPGIRPP induce spliceosome autoimmunity. *J Exp Med* 181, 453–461.
10. Laemmli, U.K. (1970) Cleavage of structural proteins during assembly of the head of bacteriophage T4. *Nature* 227, 680–685.
11. Kurien, B.T., Jackson, K., and Scofield, R.H. (1998) Immunoblotting of multiple antigenic peptides. *Electrophoresis* 19, 1659–1661.
12. Kurien, B.T. and Scofield, R.H. (1998) Heat mediated quick Coomassie blue protein staining and destaining of SDS-PAGE gels. *Indian J Biochem Biophys* 35, 385–389.

Chapter 50

Detection of La/SS-B by Western Blot Using Nanogold-Tagged Antibodies and Silver Enhancement

Shannon Maier, Sherry Hubbell, and R. Hal Scofield

Summary

Immunogold staining with silver enhancement is a versatile, sensitive and specific method for immunodetection of diverse protein antigens separated by sodium dodecyl sulfate polyacrylamide gel electrophoresis and transferred to nitrocellulose or polyvinylidene difluoride membranes. "Next-generation" antibodies tagged with nanogold particles have a wide scope of use including but not limited to immunohistochemistry, western blotting, electron microscopy, fluorescent activated cell sorting procedures, and cell isolation and migration studies. Herein, we describe the use of a nanogold-tagged anti-mouse IgG secondary antibody and silver enhancement methodologies coupled with antigen-specific unlabeled primary antibodies for the detection of the La/SS-B autoantigen by western blotting as a useful alternative to chemiluminescent and enzymatic detection methods.

Key words: Western blotting, Nanogold, Silver enhancement, Autoantigen, La/SS-B

1. Introduction

La/SS-B is a 48-kD protein ubiquitously present within the nucleus and cytoplasm of cells and exists as part of a ribonuclear protein (RNP) complex with Ro/SS-A and small cytoplasmic RNAs (YRNAs) *(1)*. Expressed as a heterodimer containing a Mr28 amino-terminal and a Mr23 carboxy-terminal domain, the N-terminal domain contains an RNA-binding and recognition motif (RRM) and the C-terminal domain contains a second RRM, an ATP-binding site, and a nuclear localization signal (NLS). The C-terminus of the protein can be cleaved within a linker region lying between the two domains, causing its translocation

Biji T. Kurien and R. Hal Scofield (eds.), *Methods in Molecular Biology, Protein Blotting and Detection, vol. 536*

DOI: 10.1007/978-1-59745-542-8_50

into the cytoplasm *(2, 3)*. It has been proposed that La has chaperone activity in the nucleus due to its ability to recognize and bind RNA. In the cytoplasm, the La protein is involved in initiation, correction, and efficient termination of transcription for newly synthesized RNA Polymerase III transcripts *(4, 5)*. Western blotting of this protein is useful in the detection of autoantibodies against the La protein in serum of patients with diseases such as systemic lupus erythematosus (SLE) and Sjögren's syndrome (SS).

The concept of using colloidal gold particles as a detection method for proteins was based on initial discoveries in the field of photography during the nineteenth century, in which precipitation of silver ions was catalyzed by the presence of colloidal gold particles in the presence of hydroquinone *(6, 7)*. Initial staining using colloidal gold particles suitable for visualization by the naked eye required the use of larger quantities and particles of diameter size ranging from 10–250 nm *(8)*. Silver enhancement of colloidal gold staining was originally described by Danscher *(9)*, in which gold particles catalyze the reduction of silver ions to metallic silver leading to enhanced visualization *(8)*. Using the premise behind autometallography, in which silver ions are precipitated by catalysis around a gold nucleus in the presence of a reducing agent, smaller particles from 1 to 10 nm (nanoparticles) could be used for detection *(6, 8)*. These nanogold particles were subsequently conjugated to antibodies that could directly target antigen or act as secondary labeling systems when used with unlabeled primary antibodies.

Immunogold antibody staining with silver enhancement is a versatile, sensitive, and specific method that has been used in western blotting to detect diverse types of protein antigens that have been separated by polyacrylamide gel electrophoresis and transferred to nitrocellulose (NC) or polyvinylidene difluoride (PVDF) membranes *(6, 8)*. Traditional methods of total protein detection in electrophoresis and western blotting such as Coomassie Blue, fast green, and colloidal gold staining are either unspecific or less sensitive with regard to antigen detection. Sensitivity and specificity in protein detection has been improved through the use of antigen-specific antibodies coupled with secondary antibodies conjugated to a chromogenic or chemiluminescent reporter. Unfortunately, these antibody systems have limited utility and are used primarily for immunohistochemistry and western blotting only. *(6–8)*. Conversely, "next-generation" antibodies tagged with nanogold particles have a wide scope of use including, but not limited to, immunohistochemistry, protein blotting, electron microscopy, fluorescent activated cell sorting procedures, and cell isolation and migration studies *(10–12)*. Herein, we describe the use of 15-nm nanogold-tagged anti-mouse IgG and silver enhancement methodologies coupled with antigen-specific unlabeled primary antibodies for the detection of autoantigens such as La/SS-B by protein blotting. Using this methodology, protein blotting can be performed with sensitivity and specificity that rivals that of chromogenic and chemiluminescent detection systems without the confounding issue of light sensitivity or fading.

2. Materials

2.1. Proteins

1. 6× His-tagged human La cDNA fusion constructs containing the endogenous human promoter, regulatory sequences, and exon coding sequence were obtained from Dr. Darise Farris (Oklahoma Medical Research Foundation, Oklahoma City, OK, USA).

2.2. Western Blotting for La/SS-B Using La-Specific SW5 Monoclonal and Donkey Anti-Mouse Nanogold-Tagged Antibodies

1. Transfer buffer (10×): 0.25 M Tris and 1.9 M glycine (*see* **Note 1**). Dilute 100-mL 10× transfer buffer in 900-mL water to prepare working solution. Store at 4°C.
2. Nitrocellulose membrane (Millipore, Bedford, MA, USA) and 3MM CHR chromatography paper (Biometra, Goettingen, Germany).
3. Tris-buffered saline (TBS; 10×): 0.05 M Tris–HCl, 1.5 M NaCl, pH 7.4. Dilute 100-mL 10× TBS in 900-mL water to prepare working solution. Store at 4°C.
4. TBS with Tween-20 (TBST): Supplement 1× working stock of TBS with 0.1% Tween-20.
5. Blocking buffer: 5% (w/v) nonfat dry milk in TBST.
6. Primary antibody: SW5 monoclonal antibody with specificity to the La/SS-B protein (kindly provided by Dr. Darise Farris).
7. Primary antibody dilution buffer: 5% (w/v) nonfat dry milk in TBST.
8. Secondary antibody: Donkey anti-mouse 15-nm IgG nanogold conjugated antibody (Electron Microscopy Sciences, Molecular Probes, Eugene, OR, USA) (*see* **Note 2** and consult the product information sheet).
9. Secondary antibody incubation buffer: 0.1% Aurion acetylated BSA (Molecular Probes, Eugene, OR, USA) in 1× PBS.
10. Silver enhancement: LI silver enhancement kit (Molecular Probes) containing enhancer reagent and initiator reagent. Store at 4°C (*see* **Note 3** and consult the product information sheet). Working solution is prepared by mixing equal amounts of the enhancer and developer immediately before use.

3. Methods

3.1. SDS-PAGE of La/SS-B Protein

1. *SDS-PAGE*: Carry out SDS-PAGE according to Cleveland et al. *(13)* (*see* Chapter Non-electrophoretic bi-directional transfer of a single SDS-PAGE gel with multiple antigens to obtain twelve immunoblots). La and Plus Protein dual-color protein standards (BioRad Laboratories, Hercules, CA, USA) were separated on 4–20% precast (15 wells) Tris-glycine polyacrylamide gels (NuSep, Ltd., Frenchs Forest NSW 2086, Australia).

3.2. 6× His La/SS-B Transfer to Nitrocellulose and Immunostaining

1. Prepare 1× transfer buffer by diluting 100 mL of 10× transfer buffer with 900-mL water.
2. Wet a piece of nitrocellulose membrane that has been cut to fit the gel (~10 cm × 8 cm) with water followed by equilibration in 1× transfer buffer ≥15 min.
3. Remove the electrodes from the power supply, disassemble the electrophoresis unit, and remove the precast gel. Discard the used electrophoresis buffer. Using a razor blade carefully cut the adhesive that holds the two precast gel plates together.
4. Using a spatula or other thin devices, gently pry apart the two plates making sure that the gel remains on the lower plate. Using either the razor blade or spatula, remove the well dividers and score the corner of the gel for orientation. Carefully transfer the gel into a tray containing 1× transfer buffer and equilibrate the gel for 15 min.
5. Briefly equilibrate the fiber pads and four 3MM filter paper squares (~10 cm × 8 cm) in 1× transfer buffer.
6. In the transfer buffer, carefully assemble the transfer cassette from anode (black) to cathode (red) by stacking a fiber pad, two equilibrated 3MM filter paper squares, the gel, the nitrocellulose membrane, two additional equilibrated 3MM filter paper squares, and the other fiber pad.
7. Assemble the transfer unit by placing the cassette into the transfer tank by aligning the anode (black) of the cassette with the anode of the tank (*see* **Note 4**). Add a prefrozen gel tray or Polar Ice® pack into the transfer tank and fill the tank with 1× transfer buffer.
8. Finish assembling the transfer tank and place the electrodes in a power supply. Transfer the blot at 100 V for 1 h.
9. Remove the electrodes from the power supply, disassemble the transfer tank, remove the gel–membrane sandwich, and discard the used transfer buffer.
10. Remove the membrane from the sandwich and immediately cover the membrane in a blocking solution. Colored

bands in the standards lane typically indicate that the transfer was successful. Transfer of proteins can also be verified by staining the membrane 5–10 min with reversible dilute fast green stain that can be removed by rinsing the membrane in TBST.

11. Block unspecific binding sites on the nitrocellulose membrane by incubating in the blocking solution overnight at 4°C.
12. Discard the blocking buffer and rinse the membrane once in cold TBST for 5 min with continuous rocking at RT.
13. The membrane is then incubated with a 1:40 dilution of SW5 supernatant containing the anti-La/SS-B antibody in antibody dilution buffer for 1 h followed by three 5-min washes in TBST with continuous rocking at RT. Discard the primary antibody solution.
14. Nanogold-tagged donkey anti-mouse IgG is diluted 1:10 or 1:20 in secondary antibody incubation buffer, the membrane is incubated for 1 h in incubation buffer at RT, and the secondary antibody solution is discarded. The membrane is rinsed three times for 5 min with incubation buffer, two times for 5 min with 1× PBS, and two times with water to remove all residual chloride ions that may increase background.
15. LI silver enhancement buffer is prepared by combining equal parts of the enhancer (solution A) and the initiator (solution B) immediately before use. The membrane is then added to the enhancement buffer and incubated until the specimen is visible as a dark brown to black band on the membrane (5–20 min). If the bands remain light following the first incubation, the membrane can be incubated a second time in freshly mixed LI silver developer until the bands intensify (5–25 min).
16. Stop the LI silver enhancement by rinsing in water for 5 min three times, remove the membrane from the water, and place it in a seal-a-meal bag for scanning or imaging (*see* **Note 5**). An example of the expected results produced is shown in **Fig. 1**.

4. Notes

1. All solutions described herein were prepared using 18.2-megaohm water, subsequently referred to as "water" throughout this manuscript.
2. *Nanogold* particles degrade when exposed to concentrated thiols such as b-mercaptoethanol or dithiothreitol and may not be stable at temperatures over 50°C. Optimal results are obtained when the particles are used between refrigerator

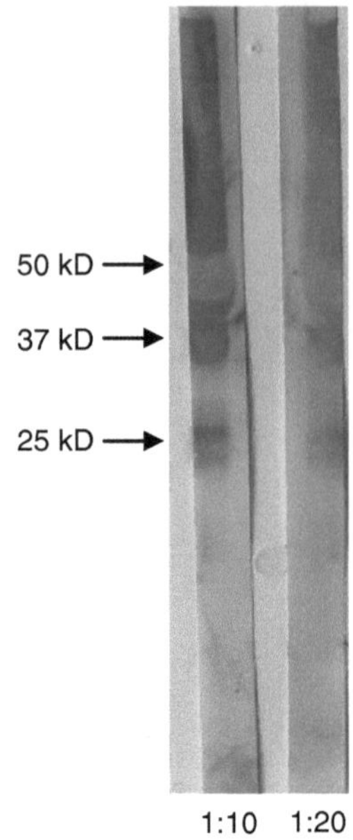

Fig. 1. Western blot of 2-mg La/SS-B protein stained with a 1:10 or 1:20 dilution of a donkey anti-mouse NANOGOLD antibody with silver enhancement. The observed bands correspond to the La/SS-B protein (48 kD), its homodimers (around 25 kD) and its degradation products. *Black arrows* designate locations of molecular weight markers.

and room temperatures. Incubations above these temperatures such as 37°C should be avoided.

3. The enhancer and initiator solutions must remain at 2–8°C and must not be frozen, exposed to extreme heat or light, and not be in contact with metallic objects that may induce silver precipitation. The stability of the enhancement solution is greatly dependent on temperature and time. Therefore, silver enhancement must be optimized by the user to optimize specific enhancement without increasing background.
4. Improper placement of the cassette into the transfer tank can lead to the loss of proteins due to charge reversal. Be sure to load the cassette and assemble the unit properly by matching the black plate of the sandwich holder with the anode (black) side of the transfer tank.
5. Membranes can be imaged using a computer scanner and saved to disk for easy data storage. Membranes remain stable in seal-a-meal bags and can be stored for prolonged times without loss of intensity.

References

1. Wolin, S. L. and Steitz, J. A. (1984) The Ro small cytoplasmic ribonucleoproteins: identification of the antigenic protein and its binding on the Ro RNAs. *Proc Natl Acad Sci USA*. 81, 1996–2000.
2. Dong, G., Chakshusmathi, G., Wolin, S. L., and Reinisch, K. M. (2004) Structure of the La motif: a winged helix domain mediates RNA binding via a conserved aromatic patch. *EMBO J.* 23, 1000–1007.

3. Chambers, J. C., Kenan, D., Martin, B. J., and Keene, J. D. (1988) Genomic structure and amino acid sequence domains of the human La autoantigen. *J Biol Chem.* 263, 18043–18051.
4. Maraia, R. J. (1996) Transcription termination factor La is also an initiation factor for RNA polymerase III. *Proc Natl Acad Sci USA.* 93, 3383–3387.
5. Maraia, R. J., Kenan, D. J., and Keene, J. D. (1994) Eukaryotic transcription termination factor La mediates transcript release and facilitates reinitiation by RNA polymerase III. *Mol Cell Biol.* 14, 2147–2158.
6. Moeremans, M., Daneels, G., De Raeymaeker, M., De Wever, B., and De Mey, J. (1987) The use of colloidal metal particles in protein blotting. *Electrophoresis.* 8, 403–409.
7. Lackie, P. M. (1996) Immunogold silver staining for light microscopy. *Histochem Cell Biol.* 106, 9–17.
8. Fowler, S. J. (1994) The detection of proteins on blots using gold or immunogold. *Methods Mol Biol.* 32, 239–255.
9. Danscher, G. (1981) Histochemical demonstration of heavy metals. A revised version of the sulphide silver method suitable for both light and electronmicroscopy. *Histochemistry.* 71, 1–16.
10. Molecular Probes. (2001) LI Silver Enhancement Kit.
11. Molecular Probes. (2001). NANOGOLD and FluoroNanogold Conjugates.
12. Sun, X. J., Tolbert, L. P., and Hildebrand, J. G. (1995) Using laser scanning confocal microscopy as a guide for electron microscopic study: a simple method for correlation of light and electron microscopy. *J Histochem Cytochem.* 43, 329–335.
13. Cleveland, D. W., Fischer, S. G., Kirschner, M. W., and Laemmli, U. K. (1977) Peptide mapping by limited proteolysis in sodium dodecyl sulfate and analysis by gel electrophoresis. *J Biol Chem.* 252, 1102–1106.

Chapter 51

Imaging Systems for Westerns: Chemiluminescence vs. Infrared Detection

Suresh T. Mathews, Eric P. Plaisance, and Teayoun Kim

Summary

Western blot detection methods have traditionally used X-ray films to capture chemiluminescence. The increasing costs for film, reagents, and maintenance have driven researchers away from darkrooms to more sensitive and technologically advanced digital imaging systems. Cooled charge coupled devices (CCD) cameras capture both chemiluminescence and fluorescence images, with limitations for each detection method. Chemiluminescence detection is highly sensitive and relies on an enzymatic reaction that produces light, which can be detected by a CCD camera that records photons and displays an image based on the amount of light generated. However, the enzymatic reaction is dynamic and changes over time making it necessary to optimize reaction times and imaging. Fluorescent detection with a CCD camera offers a solution to this problem since the signal generated by the proteins on the membrane is measured in a static state. Despite this advantage, many researchers continue to use chemiluminescent detection methods due to the generally poor performance of fluorophores in the visible spectrum. Infrared imaging systems offer a solution to the dynamic reactions of chemiluminescence and the poor performance of fluorophores detected in the visible spectrum by imaging fluorphores in the infrared spectrum. Infrared imaging is equally sensitive to chemiluminescence and more sensitive to visible fluorescence due in part to reduced autofluorescence in the longer infrared wavelength. Furthermore, infrared detection is static, which allows a wider linear detection range than chemiluminescence without a loss of signal. A distinct advantage of infrared imaging is the ability to simultaneously detect proteins on the same blot, which minimizes the need for stripping and reprobing leading to an increase in detection efficiency. Here, we describe the methodology for chemiluminescent (UVP BioChemi) and infrared (LI-COR Odyssey) imaging, and briefly discuss their advantages and disadvantages.

Key words: Chemiluminescence, Infrared, UVP, LI-COR, Odyssey, Western blotting

Biji T. Kurien and R. Hal Scofield (eds.), *Methods in Molecular Biology, Protein Blotting and Detection, vol. 536*

DOI: 10.1007/978-1-59745-542-8_51

1. Introduction

Over the last decade, the use of cooled, charge coupled device (CCD) cameras to capture chemiluminescence from western blots has become increasingly popular. Increasing costs for film, reagents, and maintenance have forced researchers to embrace newer, more sensitive, and technologically advanced digital imaging systems. CCD cameras have the ability to capture both chemiluminescence and fluorescence. Cooling CCD cameras to –20°C increases exposure times to over 5 min without generating a grainy picture and cooling to –40°C can integrate exposures for an hour without significant loss of image quality due to "hot pixels." Most imagers offer a wide dynamic range (3–4 orders of magnitude) generating a high-quality image compared with the limited linear dynamic range of film (1.5 orders of magnitude). Additionally, the available image analysis software has transformed the once frustrating western blotting of yester years to that which can be imaged accurately and repeatedly with a one-touch, preset or user-defined setting. Although chemiluminescence is one of the most frequently used protein detection methods in western blotting due to its high sensitivity (**Fig. 1**), it is

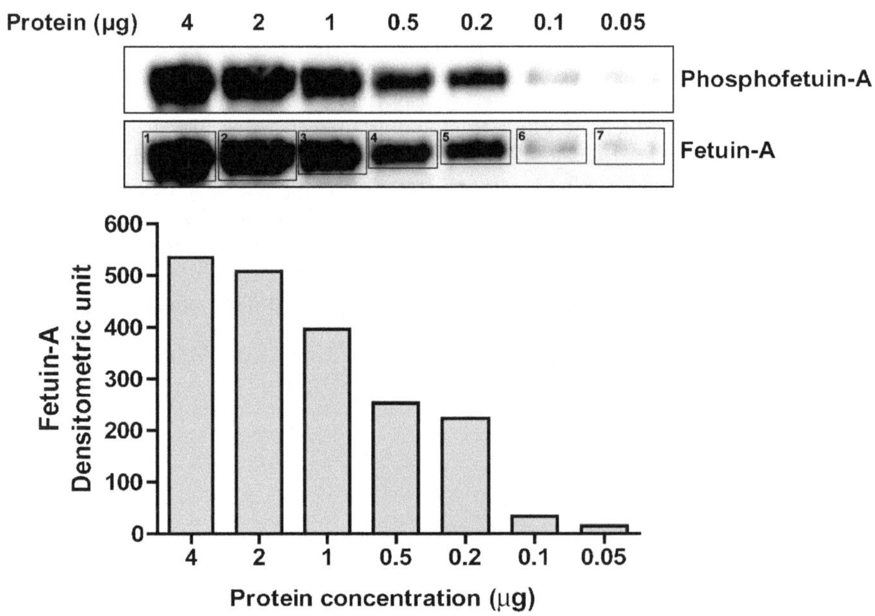

Fig. 1. Chemiluminescence imaging using cooled CCD. Purified total (*bottom*) and phosphorylated (*top*) forms of recombinant, human fetuin-A, a liver-secreted glycoprotein and inhibitor of insulin receptor tyrosine kinase, were separated by SDS-PAGE and transferred to a nitrocellulose membrane and immunoblotted with appropriate antibodies. Chemiluminescent images were captured on a UVP Biochemi Imaging System and quantitated (*bar graph*) with LabWorks software. The linear detection capability for fetuin-A ranged from 50 to 2,000 ng.

not without limitations. The light-producing enzymatic reactions of chemiluminescent detection are dynamic and fade quickly over time. Thus, obtaining images within an optimum time is critical, failing which quantitation accuracy may be compromised. To address this issue manufacturers currently offer extended-duration chemiluminescent substrates, which continue to generate stable chemiluminescence over a longer time period.

Infrared detection methods became commercially available in 2001 with the introduction of the Odyssey® Infrared Imaging System by LI-COR Biosciences. The system is composed of two solid state diode lasers, which can provide light excitation simultaneously at 685 and 785 nm. Collimating lenses, optical bandpass filters, and a focusing lens focus and tune the laser beams to produce an excitation spot on the scanning surface that is magnified and split to photodiodes that convert the light to an electrical signal. Infrared detection offers solutions to the problems associated with chemiluminescence and fluorescent detection methods including static detection, low autofluorescence, equivalent or higher sensitivity than chemiluminescence, higher signal-to-noise ratio, and detection range, multiplex labeling, and greater efficiency. Fluorescent secondary antibodies detected in the infrared spectrum produce a constant amount of light. The static nature of the intensity of light generated by infrared activation improves precision and the ability to differentiate differences in signal intensity produced by antibodies bound to proteins. This allows for a more accurate quantification of protein levels compared with enzyme-labeled secondary antibodies.

One major advantage of infrared imaging over chemiluminescence is the ability to detect two proteins simultaneously in color *(1)*. For example, it is often necessary to quantify total and phosphorylated forms of intracellular proteins to investigate cellular signaling pathways. Infrared-labeled antibodies available for the 680 and 800 nm spectrum make it possible to view total and phosphorylated proteins separately by manipulating individual wavelength channels in the same band. This permits simultaneous viewing of the effects of a treatment on protein phosphorylation without the need for membrane stripping and reprobing (**Fig. 2**). Likewise, the production of light generated from fluorophores detected in the infrared spectrum not only improves quantification and accuracy, but facilitates normalization and comparative analysis months to years later without a loss of signal if membrane imaging is desired for future analyses. Therefore, infrared imaging has the potential to dramatically improve the efficiency of western blotting methods.

Here, we describe methodology to quantitate proteins on western blots using chemiluminescence (UVP BioChemi, UVP LLC, Upland, CA) and infrared imaging (LI-COR Biosciences, Lincoln, NE).

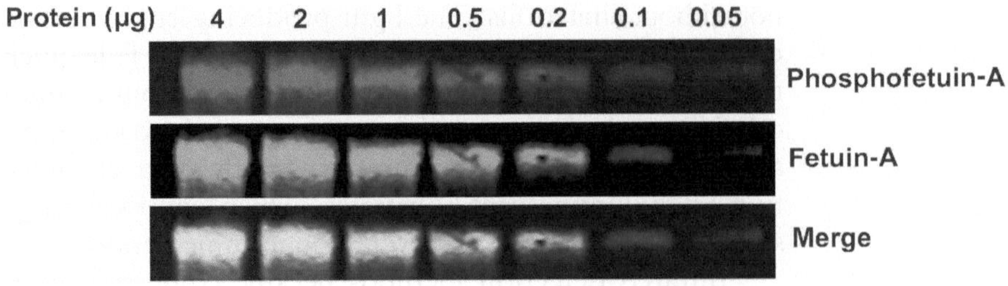

Fig. 2. Infrared imaging using LI-COR Odyssey® Imaging System. Purified total (*middle*) and phosphorylated (*top*) forms of recombinant, human fetuin-A, a liver-secreted glycoprotein and inhibitor of insulin receptor tyrosine kinase, were separated by SDS-PAGE and transferred to a nitrocellulose membrane and immunoblotted with appropriate antibodies. Overlaid image (*merge*), 700 channel (*red*), and 800 channel (*green*) were acquired using the Odyssey 2.0 software. LI-COR Odyssey® Imaging System allows the user to simultaneously detect phosphorylated and total forms of a protein without the need for stripping and reprobing (*see Color Plates*).

2. Materials

2.1. Chemiluminescence Imaging: UVP BioChemi

1. Western blot wash buffer: 10 mM Tris–HCl, pH 7.4, 150 mM NaCl, 0.1% Tween-20.
2. Blocking buffer: 5% Nonfat dry milk powder or 1% BSA diluted in western blot wash buffer.
3. Diluted primary antibody with blocking buffer: Optimum dilution depends on the antibody of interest (ranging from neat to 1:5,000).
4. Diluted secondary antibody: 1:5,000–1:20,000 dilution in blocking buffer.
5. UVP BioChemi system with LabWorks Image Acquisition and Analysis software (UVP LLC, Upland, CA).
6. Shaker.

2.2. Infrared Imaging: LI-COR Odyssey®

1. Phosphate buffered saline (PBS, 1×) (without calcium and magnesium (Fisher Scientific, Hampton, NH, USA).
2. PBS with 0.1% Tween-20 (PBST).
3. Odyssey blocking buffer (diluted 1:1 with PBS without loss of performance) (LI-COR Biosciences, Lincoln, NE, USA).
4. Diluted primary antibody with Odyssey blocking buffer. Optimum dilution depends on the antibody of interest (typically 1:1,000 to 1:5,000). Odyssey blocking buffer can be diluted 1:1 with PBS without a loss of performance. For multiple antibody detection, *see* **Note 8** and **Subheading 3**.
5. LI-COR Odyssey Infrared Imaging system (LI-COR Biosciences, Lincoln, NE, USA).
6. Shaker.

3. Methods

SDS-PAGE (*see* Chapter "Non-electrophoretic bi-directional transfer of a single SDS-PAGE gel with multiple antigens to obtain twelve immunoblots") and protein transfer (*see* Chapter "Detection of calcium binding by Ro 60 multiple antigenic peptides on PVDF membrane using Quin-2" for transfer to PVDF) to membranes should be conducted according to the procedures of Towbin et al. *(2)*. Either nitrocellulose or PVDF membranes may be used for western blotting using the UVP BioChemi Image Analysis system.

3.1. Chemiluminescence Imaging: UVP BioChemi

3.1.1. Western Blotting

1. Following protein transfer to the nitrocellulose or PVDF membrane, gently soak the membrane in PBS for 10 min to remove transfer buffer.
2. Block the membrane in the appropriate blocking buffer for 1 h at room temperature (RT) or overnight at 4°C if necessary with gentle rotation or shaking. The blocking buffer must cover the membrane (a minimum of 0.4 mL/cm^2 is recommended).
3. Dilute the primary antibody of interest in blocking buffer (1:1,000 to 1:5,000 are generally good starting points).
4. The primary antibody should be incubated for 60 min at RT or for longer periods (12–16 h) at 4°C. Optimum incubation times depend on the protein of interest and antibody dilutions used.
5. Following incubation with the primary antibody, wash the membrane five times for 5 min each wash, at RT with western blot wash buffer. The membranes should be fully covered with PBST and shaken gently to facilitate the removal of unbound antibody.
6. Incubate the membrane with secondary antibody diluted in blocking buffer (1:5,000 to 1:20,000 are typical ranges used). We have generally found that higher dilutions (1:10,000 or greater) provide maximum sensitivity with reduced unspecific binding.
7. The secondary antibody should be incubated for 60–120 min at RT with gentle shaking. Incubations longer than 120 min may increase background noise.
8. Following incubation with the secondary antibody, wash the membrane five times for 5 min each with western blot washing buffer.
9. Place the membrane on Saran- or Glad-wrap, add chemiluminescent substrate (prepared fresh by mixing luminol and peroxide containing reagents), and follow manufacturer's recommendation for length of incubation with substrate.

3.1.2. Image Acquisition

1. Drain excess substrate, and then place the membrane in a sheet protector.
2. Next, open the BioChemi Image analyzer dark box and lower metal tray onto light box (*see* **Note 1**).
3. To make sure that the captured chemiluminescence image is in focus, place a sheet of paper with graphics, words, or a combination of words and graphics.
4. Click the Video/digital capture icon on toolbar, which opens the Live Workspace Preview. Now focus the image by manually adjusting the focus ring on the camera. Close Live Workspace preview on completion.
5. Next, position the membrane in the middle of the metal tray inside the UVP Biochemi imager, and then close the imager.
6. Set Emission filter on the imager to the Clear position.
7. Turn off power to the imager. Then click Acquire and choose the Video/digital option (*see* **Note 2**).
8. To take a series of images to decide the best exposure times, click Integration tab in camera menu. Choose Sequential Integration. This allows the user to capture images sequentially, in an incremental manner. Since the capture process is automatic, this does not require the user to be present. Once completed, the user may choose the best captured image.
9. On sequential settings, choose irregular intervals; choose western Chemi option, and use default settings or make appropriate changes on sequential integration timings. Click OK.
10. Click capture in the Camera tool bar. This begins the capture process. Images captured are sequentially added to the Desktop.
11. Once completed, an image of appropriate band intensity may be chosen and saved.
12. Alternatively, a single image can be captured from the Video/digital option, for a defined exposure time, and saved (*see* **Note 3**).

3.1.3. Enhancement of Chemiluminescent Images

1. Under the Edit menu, click Display Range, which is a graphical representation of the total pixels in the image and their gray scale values.
2. Next, click Invert button, to make the bands appear dark on a light background.
3. In the Display Range, slide the green and blue bars from the right side of the graph to the left until you can see the blot on the image.

4. Next, click on the Z+ button to show more detail in the specific area of the image where pixels are concentrated.
5. Then slide both left and right bars to further enhance the blot.
6. Next, click Process menu for sharpening of the image. Hot pixels can be substracted/removed by clicking the Starfield subtraction button.
7. Next, click Image Enhancement icon in LabWorks and then click Apply to make the changes in the image permanent (*see* **Note 4**).

3.1.4. Calibrating Chemiluminescent Blots with Visible Light Protein Markers

1. Acquire an image of the visible protein markers on the blot using overhead epi-white lighting of the Biochemi system darkroom (black box); optimize this image using the Display Range feature of LabWorks, by clicking the Edit menu, and then save the image on the computer.
2. Now, acquire an image of the chemiluminescence at the same settings as the image of the visible light protein markers, and save the image. Do *not* move the membrane or adjust the zoom setting on the lens.
3. Optimize chemiluminescence image using the Display Range feature of LabWorks software, by clicking the Edit menu.
4. To make both images (visible protein marker image and chemiluminescence image) look similar, invert the image of the chemiluminescent blot so that bands appear dark in front of a light background.
5. Click Image Enhancement icon in LabWorks and then click Apply to make the changes in the image permanent.
6. Next, highlight the image of visible light marker, click Edit and Copy from LabWorks software menu (*see* **Note 5**).
7. Click the image of chemiluminescent blot and select Edit and Paste. This overlays the visible light markers onto the image of the chemiluminescent blot precisely orienting the markers as it appears on the membrane.

3.1.5. Molecular Weight Calculation

The LabWorks software can be used to calculate molecular weight of the protein band detected by chemiluminescent imaging.

1. In the LabWorks software, click 1-D gel tab. Then click "show toolbar." Next, click Lanes. This will automatically detect all lanes.
2. Next click Molecular weight standards in 1-D gel tool bar. Click New and then enter information about your protein markers, name, units, and type of material (protein) and diffusion (default is logarithmic).

3. Click Add, and then key in molecular weight of the protein markers starting with the highest molecular weight. When done, click Add again. Repeat this procedure till all molecular weights have been added.
4. Next, click Auto-locate in the Molecular Weight Standards tool bar. Software automatically detects bands and adds molecular weight standards to it. Based on the intensity of the protein marker bands, some molecular weight may not be accurately labeled.
5. To manually correct this, click Locate in the Molecular Weight Standards tool bar, and then click Detach All. Now the band position indicator can be moved to appropriate position on the image where the molecular weight markers have migrated. Once the position of all markers is confirmed click OK on the Molecular Weight Standards tool bar.
6. To obtain the results of the densitometric analysis, click Results on the open 1-D gel toolbar. Results can be exported to a Microsoft Excel file, by clicking File, then DDE to Excel.

3.1.6. Quantitation of Western Blots: Densitometric Analysis

1. In the LabWorks software menu, click Tools tab, and then click Area Density.
2. Next, click Define Region. This allows the user to mark an area of interest (chemiluminescent protein band) in the image. On the tool bar choose between rectangle, elliptical, and free-form regions by selecting the area of interest icon (AOI) in the main tool bar. From experience we have found that rectangle size works best for quantitation of western blots (**Fig.1**).
3. If the bands appear irregular, select Irregular Shape icon from the LabWorks main menu.
4. Click and draw a rectangle around the protein band of interest in the first lane. Make the first rectangle big enough to accommodate the biggest band on the image. To retain the same size for all bands, copy the first rectangle, by right clicking on the mouse.
5. A copy of the first rectangle is made. Now click and drag to the band of interest in the second lane and drop.
6. Right click again to copy another rectangle for the next band. Repeat the procedure until an area of interest has been defined for all bands. Once completed, click the End button in the Area Measurement window.
7. This opens up the Area Density Tool, which gives total density, mean density, mean background, total raw density, and minimum intensity of the analyzed bands.

8. Density measurements can be exported as a Microsoft Excel file by clicking File, and then choosing DDE to Excel option.

3.1.7. Exporting Images to PowerPoint or Word Files

The UVP Biochemi Chemiluminescent Imager captures images as 12-bit or 16-bit files. These 12-bit and 16-bit image files cannot be viewed in other imaging softwares, or be imported into Microsoft PowerPoint or Microsoft Word, and need to be converted to an 8-bit file. This is done as follows:

1. Highlight the image, then click Edit and choose "Convert to."
2. Next, click Gray Scale 8 and then click Convert. This converts the image to an 8-bit file. However, the user will need to redo the image enhancements (Invert image, Display range, etc.).
3. Next, click the Image Enhancement icon in LabWorks and choose Apply to make the changes in the image permanent, and save the image.

3.2. Infrared Imaging: LI-COR Odyssey®

3.2.1. Western Blotting

1. Following protein transfer to the nitrocellulose or PVDF membrane, gently soak the membrane in PBS for 10 min to remove transfer buffer (*see* **Note 6**).
2. Block the membrane in Odyssey blocking buffer (1:1 with PBS) for 1 h at RT or overnight at 4°C if necessary by applying gentle rotation or shaking (*see* **Note 2**). The blocking buffer must cover the membrane (a minimum of 0.4 mL/cm^2 is recommended) (*see* **Note 7**).
3. Dilute the primary antibody/antibodies of interest in blocking buffer (1:1 in PBS) (*see* **Notes 7** and **8**). Primary antibody dilutions ranging from 1:1,000 to 1:5,000 are generally good starting points. Add 0.1% Tween-20. Tween-20 added to the primary antibody solution decreases background staining and increases protein renaturing, thereby improving antigen–antibody binding *(3)*.
4. The primary antibody should be incubated for 60 min at RT or for longer periods (12–16 h) at 4°C. Optimum incubation times depend on the protein and antibody of interest.
5. Following incubation with the primary antibody, wash the membrane four times for 5 min each at RT with PBST. The membranes should be fully covered with PBST and shaken gently to facilitate the removal of unbound antibody (*see* **Note 8**).
6. Infrared (fluorescent-labeled) secondary antibody is diluted in blocking buffer. As stated previously, blocking buffer can be diluted 1:1 in PBS. Dilutions in the range of 1:5,000 to 1:25,000 are typical ranges used. We have generally found that

lower concentrations (1:20,000) provide maximum sensitivity with reduced unspecific binding. However, each antigen–antibody relationship will possess specific binding affinities that may alter incubation concentrations. Again, 0.1% Tween-20 can be added to the diluted antibody as performed previously for the primary antibody (*see* **Note 9**).

7. The secondary antibody should be incubated for 30–60 min at room temperature with gentle shaking. Incubations longer than 60 min may increase background noise. From **step 7** onward throughout these procedures, it is important to protect the secondary antibody incubation from light to avoid loss of signal.
8. Following incubation with the secondary antibody, wash the membrane four times for 5 min each with PBST.
9. The membrane should be rinsed with PBS to remove Tween-20 prior to imaging.

3.2.2. Infrared Image Acquisition

1. The membranes should be placed facedown with the top of the membrane toward the front of the imager at the 0,0 (*x*,*y*) coordinates. This will allow the user to specify a scan boundary that is 1 cm larger than the membrane in all dimensions, which minimizes the truncation of annotations made on the scan. Since scanning starts at the lower left hand corner of the scanning surface and progresses first across the *X*-axis and then steps up incrementally to the *Y*-axis, membranes are generally easier to analyze if the sides are parallel to the *X*/*Y* axes scans. However, the position of the membrane on the scanning surface is not detrimental since images can be flipped or rotated following the scan. Always clean the scanning surface with deionized water and a dust-free towel or cloth before imaging. Dust and other particulates trapped on the scanning surface can be transferred to the membrane and captured on the image. Also remove all bubbles underneath the membrane by using a roller device provided by the manufacturer (*see* **Note 10–12**).
2. Imaging may be started from the instrument's front panel, an internet browser, or from software provided by the manufacturer. Procedures described in this section are for 2.1 software operation only.
3. Start the software and choose a new project or open an existing project folder. Enter a name for the scan to be placed in the project folder. The "Preset" for the instrument should be set to "Membrane." Resolution can be set to 12, 42, 84, 169, or 137 mm – 169 mm is used for typical scans of membranes. Quality refers to how much detector signal is processed for a given area on a membrane to form a pixel on the image.

Medium is typical. Focus Offset should always be set to 0. Focus Offset accomadates additional imaging strategies such as microplates. The "Channels" check boxes are used to specify whether to detect the 700 channel dye, 800 channel dye, or both. Intensity is typically set to 5.0 for membranes. The scan area is drawn by clicking and dragging the rectangle on the scan grid. Click "Start Scan" to send the scan parameters to the imager and to start the scan (*see* **Note 13**).

3.2.3. Enhancement of Infrared Images

1. During the course of a scan or following its completion, the image generated can be rotated, cropped, and adjusted. Adjustments to the brightness, contrast, or sensitivity can be made in either or both of the 700- or 800-nm channels to optimize the image. If bands are missing on an image, increase the sensitivity using the Linear Manual Sensitivity slider. If bands are dim, use the Brightness and Contrast sliders to adjust.
2. In most circumstances, the default settings in the scan console are optimal. Variation from the default settings is typically related to insufficient protein or primary and secondary antibody dilutions. The high sensitivity of the Odyssey® imager permits the use of primary antibodies in the range of 1:5,000 to 1:10,000 and secondary antibodies in the range of 1:15,000 to 1:25,000.

3.2.4. Sizing Bands and Quantification

1. Acquire an image of the visible protein markers and samples using the Odyssey® Imager and software as described earlier.
2. Sizing begins by adding lanes to the image with both image channels overlaid allowing the lanes to be added to both images simultaneously and in identical positions.
3. Click the "Add Lane" tool in the left toolbar. Move the cursor above the highest MW marker at the top of the first lane. Center the cursor in the lane and pull the cursor down below the band with the lowest molecular weight and double click. Always check each lane to be sure that it is centered over the lane in the image and is wide enough to enclose all the bands in the lane.
4. Add other lanes by copying and pasting unless some curves have exaggerated curvature. Another option is to click the "Add Multiple Lanes" tool, which allows the number of additional lanes required. All lanes will be marked, and bands should be enclosed by a rectangular band marker. In some instances, bands that are faint will not be found during the automatic lane finding. The "Add Band" marker tool may be used to add a band. Bands may also be deleted by pressing Delete on the keyboard.

5. Once the molecular weight markers have been identified, the size of each marker can be entered through the "Analyze" dropdown and by clicking "Edit Size Standard." Assigning the size standards permits the user to develop a molecular weight line for protein identification in samples with multiple bands in the same lane.
6. Adding lanes to the membrane not only allows quantification of the molecular weight marker, but also allows for accurate densitometric analysis of each band. By simply clicking on the center of the box, the densities of each band can be determined.

4. Notes

1. Turn on power to cooled CCD camera ~10–15 min before image acquisition. This allows the CCD camera to fully cool to −28°C reducing "hot pixels."
2. Since chemiluminescence imaging is done in the dark, one common mistake, often encountered, is *forgetting to turn off* the white light on the imaging system. Further, the imaging system has an built-in safety mechanism that turns the UV-light off whenever the door to the imager is opened. The user may not realize that the UV-lights are on until the image is being captured. To make sure that all lights are turned off, a "safe lab practice" may be to turn the power to the UVP Bio-chemi system off completely.
3. A captured western blot image may not be typically visible without software enhancement. This is because of the low light emission from the chemical reaction and not because of a faulty system or camera. Sometimes, the image may look completely black and may be misleading to first-time users. The Image Enhancement tool needs to be used to enhance low light chemiluminescent images.
4. Because of the voluminous nature of a typical western blot lab, it sometimes becomes necessary to annotate chemiluminescent images. This is easily done by clicking Edit on the LabWorks menu and choosing Annotate option. The user can select the alphabet tool bar and using the mouse can now click on the desired location on the image to annotate. The desired lane-identification label or molecular weight size can now be keyed into the Annotation Object Properties tool bar. The software offers the ability to change the font, font size, or color of label.

5. When calibrating visible light protein markers with chemiluminescent blot for estimation of molecular weight on protein bands, it is extremely convenient to just draw a rectangle around the marker lane (make sure that this rectangle covers the entire area from the top to the bottom of the image) and overlay this image (rather than the entire white light image of the blot) onto the chemiluminescent blot.
6. Following the transfer, handle the membrane carefully and mark the side of the membrane where the transfer occurred with a pencil. Markings with an ink pen or markers can fluoresce and will be observed on the imaged blot. Be sure to mark the membrane on the top side of the membrane where the protein transfer occurred.
7. Although we have found that membrane blockers such as nonfat dry milk are suitable for use with anti-rabbit and anti-mouse secondary antibodies with the Odyssey® System, we have found that blocking buffer available by the manufacturer provides less background noise and unspecific labeling for anti-goat secondary antibodies. The diluted blocking buffer (1:1) with PBS can be used up to three times if stored at 4°C without a loss of performance before disposal. It should also be noted that the membrane *should not* be exposed to Tween-20 prior to blocking. Exposure to Tween-20 can cause strong background fluorescence on the blot. An important mistake that should be highlighted here is to ensure that the rinse immediately following transfer should be done in PBS only. Often times during the course of a western blot it becomes necessary to discontinue the assay due to time constraints. Blocking can be conducted overnight at 4°C if necessary.
8. The detection of two separate antibodies in the 700 and 800 channels of the instrument requires primary antibodies from different host species (rabbit, mouse, or goat IgG).
9. Diluted secondary antibodies can be saved and reused usually up to three times. The decision to reuse the secondary antibody should be based on the length of time between assays. As expected, secondary antibody used three times over the course of 3 days yields similar labeling sensitivity. In contrast, secondary antibody used three times over the course of 3 weeks will often yield disparate differences in the intensity of bands. Along these lines, it is also critical that the diluted secondary antibody be protected from light to maintain performance. Finally, 0.01% Tween-20 can be used to reduce membrane background.
10. Scanning on the Odyssey can be conducted using either the 700 or 800 infrared channels or both depending on the

protein(s) of interest. It is important to note that the scanning surface of the Odyssey should be cleaned with distilled water prior to use and removed with lent-free wipes to decrease particulate on the imaging surface. Also, remember that the blot must be protected from light and from contact with particulates such as dust, which will be imaged if transferred. Once the membrane is imaged, it may be stored dry for months protected from light. If there is any possibility that the image may need to be stripped and reprobed then store the membrane in PBS protected from light since stripping agents are generally ineffective when membranes are dry.

11. Membranes may be stored dry or in PBS buffer at 4°C.
12. Imaging is conducted by placing the blot facedown and inverted on the scanner surface of the imaging system.
13. "In-cell" western assays can detect up to two proteins simultaneously within the cellular environment without the need for generating cell lysates. While the technique is based on standard immunocytochemical methods, infrared detection enables sensitive and quantitative analysis of protein signaling pathways in cultured cells without reduced interference from cell, plate, and treatment autofluorescence *(4)*. Cells may be grown or transferred to plates (including 96-well microtiter plate) prior to experimentation. The cells are then fixed (usually with 4% paraformaldehyde), blocked with Odyssey blocking buffer or nonfat dry milk, and then incubated with primary and secondary antibodies, respectively, for the target protein(s).

Acknowledgement

S.M. has received grant support from the American Diabetes Association (7-04-JF-36) and the Auburn University Biogrant Program (2-13820).

References

1. Picariello, L., Carbonell Sala, S., Martineti, V., Gozzini, A., Aragona, P., Tognarini, I., et al. (2006) A comparison of methods for the analysis of low abundance proteins in desmoid tumor cells. *Anal Biochem.* 354, 205–212.
2. Towbin, H., Staehelin T., and Gordon J. (1979) Electrophoretic transfer of proteins from polyacrylamide gels to nitrocellulose sheets: procedure and some applications. *Proc Natl Acad Sci USA* 76, 4350–4354.

3. Zampieri, S., Ghirardello, A., Doria, A., Tonello, M., Bendo, R., Rossini, K., et al. (2000) The use of Tween-20 in immunoblotting assays for the detection of autoantibodies in connective tissue diseases. *J Immunol Methods* 239, 1–11.

4. Chen, H., Kovar, J., Sissons, S., Cox, K., Matter, W., Chadwell, F., et al. (2005) A cell-based immunocytochemical assay for monitoring kinase signaling pathways and drug efficacy. *Anal Biochem.* 338, 136–142.

Chapter 52

A Fluorescent Codetection System for Immunoblotting and Proteomics Through ECL-Plex and CyDye Labeling

Ciara A. McManus[†], Pamela M. Donoghue, and Michael J. Dunn

Summary

The qualitative and quantitative capabilities of 2-D electrophoresis and its use in widespread proteome analysis has been revolutionized over the past decade with the introduction of differential gel electrophoresis commonly known as DIGE. This highly sensitive CyDye protein labeling technique now attempts to advance conventional western blotting by the combination of DIGE labeling with the recently developed ECL-Plex CyDye conjugated secondary antibodies. The ability of this method to simultaneously visualize the total protein expression profile as well as the specific immunodetection of an individual protein species will significantly aid protein validation following 2-D gel separation by confirming the exact location of proteins of interest. This simple, rapid, and reproducible technique is demonstrated by 1-D and 2-D electrophoresis through the detection of the small 27-kDa heat shock protein (hsp 27), a protein known to be expressed in the human heart, from a complex cardiac protein extract.

Key words: CyDye immunoblotting, ECL-Plex, Western blotting, Codetection

1. Introduction

Developed by Towbin et al. *(1)*, western blotting is deemed widely as an extremely robust means of identifying and/or verifying specific interactions between immobilized proteins and antibodies in one-dimensional gel electrophoresis (1-DE). However, applying this technique to two-dimensional gel electrophoresis (2-DE) is somewhat cumbersome. Advantages of 2-DE over 1-DE lie in its ability to simultaneously display up to 2,000 proteins on a single gel *(2)*. Although the actual immunoprobing of a 2-DE gel is relatively straightforward, difficulties arise in the

[†]Ciara A.McManus and Pamela M.Donoghue Contributed equally to the Work.

Biji T. Kurien and R. Hal Scofield (eds.), *Methods in Molecular Biology, Protein Blotting and Detection, vol. 536*

DOI: 10.1007/978-1-59745-542-8_52

ability to unequivocally determine which spot on the immunoblot corresponds to which spot on the gel.

There are a considerable number of both pre- and poststaining methods available, none of which overcome the problem. For example, gel-staining techniques such as Coomassie Brilliant Blue and silver stain along with fluorescence-based methods such as SYPRO-Ruby and Deep Purple give a good overall impression of the total protein profile, but many are not compatible with downstream immunoblotting analysis. Similarly, general protein stains for blot membranes such as Amido black, Ponceau-S, Fast green FC, Colloidal gold, and India ink are also lost during antibody incubation *(3)*.

We describe here a method that combines visualization of the total protein profile and the immunoblotting of a specific individual protein, addressing the challenges associated with spot correlation between 2-DE gels and membranes. It is based upon the recently developed ECL Plex-CyDye immunoblot detection system (GE Healthcare, Amersham, Bucks, UK). Protein samples are labelled with the charge-matched 'minimal labeling' cyanine CyDye reagents (GE Healthcare), Cy3 and Cy5, prior to 1-D or 2-D gel electrophoresis. Only brief details of the 2-DE methodology are given in this chapter, but a detailed protocol can be found elsewhere *(4)*. Separated proteins are then transferred onto membranes by semidry electroblotting and can be subsequently be probed with a specific primary antibody of choice. Membranes are then incubated with ECL Plex fluor-labelled species-specific Cy3 or Cy5 secondary antibodies (GE Healthcare) and then scanned using a fluorescent imager at their corresponding specified wavelengths. Thus, this highly sensitive method allows the simultaneous viewing of the total protein profile expression along with the specific immunoreactive protein(s).

A stepwise increase in protein loading (150, 300, and 600 μg) was utilized to demonstrate both the robustness and reproducibility of the technique in 2-DE. Similarly, scanning at these different loadings showed no interference between increased protein loads and calculated fluorescence *(5)*.

CyDye immunoblotting also resolves several of the difficulties associated with DIGE protein labeling. This technique, although highly sensitive, does result in a small but significant shift in pI during 2-D gel electrophoresis; as a result, significant spot comparison between DIGE and non-DIGE 2-D gels can prove extremely difficult *(6)*. The combination of DIGE and ECL Plex labeling overcomes this problem by displaying the exact location of the protein of interest in the overall protein profile considerably aiding spot picking for mass spectrometry identification.

With the ECL Plex system, it is feasible that multiple protein targets can be detected simultaneously since multiple species-specific secondary antibodies labelled with different CyDyes are currently available providing an added quantitative capability to this approach (http://www4.gelifesciences.com/aptrix/upp00919.nsf/Content/DD5A06BB56FC99CAC12571DB00811F1B/$file/28904234AA.pdf).

2. Materials

2.1. Tissue Sample Preparation

1. Liquid nitrogen canister Dilvac Dewar flask 1 L (Keison, Chelmsford, UK).
2. Pestle and mortar (*see* **Note 1**).
3. DIGE compatible lysis buffer: 9.5 M urea (w/v), 2% CHAPS (w/v), and 20 mM Tris–HCl (w/v) pH 8.0–8.5. Stored in aliquots at −20°C. All reagents are available from Sigma-Aldrich, St. Louis, MO, USA (*see* **Note 2**).
4. Hand-held sonicator: Microson XL 2000 (Misonix, Farmingdale, NY, USA) (*see* **Note 3**).
5. Protease inhibitor cocktail: Roche complete protease inhibitor cocktail tablets (Roche Diagnostics GmbH, Mannheim, Germany).
6. Bradford protein-binding reagent: Bio-Rad Protein Assay (Bio-Rad laboratories GmbH, Munchen, Germany).

2.2. DIGE Labeling

1. CyDye stock solution: Cy3 and Cy5 CyDyes (GE Healthcare) supplied as 25 ng of lyophilized powder resuspended 1:1 in dimethylformamide (DMF) (*see* **Note 4**) (Sigma-Aldrich, UK). Dyes are used immediately or stored at −80°C in the dark for up to 3 months.
2. CyDye working solution: Dye working solutions are made up to the desired concentration in DMF, and can be stored at −80°C for up to 2 weeks in the dark.
3. Labeling stop buffer: 10 mM lysine. Can be stored at RT.

2.3. 1-D SDS-Polyacrylamide Gel Electrophoresis

1. Bio-Rad mini-Protean minigel system, Bio-Rad laboratories, Hercules, CA, USA.
2. 4–20% Precast Precise Protein gels (Pierce, Rockford, IL, USA).
3. Sample buffer: CyDye labeled protein samples were combined with an equal volume of Laemmeli sample buffer (Bio-Rad Laboratories).
4. Running buffer: 1 M Tris Base, 1 M HEPES, and 1% SDS.

2.4. Western Blotting for HsP-27

1. Hybond nitrocellulose membranes (Amersham Biosciences, Bucks, UK).

2. Protean blotting system (Bio-Rad laboratories).
3. Transfer buffer: 25 mM Tris, 192 mM glycine, and 20% methanol.
4. Membrane blocking buffer: 0.01 M Phosphate buffered saline (PBS) containing 5% bovine serum albumin (BSA) (Sigma-Aldrich).
5. Primary antibody: Monoclonal anti-Hsp 27 antibody (SPA-800) (Stressgen, Bioscience, Cambridge, UK), 1:800 dilution in PBS Tween-20 (PBST), 0.01 M PBS, and 0.1% Tween-20 (PBST).
6. Membrane washing buffer: PBST as described earlier.
7. Secondary antibody: ECL Plex goat-α-mouse IgG, Cy 3 conjugated or goat-α-rabbit IgG, Cy 5 conjugated secondary antibody (GE Healthcare), 1:2,000 dilution in PBST (*see* **Fig. 1**; **Note 5**).
8. Membrane storage: Membranes can be stored wrapped in lint-free tissue and tin foil for up to 3 months in the dark (*see* **Note 6**).

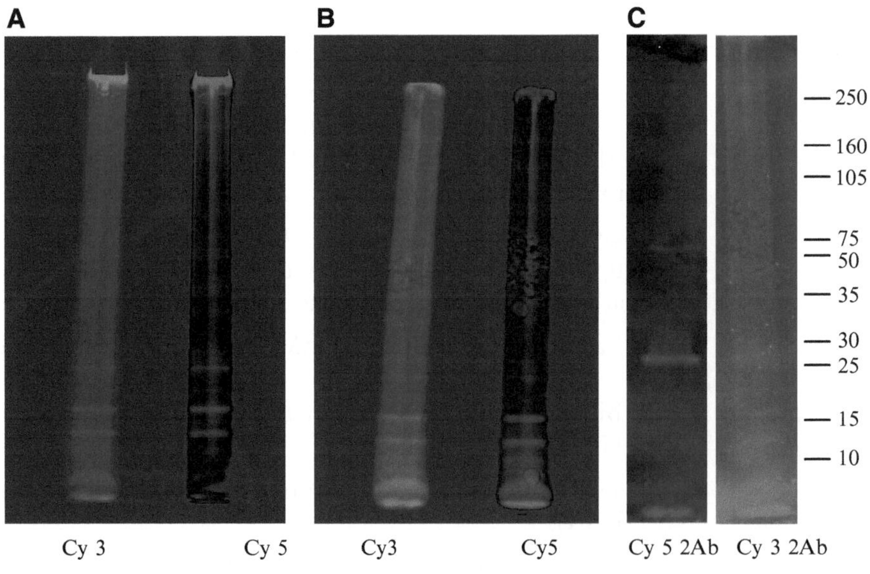

Fig. 1. ECL Plex 1-D CyDye blotting: ECL Plex human left ventricular heart fractions prelabeled with Cy3 and Cy5 CyDyes were electrophoretically separated on a 1-D 4–20% gradient gel (**A**). Following protein transfer onto nitrocellulose membranes (**B**), blots were probed with monoclonal Hsp 27 Ab and visualised with Cy5 and Cy3 secondary conjugated Abs (**C**). Flour labeled ECL Plex rainbow molecular mass markers were employed to correctly identify labeled proteins. All gels and membranes were visualized using the Typhoon variable mode imager (reproduced from *(4)* with permission from Wiley-VCH) (*see Color Plates*).

2.5. Isoelectric Focusing

1. Focusing apparatus: IPGphor, Amersham Biosciences, Bucks, UK.
2. Strip Rehydration Buffer: 8 M urea, 0.5% (w/v) CHAPS, 0.2% (w/v) DTT, and 0.2% (w/v) Pharmalyte pH 3–10.
3. 2× lysis buffer: 9.5 M urea, 2% CHAPS, 2% DTT, and 1.6% Pharmalyte pH 3–10.
4. Equilibration buffer: 6 M urea, 30% (v/v) glycerol, 2% (w/v) SDS, 0.05 M Tris–HCl, pH 8.8, and 0.01% (w/v) bromophenol blue; aliquot and store at –20°C.
5. 1% (w/v) DTT, 4.8% (w/v) iodoacetamide (*see* **Note 7**).

2.6. 2-D Electrophoresis (2-DE)

1. 2-D gel system: BioRad Protean Plus Dodeca Cell (Bio-Rad laboratories).
2. Protogel 30% (w/v) acrylamide/methylene bisacrylamide solution (37:5:1 ratio) (National Diagnostics, Atlanta, GA, USA) (*see* **Note 8**).
3. Protogel resolving buffer (4×): 0.375 M Tris–HCL and 0.1% SDS, pH 8.8 (National Diagnostics, Atlanta, GA, USA).
4. Ammonium persulphate (APS): 10% solution in ddH_2O.
5. *N,N,N,N'*-Tetramethyl-ethylenediamine (TEMED) (Sigma-Aldrich).
6. 2-D running buffer: 25 mM Tris, 192 mM glycine, and 0.1% SDS.
7. Water saturated isobutanol: Add equal volumes of isobutanol and ddH_2O to a glass bottle and shake; use upper phase. Store at RT.
8. Agraose sealing solution: 1% agarose powder dissolved in 2-D running buffer.

2.7. 2-D Western Blotting.

1. Gel equilibration buffer: 20 mM Tris base, 150 mM glycine, pH 8.3.
2. Hybond nitrocellulose membranes (GE Healthcare) (*see* **Note 11**).
3. Transblot semidry transfer cell (Bio-Rad laboratories).
4. For primary, secondary, membrane blocking materials, and storage material *see* **Subheading 2.4**, **items 4–8**.

2.8. 1-D and 2-D Typhoon Scanning

1. Scanner: Typhoon variable mode imager 9400 (GE Healthcare).
2. Scanner settings:
   (a) Acquisition mode: Fluorescence.
   (b) Laser: Cy3 580 Bp 30, Cy5 670 BP 30.

(c) Pixel size: 100 μm.

(d) Focal plane: Gel, +3 mm and press sample, membrane, platen, and do not press.

(e) Sample.

(f) DIGE file naming format.

(g) Image quality control: ImageQuant tools.

3. Methods

3.1. Tissue Sample Preparation and Lysis

1. Grind human left ventricular heart tissue (0.2 g) to a fine powder with the aid of liquid nitrogen and a pestle and mortar (*see* **Note 1**).
2. Immediately add ground tissue (0.25 g) to DIGE-compatible lysis buffer (1 mL) and incubate at RT for 1 h.
3. Sonicate the protein samples on ice for 3 × 10 s bursts. Following centrifugation at 14,000 × g_{av} for 30 min, collect the protein fractions, aliquot, and store at −80°C until further use.
4. The Bradford dye binding assay is used to determine the protein concentration of the samples *(7)*.

3.2. Minimal CyDye Labeling

1. Minimal CyDye labeling (GE Healthcare) with both Cy 3 and Cy 5 is performed at a concentration of 25 μg of protein/200 pmol of CyDye for 1-DE. Because of the separation capabilities of 2-DE, a minimum protein loading of 150 μg/1,000 ppmol of CyDye is required for 2-DE analysis. However, increased protein loadings of 300 and 600 μg were also labeled at the same ratio (*see* **Note 14**).
2. Remove the CyDyes from −20°C and allow to warm to RT for 5 min. Reconstitute the CyDye in the specified volume of DMF as supplied with the dye by pipetting the DMF down the side of the tube and vortex briefly to ensure total resuspension (*see* **Note 4**).
3. Spin for 30 s at 12,000 × *g*. CyDye stock solutions are stable for up to 3 months at −80°C.
4. Add 1 μL of stock solution CyDye to 4 μL of DMF to make a 200 pmol working solution. This is stable at −20°C for 1 week. Protein to dye ratio is both user- and sample specific and thus should be optimized for each experiment.
5. Check pH of the protein sample to be labeled by spotting 1 μL of sample onto a pH indicator strip. Ensure that the sample pH is 8.5 prior to labeling (*see* **Note 10**).

6. Add the equivalent protein sample required for labeling to a microfuge tube and add 1 μL of working Cy3 or Cy5 dye solution. Mix by vortexing, spin briefly at 12,000 × *g*, and incubate samples on ice for 30 min in the dark.
7. The labeling reaction is terminated by the addition of 1 μL of 10 mM lysine. Spin briefly and incubate the samples for further 10 min in the dark.
8. Labeled protein fractions can be either utilized immediately or otherwise stored at −20°C.

3.3. 1D SDS PAGE

1. For analyses of protein samples by 1-D SDS-PAGE, samples are combined with an equal volume of Laemmli buffer. One-dimensional SDS-PAGE is carried out using the Bio-Rad minigel system on 4–20% gradient gels (Pierce, IL, USA) according to Laemmli et al. *(8)* (*see* Chapter "Non-electrophoretic Bidirectional Transfer of a Single SDS-PAGE Gel with Multiple Antigens to Obtain 12 Immunoblots"). However, any gel system can be utilized.
2. Combine prelabeled protein samples with an equal volume of Laemmli sample buffer (Bio-Rad). Using the Bio-Rad mini-Protean III system, run the gel at 100 V for 1 h or until the dye front has migrated off the end.

3.4. 1-D Western Blotting

1. Transfer SDS-PAGE gels electrophoretically onto Hybond nitrocellulose membranes (GE Healthcare) according to Towbin et al. *(1)* using a Protean blotting system (Bio-Rad) at 100 V for 1 h at 4°C in 1-D transfer buffer.
2. Block the membranes for 1 h at RT in PBS containing 5% BSA and incubate overnight in monoclonal anti-Hsp 27 antibody (1:800 dilution) in PBST. Ensure that the membranes are protected from light.
3. Wash the membranes twice in PBST and then incubate in either ECL Plex goat-α-mouse IgG, Cy 3 or goat-α-rabbit IgG, Cy 5 conjugated secondary antibody (1:2,000 dilution) for 1.5 h in PBST protected from light.
4. After incubation in secondary antibodies, wash the blots twice in PBST for 10 min each, followed by further two washes in PBS while protected from light.

3.5. 2-DE: First-Dimension Isoelectric Focusing

1. Use 24 cm immobilized pH 4–7 IPG strips to perform first-dimension isoelectric focusing (IEF) separation.
2. Combine 2-DE samples (at loading concentrations of 150, 300, and 600 μg of human heart left ventricle) initially with an equal volume of 2× lysis buffer and then dilute in rehydration solution to a total volume of 450 μL. Rehydrate strips overnight at RT using an in-gel rehydration method *(9)*.

3. The following day, focus the rehydrated strips at 0.05 mA/IPG strip for 72,000 Vh at 20°C.

3.6. 2-DE: Second Dimension SDS-PAGE

1. Following IEF, equilibrate the strips in equilibration buffer with the addition of 1% (w/v) DTT for 15 min at RT. Subsequently, equilibrate the strips in the same buffer without DTT but with the addition of 4.8% (w/v) iodoacetamide for 15 min at RT *(10)*.
2. Perform the second dimension using 1-mm thick large format (26 × 21 cm) 12% SDS-PAGE gels overnight on a BioRad Protean Plus Dodeca Cell system at 1 W/gel at 15°C. The 2-DE tank should be covered to protect the fluorescent dyes from exposure to light. The gels are run until the dye front has just migrated off the lower end of the gels.

3.7. 2-DE Semidry Western Blotting

1. Immediately after 2-DE, equilibrate the gels in 250 mL/gel 2-DE transfer equilibration buffer with gentle agitation at RT for 30 min (*see* **Note 7**).
2. Soak filter paper and Hybond nitrocellulose in 2-DE transfer equilibration buffer prior to loading onto the semidry apparatus (*see* **Note 12**).
3. Wet the anode plate with deionized water and stack six sheets of soaked filter paper onto the plate. Remove any air bubbles in the filter paper by rolling a 10-mL pipette over the sheets one at a time. Place the nitrocellulose on top and remove any air bubbles.
4. Place the gel on top of the nitrocellulose and take note of the orientation. Add the remaining six sheets of filter paper and remove any air bubbles (*see* **Note 11**).
5. Wet the cathode plate with deionized water and connect the supply.
6. Transfer the separated proteins for 2 h at 0.8 mA/cm^2 of gel area.
7. Block the nitrocellulose membranes in 250-mL blocking buffer for 2 h at RT. Incubate the membranes overnight in a minimum of 50 mL of monoclonal anti-Hsp 27 antibody (1:800 dilution) in PBST, ensuring that the membranes are protected from light at all times (*see* **Note 5**).
8. Wash the membranes twice in 250 μL of PBST and then incubate in either 50 mL of ECL Plex goat-α-mouse IgG, Cy 3 or goat-α-rabbit IgG, Cy 5 conjugated secondary antibody (1:2,000 dilution) for 1.5 h in PBST, again protected from light (*see* **Note 9**).

9. After incubation in secondary antibodies, wash the membranes twice in 250-mL PBST for 10 min each, followed by a further two washes in PBS while protected from light.

3.8. 1-DE and 2-DE Scanning

1. Visualize all gels, membranes, and immunoblots by scanning at each stage (i.e., scan gels after 1-D or 2-D electrophoresis and membranes after transfer, to ensure transfer efficiency, and immunoblots after antibody probing) using the Typhoon variable mode imager 9400.
2. Allow the scanner to warm up for at least 30 min prior to scanning.
3. Following 1-DE or 2-DE separation, allow the gels to warm to RT. Wash the outer plates with deionized water and clean with EtOH. It is recommended that the gels are scanned while still assembled in the plates.
4. To scan membranes, place the wet membrane facedown and roll a 10-mL pipette over the membrane to remove any air bubbles.
5. The maximum pixel volume should be between 40–60 μm in order to avoid spot saturation (*see* **Note 15**).
6. For scanner settings please refer to **Subheading 2.8**.

4. Notes

1. The frozen tissue must be ground into a fine powder using the liquid nitrogen and the pestle and mortar to ensure maximum protein yield. This process can be simplified by using a large pestle and a small mortar bowl, minimizing tissue loss from the mortar during grinding, and by rapidly grinding the tissue once the added liquid nitrogen has burnt off when the frozen tissue is at its hardest.
2. DIGE-compatible lysis buffer contains no reducing agents such as dithoithreitol (DTT) used in conventional lysis buffer as these reagents interfere with CyDye labeling. Aliquots of this buffer should not be refrozen once thawed.
3. Sonication should always be carried out on ice to prevent overheating of the sample and in turn protein degradation. Protease inhibitors are added to stock DIGE-compatible lysis buffer and aliquoted accordingly.

4. DMF must be anhydrous and unopened prior to CyDye stock reconstitution and labeling. It is recommended to use a fresh bottle of DMF for every new batch of dyes.
5. CyDye labeled secondary antibodies must be stored in the dark. During incubation membranes can be wrapped in tin foil to exclude light.
6. Membranes can be stored in lint-free tissue and tin foil to exclude light.
7. DTT and iodoacetamide are added fresh to thawed aliquots of equilibration buffer. These buffers cannot be reused/refrozen.
8. Unpolymerized acrylamide is a neurotoxin and should be handled as a hazardous substance.
9. The sensitivity of the goat-anti-mouse IgG Cy3 secondary is somewhat inferior to its anti-mouse Cy5 counterpart; therefore, the Cy3 prelabeling followed by the anti-mouse Cy5 secondary combination was used in all subsequent analysis.
10. Protein sample must be adjusted to pH 8.5 to ensure optimal CyDye labeling; this can be achieved by the addition of DIGE lysis buffer.
11. The format size of the semidry blotting system may differ to that of the gel size. Gel trimming can be performed before transfer by cutting away unwanted areas of the gel while the gel is still on the glass plate; this results in a better gel-membrane fit.
12. It is necessary to separately soak the nitrocellulose and filter paper in equilibration buffer to avoid the transfer of artifacts to the membrane.
13. As seen in **Fig. 2**, total protein expression is overlaid with the immunolabelled Hsp27 protein spots. These images can also be viewed separately or simultaneously by selecting the specific fluor channel for each CyDye.
14. Scanning of 150, 300, and 600-μg gels (n = 3) using the Typhoon 9400 showed no interference between increased protein loads and calculated fluorescence as pixel volume was adjusted within a common range (40,000–60,000 pixels) for all scans.
15. Adjusting the pixel volume for each scan can be achieved by reducing/increasing the PMT setting for each wavelength.

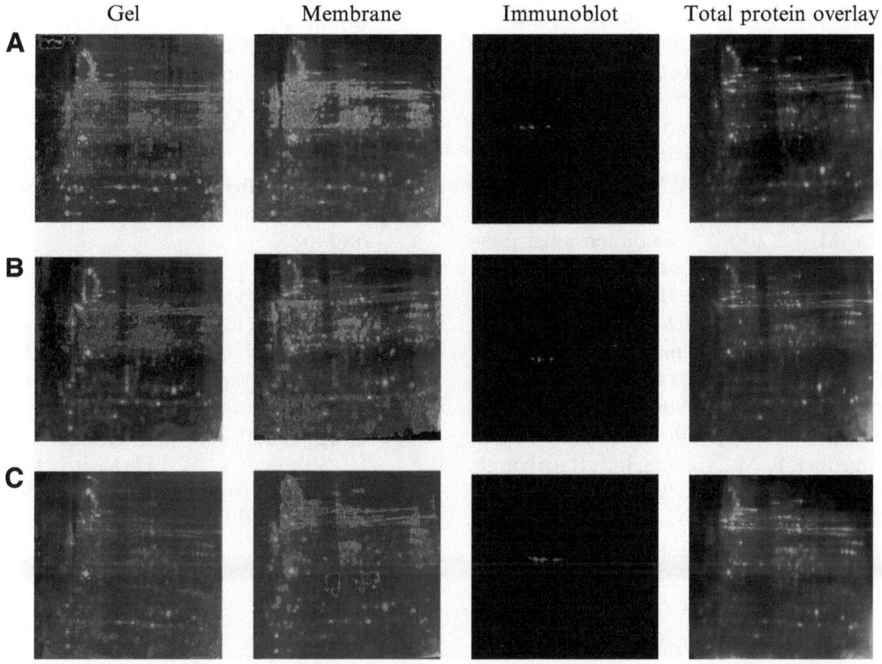

Fig. 2. ECL Plex 2-D CyDye blotting: Human left ventricle protein fractions [150 μg (**A**), 300 μg (**B**), and 600 μg (**C**)] were primarily CyDye (Cy3) labeled, focused on pH 3–10 NL IPG strips, and subjected to 2-D gel electrophoresis using 12% SDS polyacrylamide gels, column 1 (**A**, **B**, and **C**). Separated proteins were transferred onto nitrocellulose membranes, column 2. Primary incubation with monoclonal Hsp 27 antibody was followed by decoration with CyDye (Cy5) conjugated secondary antibodies, column 3. Total protein expression combined with specific Hsp 27 protein detection is overlaid and visualized at specific wavelengths for both Cy Dyes, column 4 (reproduced from *(4)* with permission from Wiley-VCH) (*see Color Plates*).

Acknowledgements

This material is based upon works supported by the Science Foundation Ireland under Grant No. 04/RP1/B499. The sample of human ventricular heart tissue was kindly provided, with ethical approval, by Prof. Marlene Rose (Imperial College, London, UK).

References

1. Towbin, H., Staehelin, T., and Gordon, J. (1979) Electrophoretic transfer of proteins from polyacrylamide gels to nitrocellulose sheets: procedure and some applications. *Proc Natl Acad Sci USA* 76, 4350–4354.
2. Gorg, A., Towbin, H., Staehelin, T., and Gordon, J. (1979) Electrophoretic transfer of proteins from polyacrylamide gels to nitrocellulose sheets: procedure and some applications. *Proc Natl Acad Sci USA* 76, 4350–4354.

2. Gorg, A., Weiss, W., and Dunn, M. J. (2004) Current two-dimensional electrophoresis technology for proteomics. *Proteomics* 4, 3665–3685.
3. Patton, W.F., in Pennington, S.R., Dunn, M.J. (Eds), (2001) *Proteomics: From Protein Sequence to Function*, BIOS Scientific, Oxford 65–86.
4. Dunn, M. J. (2007) Two-dimensional polyacrylamide gel electrophoresis for cardiovascular proteomics. *Methods Mol Biol.* 357, 3–13.
5. Donoghue, P. M., McManus, C. A., O'Donoghue, N. M., Pennington, S. R. (2006) CyDye immunoblotting for proteomics: co-detection of specific immunore active and total protein profiles. *Proteomics* 6, 6400–6404.
6. Leimgruber, R. M., Malone, J. P., Radabaugh, M. R., LaPorte, M. L., Violand, B. N., and (2002) Development of improved cell lysis, solubilization and imaging approaches for proteomic analyses. *Proteomics* 2, 135–144.
7. Bradford, M. M. (1976) A rapid and sensitive method for the quantitation of microgram quantities of protein utilizing the principle of protein-dye binding. *Anal Biochem.* 72, 248–254.
8. Laemmli, U. K. (1970) Cleavage of structural proteins during the assembly of the head of bacteriophage T4. *Nature* 227, 680–685.
9. Rabilloud, T., Valette, C., and Lawrence, J. J. (1994) Sample application by in-gel rehydration improves the resolution of two-dimensional electrophoresis with immobilized pH gradients in the first dimension. *Electrophoresis* 15, 1552–1558.
10. Gorg, A., Postel, W., Gunther, S., Weser, J., Strahler, J. R., Hanash, et-al. (1988) Approach to stationary two-dimensional pattern: influence of focusing time and immobiline/carrier ampholytes concentrations. *Electrophoresis* 9, 37–46.

Chapter 53

Visualization of Unstained Protein Bands on PVDF

Jun Park, Masaharu Mabuchi, and Ajay Sharma

Summary

In 1988, two separate investigators reported a novel method of detecting unstained protein bands on polyvinylidene fluoride (PVDF) membranes using white light transillumination. This simple method exploits the intrinsic hydrophobicity of PVDF membrane, which enables the visual observation of transferred protein bands due to differential wetting patterns between protein bands and the membrane itself. This method applies only to hydrophobic PVDF membranes, because hydrophilic membranes such as nitrocellulose wet out completely, rendering the protein bands invisible by transillumination. Transilluminational protein visualization can detect submicrogram quantities of proteins while circumventing the use of protein stains, which can potentially interfere with downstream analysis such as N-terminal sequencing. In this chapter, we demonstrate efficient transilluminational protein visualization on a recently introduced low-fluorescence PVDF membrane, normally used for downstream fluorescent immunodetection.

Key words: Transillumination, PVDF, Fluorescence

1. Introduction

Hydrophobicity of polyvinylidene fluoride (PVDF) membranes is the key contributing factor to its high protein-binding capacity and its wide use in western blotting applications. While optimal for traditional western blotting using chemiluminescent detection, conventional PVDF membranes exhibit high background in fluorescent western blot applications. A recently introduced, low-fluorescence PVDF membrane (Immobilon®-FL membrane) retains the high protein-binding capacity of the original PVDF membrane while exhibiting substantially reduced fluorescence background, making it an ideal membrane for fluorescent

Biji T. Kurien and R. Hal Scofield (eds.), *Methods in Molecular Biology, Protein Blotting and Detection, vol. 536*

DOI: 10.1007/978-1-59745-542-8_53

applications. Initial protein detection by transillumination prior to fluorescent immunodetection can streamline optimization of the immunodetection process; however, the visualization of unstained protein bands via transillumination has been previously demonstrated using only conventional PVDF membranes. We now show similar results with the new Immobilon-FL membranes.

PVDF membrane's hydrophobicity causes the membrane to appear opaque when transilluminated with white light. This opacity can change dramatically when a sufficient amount of organic solvent (such as methanol) is added to wet the membrane. In general, complete wetting of the Immobilon-FL membrane, achieved with ~50% methanol, will render the membrane translucent. Interestingly, when the membrane is coated with protein, the methanol concentration required to wet the membrane is reduced. This reduction occurs because adsorbed proteins locally decrease the membrane's inherent hydrophobicity, increasing its propensity for wetting. Electroblotted, dried PVDF membrane can be placed in 20–40% methanol (which, under normal circumstances, will not wet out the membrane), and selective wetting will occur in areas corresponding to protein bands. Transillumination then reveals unstained hydrophilic protein bands, which appear translucent against the opaque background of the hydrophobic membrane *(1, 2)* (**Fig. 1a**). By this method, even submicrogram quantities of proteins are visible *(1)*.

The transilluminational protein visualization method offers several advantages over other detection methods. First, the method is simple and rapid. Wetting of the protein bands occurs almost instantaneously in the appropriate amount of methanol. No special dyes or washing buffers are required, and the overall quality of electroblotting can be assessed rapidly. Furthermore, unlike methods using protein dyes, there is no danger of "overstaining" with transillumination because prolonged incubation with the proper concentration of methanol does not increase background "staining." Third, transillumination can be performed as many times as needed without the loss of detection sensitivity, since no membrane washing is required. After visualization, blots can be simply stored dry until the actual immunodetection or N-terminal sequencing. Finally, the method does not interfere with downstream applications. The visualized bands can be recorded using pin marks, and blots can be subsequently analyzed for specific proteins by N-terminal sequencing or immunodetection.

The sensitivity of this method critically depends on the concentration of the wetting reagent. For example, when methanol concentration is too low, protein bands will not wet out optimally, reducing the sensitivity and the overall quality of the results (**Fig. 1a**). Conversely, a higher-than-optimal concentration of methanol will wet out the membrane, and differential

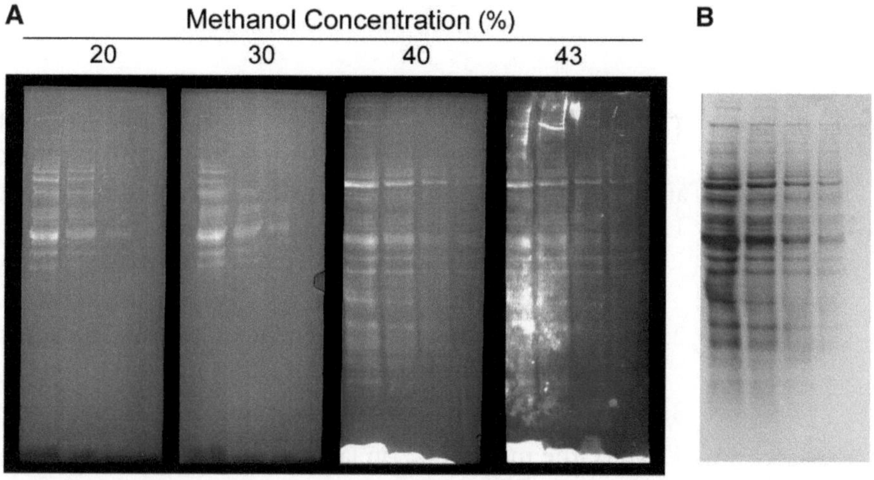

Fig. 1. Visualization of unstained protein bands using transillumination. (**A**) Shown are transilluminated Immobilon-FL membranes treated with the indicated concentration of methanol. Serially diluted total A431 cellular lysates (20, 10, 5, and 2.5-μg total protein per lane, from left to right) were separated electrophoretically and electroblotted onto Immobilon-FL membranes. Membranes were stored dry at room temperature and, on the day of analysis, treated with the indicated concentration of methanol. The boundaries of the membranes were layered with black plastic sheets to clearly demarcate the membranes. Protein bands appear translucent against the opaque background of the membrane due to the differential wetting. Notice the effect of methanol concentration on the visualization sensitivity. With increasing methanol concentration, visualization becomes more sensitive. When too much methanol (43%) is used, certain areas of membrane become completely translucent (white areas) and the visualization of protein bands is compromised as differential wetting patterns are masked. (**B**) The same FL membrane shown in panel (**A**) (40% methanol) was stained with Coomassie Blue. Similar detection sensitivity can be seen between transillumination and Coomassie stains.

wetting patterns will be lost. In general, 20–30% methanol is recommended, since, at this concentration range, methanol is unable to wet out the membrane. However, if visualization requires higher sensitivity, the membrane can be treated with gradually increasing concentrations of methanol (in 3% increments) up to 40%. In the optimal methanol concentration, it is possible to achieve visualization sensitivity similar to that of Coomassie stain (**Fig. 1b**) *(1)*.

2. Materials

1. Methanol 20–40% (v/v).
2. Low-fluorescence PVDF membrane (Immobilon-FL membrane, Millipore Corp, MA, USA).
3. Shallow tray large enough to hold membrane.
4. White light box.
5. Black paper.

3. Methods *(3)*

3.1. Membrane Drying Methods

The membrane must be completely dried before the transillumination procedure. Rinse the blotted membrane with Milli-Q® water and dry it using one of the following methods (*see* **Note 1**).

1. Soak membrane in 100% methanol for 10 s. Remove from methanol and place on piece of filter paper until dry (approximately 15 min).
2. Place the membrane between two sheets of filter paper and place in a vacuum chamber for 30 min.
3. Place the membrane on filter paper and incubate at 37°C for 1 h.
4. Place the membrane on lab bench and let it dry at room temperature (RT) for 2 h.

3.2. Visualization by Transillumination

1. Dry the blotted membrane completely as described earlier in **Subheading 3.1** and place it in a shallow tray.
2. Fill the tray with enough methanol (20–40%) to cover the membrane.
3. Incubate at RT for 5 min (*see* **Note 2**).
4. Place the wetted membrane on a light box and mask the areas around the membrane with a sheet of black paper.
5. The protein bands appear as clear areas against an opaque background.
6. For a permanent record, photograph the wet membrane.
7. Once the entire membrane has dried, it will revert to its original appearance and differential wetting patterns will disappear.

4. Notes

1. It is important to have a completely dry membrane. Overdrying will not interfere with the visualization.
2. Longer incubation does not interfere with differential wetting patterns. Protein bands can be visualized any time following differential wetting without the risk of overstaining.

Acknowledgement

The authors would like to thank Dr. Elena Chernokalskaya for her advice and critically reviewing the manuscript.

References

1. Reig, J.A. and Klein, D.C. (1988) Submicrogram quantities of unstained proteins are visualized on polyvinylidene difluoride membranes by transillumination. *Appl Theor Electrophor* 1, 59–60.
2. Walsh, M.J., McDougall, J., and Wittmann-Liebold, B. (1988) Extended N-terminal sequencing of proteins of archaebacterial ribosomes blotted from two-dimensional gels onto glass fiber and poly (vinylidene difluoride) membrane. *Biochemistry* 27, 6867–6876.

Chapter 54

Improved Protein Detection Using Cold Microwave Technology

Aaron G. Smith, Jyothi Jayaram, Carol B. Johnson, E. Ann Ellis, Stanislav Vitha, Ellen W. Collisson, and Andreas Holzenburg

Summary

Protein screening/detection is an essential tool in many laboratories. Owing to the relatively large time investments that are required by standard protocols, the development of methods with higher throughput while maintaining an at least comparable signal-to-noise ratio would be highly beneficial to many researchers. This chapter describes how cold microwave technology can be used to enhance the rate of molecular interactions and provides protocols for dot blots, western blots, and ELISA procedures permitting a completion of all incubation steps (blocking and antibody steps) within 45 min.

Key words: Cold microwave technology, Microwave, Protein detection, Dot blot, Western blot, ELISA

1. Introduction

Protein screening is a time-intensive, yet indispensable technique required on a routine basis in many laboratories. Conventional protein detection and blotting protocols call for extensive incubation times *(1)* in order to optimize antigen–antibody interactions in conjunction with appropriate visualization tools (**Table 1**). Most of these interactions can be significantly improved through the use of microwave technology. This technology greatly enhances the rate of antibody–antigen interaction, thereby significantly reducing incubation times. Microwave-assisted (MWA) research had its origins in histological studies over 30 years ago *(2)*, but

Biji T. Kurien and R. Hal Scofield (eds.), *Methods in Molecular Biology, Protein Blotting and Detection, vol. 536*

DOI: 10.1007/978-1-59745-542-8_54

Table 1
Comparison of the conventional and the MWA dot and western blot protocols

	Conventional	MWA
Blocking	Overnight at 4°C (or 2 h at 21°C)	6-min cycle
Primary antibody	Overnight at 4°C	6-min cycle
Wash (TBS + 2% Tween)	4 Times for 10 min each (21°C)	4 Times for 1 min each
Secondary antibody	1 h at 21°C	6-min cycle
Wash (TBS + 2% Tween)	2 Times for 10 min each (21°C)	2 Times at 1 min each
Wash (TBS)	2 Times for 10 min each (21°C)	2 Times at 1 min each

The 6-min cycle consists of 2 min on, 2 min off, and 2 min on at 37°C and a calibrated power of 220 W. Wash steps are at 37°C and a calibrated power of 220 W

remained enigmatic until a thorough examination into the variables allowed for improved sample reproducibility *(3)*. Since that time, a wealth of diverse protocols have been developed using MWA such as decalcification of bone *(4)*, resin polymerization for electron microscopy *(5)*, and antibody labeling of fixed biological tissue *(6)*.

However, previous western blotting and ELISA MWA studies *(7–9)* were compromised by the use of a heterogeneous microwave field and uncalibrated power settings. In essence, these conventional MWA protocols were based on using a microwave apparatus that exerted uncontrolled heat radiation in addition to microwave radiation (**Table 2**). It is thought that pure microwave radiation is different from microwave heating *(10–12)*. While the latter may cause a linear acceleration of molecules, the former could be thought of as introducing high-frequency molecular inversions (flipping). To eliminate or control the heat component, Ted Pella, Inc., developed the ColdSpot® technology *(13)*, which allows for meticulous control of sample temperature during MWA applications. The MWA protocols described in this chapter are therefore referred to as using cold-microwave technology. Protocols assisted by cold-microwave technology employ identical chemistry when compared with conventional techniques and may therefore be adapted for a wide range of other screening/visualization applications.

The most dramatic effect incurred by using cold-microwave technology is significant timesaving, cutting blot development or ELISA processing times down from ≥24 h to 45 min.

Table 2
Comparison of the conventional and MWA ELISA protocols

	Conventional	MWA
Antibody	RT for 2 h	6-min cycle
Blocking	Overnight at 4°C	6-min cycle
Wash	Twice at RT	1-min cycle
Antigen	2 h at RT	6-min cycle
Wash	Four times	1-min cycle
Biotinylated antibody	30 min at RT	6-min cycle
Wash	Thrice	1-min cycle
Streptavidin–HRPO conjugate	30 min	6-min cycle
Wash	Five times	1-min cycle
Substrate	20 min at RT	20 min at RT

The 6-min cycle consists of 2 min on, 2 min off, and 2 min on at 30°C and a wattage setting of 250 W. Wash steps are done at the same microwave settings for 1 min.

2. Materials

2.1. SDS-Polyacrylamide Gel Electrophoresis (SDS-PAGE) and Electrophoretic Transfer

1. Mini-PROTEAN 3 system (Bio-Rad, Hercules, CA, USA) used for protein separation.
2. 0.2-μm Nitrocellulose (Bio-Rad, Hercules, CA, USA) was used for electrophoretic transfer at 1 A for 50 min in 25 mM Tris–glycine/20% methanol (pH 8.3) using a Genie blotter (IdeaScientific, Minneapolis, MN, USA).

2.2. MWA Dot Blotting Protein Detection

1. PELCO BioWave® Pro Laboratory Tissue Processing System equipped with PELCO ColdSpot® Pro (Ted Pella, Inc., Redding, CA) to increase rate of antibody–antigen interactions (**Fig. 1**).
2. PELCO SteadyTemp™ (Ted Pella, Redding, CA) load cooler for maintaining the temperature of the Coldspot®.
3. Polypropylene Petri dishes (50-mm diameter × 12-mm deep (Ted Pella, Redding, CA, USA).
4. 10× Tris-buffered saline (TBS): 200 mM Tris, pH 7.6 and 1.37 M NaCl. Dilute in water to make 1× TBS.
5. Blocking buffer: 0.2% Tween-20 and 2% cold water fish gelatin (Sigma, St. Louis, MO, USA) diluted in 1× TBS.

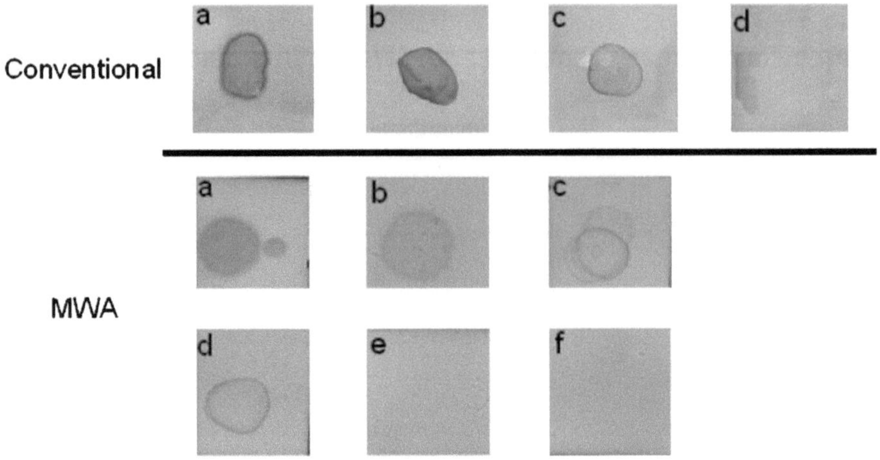

Fig. 1. Conventional and MWA dot blots in comparison. Samples from a conventional dot blot (*top row*) show whole cell lysate (**a**), peak fractions containing different amounts of the assayed protein (**b**, **c**), and a late fraction (after peak elution) as a negative control (**d**). Samples from the MWA dot blot (*middle and bottom rows*) show whole cell lysate run at 37°C (**a**) and 21°C (**b**), peak fractions containing different amounts of the assayed protein (**c**, **d**), and two negative controls, that is, a late fraction (after peak elution) (**e**) and a nonreactive protein (**f**). This figure appeared in *(12)*, Copyright © 2008 by Elsevier, Inc.

6. TBST (TBS-Tween-20) buffer: 1× TBS containing 0.2% Tween-20.
7. Next follows a suitable visualization reaction, e.g., the alkaline phosphatase reaction. Reagents for this reaction are as follows:
   (a) 10× chromogenic reaction buffer: 1 M NaCl, 50 mM $MgCl_2$, 1 M Tris–HCl, pH 8.8, store at RT, good for ~6 months.
   (b) NBT solution (3.75% w/v nitro blue tetrazolium in 100% methanol) (Research Products International, Mount Prospect, IL, USA): Store at –20°C and protect from light.
   (c) BCIP solution (5% w/v 5-Br-4-Cl-3-indolyl-phosphate in water) (Research Products International): Store at –20°C and protect from light.
9. Primary and secondary incubation steps using, e.g., anti c-myc as primary and alkaline phosphatase goat-anti mouse as secondary (Invitrogen, Carlsbad, CA, USA).

2.3. MWA Western Blotting Protein Detection

1. PELCO BioWave® Pro Laboratory Tissue Processing System equipped with PELCO ColdSpot® Pro to increase rate of antibody–antigen interactions.
2. PELCO SteadyTemp™ load cooler for maintaining the temperature of the ColdSpot.
3. Miniblotting containers (9.0 cm × 6.4 cm × 2.1 cm, Research Products International).

4. 10× TBS: 200 mM Tris, pH 7.6 and 1.37 M NaCl. Dilute in water to make 1× TBS.
5. Blocking buffer: 0.2% Tween-20 and 2% cold water fish gelatin diluted in 1× TBS.
6. TBST (TBS-Tween-20) wash buffer: 1× TBS and 0.2% Tween-20.
7. Next follows a suitable visualization reaction, such as, e.g., the alkaline phosphatase reaction. Reagents for this reaction are as follows:
   (a) 10× chromogenic reaction buffer: 1 M NaCl, 50 mM $MgCl_2$, 1 M Tris–HCl, pH 8.8, store at room temperature (RT), good for ~6 months.
   (b) NBT solution: 3.75% w/v.
   (c) BCIP solution: 5% w/v.
8. Primary and secondary incubation steps using, e.g., anti c-myc as primary and alkaline phosphatase goat-anti mouse as secondary.
9. Prestained molecular weight markers: Precision Plus protein standards (Bio-Rad).
10. Denton DV-502 vacuum pump stand (Denton Vacuum USA, Moorestown, NJ, USA) for glow discharging.

2.4. MWA ELISA

1. PELCO BioWave® Pro Laboratory Tissue Processing System equipped with PELCO ColdSpot® Pro to increase rate of antibody–antigen interactions.
2. PELCO SteadyTemp™ load cooler for maintaining the temperature of the ColdSpot®.
3. Immulon 2HB strips (Thermolab systems, Franklin, MA, USA) that can each support 12 wells.
4. 10× phosphate buffered saline (PBS): 56 mM Na_2HPO_4, 10.6 mM KH_2PO_4 and 1.54 M NaCl. Dilute in water to make 1× PBS, pH 7.2.
5. PBST (PBS-Tween-20) wash buffer: 1× PBS and 0.05% Tween-20.
6. Bovine serum albumin.
7. Triton X-100.
8. Dulbecco's Modified Eagle's Medium (DMEM) and fetal bovine serum (FBS) for antibody dilution (Gibco/BRL, Gaithersburg, MD, USA).
9. Detection reagents used: Primary FIV-2D4 monoclonal antibody *(14)* to coat the ELISA plates and capture the antigen. Biotinylated monoclonal anti-mouse secondary antibody-FIV-4F2. Hybridomas for the monoclonal antibodies were

obtained from American Type Culture Collection. Streptavidin–horseradish peroxidase conjugate for final detection (KPL Protein Research Products, Gaithersburg, MD).

10. 3,3′,5,5′-Tetramethylbenzidine (TMB) (KPL Protein Research Products, Gaithersburg, MD, USA) as a substrate for horseradish peroxidase.
11. TMB stop solution (KPL).

3. Methods

The protocols described here were developed using a readily available protein as antigen and alkaline phosphatase (dot and western blots) or horseradish peroxidase (ELISA) in conjunction with appropriate primary and secondary antibodies (*see* **Subheading 2**) for visualization. The protocols were used on both epitope tag-specific and sequence-specific antibodies. The results were comparable in both cases. This protocol is open for variations in terms of the utilization of other visualization reagents.

3.1. SDS-PAGE and Electrophoretic Transfer

1. SDS-PAGE was carried out according to Laemmli *(15)* (*see* Chapter "Non-electrophoretic bi-directional transfer of a single SDS-PAGE gel with multiple antigens to obtain twelve immunoblots").
2. Electrophoretic transfer was carried out according to Towbin et al. *(16)* (*see* Chapter "Detection of La/SS-B by Western Blot Using Nanogold-Tagged Antibodies and Silver Enhancement").

3.2. MWA Dot Blotting Protein Detection

1. Cut a piece of membrane suitable for use and place in a polypropylene Petri dish.
2. Spot the membrane with 2–5 μL of protein to be assayed (~1 mg/mL) and allow 15–20 min for sample binding.
3. Place the membrane in Petri dish and cover with 1–2 mL of blocking buffer.
4. Place the dish into microwave for a 6-min cycle (2 min on, 2 min off, 2 min on). The microwave was set to 250 W corresponding to a calibrated output wattage of 220 W and the ColdSpot® was set to 37 °C. These settings were used for all MWA steps (*see* **Notes 1** and **2**).
5. Pour off blocking buffer and immediately replace with primary antibody solution (antibody diluted in blocking buffer) (*see* **Note 3**). Incubate for a 6-min cycle (2 min on, 2 min off, 2 min on) in the microwave.

6. Pour off primary antibody solution and wash four times with TBST for 1 min each in the microwave.
7. Pour off TBST and immediately replace with secondary antibody solution (antibody diluted blocking buffer) for a 6-min cycle (2 min on, 2 min off, 2 min on).
8. Pour off secondary antibody solution and wash two times with TBST for 1 min each in microwave.
9. Pour off TBST and wash two times with 1× TBS for 1 min each (microwave).
10. Develop the blot as you would for conventional protocol (*see* **Note 4**).

3.3. MWA Western Blotting Protein Detection

1. Glow-discharge (*see* **Notes 5** and **6**) a miniblotting container.
2. After electrophoretic transfer, place membrane in the glow-discharged container and cover with 5 mL of blocking buffer (*see* **Notes 7** and **8**).
3. Place the container into microwave for a 6-min cycle (2 min on, 2 min off, 2 min on). The microwave was set to 250 W corresponding to a calibrated output wattage of 220 W, and the ColdSpot® was set to 37 °C. These settings were used for all MWA steps (*see* **Notes 1** and **2**).
4. Pour off the blocking buffer and immediately replace with primary antibody solution (antibody and blocking buffer). Incubate for a 6-min cycle (2 min on, 2 min off, 2 min on) in the microwave.
5. Pour off primary antibody solution and wash four times with TBST for 1 min each in the microwave.
6. Pour off TBST and immediately replace with secondary antibody solution (antibody and blocking buffer). Incubate for a 6-min cycle (2 min on, 2 min off, 2 min on) in the microwave.
7. Pour off secondary antibody solution and wash two times with TBST for 1 min each in the microwave.
8. Pour off TBST and wash two times with 1× TBS for 1 min each in the microwave.
9. Develop the blot as you would for conventional protocol.

3.4. MWA ELISA

1. Coat Immulon 2HB ELISA strips with 50-μL mouse monoclonal anti-FIV antibody (FTLV-2D4 MAb) at a concentration of 20 μg/mL (made in PBS pH 7.2) (*see* **Note 9**).
2. Add 300 μL of a 3% (w/v) BSA in PBS, pH 7.2 to the plates and block in the microwave for a 6-min cycle (2 min on, 2 min off, 2 min on). The microwave was set to 250 W and 30°C, and these settings were used for all subsequent MWA steps.

3. Wash the plate twice for 1 min with PBS containing 0.05% (v/v) Tween-20 (PBST).
4. Add 10 μL of Triton-X 100 per well and subsequently add 100 μL of various dilutions of the protein solution diluted in growth buffer to the wells.
5. Wash the plate four times for 1 min with PBST.
6. Add 100 μL of primary antibody diluted in PBS containing 3% FBS and incubate in microwave for a 6-min cycle (2 min on, 2 min off, 2 min on).
7. Wash the plate three times with PBST.
8. Add 100 μL of a 1:1,000 dilution of secondary antibody diluted in PBS containing 3% FBS and incubate in microwave for a 6-min cycle (2 min on, 2 min off, 2 min on).
9. Wash the plate five times for 1 min with PBST.
10. Incubate for 20 min with TMB solution.
11. Terminate the reaction with TMB stop solution after incubating the reaction for 20 min at RT (*see* **Note 10**).
12. Read O.D. at 405 nm.

4. Notes

1. MWA steps are carried out without a lid on the Petri dish or blotting container.
2. Calibrated output wattage settings are obtained as follows: Place a glass beaker containing exactly 1 L of water inside the cavity while the ColdSpot® is disconnected and incubate for 2 min at a nominal power setting of 250 W while monitoring the temperature using a digital temperature gauge. The temperature difference prior to and after the incubation (ΔT) multiplied by 35 gives the true output wattage.
3. In our experience, an approximately twofold higher antibody concentration was necessary to attain signal intensities similar to those obtained with conventional methods. For example, a 1:2,500 dilution was used for MWA blots vs. a 1:5,000 dilution used in the conventional protocol.
4. Blot development was not carried out using the microwave in order to prevent overdevelopment and introduction of unnecessary background. However, an approximately twofold background reduction is observed when conventional development is combined with all other steps carried out in the MWA mode.

5. Glow-discharging (in air) for 10 s at reduced atmospheric pressure using, e.g., the vacuum pump stand Denton DV-502 (Denton Vacuum USA) renders the surface of the containers hydrophilic and allows the use of smaller reagent volumes. Volumes used in this protocol are half of those used for non-glow-discharged containers. Containers that have been treated in such a manner should be used within 20 min as they are extremely susceptible to air-borne contaminants.
6. There are several methods of rendering surfaces more hydrophilic. However, glow-discharging in air ensures that a desired degree of hydrophilicity is achieved without introducing any additional chemicals. Since glow-discharging takes place at low vacuum levels requiring only, e.g., a rotary pump, an inexpensive home-built system could be a worthwhile investment and be substantially cheaper than buying a commercial system *(17)*.
7. Because of the short incubation times employed in this protocol, the prestained protein standards are retained on the western blot membrane (**Fig. 2**). This needs to be viewed in contrast to conventional blots where the standards are lost.
8. Membrane must remain submerged during MWA steps, or rampant background is observed during blot development.

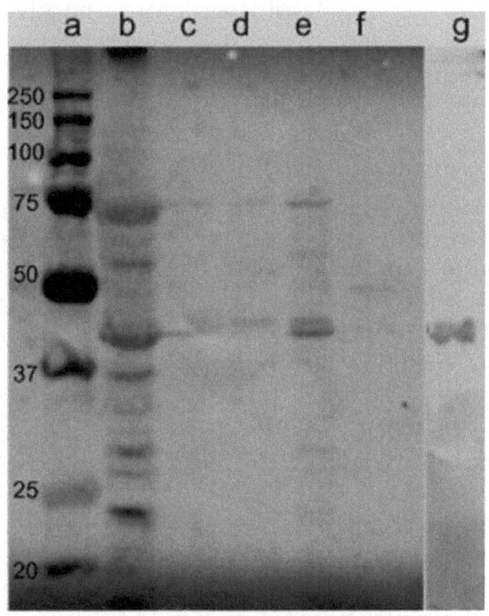

Fig. 2. MWA western blot. *Lanes* a–g have been individually loaded as follows: (**a**) prestained protein standards (highlighted in kDa), (**b**) whole cell lysate, (**c**) late fraction (after peak elution), (**d**, **e**) peak fractions containing different amounts of the assayed protein, (**f**) nonreactive protein as negative control (with a mass of 12.5-kDa), (**g**) conventional western blot (positive control). This figure appeared in *(12)*, Copyright © 2008 Elsevier, Inc.

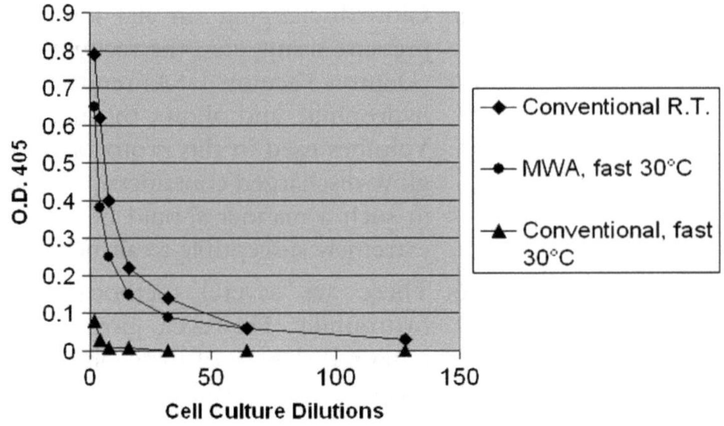

Fig. 3. Comparison of signal intensities for conventional ELISA (*diamonds*), MWA ELISA (*circles*), and a conventional protocol using times from MWA (*triangles*). Results presented here are after background subtraction from negative control (DMEM without antigen). The values presented here are the averages of three independent readings.

9. The Immulon 2HB strips gave us a graded decrease in signal with increasing dilutions of the antigen (Fig. 3). Similar results were not obtained when the ELISA was performed with 96-well microtiter ELISA plates from Sumitomo Bakelite Co. Ltd.
10. The stop solution was added after 20-min incubation of the enzyme with the substrate. As a general rule, add the stop solution immediately after a color reaction develops in the negative control.

References

1. Finney, M. (1998) Nonradioactive methods for visualization of protein blots. In: Pound J.D., ed. *Immunochemical Protocols*, Humana Press, Totowa, NJ, 207–216.
2. Mayers, C.P. (1970) Histological fixation by microwave heating. *J Clin Pathol.* 23, 273–275.
3. Giberson, R.T. and Demaree, R.S., Jr. (1995) Microwave fixation: Understanding the variables to achieve rapid reproducible results. *Microsc Res Tech.* 32, 246–254.
4. Madden, V.J. and Henson, M.M. (1997) Rapid decalcification of temporal bones with preservation of ultrastructure. *Hear Res.* 111, 76–84.
5. Demaree, R.S., Jr., Giberson, R.T., and Smith, R.L. (1995) Routine microwave polymerization of resins for transmission electron microscopy. *Scanning* 17 (Suppl 5), 25–26.
6. Chicoine, L. and Webster, P. (1998) The effect of microwave irradiation on antibody labeling efficiency when applied to ultrathin cryosections through fixed biological material. *Microsc Res Tech.* 42, 24–32.
7. Van Dorp, R., Kok, P.G., Marani, E., Boon, M.E., and Kok, L.P. (1991) ELISA incubation time can be reduced by 2.45-GJz microwaves. *J Clin Lab Immunol.* 34, 87–96.
8. Li, W., Murai, Y., Okada, E., Matsui, K., Hayashi, S., Horie, M., and Takano, Y. (2002) Modified and simplified western blotting protocol: Use of intermittent microwave irradiation (IMWI) and 5% skim milk to improve binding specificity. *Pathol Int.* 52, 234–238.
9. Toyokuni, S., Kawaguchi, W., Akatsuka, S., Makoto, M., and Hiai, H. (2003) Intermittent microwave irradiation facilitates antigen–antibody reaction in western blot analysis. *Pathol Int.* 53, 259–261.
10. Galvez, J.J., Giberson, R.T., and Cardiff, R.D. (2004) Microwave mechanisms – the energy/heat dichotomy. *Microsc Today* 12, 18–23.

11. Sanders, M.A., Anderson, T.E., and Giberson, R. (2006) Microwave methods – evidence to support a microwave effect. *Microsc Microanal Proc.* 12 (Suppl 2), 296 CD.
12. Smith, A.G., Johnson, C.B., Ellis, E.A., Vitha, S., and Holzenburg, A. (2008) Protein screening using cold microwave technology. *Anal Biochem.* 375, 313–317.
13. Sanders, M.A. (2002) Recent advances in microwave-assisted specimen processing: Keeping it cool. *Microsc Microanal Proc.* 8 (Suppl 2), 158–159.
14. Tilton, G.K., Connor, T.P., Jr., Seymour, C.L., Lawrence, K.L., Cohen, N.D., Andersen, P.R., and Tonelli, Q.J. (1990) Immunoassay for detection of feline Immuno-deficiency virus core antigen. *J Clin Microbiol.* 29, 898–904.
15. Laemmli, U.K. (1970) Cleavage of structural proteins during the assembly of the head of bacteriophage T4. *Nature* 227, 680–685.
16. Towbin, H., Staehelin, T., and Gordon, J. (1979) Electrophoretic transfer of proteins from polyacrylamide gels to nitrocellulose sheets: Procedure and some applications. *Proc Natl Acad Sci USA* 76, 4350–4354.
17. Aebi, U. and Pollard, T.D. (1987) A glow discharge unit to render electron microscope grids and other surfaces hydrophilic. *J Electron Microsc Tech.* 7, 29–33.

Chapter 55

TLC Blot (Far-Eastern Blot) and Its Applications

Takao Taki, Tania Valdes Gonzalez, Naoko Goto-Inoue, Takahiro Hayasaka, and Mitsutoshi Setou

Summary

A simple method for transfer of lipids including phospholipids, glycolipids, and neutral lipids from a high-performance thin-layer chromatography (HPTLC) plate to a polyvinylidene difluoride (PVDF) membrane, called TLC blot (far-eastern blot), is presented. Lipids separated on a HPTLC plate are blotted quantitatively. This procedure made it possible to purify individual lipids from a blotted membrane in a short time. Binding study, immunodetection, and mass spectrometric analysis are available for PVDF membrane. Furthermore, the world of molecular species imaging is opened by a scanning analysis with a combination of TLC blot and matrix-assisted laser desorption/ionization quadrupole ion trap time-of-flight mass spectrometer (TLC-Blot/MALDI-TOF MS).

Key words: Thin-layer chromatography, TLC blot, Far-eastern blot, Glycosphingolipids, Phospholipids, PVDF membrane, Imaging mass spectrometry, MALDI-TOF MS, Immunostaining, Lipidomics

1. Introduction

Recently, much attention is focused on lipids and their metabolites. From studies on signaling and receptor-mediated transactivations, lipids in membranes are known to be engaged in the mediators or stimulators of cell behavior including cell growth, adhesion, differentiation, as well as apoptosis. Furthermore, "lipid raft," a microdomain of cell membrane obtained as detergent insoluble, is a new biological topic in biology *(1–3)*. Cholesterol and sphingolipids including gangliosides are major lipid components in the microdomain, and lipid rafts are involved in playing an important role in controlling cell signaling. From this scientific background,

Biji T. Kurien and R. Hal Scofield (eds.), *Methods in Molecular Biology, Protein Blotting and Detection, vol. 536*

DOI: 10.1007/978-1-59745-542-8_55

analysis of lipids and their metabolites become more than ever important for understanding the biological mechanism.

Thin-layer chromatography (TLC) is a classical but still quite powerful analytical method for all kinds of lipids. Lipid compositions of tissue extracts, quantitative determination, and their chemical identification can be analyzed from their mobilities in combination with chemical reactions and immunostaining on a TLC plate. Major advantages are as follows: (a) Method and equipment are simple. (b) Most of the lipids can be separated by selecting the solvent system. (c) Simultaneous visual comparison of many samples can be performed on one plate. Availability of TLC is further improved by transferring the separated lipids from the TLC plate to a plastic membrane. Difficulties in handling lipids on the silica gel-coated plate have been overcome. In 1984, Towbin et al. reported a transfer method for glycosphingolipids from a TLC plate to a nitrocellulose membrane *(4)*. However, the transfer efficiency was not good enough and the membrane was not adequate for chemical detection, because nitrocellulose membrane is unstable with acid, heat, or organic solvent treatment. We examined many kinds of plastic membranes for transferring lipids and finally, PVDF membrane was found to be the best with high capacity for binding amphiphilic molecules, and stable against heat, acid, and organic solvent.

We took advantage of this powerful separation property of TLC and developed a transfer technology for lipids from an high-performance thin-layer chromatography (HPTLC) plate to a PVDF membrane, named TLC blot (far-eastern blot) *(5)*. The transfer procedure is simple and rapid with high efficiency. Established lipid detection procedures such as chemical reagent, immunostaining, enzyme assay, and binding study are available for the blotted membrane *(5–10)*. The transferred lipids on the membrane are stable and can be eluted with small amounts of methanol resulting in purification of individual lipid components without multicolumn chromatographic processes. Furthermore, imaging analysis of lipid molecular species on the PVDF membrane with ion-trap-equipped MALDI-TOF mass spectrometer (TLC-Blot/MALDI-TOF MS) opened a new gate of the lipidomics world.

2. Materials for TLC

2.1. Equipment

1. TLC tank.
2. Spray bottle.
3. UV light (315 nm).
4. Safety glass.
5. Syringe (10 μL).
6. Dryer.

2.2. Reagents

1. HPTLC plate (10 × 10 cm, Merck).
2. Lipid sample: Total lipids can be extracted from tissue or culture cells with 20 vol. of $CHCl_3$:MeOH (2:1, v/v). For example, 100 mg of sample is mixed with 2 mL of $CHCl_3$:MeOH, 2:1 for 1 day in a cold room. Most lipids can be extracted to a $CHCl_3$-rich lower fraction. Tissue residue is removed by filtration. Lipid extract is adjusted to 2 mL by the addition of methanol. An aliquot (5–20 μL) of the lipid extract is concentrated under N_2 gas and applied to the HPTLC plate.
3. Spray reagent.

2.2.1. Primulin Reagent: For All Lipid Detection

1. Stock solution: 100 mg of primulin is dissolved in 100 mL of water.
2. Spray reagent: 1 mL of the stock solution is diluted in 100 mL of a mixture of acetone:water (4:1, v/v).
3. The HPTLC plate is sprayed with primulin reagent and the separated lipid bands are detected under UV light (315 nm) in a dark room. UV light glass should be used for protection of the eyes.

2.2.2. Molybdenum Blue Reagent: For Phospholipid Detection

1. Molybdenum three oxide 4.01 g is dissolved in 25 N sulfuric acid by heating (A solution). Molybdenum powder 180 mg is added to 50 mL of A solution and the solution is heated for 15 min. After cooling, a clear upper solution was obtained by decantation (B solution).
2. Spray reagent: The same volumes of A and B solution are mixed and the mixture is added to two volumes of water.
3. The HPTLC plate is sprayed with the molybdenum blue (Dittmer reagent). After 5–10 min, a blue color remains on the bands of phospholipids.

2.2.3. Orcinol Reagent: For Glycolipid Detection

1. Orcinol 200 mg in 11.4 mL of sulfuric acid and then diluted in 100 mL of water. The HPTLC plate is sprayed with the orcinol reagent, and then the plate is heated on a hot plate at 105°C for 5–10 min. Glycolipids are stained a purple color.

2.2.4. Resorcinol Reagent: For Detection of Sialic Acid-Containing Lipids

1. Resorcinol 200 mg is dissolved in 10 mL of water. Eighty mL of conc. HCl and 0.25 mL of 0.1 M copper sulfate are added to the resorcinol solution and finally made up to 100 mL with water.
2. The HPTLC plate is sprayed with the resorcinol reagent and covered with a glass plate. The plate covered with a glass plate is heated on a hot plate at 105°C for 5–10 min. Sialic acid-containing lipids are detected as blue-purple color bands.

2.2.5. Ninhydrin Reagent: For Amino Group-Containing Lipid Detection

1. Ninhydrin 300 mg is dissolved in 100 mL of 3% acetic acid-containing butanol.
2. The HPTLC plate is heated at 100°C for 10 min, and then the ninhydrin reagent is sprayed onto the plate. Amino group-containing lipids are stained a pink color.

2.3. Solvent Systems for Developing

2.3.1. For Neutral Lipids

1. Hexane:diethylether (70:30, by vol.).
2. Petroleum ether:ditheylether:acetic acid (70:30:1, by vol.).

2.3.2. For Ceramides

1. $CHCl_3$:MeOH (95:5, by vol.).

2.3.3. For Total Phospholipids

1. Methylacetate:*n*-propanol:$CHCl_3$:MeOH:0.25%KCl (25:25:25:10:9 by vol.).
2. $CHCl_3$:MeOH:0.2% $CaCl_2$ (65:25:4, by vol.).

2.3.4. For Gangliosides

1. $CHCl_3$:MeOH:0.2% $CaCl_2$ (60:40:9, by vol.)
2. *n*-propanol:0.25% KCl (3:1, by vol.).

2.4. Reagent and Equipment for TLC Blot (Far-Eastern Blot)

1. Blotting solvent mixture: Isopropanol: 0.2% $CaCl_2$: MeOH (40:20:7, by vol.).
2. Thermal blotter (AC-5970, ATTO Co., Ltd.).

2.5. Reagents for TLC Immunostaining

1. 0.1% Bovine serum albumin (BSA)/phosphate-buffered saline pH 7.4 (PBS).
2. 0.1% BSA/1% reconstituted powdered milk in PBS.
3. First antibody solution: Anti-sialyl α2–6paragloboside solution; 1-μg mouse IgM-type monoclonal antibody/mL 0.1% BSA/PBS.
4. Second antibody solution: Horseradish peroxidase-conjugated anti-mouse IgM antibody solution. A 500–1,000 diluted with 0.1% BSA and 1% reconstituted powdered milk in PBS is used.
5. HPTLC plate: 10 cm × 10 cm silica gel 60, 0.25-mm thickness (Merck, Darmstadt, Germany).
6. Peroxidase detection reagent: 4-Chloro-1-naphthol solution (100 μL of 3% 4-chloro-1-naphthol in ethanol in 10 mL of Tris–HCl, pH 7.5 plus 10 μL of 30% H_2O_2). This reagent should be prepared just before use.
7. PVDF membrane: Clear blot membrane -P (ATTO Co. Ltd.).
8. Teflon Membrane: PTFE membrane AC 5973 (ATTO Co Ltd.).

9. Glass fiber filter: AC 5972 (100 mm × 100 mm) (ATTO Co. Ltd.).
10. Parafilm.
11. 0.4% Polyisobutylmethacrylate (PIM solution: 2.5% polyisobutylmethacrylatein $CHCl_3$ is diluted to 0.4% with hexane).
12. 0.1% BSA and 1% reconstituted powdered milk in PBS.

2.6. Materials for Mass Spectrometry

1. AXIMA-QIT mass spectrometer (Shimadzu; Kyoto Japan).
2. 2,5-Dihydroxy benzoic acid (DHB).

3. Methods

3.1. TLC

1. Before developing the plate, the TLC tank should be saturated with developing solvent.
2. One cm from the bottom, widths of 5 mm are marked with a soft pencil (6B).
3. The lipid sample diluted in 2–5 μL $CHCl_3$:MeOH (2:1) is carefully and evenly applied on these pencil marks.
4. After applying the sample, the HPTLC plate is heated by a dryer for 1 min and then developed in a TLC tank with the preferred developing solvent.
5. When the solvent front reaches about 7–8 cm from the bottom, the plate is taken and the solvent on the plate is removed with a dryer.

Figure 1 shows typical separation profiles of total lipid extracts of culture cells and brain tissue using two different solvent systems.

3.2. TLC Blot (Far-Eastern Blot) Procedure (5)

1. The developed HPTLC plate is sprayed with primulin reagent.
2. After drying the plate, the separated lipid bands are visualized under UV light (315 nm). Use UV-cut glass for protection of the eyes.
3. The lipid bands are marked with a soft red-color pencil (so as not to damage the surface of the HPTLC plate).
4. The plate is dipped in a blotting solvent for 10 s.
5. Take the plate and cover the plate with a PVDF membrane immediately. Then, Teflon membrane and glass filter paper are mounted.
6. This assembly is transferred to a TLC thermal blotter that is heated at 180°C in advance. If this apparatus is not available, 180°C heated iron is available for blotting.

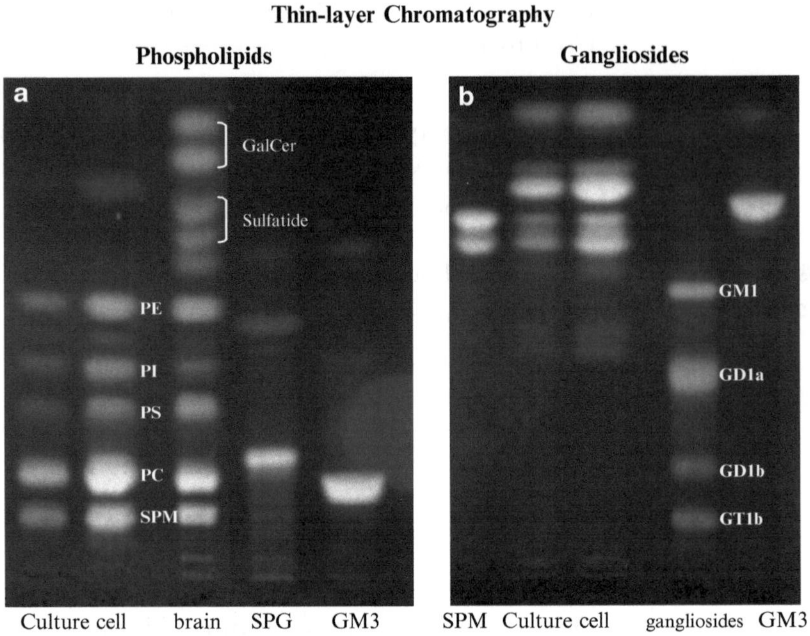

Fig. 1. Thin-layer chromatography (TLC) of phospholipids (**a**) and gangliosides (**b**). (**a**) Phospholipids were developed on the HPTLC plate with a solvent mixture of methylaetate:*n*-propanol:$CHCl_3$:MeOH: 0.25% KCl (25:25:25:10:9 by vol.). *SPG* sialylparagloboside, *GM3* sialyllactosylceramide. (**b**) Gangliosides were developed on the HPTLC plate with a solvent mixture of $CHCl_3$:MeOH:0.2% $CaCl_2$ (60:40:9, by vol.). *SPM* sphingomyelin. Primulin reagent was sprayed for detection.

7. Press the assembly for 30 s and take the PVDF membrane. Lipids separated on the HPTLC plate are transferred with a red pencil marker. You will find the lipids with the blotted pencil marks.
8. The PVDF membrane is washed with water to remove primulin reagent.
   The whole procedure is shown in **Fig.2**.

3.3. Application of TLC Blot (5)

3.3.1. Immunostaining on PVDF Membrane

Figure 3 shows immunostaining of sialylα2–6paragloboside. Various amounts of antigen glycolipid are separated on two HPTLC plates. One plate is used for immunostaining on a HPTLC plate. The other plate is used for TLC blot followed by immunostaining on PVDF membrane.

3.3.2. Immunostaining on the HPTLC Plate

1. The HPTLC plate is immersed in 0.4% of PIM solution for 30–60 s.
2. After drying the plate, the first antibody dissolved in 1% BSA/PBS is mounted by a Pasteur pipette. The antibody solution is spread on the plate with Parafilm.
3. The first antibody solution mounted on the plate is covered with Parafilm, and the plate is placed in a humidified box

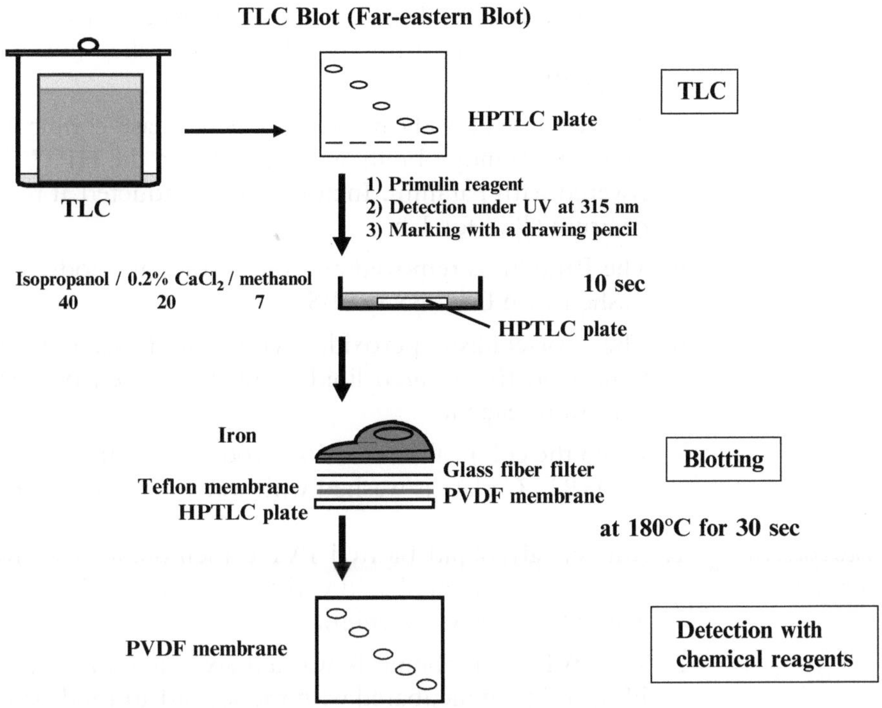

Fig. 2. Scheme of TLC blot (far-eastern blot) (reproduced from *(10)* with permission from Elsevier Ltd.).

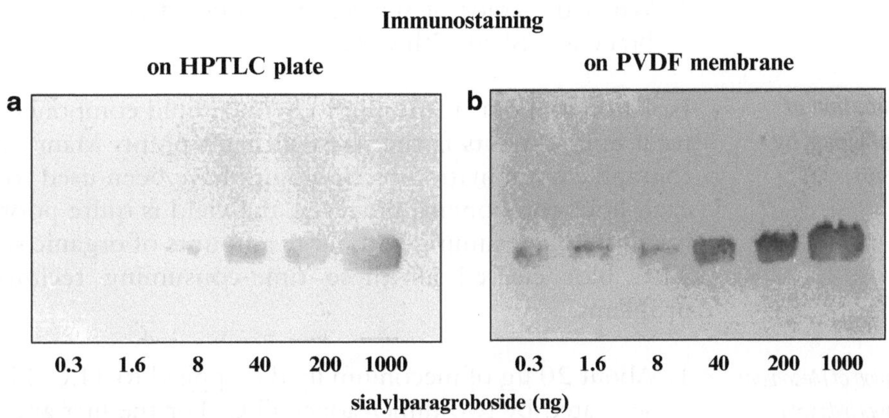

Fig. 3. Immunostaining of antigen lipid on the HPTLC plate and PVDF membrane. (**a**) Immunostaining of antigen glycosphingolipid (sialylα2–6paragloboside) on the HPTLC plate. (**b**) Immunostaining of antigen glycosphingolipid on the TLC-blotted PVDF membrane. The same amounts of antigen glycosphingolipid were developed on two HPTLC plates. One plate was used for TLC immunostaining. The other was used for TLC blot, and immunostaining was performed on the blotted PVDF membrane. The sensitivity of staining is much higher on the PVDF membrane, because the lipids are transferred to one side of the PVDF membrane, resulting in concentrating the lipids by blotting (reproduced from *(5)* with permission from Elsevier Ltd.).

with water. After leaving at 4°C overnight, the Parafilm is removed and the antibody solution is washed out with 0.1% BSA/PBS.

4. The second antibody, horseradish peroxidase-conjugated anti-mouse immunoglobulin, is mounted, and the HPTLC plate is covered with Parafilm. Incubation is conducted at room temperature (RT) for 1 h.
5. The Parafilm is removed and the second antibody solution is washed with 0.1% BSA/PBS.
6. The horseradish peroxidase-conjugated second antibody bound on the antigen lipid band is detected by peroxidase detection reagent.
7. When the color of the reaction product appears, the surface of the HPTLC plate is washed with water carefully (**Fig.3a**).

3.3.3. Immunostaining on PVDF Membrane (5)

1. Antigen glycolipid-blotted PVDF membrane is washed with water, and then incubated with the first antibody solution in a plastic bag at 4°C overnight.
2. The PVDF membrane is washed six times with 0.1% BSA/PBS, and then incubated with the second antibody solution in a plastic bag for 1 h at RT.
3. The PVDF membrane is washed six times with 0.1% BSA/PBS, and then incubated with horseradish peroxidase detection reagent (**Fig.3b**).
4. When the color of the reaction product appears, the membrane is washed with water.

3.4. Purification of Individual Lipids by TLC Blot (6)

Isolation and purification of individual lipid components from total lipid extracts is the most difficult point. Many chromatographic combination technologies have been used to isolate each lipid component; however, the yield is quite poor and it is still time consuming with large amounts of organic solvents. TLC blot cleared all these time-consuming technological problems.

3.4.1. Isolation of Meconium Glycosphingolipids with TLC Blot

1. About 20 μg of meconium lipid is applied to TLC. Lipids are separated by two-dimensional TLC. For the first and second solvent system, the same solvent system, $CHCl_3$:MeOH:0.2% $CaCl_2$ (60:40:8, by vol.), is used.
   First dimension: A lipid sample dissolved in 5 μL of $CHCl_3$:-MeOH (2:1) is applied on a HPTLC plate with a 3-mm width.
2. The plate is developed until the solvent top reaches 8 cm from the bottom. The plate is taken and dried completely, and then developed again to 8 cm.

3. After drying the plate completely, the plate is turned 90° and developed with the same solvent system as the first dimension.
4. After drying the plate, primulin reagent is sprayed onto the plate and the separated lipids are made visible under UV light (**Fig.4a**).
5. Visualized bands are marked with a soft drawing pencil.
6. The HPTLC plate is employed for TLC blot. Separated lipid bands are transferred with the pencil marks (**Fig.4b**).
7. Individual lipids marked with a pencil are separated with a cutter knife and transferred to a small glass test tube containing 50 μL of MeOH.
8. The lipid extracted with methanol from the PVDF membrane is applied to the HPTLC plate to confirm the separation using the same developing solvent system.
9. The separated lipids on the HPTLC plate are made visible by orcinol reagent. (**Fig.4c**). Twenty lipid bands separated by the HPTLC plate are successfully purified by the one-step TLC blot procedure.

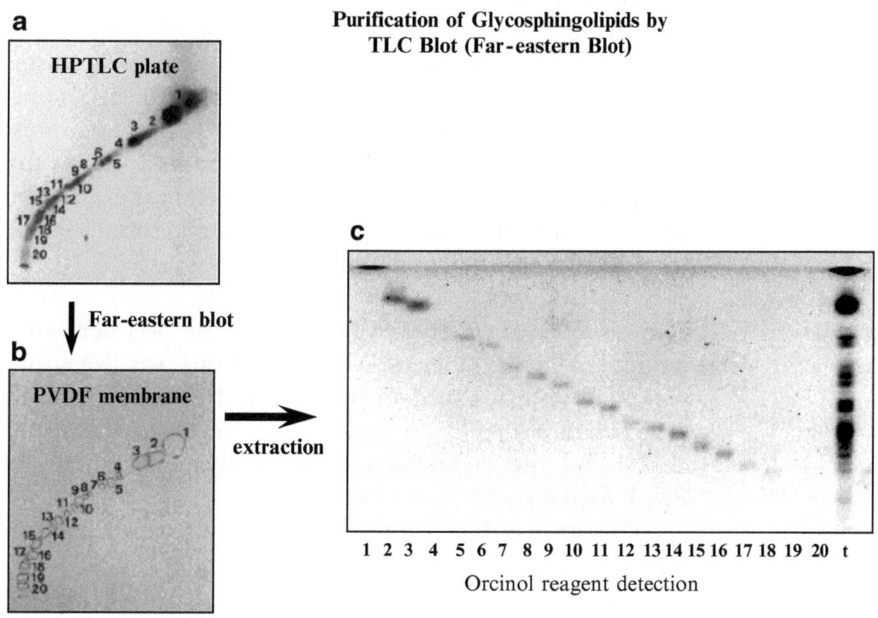

Fig. 4. Purification of glycosphingolipids by TLC blot (far-eastern blot): Glycosphingolipids of meconium (20 μg) were developed with two-dimesional TLC. (**a**) Separated lipids were made visible under UV light after spraying the plate with primulin reagent and marking with a soft drawing pencil (6B). (**b**) The separated lipids were transferred to a PVDF membrane together with pencil marks. (**c**) The marked areas were cut and used for lipid extraction with methanol. Each extract was applied to TLC and made visible by oricinol reagent (reproduced from *(6)* with permission from Elsevier Ltd.).

3.5. Application to Lipidomics with a Combination of TLC Blot and Ion-Trap-Equipped MALDI-TOF MS (TLC Blot/MALDI-TOF MS)

Lipids transferred to a PVDF membrane have been used for mass spectrometric analysis directly by secondary ionization mass spectrometric analysis *(7, 8)*. Combination with TLC-blot and MALDI-TOF MS (TLC-BLOT/MALDI-TOF MS) provides both the molecular mass and the structural information of glycosphingolipids, e.g., sugar sequences, without purification processes. Very recently, remarkable advances in MALDI-TOF MS ion trap instruments have enabled analysis with high mass accuracy and sensitivity. Taking advantage of this equipment, we can directly analyze lipids transferred on the PVDF membrane.

3.5.1. TLC Blot/MALDI-TOF MS

MALDI-TOF MS analyses are performed using an AXIMA-QIT mass spectrometer (Shimadzu; Kyoto Japan) in negative and positive ion mode.

Ionization was performed with a 337-nm pulsed N_2 laser. The ion trap chamber was supplied with two separate, independent gases: helium and argon. A continuous flow of helium gas was used for collisional cooling. Pulsed gas, argon, was used to enhance ion cooling and collision-induced fragmentation. Precursor and fragment ions obtained by collision-induced dissociation (CID) were ejected from the ion trap and analyzed by a reflectron TOF detector. The mass spectra were assembled from 100 to 300 accumulations of the profile obtained by two laser shots. 2,5-DHB solution (20–30 μL) was deposited sufficiently to lipids on the PVDF membrane. The crystallization process was accelerated under a gentle stream of cold air. Then, pressing the PVDF membrane thoroughly is an important process for achieving good mass spectra. The PVDF membrane was completely attached to a MALDI sample plate with electrified double-adhesive tape to reduce charge-up of the plate.

3.5.2. Imaging Mass Spectrometry

The raster scan on the PVDF membrane was performed automatically. The number of laser irradiations was five shots in each spot. The interval of data points was 100 μm, giving a total of 2,061 data points. Two-dimensional ion density maps were created using image reconstruction software (BioMap, Novartis, Basel, Switzerland).

Phospholipids separated on a HPTLC plate are transferred to a PVDF membrane.

The pencil-marked area is analyzed with AXIMA-QIT mass spectrometer.

Fig.5a shows a typical MS profile of phosphatidylcholine (soy bean lecithin). The distribution of fatty acid molecular species of phosphatidylcholine on the PVDF membrane is scanned and exhibited as a fingerprinting profile (**Fig.5b**).

Figure 6 shows the TLC blot (far-eastern blot) and its application.

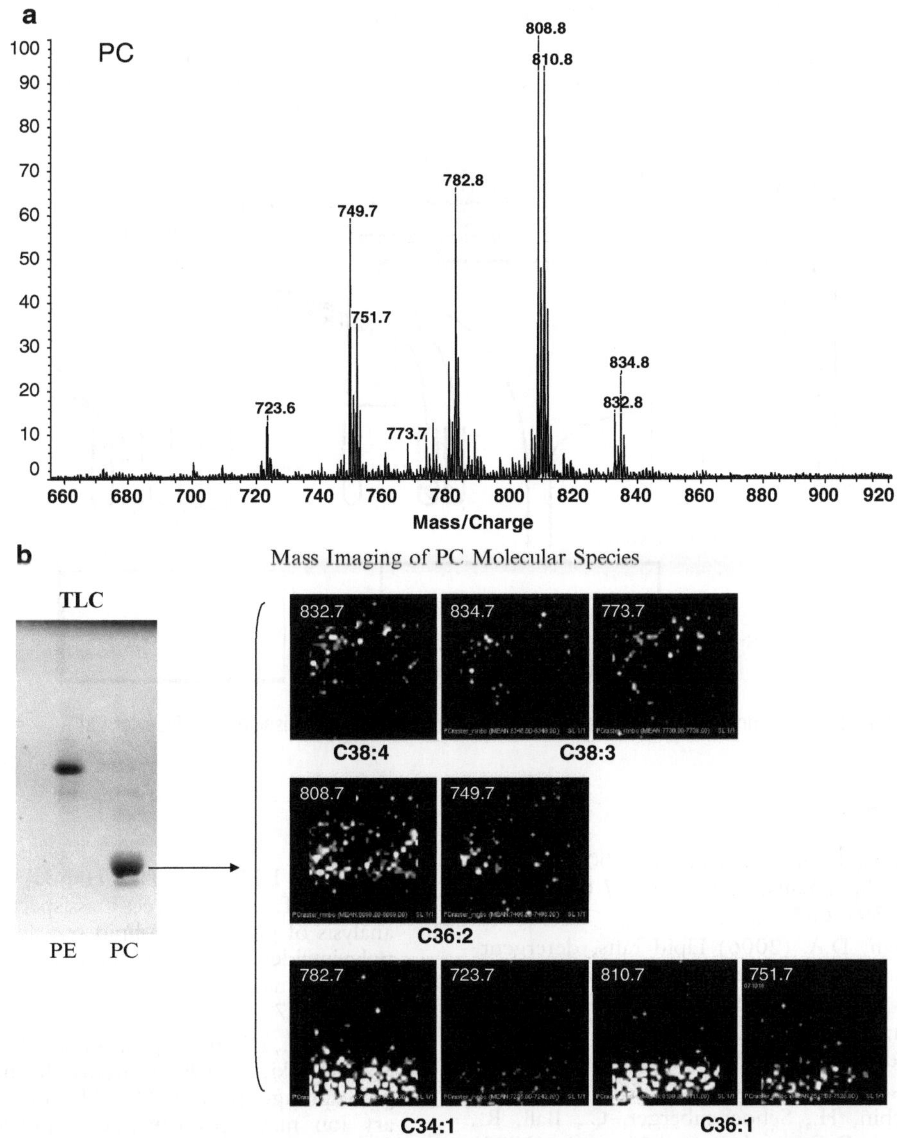

Fig. 5. TLC blot/MALDI-TOF MS of phosphatidylcholine and its imaging analysis. (**a**) Mass spectrometric analysis of phosphatidylcholine on a transferred membrane. *PE* phosphatidylethanolamine, *PC* phosphatidylcholine. (**b**) Mass imaging of phosphatidylcholine. The distribution of molecular species of phosphatidylcholine band transferred from the HPTLC plate to a PVDF membrane is shown. C38:4 means a possible combination of fatty acid constituents (C18:0 and C20:4) of phosphatidylcholine: C38:3 (C18:2 and C20:1) or (C18:1 and C20:2); C36:2 (C18:0 and C18:2) or (C18:1 and C18:1); C34:1 (C16:0 and C18:1) or (C16:1 and C18:0); and C36:1 (C18:0 and C18:1), respectively. Shorter-chain fatty acid-containing phosphatidylcholine moved slower than those containing longer chains with highly unsaturated fatty acids.

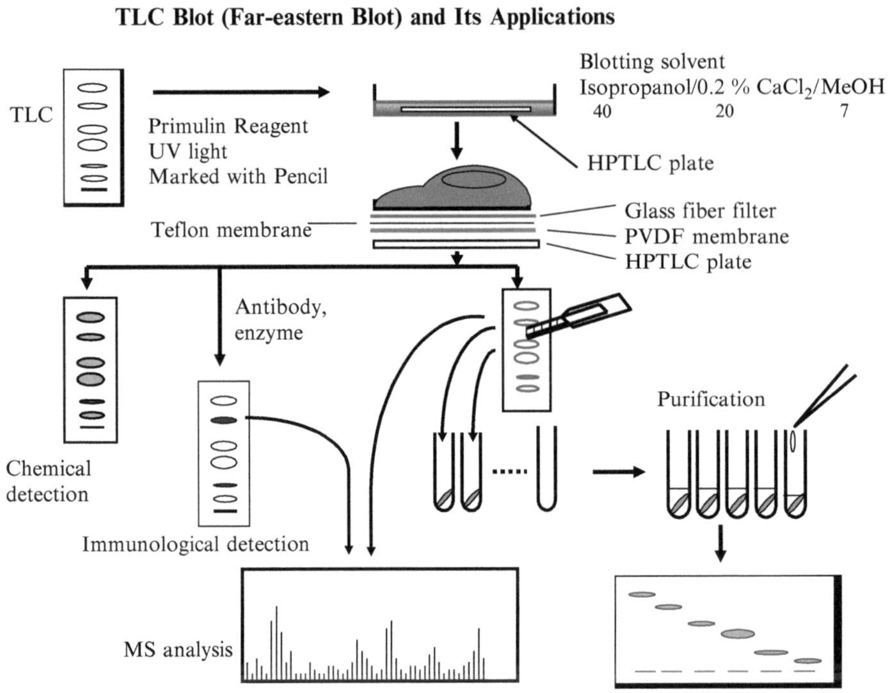

Fig. 6. Sheme of TLC blot and its applications (reproduced from *(10)* with permission from Elsevier Ltd.).

References

1. Simons, K. and Ehehalt, R. (2002) Cholesterol, lipid Rafts, and disease. *J Clin Invest.* 110, 597–603.
2. Brown, D.A. (2006) Lipid rafts, detergent-resistant membranes, and raft targeting signals. *Physiology* 21, 430–439.
3. Hanzal-Bayer, M.E. and Hancock, J.F. (2007) Lipid rafts and membrane traffic. *FEBS Lett.* 581, 2098–2104.
4. Towbin, H., Schoenenberger, C., Ball, R., Braun, D.G., and Rosenfelder, G. (1984) Glycosphingolipid-blotting: An immunological detection procedure after separation by thin layer chromatography. *J Immunol Methods* 72, 471–479.
5. Taki, T., Handa, S., and Ishikawa, D. (1994) Blotting of glycolipids and phospholipids from a high-performance thin-layer chromatogram to a polyvinylidene difluoride membrane. *Anal Biochem.* 221, 312–316.
6. Taki, T., Kasama, T., Handa, S., and Ishikawa, D. (1994) A simple and quantitative purification of glycosphingolipids and phospholipids by thin-layer chromatography blotting. *Anal Biochem.* 223, 232–238.
7. Taki, T. Ishikawa, D., Handa, S., and Kasama, T. (1995) Direct mass spectrometric analysis of glycosphingolipid transferred to a polyvinylidene difluoride membrane by thin-layer chromatography blotting. *Anal Biochem.* 225, 24–27.
8. Kasama, T., Hisano, Y., Nakajima, M., Handa, S., and Taki, T. (1996) Microscale analysis of glycosphingolipids by TLC blotting/secondary ion mass spectrometry: a novel blood group A-active glycospingolipid and changes in glycosphinolipid expression in rat mammary tumor cells with different metastatic potentials. *Glycoconjugate J.* 13, 461–469.
9. Isobe, T., Naiki, M., Handa, S., and Taki, T. (1996) A simple assay method for bacterial binding to glycosphingolipids on a polyvinylidene difluoride membrane after thin-layer chromatography blotting and *in situ* mass spectrometric analysis of the ligands. *Anal Biochem.* 236, 35–40.
10. Taki, T. and Ishikawa, D. (1997) TLC Blotting: Application to microscale analysis of lipids and as a new approach to lipid-protein interaction. *Anal Biochem.* 251, 135–143.

Chapter 56

A Brief Review of Other Notable Protein Detection Methods on Blots

Biji T. Kurien and R. Hal Scofield

Summary

Several methods have been used for detecting proteins on membranes. These include the use of quantum dot luminescent labels, oxyblot immunochemical detection, polymer immunocomplexes, "coupled" probing approach, in situ renaturation of proteins for detecting enzyme activities in crude or purified preparations, immunochromatographic assay, western-phosphatase assay, and the use of Congo red dye, a cosmetic color named Alta, Pro-Q Emerald 488 dye, or amine-reactive dye in combination with alkaline phosphatase and horseradish peroxidase–antibody conjugates for the simultaneous trichromatic fluourescence detection of proteins. Several methods have been used to improve the detection of proteins on membranes, including glutaraldehyde treatment of nitrocellulose blots, elimination of keratin artifacts in immunoblots probed with polyclonal antibodies, and the washing of immunoblots with excessive water and manipulation of Tween-20 in wash buffer. These methods are briefly reviewed in this chapter.

Key words: Protein detection, Membranes, Background reduction, Dyes, Quantum dot luminescent labels

1. Two-Dimensional Oxyblot

Oxidative modification of proteins by free radicals has been implicated in several diseases including Alzheimer's disease (AD) *(1, 2)*. Protein carbonyl formation is considered to be a detectable marker of protein oxidation and it is increased in AD. The level of carbonyls is higher in areas where the histopathology of the disease is more pronounced. Formation

Biji T. Kurien and R. Hal Scofield (eds.), *Methods in Molecular Biology, Protein Blotting and Detection, vol. 536*

DOI: 10.1007/978-1-59745-542-8_55

of carbonyls is thought to be due to reactive oxygen species (ROS)-mediated oxidation of amino acid side chains or by covalent binding to lipid peroxidation products or glycoxidation *(3)*. Unless the oxidative modification process brings about protein aggregation resulting in deposition of proteolysis-resistant protein aggregates *(4)*, oxidized proteins are more susceptible to degradation by specific proteases *(5)*. Increased production of ROS and oxidative modification of proteins in the brain has been noted in AD pathology *(2)*, thus suggesting the involvement of protein oxidation in the neurogenerative processes peculiar to AD.

By coupling two-dimensional (2D) gel fingerprinting of oxidized proteins and immunochemical detection of protein carbonyls, the identification of protein targets of oxidative modification, which could help in establishing a relationship between oxidative modification and neuronal death in AD, has been achieved *(6)*. Since this procedure was laborious, resulting in identification of only a few oxidized proteins Castegna et al. *(2)* coupled 2D fingerprinting with immunological detection of carbonyls and mass spectrometric identification of proteins. Such an approach led them to identify specific protein targets of oxidative modification.

1.1. Protein Carbonyl Derivatization

To each brain sample (obtained at autopsy from AD patients), 2,4-dinitrophenylhydrazone (DNP)/HCl was added (for mass spectrometry analysis only HCl was used). Samples were precipitated with ice-cold trichloroacetic acid following a brief incubation. Samples were centrifuged and the precipitate was resolubilized in urea. DNPH-treated samples of brain proteins from AD and control subjects were used for one-dimensional (1D) and 2D immunoblotting analysis of protein carbonyls *(6)*.

1.2. Oxyblot Immunochemical Detection

The 1D and 2D gels were electrotransferred to nitrocellulose or PVDF. After blocking with bovine serum albumin, the membranes were incubated with anti-DNP polyclonal antibody. Following addition of appropriate alkaline phosphatase secondary antibody the blots were developed with NBT (nitro blue tetrazolium)/BCIP (5-bromo-4-chloro-3-indolyl phosphate) substrate. The blots were dried and scanned. Matrix-assisted laser desorption/ionization time of flight (MALDI-TOF) mass spectrometry of trypsin-digested spots from a Coomassie Blue-stained 2D gel was also carried out for protein identification *(2)*. Using this procedure the authors identified creatine kinase BB, glutamine synthase, and ubiquitin carboxy-terminal hydrolase L-1 as the targets of oxidative modification in AD.

2. Bioconjugation of Quantum Dot Luminescent Probes for Western Blot Analysis

Detection of multiple antigens is usually done by stripping and reprobing a blot with transferred protein. Krajewski et al. *(7)* showed that it is possible to detect multiple antigens on a single blot without stripping off antibodies that have been added first by employing sequential reactions. By employing multiple fluorescent probes made from small organic dye molecules it is also possible to detect multiple antigens on a single blot without stripping off antibodies *(8)* (*see* Chapters "'Rainbow' Western Blotting" and "Multiple Antigen Detection Western Blotting"). However, such probes have several limitations. Most of these problems can, however, be eliminated by the use of quantum dot (QD) luminescent labels *(9)*.

QDs are semiconductor nanoparticles (e.g., CdSe, Inp, InAs) (having diameters of 2–10 nm) whose fundamental physical properties are influenced by quantum confinement effects *(10)*. QDs display absorption and emission peaks that progressively change to longer wavelengths with increasing particle size. Compared with standard fluorescent dyes QDs have significant advantages. Dyes, for example, have narrow absorption bands, and therefore it is difficult to excite several colors with a single excitation source. In addition, the broad spectral overlap between emissions of dyes necessitates complex mathematical analysis of the data. In comparison, QDs possess a narrow, tunable, symmetric emission spectrum allowing a larger number of probes within a detectable spectral region. A single light source can be used to excite different size populations of QDs. This makes it possible to develop simpler and more cost-effective instrumentation for multiplex detection of biomolecules. Compared with organic dyes, QDs are considerably more stable against photobleaching *(11)*. This property is important, especially for imaging applications, where the high photostability of QDs permits real-time monitoring of intracellular processes over longer periods of time *(12)*. QDs show large Stokes shifts (the difference between the maximum absorbance and emission wavelengths). In contrast to the use of fluorescent proteins (e.g., green fluorescent protein), this property permits the target signals to be clearly separated from autofluorescence and thus enables the entire emission spectra to be collected.

Makrides et al. *(9)* show a novel method of conjugating antibodies (primary or secondary) to QD, allowing the easy generation of QD-based probes for the multiplex detection of proteins in western blots. They used the immunoglobulin G (IgG)-binding Z domain, which is based on the B domain of *Staphylococcus aureus* protein A. The Z-affinity tag (6.5 kDa) is highly specific for its ligand, IgG Fc, and can easily be purified by affinity chromatography using IgG-sepharose. It has been shown earlier that the divalent ZZ domain showed ten times higher affinity for its IgG ligand compared with the monovalent Z domain. The authors designed a ZZ protein fused to a peptide that is biotinylated in vivo (by biotin protein ligase, the *birA* gene product), followed by a six-histidine tag. Bacteria were used to produce the biotinylated ZZ tag and was purified over a monomeric avidin or Ni^{2+}-NTA column and attached to streptavidin-coated QDs. Such a technology enables the biospecific coupling of any antibody to the functionalized QDs *(9)*.

Proteins electrotransferred to PVDF membranes were washed with TBST (Tris-buffered saline containing 0.1% Tween-20) and then blocked. The membranes were then incubated with the diluted primary antibody in blocking buffer and washed. The membrane was then incubated with QD_{565}-ZZ or QD_{655}-ZZ nanoparticles conjugated to secondary antibody. Following washing the protein bands were visualized using long-wavelength ultraviolet irradiation *(9)*.

The authors detected two different proteins simultaneously on the same blot by probing first with primary antibodies, followed by incubation with QD_{565}-ZZ or QD_{655}-ZZ nanoparticles or both, conjugated to secondary antibodies *(9)*.

3. Simultaneous Trichromatic Fluourescence Detection of Proteins on Western Blots Using an Amine-Reactive Dye in Combination with Alkaline Phosphatase and Horseradish Peroxidase–Antibody Conjugates

Duplicate gels are often required, for general protein staining and the other for immunoblotting, for concurrently visualizing total protein profile and a specific protein by immunoblotting. It is also possible to immunodetect two antigens by stripping the antibody complexes from the original blot and reprobing with another antibody. However, changes in gel size relative in the blot occur as a result of gel shrinkage, swelling, or other artifacts of electrophoresis, making definitive identification of protein bands/spots unreliable.

Martin et al. *(13)* report the use of an improved trichromatic detection procedure using 2-methoxy-2,4-diphenyl-2(2*H*)-furanone (MDFF) for the detection of total protein profiles, a red-fluorogenic substrate 9*H*-(1,3-dichloro-9.9-dimethylacridin-2-one-7-yl) phosphate (DDAO phosphate) to detect alkaline phosphatase conjugates, and Amplex gold reagent to detect horseradish peroxidase conjugates. The authors refer to this method as the "DyeChrome Double western blot stain." Using this procedure two different targets can be identified, when using conventional enzyme conjugates or secondary antibodies, as long as primary antibodies from two different species are employed in the analysis. For instance, one primary antibody could be raised in mouse while the other in rabbit. However, by using Zenon antibody labeling technology *(13)* two antibodies from the same species could be utilized on the same blot.

Following protein blotting, membranes were equilibrated in sodium borate buffer. MDPF in sodium borate buffer was added to the blot after removing the buffer. Following several washing steps the membrane was blocked. Both primary antibodies were prepared and added to the membrane after the blocking step. Secondary antibodies were added after washing the membrane. DDAO phosphate and Amplex gold dye were added and signals were visualized using ultraviolet epi-illumination and photography.

4. A New Immunoblotting Method Using Polymer Immunocomplexes: Detection of Tsc1 and Tsc2 Expression in Various Cultured Cell Lines

In biological research the detection of protein expression is important for functional analysis of gene products. One of the most useful conventional methods for this purpose is western blotting. It is, however, sometimes difficult to get antibodies with sufficiently high titer, especially when effective antigenic regions are unknown. High concentrations of secondary antibodies and/or long exposure times are essential when using low titer antibodies. In such a scenario secondary antibodies sometimes have a tendency to bind to membranes directly, leading to nonspecific bands.

Fukuda et al. *(14)* use a polymer immunocomplex method to effectively reduce the background in immunostaining of tissue sections and also to improve the specificity and sensitivity of western blotting. The authors obtained low titer rabbit antibodies against tuberous sclerosis proteins (Tsc1 and Tsc 2). While they obtained high background binding at first using these antibodies, they have used the polymer immunocomplex method to reduce background binding by these antibodies.

In this method, polyclonal primary or antigen preabsorbed antibodies were diluted and mixed with Envision polymer (DAKO) to generate immunocomplexes of primary and secondary antibodies. This polymer is an immunological reagent in which secondary antibodies are conjugated with several horseradish peroxidase molecules (HRP) via dextran polymer. Following this, normal rabbit serum was added and mixed with antibody solutions in order to mask the "free" antigen binding sites of Fab domains of secondary antibodies. The solution was applied to blocked membranes. The authors used biotinylated species-specific secondary antibodies and HRP-conjugated streptavidin for traditional biotin–streptavidin detection following washing of the membranes.

5. Washing of Immunoblots with Excessive Water and Manipulation of Tween-20 in Wash Buffer for Reducing Background in Western Blotting

There are some important steps that need to be taken for avoiding high background staining and unacceptable results. Wu et al. *(15)* show that of all other factors that can cause high background, inefficient washing of the membrane is the main cause. In addition, they use a lower Tween-20 concentration (0.02–0.005%) in the buffer for the reaction with antibody and a higher concentration (0.05%) in the phosphate buffered saline-Tween-20 (PBST) wash buffer.

Following standard electrophoresis and blotting to presoaked Immobilon-P membrane (Millipore, Bedford, MA, USA) the membrane was blocked and incubated with primary antibody. The major changes to standard immunoblotting occur from this stage onward, mainly in membrane washing and in the composition of buffer used for diluting antibody. The membrane was rinsed five times with distilled or deionized water followed with one 5-min wash with PBST. The authors have found that five rinses are enough to remove unbound antibody and the 5-min wash with the buffer was sufficient to make the pH and ionic strength appropriate for interaction with antibody. The wash after secondary antibody incubation was also repeated similarly.

The authors report better background reduction using the chemiluminiscence technique with their method of immunodetection compared with that obtained with classical immunodetection with chemiluminiscence. There are two advantages by washing with water coupled with a single buffer wash compared with the conventional washing with buffer alone. First of all, it is possible to use large volumes of water without significantly increasing the cost or labor involved. Second, water has a lower ionic strength and therefore removes excess antibodies and other molecular contaminants more efficiently (owing to its lower ionic strength it can absorb both solute and solvent molecules). The authors have found no membrane damage or stripping of blotted proteins on account of this excessive water washing. The authors also report obtaining high-quality, reproducible protein blots with significantly lower background using this procedure. This method was found to be applicable to a variety of proteins under different experimental conditions *(15)*.

6. Elimination of Keratin Artifacts in Immunoblots Probed with Polyclonal Antibodies

Ever since silver staining of protein in sodium dodecyl sulfate (SDS) polyacrylamide gels *(16)* had become common it became clear that contaminating protein bands are often present in protein patterns. It was first thought that these bands were artifacts owing to the use of β-mercaptoethanol, since they appeared only under reducing conditions *(17)*. Skin keratins were suspected to be responsible for these undesirable bands resulting from the contamination of protein samples or the buffers used for SDS-PAGE *(18)*. This observation was consistent with the fact that, under nonreducing conditions, interchain disulfide linkages occurring between keratin molecules may prevent entry into polyacrylamide gels *(19)*. Also, these artifacts were not present when monoclonal antibodies were used. To eliminate these artifacts, the only way, until Berube et al. described the adsorption of rabbit polyclonal antiserum on skin keratins, was to take great care to avoid keratins during sample preparation or performing SDS PAGE *(19)*.

The authors prepared the keratins, used in adsorbing the sera, from human skin. They first obtained the skin specimens from the hospital they worked in and later obtained large quantities of human squamous keratin obtained from the feet from local estheticians. They froze the human keratin in liquid nitrogen and then pulverized it to a powder using a mortar and pestle. The powder was washed by centrifugation and incubated with shaking in Tris-buffered saline (TBS) containing Nonidet P40. The detergent was removed by six successive centrifugations of the keratin in TBS. Following the final spin, the keratin was frozen at −80°C until use. Preimmune or immune sera (3 volumes) were adsorbed on keratin (1 volume) for a minimum of 2 h under agitation. After this the sera were spun and recovered and used in western blotting. Thus, it was shown that the quality of immunological detection could be improved by adsorption of the rabbit polyclonal antiserum on skin keratins *(19)*.

7. A Simplified "Coupled" Probing Approach for the Detection of Proteins on Dot and Western Blots

Sundaram *(20)* describes a novel "coupled blotting" approach that used simultaneous probing of antigens on dot and western blots with primary and secondary antibodies. The author used the highly sensitive enhanced chemiluminescence (ECL) detection system, purified E7 protein of cottontail rabbit papillomavirus (CRPV), and E7-specific antibodies. The abilities of sequential primary antibody followed by secondary reagent and coupled treatments to detect E7 protein on blots were compared and it was found that there was no reduction in signal strength after coupled probing. This coupled probing procedure, involving the addition of the primary and secondary antibodies together in one incubation, has the advantage of saving hands-on time and buffer solutions compared with the standard procedure. The coupled blotting has been shown to be useful for the rapid detection of proteins without compromising quality.

8. Staining Proteins on Nitrocellulose with Congo Red Dye

Protein visualization on nitrocellulose is a procedure that is routinely carried out in enzymology, molecular biology, and protein chemistry. Staining with Ponceau S dye and amido black has been popular (*see* Chapter "Protein Stains to Detect Antigen on Membranes"). Congo red (an anionic dye) binds with carboxymethyl cellulose (CMC) but not with degraded CMC, a property that has been exploited for detection of endoglucanase activity in agar *(21)*. The observation of a blue or violet band while detecting the enzyme activity in gel, showing that it interacted with protein, encouraged Mehta and Rajput *(22)* to develop a staining method on nitrocellulose membrane using Congo red dye.

A stock solution of Congo red dye (Reidel, Germany) was made in distilled water and stored at room temperature. A working solution of this was made by diluting 1 mL of the stock solution with 9 mL of 0.2 M acetate buffer (pH 3.5; while diluting, some dye was found to become insoluble without affecting the results obtained).

The nitrocellulose membrane was immersed for 5 min at room temperature in the diluted Congo red dye immediately following transfer of proteins from sodium dodecyl sulfate polyacrylamide gel (10%). The staining was stopped by washing the membrane with distilled water. The membrane was destained with water until brown bands against a light pink background became visible. The membrane was shaken mildly during both staining and destaining. The dye was found to stain widely different proteins. The method was found to detect up to 500 ng within 10 min. The sensitivity was found to be higher than that obtained with Ponceau S. Staining of blots with India ink is capable of detecting as little as 80 ng of protein, but the staining procedure takes several hours. Therefore, for qualitative purposes Congo red dye staining of nitrocellulose offers a quick method to detect proteins. If sensitivity is a prime concern, then alternative staining procedures such as colloidal gold staining could be the method of choice *(22)*.

9. Staining of Proteins on SDS Polyacrylamide Gels and on Nitrocellulose Membranes by Alta, a Color Used as a Cosmetic

Pal et al. *(23)* describe the use of Alta, a preexisting scarlet-red stain used for cosmetic purposes, for staining gels and nitrocellulose membranes during western blot analysis. Alta is made of 0.8% Crocein scarlet (brilliant Crocein) and 0.2% Rhodamine B.

It is inexpensive and easy to use, while being almost as sensitive as Coomassie Blue R-250. This stain has been used in some parts of India by women as a cosmetic to decorate their feet. For western blot analysis, Alta (purchased from local market in Pune, India) is added in the upper tank buffer to a final concentration of 5% (v/v; 0.4% Crocein scarlet and 0.1% Rhodamine B) prior to electrophoresis. The gel was viewed on a UV-transilluminator, following SDS-PAGE and the protein profile was recorded using a gel documentation system. The gel was electrotransferred and following transfer the membrane was viewed on a UV-transilluminator as before. In addition, this membrane can be processed further for immunodetection without any interference by the background stain. The protein profile can be monitored as the gel runs and can be seen on the nitrocellulose membrane following electrotransfer. This eliminates the need to run individual gels for protein staining on the gel and for western blot analysis.

10. Renaturation of Calcium/Calmodulin-Dependent Protein Kinase Activity After Electrophoretic Transfer from Sodium Dodecyl Sulfate-Polyacrylamide Gels to Membranes

Detecting enzyme activities in crude or purified preparations has been performed by in situ renaturation of proteins following separation by sodium dodecyl sulfate-polyacrylamide gel electrophoresis (SDS-PAGE). Removal of SDS from the gel by extensive washing has been used to allow renaturation of the proteins and detection of enzyme activity in activity gels or gel overlay protocols. Electroblotting of proteins separated by SDS-PAGE to nitrocellulose or PVDF membranes prior to denaturation, renaturation, and phosphorylation in situ has been a further refinement to this approach.

Shackelford and Zivin *(24)* have used a protocol to renature calcium/calmodulin-dependent protein kinase following transfer to PVDF membranes. Following transfer to PVDF, the proteins bound to the membrane were denatured first in a denaturing buffer. The membrane is rinsed and incubated in a renaturation buffer for around 16 h (with gentle rocking). This was followed by a brief incubation with Tris–HCl. Functional assays were carried out following this. The authors have used this system to identify autophosphorylation of a subset of bound kinases. They also find that the membrane can also be used for immunoblotting, phosphoamino acid analysis, or peptide mapping following the in situ renaturation and phosphorylation procedure.

11. Detection of Glycoproteins in Polyacrylamide Gels and on Electroblots Using Pro-Q Emerald 488 Dye, a Fluorescent Periodate Schiff-Base Stain

Oligosaccharides are co- or posttranslationally attached to proteins, commonly by a number of glycosidases and glycosyltransferases. Cell surface proteins and extracellular matrix proteins are especially rich in sugar moieties. Glycosylation of proteins is vital to growth control, cell adhesiveness, cell migration, tissue differentiation, and inflammatory reaction cascades. Changes in the profiles of glycosylation are often useful indicators for the assessment of disease states. There have been only relatively few methods available for direct analysis of glycoproteins transferred to membranes.

Reacting carbohydrate groups by a periodate/Schiff's base (PAS) mechanism and noncovalent binding of specific carbohydrate epitopes using lectin-based detection systems have been the two most widely utilized methods for the detection of glycoproteins on blots. In the PAS method the carbohydrate groups are oxidized followed by conjugating with a chromogenic substrate (Alcian Blue, acid fuchsin), a fluorescent substrate Pro-Q Emerald 300 dye, dansyl hydrazine, 8-aminonapthalene-1,3,6-trisulfonate, or a biotin hydrazide or digoxigenin hydrazide tag. In the case of the chromogenic and/or fluorescent conjugates the signal is detected directly and in the case of the tags it is detected indirectly using enzyme conjugates of streptavidin or antibodies to the tags *(25)*.

Pro-Q Emerald 300 dye is a fluorescent hydrazide excitable at 300 nm that was reported by Steinberg et al. *(26)* not too long ago. This dye is suitable for the sensitive direct fluorescence detection of glycoproteins in gels or on electroblots without using enzyme amplification systems. However, this dye does not have a visible excitation peak and therefore the glycoproteins cannot be visualized using laser-based gel scanners. Therefore, a new fluorescent hydrazide dye named Pro-Q Emerald 488 with an excitation maximum of 510 nm and emission maximum of 525 nm was developed by this group. This dye is also linked to glycoproteins through the standard PAS conjugation mechanism. The glycols in glycoproteins are oxidized initially to aldehydes with the use of periodic acid. A fluorescent conjugate is generated as the dye reacts with the aldehydes on the glycoproteins. In this procedure, there is no requirement for a reduction step with sodium metabisulfite or sodium borohydride to stabilize the resulting conjugate. Differential display maps of protein glycosylation and expression levels are easily generated using computer-assisted overlay techniques *(25)*.

12. Development of Rapid One-Step Immunochromatographic Assay

In the past, most immunoassays have been performed in laboratories possessing tools and devices for analysis and this has been customarily performed by skilled personnel. Instantaneous examination of alterations in one's own physical symptoms or health status is preferred increasingly. Future health care systems may have self-tests done at home as an integral part. The pregnancy test, based on the rapid detection of human chorionic gonadotropin in urine by just adding urine to the test kit, was the first successful commercial kit.

The novel concept of immunochromatographic assay that depends on the transport of a reactant to its binding partner immobilized on the surfaces of the membrane has been used to achieve the speed and convenience of the test. The transport is brought about by the capillary action of aqueous medium through membrane pores, and therefore, this transport also separates the unbound reactant from the binding complex formed at the liquid–solid surface. The immunochromatography assay provides a way for carrying out the test without the handling of reagents (that is, permitting a one-step assay) in addition to speeding up the analytical procedure.

Consequently, the assay may be carried out at places where the specimen is collected rather than at a specialized location.

Immunochromatography can be used for both qualitative and quantitative analyses. Since the analytical system was first developed for on-site determination of pregnancy, it has been made as an on/off format without adopting a signal detector. In this current model, two antibodies binding distinct epitopes on an analyte are utilized. One antibody (detection antibody) is labeled with a signal generator (e.g., latex beads), gold colloids, and the other antibody (capture antibody) is immobilized onto solid surfaces. The labeled antibody is kept in a dehydrated state within a glass-fiber membrane in such a way that it can be dissolved immediately upon contact with an aqueous medium containing the substance to be measured (analyte). The antibody then takes part in the binding reaction to form a complex with the analyte in the liquid phase. This antibody–analyte complex moves forward in a continuous fashion until it is ultimately captured by the antibody immobilized on the surface of the nitrocellulose membrane. The membrane with its uniform pores provides a liquid–solid interface for reproducible antigen–antibody binding. The two membranes (with immunoreagents) are excised into strips and are attached in length contiguously and a cellulose membrane is present at the top to bring about a contiguous wicking that permits the immune complex to be pulled to the immobilized antibody. Within 10 min a color signal can be read, the intensity determining the amount of analyte *(27)*.

13. Unmasking of Phosphorylation-Sensitive Epitopes on p53 and Mdm2 by a Simple Western-Phosphatase Procedure

Several cellular proteins become phosphorylated either constitutively or in response to a number of regulatory signals. Sometimes, this phosphorylation can happen on a part of an epitope recognized by a monoclonal antibody, leading to decreased immunoreactivity or total lack of binding by the antibody. To solve this, the extract or the immunoprecipitated sample has been treated with phosphatase prior to SDS PAGE.

Maya and Oren *(28)* describe a simple procedure in which the phosphatase treatment is carried out on the nitrocellulose membrane following western blotting. This procedure can be used before the use of any antibody whose epitope is known to be altered as a result of phosphorylation.

The nitrocellulose membrane, following protein transfer, was rinsed briefly with double-distilled water and incubated with phosphatase buffer containing calf intestinal alkaline phosphatase. The membrane was then washed and blocked prior to normal immunodetection. The authors found better detection of specific epitopes using this procedure.

14. Glutaraldehyde Fixation of Calmodulin to Nitrocellulose to Improve Immunodetection

Low molecular weight proteins such as calmodulin can be detected more efficiently by cross-linking to nitrocellulose using glutaraldehyde. Samples are dot blotted onto nitrocellulose membrane. The membrane is then incubated with 0.2% glutaraldehyde for 15–20 min at room temperature. Following fixation the membrane is rinsed briefly with Tris-buffered saline and subjected to immunodetection *(29)*.

15. Concentrating Protein Samples for Sodium Dodecyl Sulfate-Polyacrylamide Gel Electrophoresis and Isoelectric Focusing Using Protein-Blotting Membranes

A crucial limiting factor for successfully analyzing a biological sample electrophoretically is the protein concentration. Increasing protein concentration is not easy and several methodologies have been used for this purpose. However, they are often not satisfactory, require special equipment or the use of hazardous chemicals. Liang et al. *(30)* report a simple method for concentrating dilute protein samples. Their method consists of absorbing proteins onto protein-blotting membrane strips. They incubate blotting membrane strips in dilute protein solutions to capture proteins. Then they loaded the protein-absorbed membrane strips directly into the sample wells containing a strong protein elution buffer (for either sodium dodecyl sulfate-polyacrylamide gel electrophoresis (SDS-PAGE) or isoelectric focusing (IEF), carried out according to standard techniques). In this manner the authors were able to successfully concentrate protein samples for SDS-PAGE or IEF.

16. Treatment of Nitrocellulose Blots to Improve Detection

Numerous proteins bind strongly to nitrocellulose under different experimental conditions. However, milk *(31)*, Triton X-100 *(32)*, Nonidet P-40 *(33)*, and Tween 20 *(34)* have all been shown to remove bound proteins. Milk, however, has been generally used to block unoccupied sites on the nitrocellulose membrane and thus prevent unspecific binding. Hoffman et al. *(34)* have shown that the sensitivity of nondenaturing blots can be increased by soaking the membrane in acidic buffer after the transfer. The authors showed that following exposure to acidic buffer, milk-stripping of antibody from membrane was completely eliminated.

References

1. Kurien, B.T., Hensley, K., Bachmann, M., and Scofield, R.H. (2006) Oxidatively modified autoantigens in autoimmune diseases. *Free Radic Biol Med.* 41, 549–556.
2. Castegna, A., Aksenov, M., Thongboonkerd, V., Klein, J.B., Pierce, W.M., Booze, R., et al. (2002) Proteomic identification of oxidatively modified proteins in Alzheimer's disease brain. II. Dihydropyrimidinase-related protein 2, alpha-enolase and heat shock cognate 71. *J Neurochem.* 82, 1524–1532.
3. Dean, R.T., Fu, S., Stocker, R., and Davies, M.J. (1997) Biochemistry and pathology of radical-mediated protein oxidation. *Biochem J.* 324, 1–18. Review.
4. Davies, K.J. (2001) Degradation of oxidized proteins by the 20S proteasome. *Biochimie* 83, 301–310. Review.
5. Stadtman, E.R. (1990) Metal ion-catalyzed oxidation of proteins: biochemical mechanism and biological consequences. *Free Radic Biol Med. 9, 315–325. Review. Erratum in Free Radic Biol Med.* 10, 249.
6. Aksenov, M., Aksenova, M., Butterfield, D.A., and Markesbery, W.R. (2000) Oxidative modification of creatine kinase BB in Alzheimer's disease brain. *J Neurochem.* 74, 2520–2527.
7. Krajewski, S., Zapata, J.M., and Reed, J.C. (1996) Detection of multiple antigens on western blots. *Anal Biochem.* 236, 221–228.
8. Gingrich, J.C., Davis, D.R., and Nguyen, Q. (2000) Multiplex detection and quantitation of proteins on western blots using fluorescent probes. *Biotechniques* 29, 636–642.
9. Makrides, S.C., Gasbarro, C., and Bello, J.M. (2005) Bioconjugation of quantum dot luminescent probes for western blot analysis. *Biotechniques* 39, 501–506.
10. Bruchez, M., Jr, Moronne, M., Gin, P., Weiss, S., and Alivisatos, A.P. (1998) Semiconductor nanocrystals as fluorescent biological labels. *Science* 281, 2013–2016.
11. Wu, X., Liu, H., Liu, J., Haley, K.N., Treadway, J.A., Larson, J.P., et al. (2003) Immunofluorescent labeling of cancer marker Her2 and other cellular targets with semiconductor quantum dots. *Nat Biotechnol.* 21, 41–46.
12. Dubertret, B., Skourides, P., Norris, D.J., Noireaux, V., Brivanlou, A.H., and Libchaber, A. (2002) In vivo imaging of quantum dots encapsulated in phospholipid micelles. *Science* 298, 1759–1762.
13. Martin, K., Hart, C., Liu, J., Leung, W.Y., and Patton, W.F. (2003) Simultaneous trichromatic fluorescence detection of proteins on western blots using an amine-reactive dye in combination with alkaline phosphatase- and horseradish peroxidase-antibody conjugates. *Proteomics* 3, 1215–1227.
14. Fukuda, T., Tani, Y., Kobayashi, T., Hirayama, Y., and Hino, O. (2000) A new western blotting method using polymer immunocomplexes: detection of Tsc1 and Tsc2 expression in various cultured cell lines. *Anal Biochem.* 285, 274–276.
15. Wu, M., Stockley, P.G., and Martin, W.J., II. (2002) An improved western blotting technique effectively reduces background. *Electrophoresis* 23, 2373–2376.
16. Morrisey, J.H. (1981) Silver stain for proteins in polyacrylamide gels: a modified procedure with enhanced uniform sensitivity. *Anal Biochem.* 117, 307–310.
17. Dunbar, B.S. (1987) Two Dimensional Electrophoresis Immunological Techniques, Plenum, New York, p. 74.
18. Ocha, D. (1983) Protein contaminants of sodium dodecyl sulfate-polyacrylamide gel. *Anal Biochem.* 135, 470–474.
19. Bérubé, B., Coutu, L., Lefièvre, L., Bégin. S., Dupont, H., et al. (1994) The elimination of keratin artifacts in immunoblots probed with polyclonal antibodies. *Anal Biochem* 217, 331–333.
20. Sundaram, P. (1996) A simplified 'coupled' probing approach for the detection of proteins on dot and western blots. *J Immunol Methods* 195, 149–152.
21. Béguin, P. (1983) Detection of cellulase activity in polyacrylamide gels using Congo red-stained agar replicas. *Anal Biochem.* 131, 333–336.
22. Mehta, S. and Rajput, Y.S. (1998) A method for staining of proteins in nitrocellulose membrane and acrylamide gel using Congo red dye. *Anal Biochem.* 263, 248–251.
23. Pal, J.K., Godbole, D., and Sharma, K. (2004) Staining of proteins on SDS polyacrylamide gels and on nitrocellulose membranes by Alta, a colour used as a cosmetic. *J Biochem Biophys Methods* 61, 339–347.
24. Shackelford, D.A. and Zivin, J.A. (1993) Renaturation of calcium/calmodulin-dependent protein kinase activity after electrophoretic transfer from sodium dodecyl sulfate-polyacrylamide gels to membranes. *Anal Biochem.* 211, 131–138.

25. Hart, C., Schulenberg, B., Steinberg, T.H., Leung, W.Y., and Patton, W.F. (2003) Detection of glycoproteins in polyacrylamide gels and on electroblots using Pro-Q Emerald 488 dye, a fluorescent periodate Schiff-base stain. *Electrophoresis* 24, 588–598.
26. Steinberg, T.H., Pretty On Top, K., Berggren, K.N., Kemper, C., Jones, L., Diwu, Z., et al. (2001) Rapid and simple single nanogram detection of glycoproteins in polyacrylamide gels and on electroblots. *Proteomics* 1, 841–855.
27. Paek, S.H., Lee, S.H., Cho, J.H., and Kim, Y.S. (2000) Development of rapid one-step immunochromatographic assay. *Methods* 22, 53–60.
28. Maya, R. and Oren, M. (2000) Unmasking of phosphorylation-sensitive epitopes on p53 and Mdm2 by a simple western-phosphatase procedure. *Oncogene* 19, 3213–3215.
29. Dominguez, D.C., Adams, H., and Hageman, J.H. (1999) Immunocytochemical localization of a calmodulinlike protein in Bacillus subtilis cells. *J Bacteriol.* 181, 4605–4610.
30. Liang, F.T., Granstrom, D.E., and Shi, Y.F. (1997) Concentrating protein samples for sodium dodecyl sulphate-polyacrylamide gel electrophoresis and isoelectric focusing using protein-blotting membranes. *J Chromatogr A* 764, 143–150.
31. Spinola, S.M. and Cannon, J.G. (1985) Different blocking agents cause variation in the immunologic detection of proteins transferred to nitrocellulose membranes. *J Immunol Methods* 81, 161–165.
32. Batteiger, B., Newhall, W.J., Newhall, V., and Jones, R.B. (1982) The use of Tween 20 as a blocking agent in the immunological detection of proteins transferred to nitrocellulose membranes. *J Immunol Methods* 55, 297–307.
33. Lin, W. and Kasamatsu, H. (1983) On the electrotransfer of polypeptides from gels to nitrocellulose membranes. *Anal Biochem.* 128, 302–311.
34. Hoffman, W.L., Jump, A.A., and Ruggles, A.O. (1994) Soaking nitrocellulose blots in acidic buffers improves the detection of bound antibodies without loss of biological activity. *Anal Biochem* 217, 153–155.

Chapter 57

Sending Secret Messages on Nitrocellulose Membrane and the Use of a Molecular Pen for Orientation in ECL Membrane Assays

Biji T. Kurien

Summary

Invisible ink and writing secret messages have been part of man's fantasy, having proven useful in clandestine and high sensitivity areas. Security inks, made up of invisible materials that give printed images that cannot be photocopied, that can be read only under special environments have become important. An ink formulation based on silicon (IV) 2,3-naphthalocyanine bis(trihexylsilyloxide) as colorant, invisible to the naked eye but infrared readable, has been described earlier. Biometric DNA ink has also been developed for security authentication. In lighter vein, many budding scientists and others have often experimented with writing secret messages on paper, either for purposes of fun or actually sending secret messages to friends. It involved the use of lemon juice, milk, or other solutions that could be used with a dip pen, brush, or a fountain pen to write invisible messages on a blank white paper. Words turn up as though by magic when the paper is exposed' to heat in one form or the other. Here, an attempt is made to end this book on a slightly humorous note by showing that invisible messages can be written on nitrocellulose membranes (but not on polyvinylidene difluoride membranes) using an appropriately diluted horseradish peroxidase/alkaline phosphatase anti-IgG conjugate (rabbit, mouse, or human anti-IgG). The message is written on the membrane, preferably with a fountain pen, and the membrane is allowed to dry. Regular detection with enhanced chemiluminiscence (ECL) plus or nitro blue tetrazolium/5-bromo-4-chloro-3-indolyl phosphate systems is used to unravel the secret message. In addition, this method could be used to mark nitrocellulose membranes for orientation purposes using ECL detection system and thus can eliminate the use of autoradiography pens.

Key words: Nitrocellulose membrane, Secret message, Invisible ink, Molecular pen, ECL detection, Autoradiography pens

Biji T. Kurien and R. Hal Scofield (eds.), *Methods in Molecular Biology, Protein Blotting and Detection, vol. 536*

DOI: 10.1007/978-1-59745-542-8_57

1. Introduction

Dr. Harry Towbin recollects his memories associated with his finding of protein transfer thus "I felt like a child that reveals secret messages written in invisible ink by holding a sheet of paper over a flame" (*see* Chapter "Origins of protein blotting"). Writing secret messages has been a fancy of several youngsters.

Lemon juice, milk, and other solutions have been used as invisible ink and applied to plain white paper using fountain pens, dipping pens, or even toothpicks. The brown-colored messages appear as if by magic, when the paper is exposed to heat. Other types of invisible inks are available that appear only under the use of specific conditions (use of ultraviolet or infrared lights).

Security inks, made up of invisible materials that give printed images that cannot be photocopied, that can be read only under special environments have become important. Kishimura et al. have reported the use of a novel photoluminescent ink for rewritable media that emits phosphorescence dichroically owing to a structural bistability of the self-assembled luminophore *(1)*. Using conventional thermal printers long-lasting images have been developed (readable only by exposure to ultraviolet light) that are thermally erasable for rewriting *(1)*. An ink formulation based on silicon (IV) 2,3-naphthalocyanine bis(trihexylsilyloxide) as colorant, invisible to the naked eye and infrared (IR) readable, has been described earlier *(2)*. In contrast to the currently used mechanism of excitation and fluorescence detection, this ink formulation is based on absorption and reflection of light in the infrared (absorbing strongly at 790 nm with highly transmitting characteristics in the visible).

Biometric DNA ink has also been developed for security authentication. Hashiyada *(3)* describes the development of DNA ink, using synthetic DNA mixed with printing inks. DNA fragments (single-stranded) encoding a personalized set of short tandem repeats (STR) were made. Owing to the high discrimination power of STRs, it has become possible to use them for biometric personal authentication systems.

Here it is shown, using diluted horseradish peroxidase conjugate as the invisible ink, that it is possible to write secret messages on nitrocellulose membrane (and not on PVDF membrane) that can be easily developed by ECL (**Fig. 1**) or NBT/BCIP (**Fig. 2**). The entire procedure can be carried out within 10 min in the case of development with NBT/BCIP. The fountain pen filled with the conjugate can be used as a substitute for autoradiograph pens that are available for purchase.

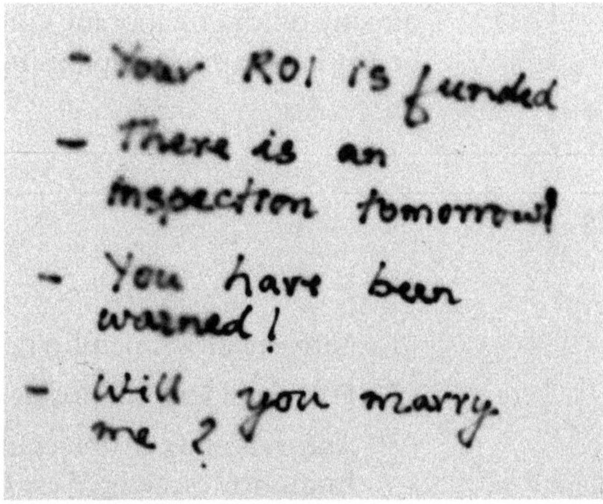

Fig. 1. Sample messages written on nitrocellulose with the invisible ink and developed using ECL plus. The invisible ink was concocted by diluting horseradish peroxidase conjugated (1:500) anti-rabbit IgG. Molecular weight markers on a nitrocellulose membrane can be marked directly on a membrane as dashes shown on the left. The messages were written with a fountain pen containing the invisible ink and developed directly using ECL plus.

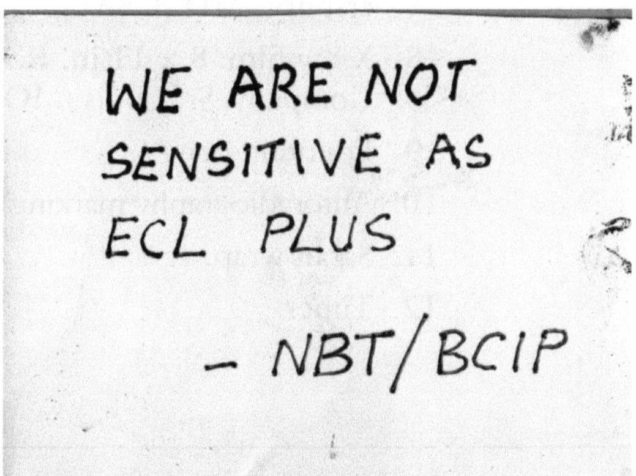

Fig. 2. Sample messages written on nitrocellulose with the invisible ink and developed using NBT/BCIP. The invisible ink was concocted by diluting alkaline phosphatase conjugated (1:500) anti-rabbit IgG. The message was written with a fountain pen containing the invisible ink and developed directly using NBT/BCIP. The whole procedure could be done in 7 min or less.

Thus, any horseradish peroxidase or alkaline phosphatase conjugate could be used as an invisible ink to send secret messages or as a marker for orientation purposes in ECL detection assays. The advantage in using a fountain pen filled with this invisible ink is that molecular weight markers can be marked on the membrane directly (either the molecular weight in kD or just as a dash spanning the width of the marker). The mark can be made directly on the membrane compared with the mark that needs to be made on a sticky tape using autoradiography pens.

2. Materials

1. Nitrocellulose membrane (Bio-Rad, Hercules, CA, USA).
2. Polyvinylidene difluoride (Bio-Rad).
3. Anti-rabbit IgG horseradish peroxidase conjugate (Jackson ImmunoResearch, West Grove, PA, USA).
4. Anti-human IgG horseradish peroxidase conjugate (HRP) (Jackson ImmunoResearch).
5. Anti-rabbit IgG alkaline phosphatase conjugate (Jackson ImmunoResearch).
6. ECL plus kit (Amersham/GE Healthcare, Piscataway, NJ, USA).
7. NBT/BCIP (Nitro blue tetrazolium/5-bromo-4-chloro-3-indolyl phosphate) (KPL Protein research products, Gaithersburg, MD, USA).
8. X-ray film: 8 × 11 in. Kodak Biomax Film (Sigma Chemical Company, St. Louis, MO, USA).
9. Fountain pen.
10. Autoradiography marking pens (Midsci, St. Louis, MO, USA).
11. Saran wrap.
12. Timer.

3. Methods

3.1. Writing the Message

1. Cut a nitrocellulose or PVDF membrane to the required size.
2. Remove the top flap covering the membrane.
3. Dilute conjugate 1:500 with water in a microcentrifuge tube (*see* **Note 1**).
4. Spin the microcentrifuge tube at 10,000 × g for 1 min (*see* **Note 2**).
5. Save the supernatant. Aspirate into a fountain pen using the self-filler (*see* **Note 3**).
6. Write message on the membranes (*see* **Note 4**). The writing will diffuse well into the nitrocellulose membrane but not into the PVDF membrane (*see* **Note 5**).

7. Let the membrane dry. It should dry within 5 min on nitrocellulose membrane (*see* **Note 6**).

3.2. Marking Molecular Weight Markers with the Invisible Ink

1. Run sodium dodecyl sulfate polyacrylamide gels (*see* Chapter "Non-electrophoretic bi-directional transfer of a single SDS-PAGE gel with multiple antigens to obtain twelve immunoblots") with molecular weight standards.
2. Transfer to nitrocellulose membrane (*see* Chapter "Detection of La/SS-B by Western Blot Using Nanogold-Tagged Antibodies and Silver Enhancement").
3. If using ECL plus system to detect antigens, fill up the fountain pen with the invisible ink (1:500 dilution of HRP conjugate). Mark each molecular weight marker (along the width of each marker; *see* **Fig. 1**) as a thin dash (*see* **Note** 7) (normally the membrane is fixed to a flat surface such as a used X-ray film via sticky labeling tapes, and orientation marks are made on these tapes using an autoradiography marking pen) (*see* **Note 8**). Let the membrane dry.
4. Develop the membrane with ECL plus (*see* **Subheading 3.3**).

3.3. Detection with ECL Plus

1. Make sure that the X-ray machine is warmed up prior to use. Otherwise, one would have to wait 15 min for the machine to warm up.
2. Take the membranes, Saran wrap, pipettes, pipette tips, container for membrane, X-ray film, glass tube (10 mL), and timer to the X-ray developing room.
3. Turn on the regular lamp in the X-ray development room.
4. Mix the ECL plus solution A and B (40:1) in a 10 mL glass tube or in a small container that is small enough to fit the membrane.
5. Place the nitrocellulose membrane in the container and swish the solution continuously over the membrane for 5 min (*see* **Note 9**).
6. At the end of 5 min the excess ECL plus solution is removed.
7. Place Saran plastic wrap, roughly three times the size of the membrane on the workbench.
8. Place the membrane in the middle of the Saran wrap. The right side end of the plastic wrap is folded over the membrane to the left (the entire membrane should be covered with Saran wrap) (*see* **Note 10**).
9. Ensure that the surface of the workbench, in the X-ray developer room, is not moist.
10. Turn off the regular light and turn on the safety light (red light). Close eyes for a few moments prior to turning off the

regular light and also after turning it off so that one's eyes will get acclimatized to the darkness faster.

11. Open film box, retrieve a film, and close the box (*see* **Note 11**).
12. Expose the film to the membrane. A 30 s exposure gave good results for 1:500 dilution of the conjugate (*see* **Note 12**).
13. Develop the film. Place the film against the left wall of the feeder of the X-ray developer. This way the film will travel in a straight line as it slides into the machine.
14. Re-expose another film (for 1 min) to the membrane if signal is not strong enough (*see* **Note 13**).
15. Repeat the ECL plus and X-ray film exposure for the PVDF membrane (*see* **Note 14**).
16. Scan to save image (*see* **Fig. 1**).

3.4. Detection with NBT/BCIP

1. Once the message is written with the alkaline phosphatase conjugate (1:500 dilution), allow it to dry. It should take less than 5 min for this.
2. There is no need to wash the membrane.
3. Add 10 mL of NBT/BCIP mixture and swish it over the membrane (*see* **Note 15**).
4. The message should appear in a few minutes. Wait till the development is complete and then rinse with deionized water (*see* **Note 16**).
5. Scan to save image (*see* **Fig. 2**).

4. Notes

1. The conjugates were diluted 1:100, 1:500, and 1:1,000 for rabbit and human horseradish peroxidase conjugates, and it was found that 1:500 gave good results. The 1:1,000 dilution also gave good results but required slightly longer exposure timings. There was significant difference between the rabbit and human conjugates. No buffers were checked as diluent.
2. The timer was set at 1 min and 15 s to obtain a 1-min centrifugation at 10,000 × *g*. The sample was spun to ensure that the fountain pen does not get clogged from any precipitate that might be found in the conjugates. However, in subsequent experiments the samples were used without centrifugation and there was no clogging of the fountain pen.
3. The fountain pen that we used did not aspirate sufficiently enough and therefore it had to be filled multiple times just

to ensure continuous flow of the conjugate. Therefore, when selecting a pen it would be a good idea to first check whether it could pick up enough fluid (if the fountain pen is a self-filler/suction filler).

4. Try to write with an even speed. Writing fast does not work very well. Writing slowly ensures even flow of the conjugate (this also depends on the type of fountain pen used). If it is too slow, there would be a higher diffusion of the conjugate into the membrane. This is something that will work better with practice. **Figures 1** and **2** were done in the hurry toward the last days of editing this book. Drawing caricatures was also attempted. However, the data are not shown here.
5. It takes much longer for the conjugate to dry on the PVDF membrane. It may be possible to make the conjugate to diffuse into the PVDF membrane. Diluting the conjugate in a 20% methanol solution may work in making the invisible ink to diffuse. However, studies should be undertaken to find out the level of methanol that would be tolerated and allow the horseradish peroxidase to remain active.
6. Care needs to be taken not to use pressure to write. If the pen nib is pressed too hard against the membrane, as the words are written, the impression of the words will be seen on the membrane and the message will not be a secret anymore (it is not very easy to figure out the message, even if pen impressions are seen). Control experiments were carried out using the fountain pen filled with just water to see if any background writing would be detected (resulting from pressure marks of the fountain pen nib). The results were negative (data not shown). There is no need to wash the membrane to wash away unbound or loosely bound conjugate.
7. Identifying molecular weight is a bane in ECL plus detection system. The nitrocellulose membrane will first have to be oriented with respect to the marks on labeling tape using autoradiography marking pens. Then they will have to be carefully overlaid over the developed X-ray film and the markers are marked on the film. This problem can be averted if marks could be made directly on the molecular weight marker using the invisible ink described in this chapter.
8. Autoradiography marking pens (http://www.midsci.com/docs/opt/autoradiography_pens.html1) are expensive, costing $62 per pen. One milliliter of undiluted conjugate costs only around $80. Dilution at 1:500 (8 cents) is enough for marking membranes for a long time (we have not tested the longevity of the diluted conjugate yet). Besides, autoradiography pens are to be used on labeling tapes, whereas the invisible ink described here could be used directly on the membrane and also used to mark molecular weight markers.

9. An alternate way to save on ECL plus solution is to place the membrane on a piece of Saran wrap that is about three times the size of the membrane. Pipet 2 mL of the ECL mix onto the membrane. Aspirate the mix back into a plastic Pasteur pipette and back into the membrane. Repeat this for 5 min covering the entire area of the membrane.
10. The plastic wrap can be gently lowered on the membrane in such a way that it will not form wrinkles on top of the membrane. Even if it does form it can be pressed away with the fingers.
11. Unwrap one film from the film box under the safety light, and close the box (make sure the opening of the film bag in folded and facing the inside part of the box as the box is closed).
12. Place the left upper corner of the film over the membrane for few seconds. One can time with a timer or can time it mentally if exposure time is short. Press with hand and keep it immobile. Move the film so that the right upper corner of the film is over the membrane. Let it sit immobile for twice the time taken initially. Similarly repeat for the left lower corner and right lower corner, increasing the time of exposure two-fold each time. The film should be kept immobile each time. Another way would be to fix the membrane inside a Kodak X-ray exposure holder using Scotch tape (rather than placing the membrane on the workbench) and expose to X-ray film inside the holder. A 14 × 17-in. holder can be used if the film needs to be exposed to the membrane in four different corners (to get four different autoradiograms in one 8 × 11-in. film).
13. The exposure time would depend upon the manner in which the message is written. If it was written uniformly then a short exposure time would work. If some words are written faster than other words, then the signal would be lower in areas that were written faster (*see* **Fig. 1**). Writing at the right speed is a must to generate a visually appealing message. Again it will depend on the type of fountain pen used.
14. PVDF membrane behaves in a manner opposite to nitrocellulose membrane. The signal obtained is not good. However, if the ECL plus solution is retained on the membrane and the Saran wrap in enveloped over it the X-ray film exposed on the membrane reveals portions of the message. The signal appears as a negative against a positive background (the message appearing blank and the surrounding area covered by the ECL plus solution appearing dark). However, the clarity of the message is poor.
15. The NBT/BCIP mixture is stored at 4°C. It would accelerate the reaction if the required volume is brought to room temperature prior to adding to the membrane.

16. The loosely bound (to the membrane) AP conjugate can be seen coming off as purple particles. This could be avoided by rinsing the membrane briefly with deionized water. However, it is not necessary to wash the membrane as the message develops strongly in a minute or two.

Acknowledgements

This work was supported by NIH grant ARO1844 and Oklahoma Center for the Advancement of Science and Technology to Dr. Hal Scofield, OMRF, OKC, OK, USA. Thanks are also due to Dr. Rachna Aggarwal for the purchase of the fountain pen used in this experiment from New Delhi, India.

References

1. Kishimura, A., Yamashita, T., Yamaguchi, K., and Aida, T. (2005) Rewritable phosphorescent paper by the control of competing kinetic and thermodynamic self-assembling events. *Nat Mater.* 4, 546–549.
2. Yousaf, M. and Lazzouni, M. (1995) Formulation of an invisible infrared printing ink. *Dyes Pigments* 27, 297–303.
3. Hashiyada, M. (2004) Development of biometric DNA ink for authentication security. *Tohoku J Exp Med.* 204, 109–117.

Index

Q

R

S

Color Plates

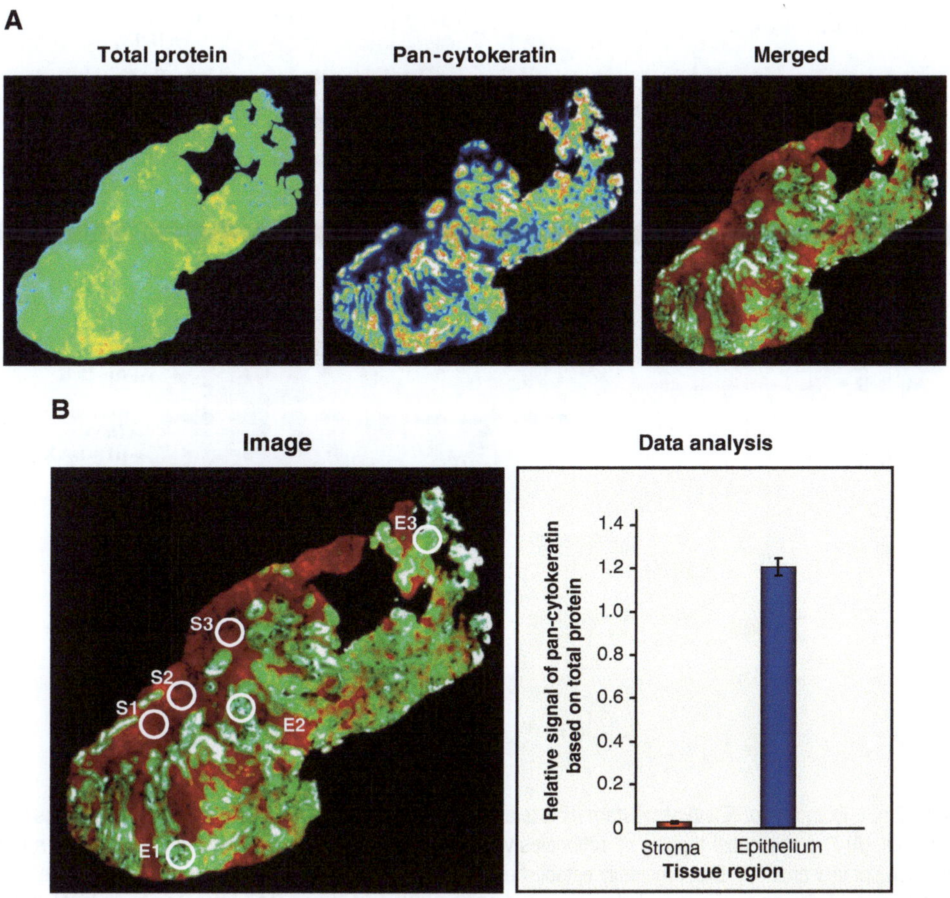

Chapter 16, Fig. 2. Representative image and quantitation of whole tissue section by multiplex tissue immunoblotting. (**A**) Prior to immunoblotting, a formalin-fixed and paraffin-embedded tissue section of a gastrointestinal tumor (5 mm thickness) was transferred to the five-membrane stack by heat-facilitated capillary transfer system. The third membrane was incubated in biotinylation solution and then was reacted with anti-pan-cytokeratin antibodies (1:200-fold diluted). After primary antibody incubation is completed, total protein and specific target signals were detected using streptavidin linked Cy5 and FITC conjugated anti-mouse IgG (1:1,000-fold diluted). Membranes were imaged with a microarray scanner. Fluorescent scans are represented in *pseudocolor*, where signal intensity is *white–red–yellow–green–blue–black* from maximum to minimum signal. (**B**) We selected three different representative stroma and epithelium tissue regions based on the H&E slide. We subsequently quantified those areas using ImageQuant 5.2 software. After normalization with total protein level, relative expressional signals were represented as a ratio. The *bar graph* shows the average ± SD of three circle areas (*S* Stroma, *E* Epithelium).

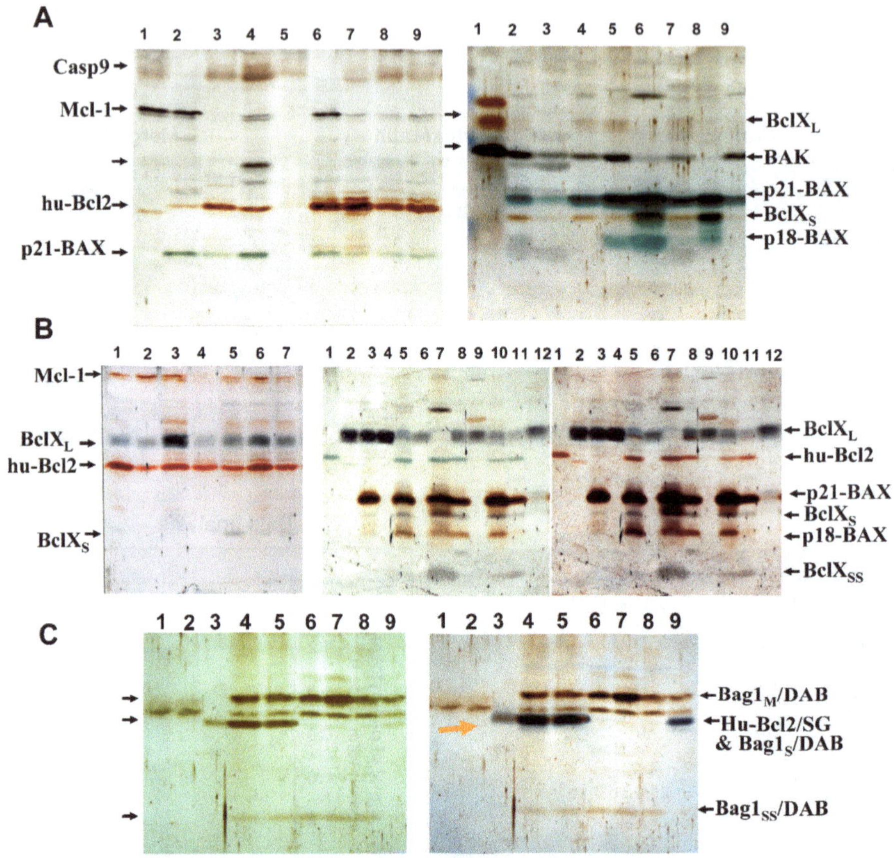

Chapter 47, Fig. 1. Examples of *Rainbow western blots* to assess the physiological content of Bcl2 family proteins in normal human tissues. (**A**) The sequential tagging of antibodies with different HRP substrates resulted in 3–4 colors specifically linked to each primary antibody. Experimentally established order of chromogen developments prevented color "overwriting" [(**B**) *left, middle*]. Changing the order led to color overwriting [(**B**) *right*, *green* TMB color was turned into *red* by AEC reagent]. Panel **C** demonstrates strong immunoreactions for two proteins with similar molecular weight, Bcl2 and Bag-1 using Vector® SG and DAB chromogens. All human samples were loaded at 100 mg total protein per lane. The cell culture lysates were normalized for protein content to apply 30–50 mg per lane.

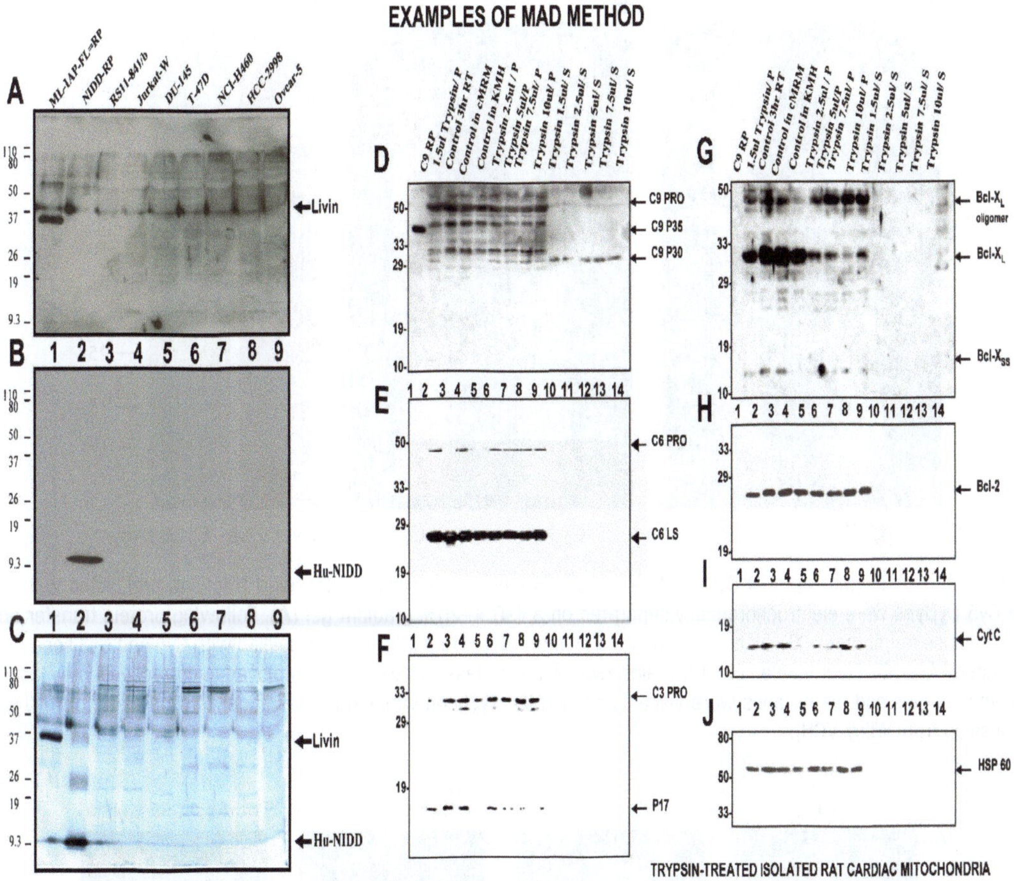

Chapter 48, Fig. 1. Panels (**A–C**) show the colorimetric (DAB/ Vector® SG) background reducing procedure for antibodies raised to apoptotic proteins Livin [panel (**A**); first ECL probing) and NIDD [panel (**B**); second probing]. The blot contains a variety of human cancer cell lines and recombinant protein positive controls. Panel (**C**) demonstrates the abundance of the cross-reacting lanes after chromatographic development. Panels (**D–J**) depict results obtained from a single membrane sequentially probed with 11 antibodies in an experiment with isolated trypsin-treated rat heart mitochondria. Selected blots are shown for antibodies specific to the following proteins: caspase-9 (1st probing), Bcl-X_L (3rd probing), active caspase-6 (4th probing), proform and active caspase-3 (5th), mouse Bcl-2 (8th), cytochrome *c* (10th) and HSP60 (11th). The results indicate early digestion of Bcl-X_L, release and activation of caspases with late changes in Bcl2 levels in mitochondrial membranes.

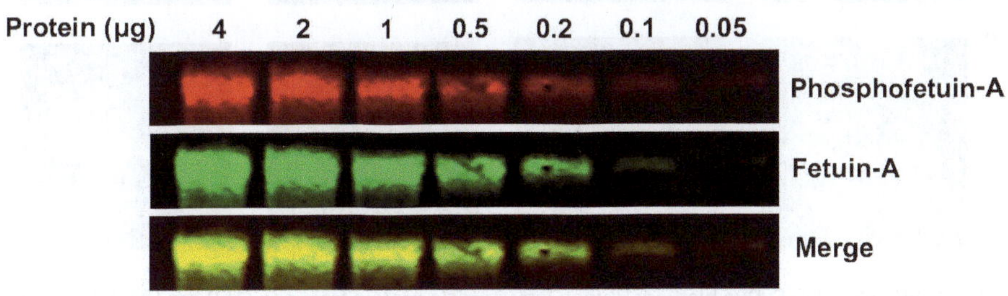

Chapter 51, Fig. 2. Infrared imaging using LI-COR Odyssey® Imaging System. Purified total (*middle*) and phosphorylated (*top*) forms of recombinant, human fetuin-A, a liver-secreted glycoprotein and inhibitor of insulin receptor tyrosine kinase, were separated by SDS-PAGE and transferred to a nitrocellulose membrane and immunoblotted with appropriate antibodies. Overlaid image (*merge*), 700 channel (*red*), and 800 channel (*green*) were acquired using the Odyssey 2.0 software. LI-COR Odyssey® Imaging System allows the user to simultaneously detect phosphorylated and total forms of a protein without the need for stripping and reprobing.

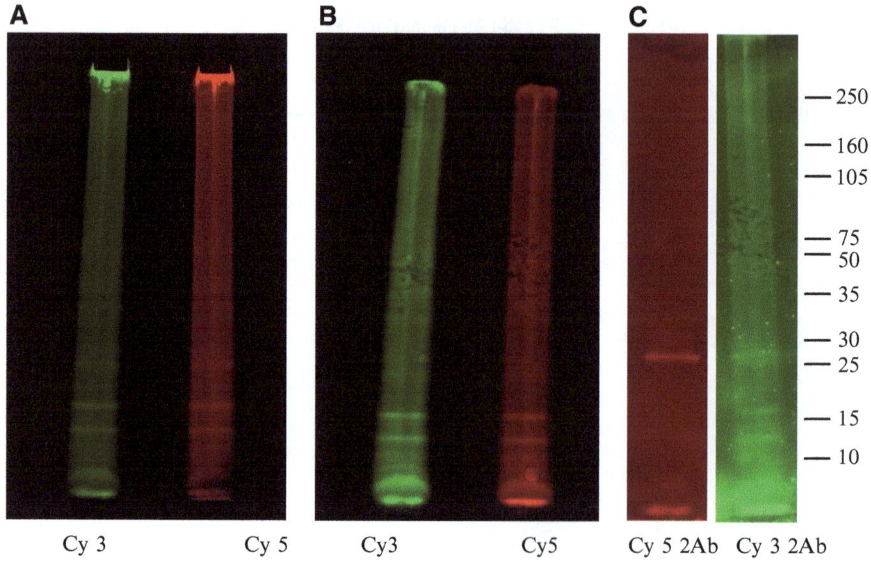

Chapter 52, Fig. 1. ECL Plex 1-D CyDye blotting: ECL Plex human left ventricular heart fractions prelabeled with Cy3 and Cy5 CyDyes were electrophoretically separated on a 1-D 4–20% gradient gel (**A**). Following protein transfer onto nitrocellulose membranes (**B**), blots were probed with monoclonal Hsp 27 Ab and visualised with Cy5 and Cy3 secondary conjugated Abs (**C**). Flour labeled ECL Plex rainbow molecular mass markers were employed to correctly identify labeled proteins. All gels and membranes were visualized using the Typhoon variable mode imager (reproduced from *(4)* with permission from Wiley-VCH).

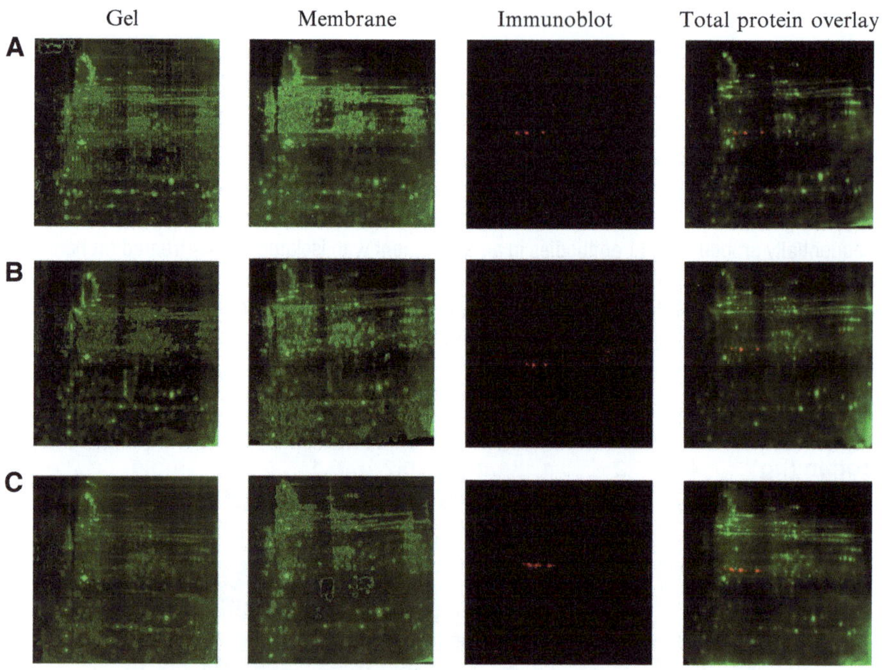

Chapter 52, Fig. 2. ECL Plex 2-D CyDye blotting: Human left ventricle protein fractions [150 mg (**A**), 300 mg (**B**), and 600 mg (**C**)] were primarily CyDye (Cy3) labeled, focused on pH 3–10 NL IPG strips, and subjected to 2-D gel electrophoresis using 12% SDS polyacrylamide gels, column 1 (**A**, **B**, and **C**). Separated proteins were transferred onto nitrocellulose membranes, column 2. Primary incubation with monoclonal Hsp 27 antibody was followed by decoration with CyDye (Cy5) conjugated secondary antibodies, column 3. Total protein expression combined with specific Hsp 27 protein detection is overlaid and visualized at specific wavelengths for both Cy Dyes, column 4 (reproduced from *(4)* with permission from Wiley-VCH).

MIX
Papier aus verantwortungsvollen Quellen
Paper from responsible sources
FSC® C105338

If you have any concerns about our products,
you can contact us on
ProductSafety@springernature.com

In case Publisher is established outside the EU,
the EU authorized representative is:
Springer Nature Customer Service Center GmbH
Europaplatz 3, 69115 Heidelberg, Germany

Printed by Libri Plureos GmbH
in Hamburg, Germany